MAGILL'S ENCYCLOPEDIA OF SCIENCE

ANIMAL LIFE

MAGILL'S ENCYCLOPEDIA OF SCIENCE

ANIMAL LIFE

Volume 3
Lemurs–Respiration in Birds

Editor
Carl W. Hoagstrom, Ph.D.
Ohio Northern University

Project Editor
Tracy Irons-Georges

SALEM PRESS, INC.
Pasadena, California
Hackensack, New Jersey

Editor in Chief: Dawn P. Dawson

Project Editor: Tracy Irons-Georges	*Photograph Editor:* Philip Bader
Copy Editor: Leslie Ellen Jones	*Production Editor:* Joyce I. Buchea
Research Supervisor: Jeffry Jensen	*Page Design:* James Hutson
Research Assistant: Jeff Stephens	*Layout:* Eddie Murillo
Aquisitions Editor: Mark Rehn	*Additional Layout:* William Zimmerman

Graphics by: Electronic Illustrators Group, Morgan Hill, Calif.
Cover Design: Moritz Design, Los Angeles, Calif.

Library of Congress Cataloging-in-Publication Data

Magill's encyclopedia of science : animal life / editor, Carl W. Hoagstrom.
    p. cm.
  Includes bibliographical references (p.).
  ISBN 1-58765-019-3 (set : alk. paper) — ISBN 1-58765-020-7 (v. 1 : alk. paper) —
ISBN 1-58765-021-5 (v. 2 : alk. paper) — ISBN 1-58765-022-3 (v. 3 : alk. paper) —
ISBN 1-58765-023-1 (v. 4 : alk. paper)
  1. Zoology—Encyclopedias. I. Hoagstrom, Carl W.

QL7 .M34 2002
590′.3—dc21

2001049799

First Printing

TABLE OF CONTENTS

MAGILL'S ENCYCLOPEDIA OF SCIENCE

ANIMAL LIFE

LEMURS

Types of animal science: Anatomy, behavior, classification, ecology, evolution
Fields of study: Conservation biology, ecology, ethology, zoology

Over fifty species of lemurs have been identified and classified in the superfamily Lermuroidea in the Primate order. They have evolved separately from East African and other primates for at least forty million years.

Principal Terms

DENTAL COMB: forward projecting lower incisors and canines that are used for grooming and feeding

DENTAL FORMULA: denotes the types of teeth in one quarter of the mouth: incisors, canines, premolars, molars

POLYGYNY: males compete for access to several females

PRIMITIVE: a feature that reflects an ancestral condition rather than those that represent more recent evolutionary changes (derived)

SEXUAL DIMORPHISM: behavioral and anatomical differences between the sexes

SUBLINGUA: fleshy plate under the tongue used to clean the dental comb

Lemurs are primitive monkeys that originally inhabited parts of North America, Eurasia, and Africa, but competition with monkeys and apes resulted in their present distribution, restricted to Madagascar. Humans arrived on Madagascar about two thousand years ago, contributing to extinctions of one third of the island's species through habitat destruction, hunting, climate change, and perhaps new diseases. Most lemurs are highly endangered.

Most lemurs are arboreal (tree-dwelling), although ring-tailed lemurs spend more time on the ground. Lemurs subsist primarily on three to four food species, consisting of a mix of leaves, fruit, buds, bark, and shoots, but favored foods can vary monthly or seasonally. The golden bamboo lemur tolerates high levels of cyanide in the bamboo shoots. The smaller, nocturnal lemurs and the aye-aye eat more insects. Several species are cathemeral (active during parts of day and night), and there are also diurnal (day-active) forms, such as ring-tailed lemurs or indris. The diurnal species may have a nocturnal past, because they share certain features with nocturnal lemurs, including seasonal breeding, female dominance, and reliance on scent for communication.

Unique Primates

Lemurs exhibit many types of social organization. They are solitary in smaller nocturnal forms, while larger diurnal or cathemeral forms live in pairs that sometimes congregate in larger groups, usually with 1:1 sex ratios. Lemurs lack sexual dimorphism, and if there is a difference, females are larger than males. In social species, females largely dominate males, to the extent that males signal submissiveness to all females independent of context. This unusual primate pattern may be due either to seasonally low food productivity and the resulting reproductive costs, or may be partly a function of extinctions of raptors that preyed on primates.

Lemurs range in size from the pygmy mouse lemur at thirty-one grams (about one ounce) to the indri at seven kilograms (sixteen pounds), although many extinct forms were larger. All lemurs have longer posterior than anterior limbs, and their anatomy reflects their ability to practice vertical clinging and leaping. This is most developed in the sifakas and indris, which generally position themselves vertically while in the trees, leap from tree to tree, and exhibit a leaping, kangaroo-like gait when on the ground.

Cheirogaleidae include the fat-tailed and greater dwarf lemurs, primates that hibernate for up to six months during the dry season. Many of the females in Cheirogaleidae, Megaladapidae, and Lemuridae carry offspring in their mouths and sometimes park them on branches when active.

The aye-aye has a high brain-to-body ratio and also has incisors that grow throughout its life (as in rodents) resulting in a dental formula of 1-0-0-3.

The teeth are used to gnaw on dead wood during searches for grubs that are then removed with a lengthy third finger.

—*Joan C. Stevenson*

See also: Apes to hominids; Baboons; Cannibalism; Chimpanzees; Communication; Communities; Evolution: Animal life; Evolution: Historical perspective; Fauna: Madagascar; Gorillas; Groups; Hominids; *Homo sapiens* and human diversification; Human evolution analysis; Infanticide; Learning; Mammalian social systems; Monkeys; Orangutans; Primates.

Bibliography

Falk, Dean. *Primate Diversity*. New York: W. W. Norton, 2000. An introductory overview.

Jolly, Alison. "Pair-bonding, Female Aggression, and the Evolution of Lemur Societies." *Folia Primatologica* 69, supplement 1 (1998): 1-13. Tries to explain the unique aspects of lemur society.

Kappeler, Peter M. "Determinants of Primate Social Organization: Comparative Evidence and New Insights from Malagasy Lemurs." *Biological Reviews of the Cambridge Philosophical Society* 72, no. 1 (February, 1997): 111-151. An exhaustive review.

Most lemurs live an arboreal lifestyle. (PhotoDisc)

Lemur Facts

Classification:

Kingdom: Animalia

Subkingdom: Bilateria

Phylum: Chordata

Subphylum: Vertebrata

Class: Mammalia

Order: Primates

Suborders: Prosimians (lemurs, lorises, and tarsiers) or Strepsirhini (lemurs and lorises only)

Superfamily: Lemuroidea

Families: Lemuridae (brown, black, crowned, red-bellied lemurs, bamboo lemurs, ring-tailed lemurs, mongoose lemurs, ruffed lemurs); Megaladapidae (sportive lemurs); Cheirogaleidae (dwarf lemurs, mouse lemurs, fork-marked lemurs); Indriidae (avahi or woolly lemurs, indris, sifakas); Daubentoniidae (aye-ayes)

Geographical location: Madagascar and adjacent Comoro Islands

Habitat: Forests (primary or secondary, dry, humid, or rain forests, evergreen or bamboo forests); sometimes also bush, scrub, or savanna edges of humid forests, spiny deserts, and tree plantations

Gestational period: From 65 days in the dwarf lemur to 175 days in the larger sifaka, indri, and aye-aye

Life span: Ranges from about nine years in the greater dwarf lemur to twenty-five to thirty years in ring-tailed, brown, and black lemurs

Special anatomy: A long snout with a sensitive, moist pad (rhinarium); incompletely fused bony eye socket; reflecting retina (tapetum); ancestral dental formula of 2:1:3:3; dental comb in the lower jaw and sublingua; unfused lower jaw and frontal bone; multiple pairs of breasts; two-section uterus; grooming claw on second toe of foot

Mittermeier, Russell A., William R. Konstant, David M. Meyers, and Roderic B. Mast. *Lemurs of Madagascar.* Washington, D.C.: Conservation International, 1994. A field guide, which also covers destruction of habitat, and conservation efforts.

Rakotosamimanana, Berthe, Hanta Rasamimanana, Jorg U. Ganzhorn, and Steven Goodman. *New Directions in Lemur Studies.* New York: Plenum, 1999. Recent studies on behavior and conservation.

Rowe, Noel. *The Pictorial Guide to the Living Primates.* East Hampton, N.Y.: Pogonias Press, 1996. Superb photos and a summary including anatomy and life history of each species.

LEOPARDS

Type of animal science: Classification
Fields of study: Anatomy, physiology, wildlife ecology

Large cats widely distributed over Africa and Asia, leopards have the greatest geographic range of any wild cat, but face severe pressure from humans within many of their natural habitats.

Principal Terms

CANINES: four pointed, elongated teeth that grasp and kill prey

CARNASSIALS: pairs of large, cross-shearing teeth on each side of jaw

CREPUSCULAR: active at twilight or before sunset

MELANISTIC: having dark coloration of skin and hair

PAPILLAE: sharp, curved projections on tongue

Male leopards vary in length from five to eight feet, including a twenty-eight to thirty-seven inch tail. They stand eighteen to thirty-two inches high at the shoulder and weigh from fifty-five to two hundred pounds, depending on subspecies and geographic area. The largest leopards tend to inhabit mountainous terrain and colder regions. On average, females are 40 percent smaller than males. Leopard coats are short and sleek in the tropics and densely furred in colder areas. Their base color varies from yellow cream in desert areas, to golden yellow in grasslands, becoming deep gold in mountains and forests. All leopards are spotted, with black spots arranged in rosettes along the back and sides. Unlike jaguars, leopard rosettes do not have a spot in the center. Leopards in dark, moist tropical forests of Asia are often melanistic and are called black panthers; dark coats may be advantageous in areas of dim light.

The name leopard is also given to two other species: the snow leopard (*Panthera uncia*), displaying cream-colored fur and grayish rosettes, that inhabits mountainous Central Asia, and the clouded leopard (*Neofelis nebulosa*) of Southeast Asia, carrying grayish fur with cloudlike blotches on its sides. Both are endangered by overhunting for their beautiful fur.

Leopard Behavior

Leopard litters average two cubs; weaned at about three months, they soon begin to accompany their mother on hunts. Cubs remain with their mothers eighteen to twenty-four months before leaving to establish their own territories. Adult leopards are solitary hunters. Males sometimes hunt with females shortly after mating but play no role in raising the young.

Leopards are stealthy nocturnal or crepuscular hunters, preferring to stalk prey in the dark or half-light. They pursue a wide variety of targets as opportunity offers, including reptiles, rodents, birds, fish, and hoofed animals. Leopards kill by biting their victims' necks, strangling them, or severing their spinal cords with canine teeth. Scissoring carnassial teeth and rasplike papillae soon clean all bones of meat. Where their habitat abuts human settlements, leopards hunt close to houses, sometimes eating pets or livestock, but rarely attacking humans. Leopards are agile climbers; in areas where they face competition from lions or hyenas, leopards carry prey twice their body weight high into trees to discourage scavengers.

Relations with Humans

Leopard populations have been decimated by intensive hunting for their prized fur and through destruction of their habitat by expanding human populations. Originally, leopards had the widest east-west range and greatest habitat tolerance of any feline species. Their range included all of Africa outside the Sahara, as well as the Middle East, India, Indonesia, China, Korea, and eastern Siberia. Leopards are now endangered in much of Asia and virtually extinct in North Africa, the Middle East, China, and Korea. The 114 Amur leopards (*Panthera pardus orientalis*) currently held in zoos greatly exceed the estimated thirty or fewer still inhabiting the wild.

International traffic in leopard skins, though banned by the Convention on International Trade in Endangered Species, continues to be a problem, yet the leopard is showing greater survival success than its feline competitors in Africa, the lion and the cheetah. Leopards persist in substantial numbers in sub-Saharan Africa, where they are protected. Perhaps 100,000 or more roam the African plains, a number greater than the estimated total population of all other great cats—lions, tigers, jaguars, and mountain lions—combined. However, even where they are valued and guarded

---

Leopard Facts

Classification:
Kingdom: Animalia
Subkingdom: Bilateria
Phylum: Chordata
Subphylum: Vertebrata
Class: Mammalia
Subclass: Eutheria
Order: Carnivora
Family: Felidae (cats)
Genus and species: Panthera pardus

Geographical location: Found over most of Africa south of the Sahara, in the Middle East and India, north to central Asia, and south to Indonesia

Habitat: Grasslands, forests, mountains, and jungles

Gestational period: 3 to 3.5 months

Life span: About fifteen years in the wild, twenty to twenty-five years in captivity

Special anatomy: Large eyes with excellent night vision; jaws adapted to seizing and gripping prey, teeth designed for tearing and slicing flesh

---

Leopards are very agile climbers and usually haul their prey into trees to keep it away from scavengers. (Digital Stock)

as tourist attractions in national parks, illegal hunters continue killing them and farmers on the edge of parks spread poison to protect their cattle and sheep from wandering predators.

Wherever human populations press on leopard habitat, often exaggerated stories of man-eaters arise. One such narrative describes the "Man-Eating Leopard of Rudyaprayag," who was accused of stalking and killing 115 pilgrims en route to a religious shrine in northern India, over a period of eight years, before being hunted down and killed.

—Milton Berman

See also: Carnivores; Cats; Cheetahs; Fauna: Africa; Fauna: Asia; Jaguars; Lions; Mountain lions; Predation; Tigers.

Bibliography

Adamson, Joy. *Queen of Shaba: The Story of an African Leopard*. New York: Harcourt Brace Jovanovich, 1980. Describes raising a three-month-old female leopard cub and releasing her in the wild, where she mated and produced a litter.

Brakefield, Tom. *Kingdom of Might: The World's Big Cats*. Stillwater, Minn.: Voyageur Press, 1993. Documents habits, natural history, biology, and challenges that recent changes in population and range pose for leopards. Extensive bibliography.

Kitchener, Andrew. *The Natural History of the Wild Cats*. Ithaca, N.Y.: Comstock, 1991. Summarizes recent biological and ecological research concerning leopard habits and social life. Twenty-three-page bibliography of scientific articles.

Lumpkin, Susan. *Big Cats*. New York: Facts on File, 1993. Written for younger readers, the work portrays leopard physical characteristics and behavior.

Sleeper, Barbara, and Art Wolfe. *Wild Cats of the World*. New York: Crown, 1995. Describes relations between leopards and humans from prehistoric times to the twentieth century.

LIFE SPANS

Types of animal science: Development, reproduction
Fields of study: Biochemistry, cell biology, developmental biology, genetics, population biology

An animal's life span, the time lapsing between its birth and death, differs greatly from species to species because of differing environmental pressures, chance variations in physical conditions, and heredity.

Principal Terms

LIFE CYCLE: the sequence of development beginning with a certain event in an organism's life (such as the fertilization of a gamete), and ending with the same event in the next generation

LIFE EXPECTANCY: the probable length of life remaining to an organism based upon the average life span of the population to which it belongs

LIFE SPAN: the maximum time between birth and death for the members of a species

METABOLISM: the biochemical action by which energy is stored and used in the body to maintain life

MORTALITY RATE: the percentage of a population dying in a year

---

Life span has two common meanings, often confused. Popularly, the term can refer to the longevity of an individual, but in biology it is more abstract, a characteristic of the entire species rather than individual members. In this sense, life span is the maximum time an individual can live, given its environment and heredity, and life expectancy is the amount of time remaining it at any point during its life span.

The great variety within the animal kingdom complicates the definition of life span. For some species, life span is essentially the same as its life cycle, the return to the same developmental stage from one generation to the next. Salmon, for instance, hatch from eggs in small streams, migrate to the sea where they reach maturity, then struggle up rivers to return to the site of their hatching in order to produce more eggs. After completing the reproduction cycle, both males and females die. For most animal species, however, an individual may produce several generations of offspring before dying. In the case of humans, individuals can live long after their fertility ends.

It is often difficult to measure the life span for specific species, and in fact, a time period is seldom definitive for all species members. Rather, scientists recognize that mortality, the percentage of individuals that die each year (or each day in some cases), increases for a population of organisms until at some age it reaches 100 percent, and all individuals in a generation are dead. Finding the life span of laboratory animals is fairly simple: A population is given the best possible living conditions, and observers wait for the last individual to die. That, presumably, is the optimal life span for the species. Much the same procedure can determine the life spans of pets and animals in zoos, except that the population to be observed is much smaller, sometimes only a single individual. Likewise, the records of thoroughbred domestic animals, born and raised in captivity, provide evidence for the maximum species life span. Most animals live in the wild, however, where investigations face a great variety of conditions and anecdotal evidence can be misleading. For example, biologists long thought that bowhead whales lived only about fifty or sixty years, but in the late 1990's, various new kinds of historical and biochemical evidence identified bowheads that lived well into their second century. Moreover, species

949

It has been theorized that very small mammals, such as mice, have shorter life spans than larger animals because their metabolisms run much faster; their bodies literally wear out more quickly. (Rob and Ann Simpson/Photo Agora)

whose members can, if need be, go into dormancy show a large variation in individuals' apparent longevity, even when all members pass through a single life cycle.

In general, however, the life span variation among species falls into a fairly narrow range of time. The shortest life spans, which last a single life cycle, can be a matter of days, while the longest last more than two hundred years. Humans enjoy the greatest longevity among primates; scientists estimate the theoretical maximum human life span to be from 130 to 150 years, more than double that of the species with the next greatest endurance, gorillas and chimpanzees. However, several kinds of invertebrates live more than two centuries.

Life Span Limiters and Extenders

While each species has a theoretical maximum life span, few if any individual animals reach it. Three general influences limit longevity: environmental pressures, variations in physiological processes, and heredity.

Most domesticated species have longer life spans, frequently two times longer, than their wild relatives, and wild animals in captivity often live longer than in their natural habitat. The gray squirrel, for example, lives for three to six years in the wild, but from fifteen to twenty years in a zoo. The reason is clearly the safer, healthier, less stressful environment. Predators are one of the biggest threats in the wild, as are fluctuations in climate that affect the availability of food and shelter. Natural calamities such as hurricanes, wild fires, and earthquakes also take their toll.

Disease and chance injury also kill off many organisms, but even if disease is absent, many physiological processes in the body appear to degenerate or stop with age. Biochemists find that individual cells age and die. Cross-connections among connective tissue, such as ligaments and cartilage, gradually reduce the body's flexibility and inhibit motion. Chemical plaques build up in brain tissue, hindering the electrochemical connection among neurons. Highly reactive oxidants, the ionized molecular byproducts of cellular metabolism, also build up with age and degrade the operation of cells' mitochondria, the generators of chemical energy, as well as other cellular organelles. Moreover, laboratory tests reveal that cells can divide only a certain number of times—about fifty for some human cells—and then they die, a limit called the Hayflick finite doubling potential phenomenon. In connection with this finding, geneticists discovered that the lengths of deoxyribonucleic acid (DNA) at the tips of chromosomes, called telomeres, shorten with age and appear to play a role in cell dysfunction and death.

Other genetic factors help determine the age limit of cells and the entire organism. Research in the late 1990's with mice and roundworms uncovered genes that regulate cell life, including one kind that causes cells to suicide if their DNA or internal structure is too damaged to function properly or if the cells turn cancerous. Loss of function by genes damaged during cell division (mitosis) or by environmental toxins can impair a cell's ability to maintain metabolism, or set loose functionless or even outrightly harmful proteins and enzymes into the blood stream and lymph system—all of which can bring on illness. Organisms that escape such effects still may face genetic disease. Most human populations, for instance, now have a longer life expectancy than ever before; because of it, neurodegenerative diseases (such as Alzheimer's disease), cardiovascular dysfunction, and immunological disorders, caused or made possible by genes, grow ever more common. Because many of these life-shortening maladies were rare or nonexistent before the twentieth century, natural selection has not had a chance to cull them from the human genome. Similar diseases crop up in domestic animals and wild animals in captivity.

Late twentieth century research also identified several ways to extend life. A sharply reduced diet extends the lives of mice and fruit flies beyond their normal life span (although for the brown trout, a richer diet is life-extending), and lower than normal temperature has the same effect on fruit flies, fish, and lizards. Discovery of the relation of telomere length to cell death and isolation of genes that order cells to suicide allowed scientists to bioengineer animals with cells whose DNA self-repaired the telomeres and genes that failed to trigger cell-suicide; the result was individuals that in some cases had life spans twice the normal length or more. However, the most pervasive life-extending method is the development of social life—colonies, herds, packs—in which individuals work together to protect, shelter, and feed themselves and rear their offspring. In the case of humans (and perhaps some whales), culture and intelligence permit the species to pass on recently acquired information from one generation to the next and even to alter the environment to lengthen collective and individual life spans.

Size and Life Span
Biologists noted long ago that large animals live longer than small animals, but why this should be true was a mystery. In 1883, Max Rubner proposed that the relation had to do with metabolism. Large animals have a smaller skin surface to body mass ratio than smaller animals; accordingly, large animals lose heat more slowly and so can maintain body functions with less energy burned per unit of mass; in other words, a slower metabolism. Scientists since found that individuals of different animal species all use about the same amount of chemical energy during their lives, twenty-five to forty million calories per pound per lifetime. (There are significant exceptions: Humans consume about eighty million calories per pound.) Because small animals must burn energy at a higher rate, the argument holds, they physically wear out faster.

In 1932, Max Kleiber derived a mathematical relationship for Rubner's proposal. According to Kleiber's law, also known as the quarter-power scaling law, as mass rises, pulse rate decreases by the one-fourth power. So elephants, which have 104 times the mass of chickens, have a pulse rate one tenth as fast. Scientists suggest that the relation results from the geometry of circulatory systems and point out that the quarter-power scaling law is pervasive in nature, but the underlying reason for it remains unknown. In any case, plenty of exceptions to the mass-life span correlation exist. With a life span of about one hundred years, box turtles outlive fellow reptiles, for example, and humans outlive all mammals (with the possible exception of some whale species) regardless of size. Exceptions also occur among domestic species living sheltered lives: Cats have longer life spans than dogs.

Theories of Life Span
A 1995 review of data from earlier animal studies suggested that heredity accounts for about 35 percent of the variation in life spans among invertebrates and mammals; 65 percent comes from unshared environmental influences. Nonetheless, theorists in the life sciences continue to debate the relative influence of genetic and other biochemical factors on the one hand and environment fac-

The Big, the Old, and the Immortal

In some species, the older an animal gets, the larger it grows. Big sharks are old sharks, as are big crocodiles and snakes. In 1912, a reticulated python was found that measured thirty-two feet, ten inches in length, the longest ever, but no one knows how old it was. The age record for snakes goes to Popeye, a boa constrictor at the Philadelphia Zoo that died in 1977 at the great age of forty years and three months. Even for reptiles, however, Popeye's life span was modest. One Madagascar radiated turtle lived 188 years.

Most species, including humans, grow to a maximum size and then stop. The longest attested life span for a person is 122 years, achieved by France's Jeanne Louise Calment, who died in 1997, followed by Japan's Shigechiyo Izumi, who died just short of 121 years in 1986. Despite theories relating large body mass to long life span, some big and small animals in the wild have about the same life expectancy at birth: fifty years for both golden eagles and most whales, for instance, or twenty-four years for rhesus monkeys compared with twenty-five years for lions.

There is some disagreement over which animal species has the longest life span. The tubeworms growing on the ocean floor near hydrocarbon-seep sites in the Gulf of Mexico are thought to be as much as 250 years old, and so have been called the longest-lived noncolonial animals without backbones. However, some mollusks of the class Lamellibranchia, such as quahogs, a variety of clam, have also been called the longest-lived animal. They, too, can lie snug in their ocean bed for 250 years.

Whether classified as animals or not, bacteria hold the record for life span. If they are not eaten by predators or killed by environmental change, these single-cell creatures are theoretically immortal. Furthermore, they do not get larger with age: They divide.

tors on the other hand. The debate derives from the premise that life spans are the product of the natural selection that ensured species' reproductive success. The proposals fall into three categories: random damage (stochastic) theory, programmed self-destruction theory, and ecological theory.

Random damage theories emphasize the wear and tear on the body that accumulates with metabolic action. It is the source of damage that differs from one theory to another. One holds that the buildup of metabolically produced antioxidants is the key factor, a spinoff of the long-standing conjecture that the faster an animal's metabolism is, the shorter its life span. A second theory focuses on proteins that change over time until their effect on the body alters for the worse, especially when the proteins are involved in cellular repair. There is, for example, the altered connective tissue that causes the cross-connections stiffening tendons and ligaments. Another such change is the glycosylation of proteins or nucleic acids, in which a carbohydrate is added. Glycosylation is involved in such age-related disorders as cataracts, vascular degeneration among diabetics, and possibly atherosclerosis A third theory points to the buildup of toxins inside cells, and a fourth concerns the potential problems that come from errors in metabolism or viral infection which slowly impair or kill cells. Fifth, the somatic mutations theory proposes that chance mutations accumulate in a person's nuclear or mitochondrial genome and induce cell death or produce proteins and enzymes that have aging effects.

Programmed death theories hold, as the name suggests, that a species' genetic heritage includes a built-in timer or damage sensor. Telomeres shorten as DNA ages until the genes at the end of chromosomes are unprotected and subject to deterioration during the splitting and gene crossover of mitosis. The genes then lose their ability to produce essential biochemicals, whose absence harms the body or leaves it defenseless against damage from infection or injury. Damage sensors can include the genes that instruct cancerous or malfunctioning cells to die. Although such genes clearly are a means to check the spread of disease, their cumulative effect may be harmful. Furthermore, scientists discovered genes that produce much more of, or less of, their metabolic products as cells age, which also contributes to the overall aging of the body.

The ecological theory draws conclusions about life span from a species' role in its environment. Small animals have faster metabolisms and live shorter lives, it is argued, because they are not likely to escape predators for very long. Therefore, they evolved to mature and reproduce rapidly. Large animals typically have more defenses against predators and can afford to take life slowly. Moreover, animals that evolve defensive armor, spines, or poison also avoid predation and live longer than related species that do not. Finally, species that evolve mechanisms to withstand environmental stress, as from extreme temperatures or food scarcity, also have long life spans.

The theories assume that the life span for individuals within a species serves the survival of the entire species. Yet even a species' days on earth are numbered. Environmental change can slowly squeeze them from their habitats, a catastrophe may wipe them out indiscriminately, or they may evolve into a new species. Scientists estimate that the average life span for a multicellular species lasts from one to fifteen million years. That average is stretched by several notable exceptions in the animal kingdom—such living fossils as crocodiles (140 million years old), horseshoe crabs (200 million), cockroaches (250 million), coelacanths (a type of fish, 400 million), and certain mollusks of the genus *Neopilina* (500 million).

—*Roger Smith*

See also: Aging; Birth; Death and dying; Demographics; Extinction; Extinction and evolutionary explosions; Genetics; Growth; Heterochrony; Mark, release, and recapture methods; Metabolic rates; Migration; Offspring care; Population analysis; Population fluctuations; Population genetics; Population growth; Predation; Reproduction; Reproductive strategies; Veterinary medicine; Wildlife management; Zoology; Zoos.

Bibliography

Austad, Steven. *Why We Age*. New York: John Wiley & Sons, 1997. Written by a leading investigator of aging, the book is primarily devoted to human concerns, but there is much discussion of theories of aging, as well as specific biological and environmental influences, pertinent to animals, especially mammals.

Bova, Ben. *Immortality*. New York: Avon Books, 1998. Although human life span is his primary topic, Bova discusses animal life spans and the biochemistry of aging in clear, engaging prose for general readers.

Finch, Caleb B., and Rudolph E. Tanzi. "The Genetics of Aging." *Science* 278, no. 5337 (October 17, 1997): 104-109. A concise, well-explained survey of the genetic influences on life span, drawing from studies conducted on animals. Best for readers who understand basic biology and genetics.

Furlow, Bryant, and Tara Armijo-Prewitt. "Fly Now, Die Later." *New Scientist* 164, no. 2209 (October 23, 1999): 32-35. The authors describe the age differences between closely related creatures of the same size to support their theory that environmental factors, such as predation, influence the evolution of life spans for species.

MacKenzie, Dana. "New Clues to Why Size Equals Destiny." *Science* 284, no. 5420 (April 6, 1999): 1607-1608. Discusses theories about the relation of size to life span in a style accessible to general readers.

Margulis, Lynn, and Karlene V. Schwartz. *The Five Kingdoms: An Illustrated Guide to the Phyla of Life on Earth*. 3d ed. New York: W. H. Freeman, 1998. A richly illustrated explanation of taxonomy with detailed information about the biology of the thirty-seven phyla in the kingdom Animalia, including discussion of life cycles and development.

Walford, Roy L. *Maximum Life Span*. New York: W. W. Norton, 1983. Delightfully written and lucid, this book, written by a pioneering researcher in aging, is somewhat out of date and focuses primarily upon human aging, but it still supplies solid information about the biological and ecological limits on animal life spans.

LIONS

Type of animal science: Classification
Fields of study: Anatomy, wildlife ecology

Lions, celebrated in myth as the King of Beasts, are the second largest of the big cats and the only members of the Felidae family to live and hunt in groups in the wild.

Principal Terms

CANINES: four elongated, pointed teeth that grasp and kill prey

CARNASSIALS: pairs of large, cross-shearing teeth on each side of jaw

CREPUSCULAR: active after sunset and in early morning

DIMORPHISM: existence of two distinct forms within a species

PAPILLAE: sharp, curved projections on tongue

Surpassed in size only by tigers, African male lions range in weight from 330 to 420 pounds, stand about forty-eight inches tall at the shoulder, and average 8 to 9.5 feet in length, including the tail. Females are smaller, weighing from 260 to 350 pounds, stand about forty inches tall, and average seven to eight feet in length. The Asian lion subspecies (*Panthera leo persica*) is similar in size, but somewhat shorter and stockier in build.

Among cats, lions are the only species to show sexual dimorphism—mature males display a distinctive mane encircling their head and shoulders; it darkens with age, giving the lion a majestic look. Coat color is normally tawny yellow to reddish brown in both sexes, with black accents on ears, tail tips, and manes. Cubs are born spotted, but by three months of age begin showing the uniform coat color of their parents.

Lion Behavior

Unlike other wild cat species, lions live and hunt in groups, called prides. (Domestic cats exhibit some similarities in group behavior.) The pride's core consists of two to twelve closely related lionesses, who assist each other in raising their cubs. Female offspring usually remain members of the group, but males are driven off before becoming sexually mature. Two to four unrelated males live with the pride, fathering the cubs, protecting the pride, and proclaiming their territory with scent marks and loud roars that can be heard for five miles. Males rarely control a pride more than three or four years before being replaced by younger, more powerful challengers.

Lions are crepuscular hunters, preferring to rest in the shade during the heat of the day; they emerge at sunset or in the early morning to pursue their prey. Lionesses do most of the killing, cooperating when stalking and ambushing victims. The preferred targets are medium to large hoofed animals such as antelopes, zebras, and wildebeests. Males rarely participate in chases unless their weight is needed to bring down large bull buffaloes; however, males claim first place at the pride's feasts. Lions are also opportunistic scavengers, stealing prey from leopards, cheetahs, and hyenas. Similar to other big cats, lions kill by biting their victims' necks, strangling them, or severing their spinal cords with sharp canine teeth. Scissoring carnassials and rasplike papillae leave little of their victims other than bones and skin.

Lions and Humans

Throughout history, lions have been called "King of Beasts" and used to symbolize royal authority. Rulers, often described as lions, wore lion skins to

Lionesses do most of the hunting for the pride. (Digital Stock)

impress subjects with their majesty and power. Hunters demonstrated bravery and prowess by killing lions. Lions have been celebrated in myth and legend from antiquity to the present; in astrology, Leo Major reigns as the fifth house of the Zodiac.

Originally, lions existed throughout southern Europe, all of Africa, the Near East, and from southwest Asia to India. As human population and agriculture expanded, hunting and habitat destruction greatly reduced lion numbers. By 100 C.E., lions had been eliminated from Europe, by the time of the Crusades they were extinct in the Near East, and before 1900 had vanished from North Africa and most of Asia. When the Nawab of Junagadh established the Gir Forest Lion Sanctuary in Gujarat, India, in 1965, less than 100 Asian lions remained in the world; a 1990 study estimated that the population had rebounded to about 250 individuals.

Estimates of the African lion population south of the Sahara Desert run as high as fifty thousand. The overwhelming majority live in national parks and reserves, where wild animals are protected as valuable national assets, attracting tourists and significant income. In the parks, lions have become so accustomed to humans that they sometimes rest in the shade of tour buses to avoid the afternoon sun. As human populations continue expanding, governments face pressure to reduce the preserves and permit agriculture and grazing to increase the food supply. However, most naturalists believe the monetary value of wild animals in producing foreign exchange will ensure the long-term survival of the African lion, even if reduced in numbers.

—*Milton Berman*

See also: Carnivores; Cats; Cheetahs; Fauna: Africa; Fauna: Asia; Groups; Jaguars; Leopards; Mountain lions; Predation; Tigers.

Bibliography

Adamson, Joy. *Born Free: A Lioness in Two Worlds.* 1960. 40th anniversary ed. New York: Pantheon, 2000. Describes raising an orphaned female lion cub and releasing her into the wild, where she mated and produced young.

Lion Facts

Classification:

Kingdom: Animalia

Subkingdom: Bilateria

Phylum: Chordata

Subphylum: Vertebrata

Class: Mammalia

Subclass: Eutheria

Order: Carnivora

Family: Felidae (cats)

Genus and species: Panthera leo

Geographical location: Once common to many areas of Europe, Africa, and Asia; now found only within protected areas in Africa south of the Sahara and in one wildlife refuge in India

Habitat: Grassy plains, savannas, and open woodlands

Gestational period: 3 to 3.5 months

Life span: About twelve to fifteen years in the wild, twenty years or more in captivity

Special anatomy: Large eyes with excellent night vision; jaws adapted to seizing and gripping prey; teeth designed for tearing and slicing flesh

Brakefield, Tom. *Kingdom of Might: The World's Big Cats.* Stillwater, Minn.: Voyageur Press, 1993. Documents habits, natural history, biology, and challenges that change in population and habitat pose for lions. Extensive bibliography.

Kitchener, Andrew. *The Natural History of the Wild Cats.* Ithaca, N.Y.: Comstock Publishing, 1991. Summarizes recent biological and ecological research concerning lion habits and social life. Twenty-three page bibliography of scientific articles.

Scott, Jonathan. *Kingdom of Lions.* Emmaus, Pa.: Rodale Press, 1992. Photos and commentary based on fifteen years observing behavior and problems of lions living in the Masai Mara National Reserve in Kenya.

Sleeper, Barbara, and Art Wolfe. *Wild Cats of the World.* New York: Crown, 1995. Describes hunting behavior of lions and their relations with humans.

LIZARDS

Types of animal science: Behavior, classification, ecology, evolution, reproduction
Fields of study: Anatomy, conservation biology, herpetology, physiology, systematics (taxonomy), zoology

More than six thousand species of reptiles have been identified and named. They are classified in the order Squamata, of which slightly more than four thousand are lizards. The remainder are snakes. Technically, snakes comprise one group of limbless lizards. Squamata is the largest order of extant reptiles.

Principal Terms

ACTIVE or WIDE FORAGING: moving about in search of prey

AUTOTOMY: loss of the tail by controlled muscle contractions resulting in breakage at intervertebral sutures

BEHAVIORAL THERMOREGULATION: maintaining relatively constant body temperature by shuttling between warm and cool microhabitats

ECDYSIS: sloughing off of old skin, which is replaced by new skin from underneath

LECITHOTROPHY: nutrition of developing offspring from yolk reserves within the egg

MATROTROPHY: nutrition of developing offspring directly from the mother; in the case of lizards, via a placenta

OVIPARITY: production of shelled eggs by females

PARTHENOGENESIS: reproduction in which unfertilized eggs develop into females genetically identical to their mothers

SIT-AND-WAIT FORAGING: sitting in one place, waiting, and attacking prey as they move

SPECTACLE: transparent scale covering the eye as a replacement for the eyelid; occurs in some lizards and all snakes

VIVIPARITY: production of live young by females

Lizards belong to the order Squamata, along with snakes. There are approximately four thousand lizard species.

Lizard Reproduction

Most lizards reproduce sexually, although some are parthenogenetic. Most lizards are polygynous, with males mating with more than a single female, although a few, such as Australian sleepy lizards (*Trachydosaurus rugosus*), are monogamous. Mating occurs after complex social behavior often involving prolonged courtship. Fertilization is internal. Males have paired intromittant organs called hemipenes, one of which is inserted in the female's cloaca during mating. Once eggs are fertilized, the female carries eggs or embryos for various periods of time. The weight of unborn offspring usually reduces the female's ability to run fast, thus affecting her ability to escape predators. Many females change their behavior while gravid to reduce the costs of reproduction. Costs of reproduction are not confined to increased predation risk; energy required for locomotion increases as well due to the added weight that females carry around while gravid.

Most lizards produce eggs (oviparity) but many produce live young (viviparity) following extended gestation periods. Females of oviparous species deposit eggs in places that are moist but not wet, such as inside rotted logs or in the ground, often under rocks. Most lizard eggs have pliable, leathery-shelled eggs, but a few, such as geckos, have hard, calcified eggshells. Lizard eggs

Lizard Facts

Classification:

Kingdom: Animalia

Subkingdom: Bilateria

Phylum: Vertebrata

Subphylum: Tetrapoda

Class: Reptilia

Subclass: Diapsida

Superorder: Lepidosauria

Order: Squamata

Suborders: Iguania (iguanids, chamaeleonids, and agamids); Scleroglossa (all other lizards and snakes); Gekkota (gekkonids, eublepharids, diplodactylids, and pygopodids); Autarchoglossa (teiids, gymnophthalmids, lacertids, xantusiids, dibamids, amphisbaenids, trogonophids, rhineurids, bipedids, scincids, gerrhosaurids, cordylids, anguids, xenosaurids, helodermatids, lanthanotids, varanids, and "snakes")

Geographical location: Every continent except Antarctica

Habitat: A vast majority live on land or in trees in all habitats (desert, savanna, temperate and tropical forest); a few are semiaquatic; one, the Galápagos iguana, dives into the ocean and feeds on algae

Gestational period: Highly variable; eggs of *Anolis* lizards hatch in several weeks, while embryos of some skinks (*Mabuya*) require a year

Life span: Variable; some small skinks and geckos live one year or less, whereas Gila monsters (*Heloderma*) and large monitors (*Varanus*) live thirty years or more

Special anatomy: All lizards have scales (like other reptiles), paired copulatory organs in males (hemipenes), and claws on their feet; most have four legs, an elongate tail, eyelids, and external ear openings; some lizards in several families have lost two or all four legs as adaptations for life underground, some have their ear openings completely covered by scales, some have short tails, and some have their eyes covered by a transparent scale (spectacle); some subterranean lizards, such as the strange worm lizards, have eyes reduced to an eyespot covered by scales

with a specialized scale on the front of their jaw, called the egg tooth. Parental care ends once offspring exit the eggs. Hatchlings are fully formed, resembling miniature adults. Females of viviparous species often provide some parental care to neonates (newborns). Most help neonates free themselves from embryonic membranes, often eating the membranes. A few, such as the large Australian sleepy lizard, engage in extended parental attention, but it does not involve feeding or grooming the young, as in birds and mammals.

A number of lizard species in several different families reproduce by parthenogenesis, a process in which females produce daughters that are genetically similar to their mothers without the involvement of males. In such species, potential population growth is extremely high because no energy is wasted on males and every individual produces offspring.

Although sex determination in most lizards is chromosomal, as in humans, some species of lizards lack sex chromosomes and have environmental sex determination. Eggs incubated at one set of temperatures produce all males, whereas eggs incubated at another set of temperatures produce all females.

The juvenile stage of most lizards is also a high mortality stage. Juvenile lizards are relatively small in body size, and, as a consequence, many predators can easily eat them. Because juveniles do not reproduce, all energy taken in is devoted to

in the nest are vulnerable to predation by many animals because they cannot move. Eggs contain yolk, which is high in energy and thus a good food source for snakes, mammals, and even some other lizards. Consequently, mortality of lizard eggs in nests is high. A vast majority of oviparous lizards do not provide parental care to eggs, but females of a few, such as five-lined skinks and glass lizards, remain with the eggs, brooding them until they hatch. Hatchlings cut slices in their eggshells

growth and maintenance. When lizards reach sexual maturity, growth slows and most energy is directed into reproduction and maintenance. Males use energy in reproductive related behaviors such as territorial defense and courtship, whereas females use energy for production of eggs.

Evolution of Sensing Systems

One of the most fascinating aspects of the natural history of lizards is the variation in relative importance of sensory systems and its consequences. Most lizards in the suborder Iguania are visually oriented, sit-and-wait predators. They capture prey by a process known as tongue prehension; basically, they stick their tongues out and carry insects into the mouth on their tongues. They are colored for camouflage, move very little, and attack moving prey from perches, to which they usually return. Their social systems usually involve territoriality, the defense of a specific area against intruders of the same species. In most instances, territorial behavior is directed toward other males.

Lizards in the suborder Scleroglossa use a combination of visual and chemical cues to locate and discriminate prey. They have well-developed vomeronasal systems; they pick up chemicals from the external environment with their tongues and bring them into the mouth, where they are passed over the vomeronasal organ (also called the Jacobson's organ) in the roof of their mouth. This organ transfers information directly to the lizard's brain, allowing it to discriminate prey on the basis of chemicals much like mammals' sense of smell. These lizards capture prey by grasping prey between their jaws. The tongue may be used to help manipulate prey within the mouth but is not used to capture prey. Most of these are active foragers; they travel around, searching for prey, many of which are not moving.

Venomous and Dangerous Lizards

Most lizards will bite in self-defense, and, because all lizards have teeth, some will break the skin. Only two lizard species in the world are venomous, the Gila monster and its close relative, the Mexican beaded lizard. Both have powerful jaws and are difficult to remove should they bite. They do not have fangs like poisonous snakes; rather, they have grooved teeth with venom glands located in the rear of the mouth. Venom moves along the grooved teeth as the lizards grind their jaws while biting. Bites are only rarely fatal. Large monitor lizards are nonvenomous, but their mouths contain high levels of bacteria, which can cause dangerous infections following bites. Some of the larger monitors, such as the Komodo dragon of the Lesser Sunda Islands, have sharp, serrated teeth and can bite completely through the

Iguanas are sit-and-wait predators that catch passing insects with their long tongues. (Digital Stock)

Some Types of Lizards

ANOLES (genus *Anolis*) are highly diverse, with more than three hundred species, all in the New World. Most are arboreal, and females produce only a single egg in each clutch.

CHAMELEONS (family Chamaeleonidae), restricted to the Old World, have prehensile tails and long protrusible tongues that they use to capture insects.

CHUCKWALLAS (genus *Sauromalus*) are herbivorous lizards of the American Southwest desert that hide in rock crevices, where they inflate their bodies, making it very difficult to remove them.

FLYING LIZARDS (genus *Draco*) of Southeast Asia have large flaps of skin along the sides of their bodies that become parachutes when erected by their ribs, allowing them to glide through the air after jumping from tree surfaces.

GECKOS (family Gekkonidae) include many species with adhesive toe pads, which allow them to climb smooth, vertical surfaces.

GLASS LIZARDS (genus *Ophisaurus*) have no legs and superficially resemble snakes. Their long tails are easily broken and often break into several pieces, each of which moves around violently, distracting predators while the lizard slowly escapes.

IGUANAS (*Iguana iguana*) are large tropical lizards that feed exclusively on plants. They have microorganisms in their digestive tracts that break down cellulose, allowing them to digest leaves.

MONITORS (genus *Varanus*) are large, intelligent lizards living in the Old World and Australia. The largest monitor is the Komodo dragon, which reaches nearly twelve feet in total length.

NIGHT LIZARDS (genus *Xantusia*) are secretive lizards that live in rock crevices and under branches of fallen Joshua trees in the desert southwest of North America. Even though they are very small in size, they are long-lived.

TEGU LIZARDS (genus *Tupinambis*) live in open habitats in South America and reach more than 3.5 feet in total length.

WHIPTAIL LIZARDS (genus *Cnemidophorus*) are unusual New World lizards in that many species are parthenogenetic, reproducing without males and producing only female clones.

leg muscles of large vertebrates. As a result, their bites are potentially life-threatening even though they rarely attack humans.

—*Laurie J. Vitt*

See also: Chameleons; Cold-blooded animals; Reptiles; Salamanders and newts; Smell; Snakes; Turtles and tortoises.

Bibliography

Auffenberg, Walter. *The Behavioral Ecology of the Komodo Monitor.* Gainesville: University Presses of Florida, 1981. The most comprehensive study of a single lizard species. Covers nearly all aspects of the biology of the world's largest lizard, including nice accounts of its interaction with humans.

Cogger, Harold G. *Reptiles and Amphibians of Australia.* 6th ed. Sanible Island, Fla.: Ralph Curtis Books, 2000. Of all regional guides, this is the best for amphibians and reptiles in general. Because Australia has extremely high lizard diversity, it is the best available guide for lizards, limited only by its restriction to Australia.

Cogger, Harold G., and Richard G. Zweifel, eds. *Encyclopedia of Reptiles and Amphibians.* 2d ed. San Diego, Calif.: Academic Press, 1998. A beautifully done desktop book with outstanding color photographs throughout. Information on all reptiles and amphibians is presented in a clear and concise form.

Pough, F. H., R. M. Andrews, J. E. Cadle, M. L. Crump, A. H. Savitzky, and K. D. Wells. *Herpetology*. 2d ed. Upper Saddle River, N.J.: Prentice Hall, 2001. This is one of two standard, comprehensive herpetology textbooks. Aimed at upper-division college undergraduates and graduate students.

Vitt, Laurie J., and Eric R. Pianka, eds. *Lizard Ecology: Historical and Experimental Perspectives*. Princeton, N.J.: Princeton University Press, 1994. A series of articles by leading researchers in lizard ecology. Oriented toward professional scientists.

Vitt, Laurie J., Janalee P. Caldwell, and George R. Zug. *Herpetology: An Introductory Biology of Amphibians and Reptiles*. 2d ed. San Diego, Calif.: Academic Press, 2001. This is one of two standard, comprehensive herpetology textbooks. Aimed at high school through college undergraduate-level students with background in biology.

LOCOMOTION

Type of animal science: Physiology
Fields of study: Anatomy, biochemistry, physiology, zoology

Locomotion, the ability to move from one place to another as needed, is widespread. It is seen in unicellular animals, which move via cilia, flagella, or like amoebas. The locomotion of multicellular animals includes crawling, walking and running, swimming, gliding, and flying.

Principal Terms

AIRFOIL: the wing of a flying animal or airplane that the provides lift and/or thrust needed for flight

ANGLE OF ATTACK: the angle at which an airfoil meets the air passing it

EUKARYOTE: a higher organism, whose cells have their genetic material in a membrane-bound nucleus and possess other membrane-bound organelles

LIFT: the force enabling flight; it occurs when an airfoil makes air passing it move so as to lower air pressure above the airfoil to values below that beneath the airfoil

LOCOMOTION: the ability of an organism to move from one place to another as needed

MITOCHONDRION: a subcellular organelle that converts foods to carbon dioxide, water, and energy

SESSILE: an organism that is not capable of moving from its point of origin

STRIATED MUSCLE: voluntary or skeletal muscle, capable of conscious enervation

Being a sessile animal, unable to move from place to place, makes life difficult for a wide variety of reasons. For example, if food is not available in a very limited area, death may come very quickly. As a result, a great many living animals, from unicellular microbes to the multi-cellular higher organisms, have developed the ability to move from one place to another as needed. This ability for self-directed movement is called locomotion.

Locomotion is under conscious (or voluntary) control. In essence, it serves to allow living creatures to move in chosen directions, toward food, away from danger, toward mates, and so on. Locomotion is essential for the optimization of the lives of animals. This can be seen in two ways: Most living animals are capable of locomotion; it is particularly well developed in large, higher organisms.

The Physiology of Locomotion

Many unicellular animals, such as protozoa, carry out locomotion via hairlike organelles called cilia and flagella, or by amoeba-like motion. Their movement is not complex. Multicellular animals, crawl, walk, run, swim, glide, or fly. The basis for these much more complex activities is cooperative operation of their skeletons and muscles, along with other processes needed to suit each organism to its environment. In addition, nerve impulse transmission coordinates locomotion.

The skeleton and the muscles act as the levers that can move to produce body motion in chosen directions and as the machines that produce the motive ability required for locomotion. Skeletons are of two types. The external, bony, jointed exoskeletons of arthropods (insects and crustaceans) surround them like medieval suits of armor. This design limits the dexterity of arthropods and the complexity of the motion possible for them. In contrast, the bony endoskeletons of vertebrates are much more complex and produce the ability for much more complicated and dexterous locomotion.

The muscle involved in locomotion can be exemplified by vertebrate striated skeletal muscle. This type of muscle is called striated because its fibers appear to be banded when viewed under light microscopes. Such muscle is under the conscious (voluntary) control of any higher organism containing it and wishing to move toward a desired place. Striated muscle is either red or white. Red striated muscle has a very good blood supply—the basis for its color—and obtains energy in the course of oxidizing foods into carbon dioxide and water. This occurs in its many mitochondria, subcellular organelles that carry out oxidative, energy-getting processes. White striated muscle has less of a blood supply and less mitochondria. It obtains most of its energy by converting food to lactic acid, without much use of oxidation or mitochondria.

All locomotion uses the contraction and relaxation of muscle or related systems, which can be thought of as muscle waves. In the unicellular eukaryotes (such as protozoa), cilia and flagella are used. These structures are cell organelles that are whiplike and identical in their makeup. However, their size and number per cell differ. One protozoan type may have on its surface hundreds of cilia, each ten to twenty microns long. Another species can have one or two flagella up to 250 microns long.

Both cilia and flagella cause movement by producing a wavelike motion transmitted all through the organelle, outward from its base. Their action uses organized protein microtubules in a network, interconnected by arms and spokes reminiscent of fibrils, and crosslinks of striated muscle. The mechanism of their action also uses a sliding mechanism initiated by a protein called dynein. Cilia and flagella are also found in multicellular eukaryotes, including humans. For example, flagella propel human sperm and are found in human lungs, where they mediate the removal of dust and dirt from the airways.

Amoeboid motion also occurs in protozoa. Here, their protoplasm flows in the direction of

Cheetahs are the fastest animals on earth, running at speeds of up to seventy miles per hour. They increase the distance covered in each step by leaping as well as running. (Corbis)

Muscle and Bone

Light microscopy of striated muscle fibers reveals a regular pattern of alternating light and dark bands that is repeated every few microns along the fibers. This banding pattern has been found to explain muscle contraction. The process operates due to the presence in the muscle of two kinds of filaments, thick filaments and thin filaments, which run lengthwise along each muscle fiber. Thick filaments are made mostly of a protein called myosin and have projecting cross-bridges. Thin filaments contain mostly the protein called actin and have no cross-bridges. When a muscle contracts, due to energy input, the cross-bridges of the thick filaments attach to the thin filaments surrounding them, pulling them so that they slide between thick filaments. This sliding action shortens the muscle fibers, and as the shortening proceeds, cross-bridges detach and reattach after walking to positions further along the thin filaments.

If a vertebrate animal is to walk, run, swim, or fly, a system of levers is required to turn its muscle contractions into locomotion. This is provided by its endoskeleton, which pulls the levers inside the body, joined to muscles. The skeletons of vertebrates are made of bones, which meet at joints that allow bending and twisting due to the action of muscles attached to the bones. A joint which only allows bending to occur (such as an elbow or knee) has flexor muscles to bend them and antagonistic extensor muscles to straighten them. Joints, such as those in the fingers, have the ability for much more freedom of movement. This is because they are moved by a larger number of muscles. Some muscles attach directly to the bones of the skeleton, but others join the bones via tendons.

desired motion, forming pseudopods, followed by forward locomotion of the cell. This type of locomotion seems to depend on cytoplasm transitioning between fluid endoplasm and gel-like ectoplasm. Pseudopod formation occurs due to the action of microfilaments of actin, one of the contractile proteins also found in striated muscle. Thus, the contraction mechanism may be similar to that in striated muscle.

The various forms of locomotion include crawling, walking and running, swimming, gliding, and flying. These locomotive processes all use muscle waves based on actin and myosin fibrils. However, the complete systems involved become more and more complex due to increases of the needs of locomotion, crawling on earth's surface, walking or running on land, swimming in lakes and oceans, and gliding or flying in the air. The increases in complexity are mostly due to the need for the involvement of more and more muscles.

Crawling, Walking, and Running

Crawling is movement on land without legs, produced by waves of muscular activity traveling along the crawler's body. Two crawling creatures are snakes and snails. When a snake crawls, it curves its body into bends around stones and ground irregularities. These bends travel backward along the snake's body in muscle waves. The waves push against the stones and irregularities and propel the snake forward. Snails, in contrast, usually crawl on a long foot covered with mucus, using waves of muscle contraction that travel along the foot. Snail locomotion depends on their interesting mucus, which resists gentle pressure as if it were a solid, but when pressed hard behaves like a liquid. The muscle waves of snail feet are designed so that foot portions moving forward exert pressure enough to liquefy the mucus, while parts that move backward do not.

All living organisms capable of walking or running have skeletons because the processes are intricate and require the use of a relatively large number of muscles in a complex fashion. Walking is locomotion at a slow pace, when there is no need for speedy movement. Running enables organisms to move much more quickly. It is done when predators threaten, or if there is an unthreatening reason to travel more quickly than usual, for instance, to obtain some choice food or a mate.

The complexity of walking and running is made clear by the coordination required. For example, when four-footed animals walk, at least one forefoot and one hind foot are kept on the ground at all times. In walking bipeds, at least one foot is on the ground at all times, and at intervals, both feet are on the ground. In running, there are times when both feet leave the ground simultaneously. Four-footed mammals walk slowly, but often attain higher speeds by trotting, cantering, or galloping. In each of these gaits, there are differences in the operation of the legs. For example, in a gallop, both forefeet are set down at one instant and both hind feet are set down at the next instant. This is not due to chance; rather, it allows the back muscles to contribute to the work done and spread the overall effort through the body.

Another aspect of the differences between walking and running is the interaction of the feet with the ground. Some animals, including humans, walk or run with the entire bottom of the foot on the ground; this is known as plantigrade locomotion. In contrast, most quadrupeds carry out these actions with only their toes on the ground, known as digitigrade locomotion. Such differences in ways in which vertebrates walk and run add to the complexity of the processes and the number of muscles used. The rate at which an animal moves also depends on its foot size. The larger the foot area that touches the ground while walking or running, the slower the animal. The complexity of running activity adds to total speed. For example, cheetahs can run at a speed of sixty-five to seventy miles per hour, partly by adding to their gait a series of leaps at chosen intervals. The added leaps help to produce the ability to move three times as fast as exceptional human athletes can run one hundred meter dashes.

Flying, Thrust, Drag, and Lift

Heavier-than-air craft, whether birds, bats, or airplanes, must conquer gravity to enter the air in controlled flight. In the case of animals who fly—as well as of airplanes—three different forces are involved. The first, called thrust, is empowered in animals by wing flapping. Thrust enables forward horizontal movement, as long it exceeds the second force, drag, caused by air viscosity.

The third aerodynamic force, lift, is the key to flight. It enables upward rise into the air, at a right angle to the direction of the forward motion. Lift in living creatures is supplied by the wings of flying animals. These wings (airfoils) are moved so the angle at which they meet air passing them (the angle of attack) makes the air flow much more rapidly past the upper airfoil surface than past its lower surface. This drops the air pressure above the airfoil to values below the air pressure under it and engenders lift needed to fly. In flying animals, the angle of attack (as in birds) is altered by wing motion adjustments.

The importance of the angle of attack of airfoils is clarified by its use and misuse by human pilots during flight. Angles of attack of up to fifteen degrees increase lift, enabling fast climbs but slowing airspeed. When the angle of attack is too steep, air currents atop airfoils decrease lift and cause aircraft to drop toward the ground in stalls. When pilot misjudgment produces a stall, the craft crashes unless the pilot quickly decreases the angle of attack. Birds do this instinctively, and so effortlessly that crashes due to bird stalls do not occur.

Swimming, Gliding, and Flying

Organisms that swim in water range from tiny ciliated or flagellate microbes to blue whales that are over one hundred feet long. Their locomotive operations are complex and use oarlike rowing, hydrofoil motion of propellers, body undulation, and jet propulsion. Oar and propeller actions have some similarities, but oars push in the direction of a desired movement, while propellers work at right angles to blade motion. Propeller action is due to lift, caused by hydrofoil action in the water. Propeller-mediated animal locomotion is more complex than walking or running, especially in the case of whale flukes. These hydrofoils, in addition to complex motion, must be tilted at angles which provide more lift than drag, under widely varying conditions.

In animals that swim by undulation, movements are made like the crawling of a snake. Here, the resistance to motion of poorly compressible water makes it easier for the swimmer to slide forward along its long axis than sideways along its short axis. The squid exemplifies those organisms that move by jet propulsion. A squid does this by drawing water into its mantle and squirting it out forcefully from an organ called a funnel. The squirt is aimed forward when a squid wants to swim backward, or backward when it seeks to swim forward.

One of the problems associated with locomotion by swimming is that most tissues in swimming organisms are more dense than water. Such organisms would sink, lacking ways to raise body buoyancy. These needs add to the complexity associated with swimming, compared with walking or running on land. Ways in which to counter sinking include swimming with all fins extended like airplane wings, as sharks do, to add buoyancy by increasing the surface area of the body. Also, swimming animals, such as the bony fish, adjust their densities to match that of the water via gas-filled swim bladders in their body cavities.

Only insects, birds, and bats are capable of locomotion by true flight. However, some other animals glide for short distances. Best known are flying squirrels, which glide from tree to tree. Also, flying fish use their large pectorals as wings of a sort. They leap out of the oceans with the fins spread and glide through the air for up to half a minute. As is the case with whale flukes in water, a glider supported by lift on its wings is slowed and pulled down by drag. Thus, most gliding organisms do not stay in the air for long Their time suspended aloft depends upon maintaining the appropriate angle of attack of their wings. They can adjust glide speed by changing these angles of attack. Increasing this angle slows them and decreasing it allows them to speed up.

Soaring, or gliding for long distances, is possible only for animals aloft in quickly rising air. Few animals other than birds and butterflies have wings capable of sustained gliding. The needed air currents are found mostly over hillsides or coastal cliffs and in thermals, which are currents of hot air rising from ground heated by the Sun. Vultures soaring to scan for carrion can travel for distances up to fifty miles. Butterflies, gulls, and albatrosses also soar for long distances. Long-distance soaring is very often used on migration trips, to conserve energy for other activities.

Ways to increase glide or soar distance include tilting the body to suit different angles of attack and moving wings forward or backward. These actions provide streamlining and move the center of gravity of the body. Gliding and soaring animals make turns by giving one wing a higher angle of attack than the other. Furthermore, gliding is usually done with legs retracted, to increase streamlining. The legs are lowered like the landing gear of airplanes when a braking effect is required. Clearly, locomotion in the air is even more difficult than swimming, involving much more complex interactions of the muscles and the skeleton.

Power, needed for most sustained flight, is obtained when flying animals flap their wings. There are two main animal flight types, high-speed flight and hovering flight. In the high-speed flight of birds, bats, and insects, the body moves forward at the same speed as the wings. The wings beat up and down as the animal moves forward through the air. Each down stroke produces lift and propels the flying animal forward and supports it in the air. The angle of attack of the upstroke is increased, so as to produce very little lift. This is because lift, acting upward at right angles to the wing's path, would slow high-speed flight.

In hovering flight, the body is stationary and the wings move very quickly, as when hummingbirds feed. Animals use two hovering techniques. Most often, their wings are kept horizontal and beat straight backward and straight forward. They turn upside down for each backstroke, adjusting the angle of attack so lift is obtained. The hum of hummingbirds is due to the frequencies with which their wings beat (twenty-five to fifty-five cycles per second). Insects buzz, whine, or drone, according to the sound produced by the higher frequencies at which their wings beat. In some cases, as in pigeons, hovering is accomplished when the wings clap together at the end of an upstroke. The clap enhances airflow into the space formed as they separate and so enhances lift. Initial wing-clap hovering in pigeons causes the sound first made when a flock flies away together.

—*Sanford S. Singer*

See also: Endoskeletons; Exoskeletons; Flight; Muscles in invertebrates; Muscles in vertebrates; Physiology; Respiration in birds; Wings.

Bibliography

Audesirk, Gerald, and Teresa Audesirk. *Biology: Life on Earth*. 5th ed. Upper Saddle River, N.J.: Prentice Hall, 1999. This solid, well-illustrated biology text contains a great deal of information that aids understanding locomotion, as well as useful references.

Dantzler, William H., ed. *Comparative Physiology*. New York: Oxford University Press, 1997. Contains several chapters which cover important aspects of the study of locomotion and provide good references.

Dickenson, Michael H., Claire T. Farley, Robert J. Full, M. A. R. Koehk, Rodger Kram, and Steven Lehman. "How Animals Move." *Science* 238, no. 5463 (April 7, 2000): 100-106. This article describes advances in integrative study of locomotion that have revealed general principles of locomotion. Extensive references.

Gamlin, Linda, and Gail Vines, eds. *The Evolution of Life*. New York: Oxford University Press, 1987. Chapter 24, on movement, is quite useful. Other chapters cover nervous system transmission and additional fundamentals.

Hickman, Pamela, and Pat Stephens. *Animals in Motion: How Animals Swim, Jump, Slither, and Glide*. Niagara Falls, N.Y.: Kids Can, 2000. This delightful juvenile book holds some very useful basics on locomotion.

LUNGFISH

Type of animal science: Classification
Fields of study: Anatomy, evolutionary science, paleontology, systematics (taxonomy)

Lungfish are freshwater members of the subclass Sarcopterygii, the fleshy- or lobed-finned fishes. As their name indicates, lungfish have lungs evolved from the swim bladder and can breathe air. Because they are among the fishes most closely related to the tetrapods, they play a crucial if controversial role in the study of vertebrate evolution.

Principal Terms

CONTINENTAL DRIFT: theory that the continents have moved slowly apart from an early landmass, explaining why many species appear to be closely related while separated by wide expanses of ocean

ESTIVATION: similar to hibernation; period of reduced activity or dormancy triggered by dry and/or hot environmental conditions

METAMORPHOSIS: pronounced developmental change in form or structure of an animal after birth or hatching, as in the transformation of tadpoles into frogs

SWIM BLADDER (AIR BLADDER): an internal organ evolved from the gut that allows a fish to regulate its vertical position in the water column (maintain its balance)

TETRAPODS: four-legged vertebrates (amphibians, reptiles, birds, and mammals)

Lungfish fossils first appear at the beginning of the Devonian period, about 400 million years ago. Once distributed around the world, lungfish now occur in only three areas (South America, Africa, and Australia) that were once adjacent but have since separated due to continental drift. South American and African lungfish are more closely related to one another. They have reduced gills and two lungs and are true air breathers, while Australian lungfish have one lung and rely mostly upon their gills for respiration, gulping air at the water surface to breathe only when dissolved oxygen is in short supply.

African and South American Lungfish
There are two related families in this order. The African family, Protopteridae, has one genus (*Protopterus*) and four species (*P. annectens, P. aethiopicus, P. dolloi,* and *P. amphibius*). They live in the rivers, lakes, and swamps of East and Central Africa and grow to 2 meters (6.5 feet) in length. The South American family, Lepidosirenidae, has only one species, *Lepidosiren paradoxa*. It lives in swampy areas of the Amazon and Paraná river basins and in the swamps of the Chaco region and grows to 1.2 meters (4 feet) in length.

During spawning season, males dig holes for the eggs and guard both eggs and young after they hatch. The South American male develops branched, gill-like structures on his pelvic fins that may supply extra oxygen to himself as he incubates and to his offspring. Juveniles hatch with adhering organs to attach to plants. They have external gills like salamander young and undergo metamorphosis. Adults live on crustaceans, mollusks, and small fishes. They survive the dry season by resting quietly in moist burrows dug in the mud. If it becomes too arid, they seal their burrows with mud, secrete a protective covering of mucus, and estivate, sometimes for several months. During estivation, metabolism slows down to conserve energy and air. Once the rainy season begins, water enters the burrow and awakens the estivating fish.

Australian Lungfish

The Australian lungfish was first discovered in 1870. This lungfish closely resembles fossil lungfishes except that its skull contains fewer, larger bones. It has remained unchanged for 100 million years or more, longer than any other vertebrate known. There is now only one living species of this order, *Neoceratodus forsteri*, and it occurs naturally in only two locations, the Burnett and Mary rivers in Queensland, Australia, although it is being introduced into other rivers in an effort to ensure its survival. During the summer, the rivers dry up, and the lungfish survive in small pools. They cannot estivate.

Spawning takes place after the rainy season starts. The eggs are laid in masses of fifty to one hundred on aquatic plants. The young breathe through the gills and skin. They do not have adhering organs or external gills and do not undergo metamorphosis. They develop pectoral fins at about fourteen days and pelvic fins at about ten weeks, gradually assuming the adult shape. They feed on algae, but adults feed on small fishes, mollusks, crustaceans, insect larvae, and some plants and grow to a size of about 1.8 meters (6 feet) and forty-five kilograms (one hundred pounds). They have broad, flat heads with the mouth underneath, small eyes, and pointed tails. The body is olive green or brown above and pinkish below.

—*Sue Tarjan*

See also: Fauna: Africa; Fauna: Australia; Fauna: South America; Fish; Lungs, gills, and tracheas.

Lungfish Facts

Classification:
Kingdom: Animalia
Subkingdom: Bilateria
Phylum: Chordata
Subphylum: Vertebrata
Superclass: Gnathostomata
Class: Osteichthyes
Subclass: Sarcopterygii
Order: Dipnomorpha or Dipnoi
Families: Lepidosirenidae (African and South American lungfish); Dipnorhynchidae, Dipteridae, Phaneropleuridae, Ctenodontidae, Conchopomidae, Sagenodontidae, Ceratodiae (Australian lungfish)
Geographical location: Africa, Australia, and South America
Habitat: Freshwater only (rivers, lakes, swamps, and marshes)
Gestational period: Australian lungfish, twenty-one to thirty days; African and South American lungfish, eleven to fifteen days
Life span: Unknown; larger fish may live to be many years old
Special anatomy: Elongated bodies, continuous rear fins composed of connected dorsal, caudal, and anal fins; Australian lungfish—laterally compressed body, large scales, two pairs of flipperlike pectoral and pelvic fins, one lung; African and South American lungfish—eel-like body with small scales, two pairs of long, thin pectoral and pelvic fins, paired lungs

---

Bibliography

Carroll, Robert L. *Vertebrate Paleontology and Evolution*. New York: W. H. Freeman, 1988. Comprehensive if technical study of vertebrate evolution. Includes discussion of the controversial role of ancestral lungfish in the evolution of tetrapods.

Long, John A. *The Rise of Fishes: Five Hundred Million Years of Evolution*. Baltimore: The Johns Hopkins University Press, 1995. Clear, detailed account of the evolution of fishes. Chapter 8, "Lungfishes: Fishes That Breathe," contains a wealth of information focusing on the early fossil record, including an excellent explanation of the origin of lungs. Lavishly illustrated.

Moyle, Peter B., and Joseph J. Cech, Jr. *Fishes: An Introduction to Ichthyology*. 4th ed. Upper

Saddle River, N.J.: Prentice Hall, 2000. Excellent introductory text. Chapters 11, "Relict Bony Fishes," and 13, "Evolution," discuss the lungfish in considerable depth.

Roush, Wade. "'Living Fossil' Fish Is Dethroned." *Science* 277 (September 5, 1997): 1436. Lively coverage of mitochondrial DNA analysis indicating that lungfish, not coelacanths, may be closer to amphibians on the evolutionary tree.

Weinberg, Samantha. *A Fish Caught in Time: The Search for the Coelacanth.* New York: HarperCollins, 2000. Popular account of the discovery of a "living fossil," the coelacanth, a lobe-finned fish related to the lungfish. Provides an informative presentation of the evolutionary debate over which one is the closest ancestor of the tetrapods.

LUNGS, GILLS, AND TRACHEAS

Type of animal science: Anatomy
Fields of study: Anatomy, histology

Gills, tracheas, and lungs are respiratory organs that are the primary sites of gas exchange in certain groups of animals. Gills are present in water-breathing animals, and tracheas and lungs are found in animals that breathe air. The structure and design of these organs are related to the particular manner of their function.

Principal Terms

BRONCHUS (pl. BRONCHI): an individual tube that is part of a lung and leads to one of the smaller lung parts

DIFFUSION: the passive movement of molecules from an area of high concentration (or pressure) to an area of lower concentration, often across a distance or a barrier

EPITHELIUM: a thin membrane that lines or coats a surface; in respiratory organs, the layer of cells that covers or lines the surface is called the respiratory epithelium

GILL: an extension or outpocketing of the body wall of an animal that creates a limblike structure, much as the fingers of a glove extend from the palm; gills are found almost exclusively in animals that breathe water

LAMELLA (pl. LAMELLAE): lamella means "platelike," and an individual lamella can be any one of several structures in the context of gas-exchange organs; these are usually found in gills

LUNG: a concave inpocketing of the body wall of an animal (in contrast to a gill); lungs occur in air-breathing animals

TRACHEA (pl. TRACHEAS): a tubular inpocketing of the body found mostly in insects and the spiders (arachnids), both of which are air breathers

The anatomy of a respiratory organ—gills, tracheas, or lungs—determines the way in which the organ functions. The design of the organ is specifically related to the way in which the animal functions in the world. Design refers to whether the structure is internal or external, large or small, and concerns some aspects of how it works. In order to understand the function of most organs and organ systems, it is necessary to know the structure.

The structure, or anatomy, can be considered from several different levels of organization. Gross structure is the size, shape, and position of the whole organ within the animal. Fine structure, on the other hand, refers to the microscopic level of organization. The microscopic structure includes the type and thickness of cells, the number of cells, and a description of the cell surface.

Common Features of Respiratory Organs

There are a few common features among gills, tracheas, and lungs in terms of both the gross structure and the fine structure. All three are organs of gas exchange and are therefore designed to permit oxygen and carbon dioxide gas to diffuse passively across this specific part of the body wall. Because one of the factors that directly affects diffusion is the total surface area, respiratory surfaces are greatly increased in order to maximize the movement of gas. Thus, the largest part of the total surface of an animal is the respiratory surface, no matter whether it is a lung, gill, or trachea. All three types of respiratory organs have openings to permit air or water to flow in (and out), tubes for conducting air or water to the exchange surface, the exchange surface itself, and the necessary support to hold the surface in place. The exceptions are those animals that have external gills with no enclosing cover.

The commonality of gas exchange function also imparts similar features to the microscopic level of structure. The epithelia of all three types of organs are thin and made of cells that are individually thin, providing less of a barrier to gas movement (although there is an exception to this in gill structure).

Gills

Gills are the most diverse and varied of all respiratory organs. Gills are found in aquatic animals, both vertebrate (animals with internal skeletons) and invertebrate (animals with external skeletons or no hard skeleton). With few exceptions, all animals with gills live in and breathe water, even if the volume of water is small and only one life stage of the animal breathes water. Most people are familiar with several types of animals that breathe water and use gills: fish, crabs, lobster, shrimp, crayfish, clams, aquatic insects, many worms (but not all), and numerous other animals. The form and function of gills have been variously modified in different animals through the process of natural selection.

The gills of many marine worms and of the marine sea hare and sea slug (both are related to snails) are thin, blind-ending tubules representing extensions of the body wall that form an elaborate treelike structure. For the most part, these are not complicated either in the pattern of blood flowing through them on the inside or in the way water is passed over them on the outside. Worms and sea slugs usually extend the gills into the water above and let the water's own movement bring the oxygen-filled water to the gills. Sea hares, however, cover the gills with a flap of skin, an extension of the body wall, and pump water over the gills by the beating action of numerous tiny cilia.

The gills of crabs are formed from a central structure, called an arch, and platelike extensions of the gill arch, called lamellae. Each lamella has a thin layer of epithelium, but because it is formed as an extension of the outside layer of the body, the epithelium is covered with a very thin layer of shell, called chitin. This is true of crabs, lobster, shrimp, and all the crustaceans. Nevertheless, the layer still permits the exchange of gases between the blood on the inside and water on the outside.

Each lamella is exposed to the water on both sides and has numerous supporting cells on the inside that keep the two sides apart so that blood may flow between. These support cells are called pillar cells, and if an observer could look down between the sides, the inside would look something like a forest. Blood comes into each lamella from a single channel, an artery, located at the top, and flows out through a vein located at the bottom of the lamella.

Not all lamella are alike even in the same crab, much less in crabs from habitats as different as the deep ocean and a tropical forest. One of the major differences is that some epithelia are specialized to transport salts actively across the gill in much the same way as the kidney does in a mammal. These gill lamella are thicker, with cells that are modified to transport sodium and chloride. Blue crabs, land crabs, and crayfish (which have podia instead of lamellae) are examples of animals with this type of specialization.

Not all crustaceans have gills that are lamellar in form; the crayfish and lobster have fingerlike projections of the gill arch. These projections are called podia and are more like the blind-ending tubules found in some worms than they are like the lamellae of crabs. The inside of the podia is divided, however, so that blood flows to the tip of the tube on one side and back to the animal on the other side of the podia.

Fish gills are not unlike those of a crab in many ways, but those of fish are more complex. Both are composed of flattened extensions of a trunklike support structure that contains the blood vessels. Gills of most teleost fish (bony fish, such as perch and bass) have further, or secondary, extensions from these, which are oriented perpendicular to the main lamella. The result is that the primary lamellae form a flat surface in a horizontal plane, and the secondary lamellae form the largest surface area, extending in a vertical plane. Some primitive fish, or those with adaptations to special conditions, have reduced secondary lamellae.

Some of the major differences between vertebrate and invertebrate gills are that fish gills are thinner, so that the layer separating the blood from the water is not as thick in fish. Fish-gill epithelia are usually about 20 percent as thick as those of the invertebrates described. Many fish have two different types of lamellae on the gill arch: primary and secondary lamellae. Many fish gills also contain specialized cells to transport salts across the border between the animal and the water, as happens in blue crabs.

Tracheas

Tracheas, found in insects, spiders, and related invertebrates, are long, thin tubes that extend from the outside of an animal to the inside, similar to the way a trachea leads to the lung in humans. The tracheal system, however, is made up entirely of branching tubules, each somewhat smaller than the one leading to it. The major tubes are called tracheas and the smaller ones are tracheoles; the smallest ones are as thin as hairs and reach within a very short distance of the cells inside the body. The outside opening to each trachea is controlled by a special structure, a spiracle, that can open and close the trachea.

Lungs

Lungs are actually the simplest type of respiratory system to describe in terms of animal type, distribution, and variations on the common form. Lungs are found only in air-breathing vertebrates, are always internalized, and have a single opening through which air flows both in and out. A single tube, the trachea, is the conducting passage for inspired and expired air. This tube branches into two smaller tubes, the bronchi, that lead to each of the individual sides, or lobes, of the lung. Tracheas and bronchi are usually, if not always, reinforced with cartilage rings for structural support to prevent collapse. Thus, animals with lungs actually have two, one on each side.

Tracheas and bronchi are not designed for gas exchange but are conducting tubes that also serve to "condition" the air before it reaches the lungs. Both tubes are lined with special cells that have hairlike projections called cilia on the surface. The cilia continually sweep small particles (such as dust) toward the outside to protect the lungs. Other cells in tracheas and bronchi secrete mucus to keep the surface moist and maintain high humidity so that air will be 100 percent humid when it reaches the alveoli.

At the end of the tubes of the lungs are the sac structures that form the gas-exchange surface. These very thin structures are the alveoli and are the site of gas movement into and out of the body. On the inside of each alveolus is an entire network of tiny blood vessels, capillaries, that supply the blood which is the source of the carbon dioxide that is exhaled and is the sink for the oxygen taken up at the alveoli.

Studying Respiratory Organs

The anatomy of gas-exchange organs is studied in whole living animals, in preserved specimens, and in sections prepared for microscopic examination. Both light microscopy and electron microscopy are used to study these organs. Light microscopy is simply the use of a microscope that requires standard lighting with the type of lens systems that have been used for decades; there are also some much more sophisticated systems that use specialized optics and lighting techniques. Electron microscopy uses beams of electrons to enable magnification of more than 100,000 times and the visualization of cells and their component parts. In some cases, miniaturized laser optics are used to examine the relatively large organs, such as lungs, while they are functioning. This application is often used in health-related studies and diagnosis of disease.

Studying the gross morphology of gills, tracheas, and lungs involves the dissection of a whole animal, either preserved or freshly deceased. Each option offers advantages, and both methods will be employed in a complete examination. Part of the dissection will be to determine qualitative features such as attachment points, spatial relations, and general appearance. A thorough study of the anatomy of an organ also includes measurements of features such as surface areas, number of parts, distances, and volumes. These are the quantitative measures, and they are applicable to all the organs. These numerical values are absolutely necessary in order to quantify the functions.

In the case of small animals, including shrews, insects, and fish the size of minnows, a dissection of the whole animal may have to be carried out with additional magnification, using a microscope designed for such work. These microscopes are called dissection microscopes and have enough distance between the lens and the object to allow dissection with small instruments by hand. In a few cases, the material is so small that microdissecting techniques developed by neurobiologists must be used. Traditional histological examination of respiratory organs has also revealed much about the design and nature of these structures. Histological examination (examining tissue structure, usually under a microscope) reveals the thickness of the epithelium, the number of layers, and some information about the nature of the material in the layers.

One of the purposes for studying the anatomy of all respiratory structures, and lungs in particular, is to improve medical applications. In veterinary medicine, knowledge of the general structure of lungs is crucial in treating diseased lungs and other respiratory problems. The use of animals in research has always been a controversial topic, but if such research is to be conducted properly, then accurate knowledge of the anatomy of the experimental animals is necessary. In this way, the animals can be kept healthy and monitored properly so that the intended purpose of the research may be realized.

Comparing Respiratory Structures

Evolutionary and taxonomic relationships are revealed in comparisons of respiratory structures in some groups of animals. Other relationships are

evident in making comparisons between very different groups. There are only subtle differences in lung structure among the air-breathing vertebrates with lungs: amphibians, reptiles, and mammals. These differences occur more in the gross structure than at the cellular level, indicating how evolutionary pressures create successful adaptations.

The differences in the gill structures of crabs, worms, and fish from various habitats give valuable insight into evolutionary changes. Land crabs and air-breathing fish have fewer gills, fewer lamellae with thicker surfaces, and special structures to hold them in position. These latter are needed in air because without the support offered by water the gill lamellae would otherwise adhere to one another and not function. Similarly, some fish that live in water that contains too little oxygen must depend on air breathing and show these same adaptations.

One of the most widely studied aspects of comparing respiratory structures in different animals is the evolutionary transition from breathing water to breathing air. The transition from water to land has always interested scientists, and this level of anatomy is part of that interest. Biologists have examined the differences between air and water breathers in many groups of animals generally, and in groups that have both, such as the crabs and the amphibians, in particular. The features of interest include the surface area, thickness, position in the body, and pattern of circulation.

Few groups of animals display such a clear example of the evolution of structure and function in the tracheal system as do insects. Insects have a respiratory system, the tracheas, that reaches nearly to the cells themselves, bringing them oxygen-laden air and removing the carbon dioxide produced by the cells. It would seem, then, that there would be little need for a circulatory system to serve that function, and indeed there is but a rudimentary one in insects.

Anatomy of respiratory structures is also studied in relation to other functions and organ systems. The circulatory system provides the blood that exchanges gases with the water or air on the outside of the animal. Thus, the anatomy of the blood vessels is an important part of the anatomy. Similarly, the control of the flow of blood to the lungs or gills is critical for respiratory system function.

The fine structure of several types of respiratory systems, but primarily gills, is part of the study of other, nonrespiratory functions. The gills of some aquatic animals play an important role in regulating the animal's salt balance. In these animals, the fine structure of the gills shows where this process takes place and what special modifications are needed.

—Peter L. deFur

See also: Adaptations and their mechanisms; Anatomy; Cell types; Circulatory systems of vertebrates; Gas exchange; Histology; Osmoregulation; Physiology; Respiration and low oxygen; Respiration in birds; Respiratory system.

Bibliography

Barnes, Robert D. *Invertebrate Zoology*. 6th ed. Philadelphia: Saunders College Publishing, 1994. This invertebrate zoology text complements the Romer and Parsons vertebrate zoology text. Barnes's book for advanced high school or college level is considered by many to be the best, and it contains excellent illustrations and a rather complete bibliography. The reader must get the information for each group from the appropriate chapter, but the index is sufficiently detailed so that this is not a problem. Most of the phyla have gills or trachea, so the reader must look carefully.

Comroe, J. H. "The Lung." *Scientific American* 214 (February, 1966): 56-68. This magazine only features a subject once, so the story on the lung is dated, but it still gives an accurate description of basic lung anatomy and includes information on function.

Klystra, J. A. "Experiments in Water Breathing." *Scientific American* 219 (August, 1968): 66-74. Somewhat dated, but the illustrations effectively compare lungs and gills; there are accompanying explanations of the function of each.

Raven, Peter H., and George B. Johnson. *Biology*. 6th ed. St. Louis: Times Mirror/Mosby, 2001. A text for college freshmen; includes much more than only the anatomy of the three respiratory systems. It does a better than average job of explaining gills, tracheas, and lung structure, and it includes some material on function, evolution, and related matters. The illustrations are clear, and there is a glossary.

Romer, Alfred S., and Thomas S. Parsons. *The Vertebrate Body*. 6th ed. Philadelphia: W. B. Saunders, 1986. This is one of the best texts dealing with comparative anatomy of vertebrates. Written for college students, this is certainly the most comparative survey and gives an excellent evolutionary perspective. The figures are good; the glossary is extensive but succinct.

Storer, Tracy I., Robert L. Usinger, Robert C. Stebbins, and James W. Nybakken. *General Zoology*. New York: McGraw-Hill, 1979. The authors are four of the finest zoologists of the past several decades, each with his own specialization to bring to the book. Every part is written by a specialist in the field, a feature of which few general textbooks can boast. Unlike most texts, the material is covered by function and by animal group. The respiratory structures are described in the section on that topic and for each group as presented. College level with illustrations.

MAMMALIAN SOCIAL SYSTEMS

Type of animal science: Behavior
Fields of study: Anthropology, ecology, ethology, zoology

Social organization in mammals ranges from solitary species, which come together only to breed, to large and intricately organized societies. Understanding the social systems of mammals is essential for effective conservation of species.

Principal Terms

BROWSER: an organism that feeds primarily on leaves and twigs of trees and shrubs rather than on grasses

CARNIVORE: a member of the meat-eating order Carnivora, which includes dogs, cats, weasels, bears, and their relatives

EUSOCIAL: a social system with a single breeding female; other members of the colony are organized into specialized classes (exemplified by bees, ants, and termites)

GRAZER: an organism that feeds primarily on grasses

PRIMATES: members of the order Primates—monkeys, apes, and their relatives

RODENT: a member of the order Rodentia—squirrels, rats, mice, and their relatives

SAVANNA: a grassland with scattered trees; some ecologists restrict the term to tropical regions

TERRITORY: an area that an animal defends against other members of the species and that often contains food, shelter, and other requirements for the individual or group

UNGULATE: a hoofed mammal from the order Artiodactyla (pigs, cattle, antelope, and their relatives) or from the order Perissodactyla (horses, rhinoceroses, tapirs, and their relatives)

All levels of social organization occur in mammals. There are solitary species, such as the mountain lion (*Felis concolor*), in which the male and female adults come together only to mate, and the female remains with her young only until they are capable of living independently. At the other numerical extreme are some of the hoofed mammals, which form herds of thousands of individuals. Other extremes might be considered in terms of specialization for social life. The most socially specialized mammal is probably the naked mole rat (*Heterocephalus glaber*) of Africa, which has a eusocial colony structure similar to that of ants, bees, and termites. Between these extremes, there are many variations. No current theory accounts for the diversity of mammalian social systems, but two broad generalizations are consistently employed to explain mammalian species' social organization. These are the environmental context in which the species exists and the mammalian mode of reproduction.

More than any other group of animals, mammals are required to form groups for at least part of their lives. Although in all sexually reproducing animals the sexes must come together to mate, mammals have an additional required association between mother and young: All species of mammal feed their young with milk from the mother's mammary glands. This group, a female and her young, is the basis for the development of mammalian social groups. In some species, the social group includes several females and their young and may involve one or more males as well.

The particular social organization adopted by a mammalian species is a response to the environmental conditions under which the species lives. The species' food supply and the distribution of that food supply are often the predominant determinants, but predators on the species are also important in determining the form of its social organization. The best way to see the effects of these factors on mammalian social structure is by example.

Primate Social Organization

The primates are the most social group of mammals. Monkeys demonstrate the importance of food supply and its distribution in determining social structure. The olive baboon (*Papio anubis*) occupies savannas, where it exists in large groups of several adult males, several adult females, and their young. Finding fifty or more animals in a group is not uncommon. Individual males do not guard or try to control specific females except when the females are sexually receptive. The group's food supply is in scattered patches, but each patch contains an abundance of food. The advantage of having many individuals searching for the scattered food is obvious: If any member finds a food-rich patch, there is plenty for all.

Predation probably also plays a role in the olive baboon's social organization. The savannas they roam have many predators and few refuges for escape. A large group is one defense against predators if hiding or climbing out of reach is not practical. Having many observers increases the chance of early detection, giving the prey time to elude

Primates, such as these chacma baboons, are the most social of mammal orders. Social structures offer increased safety from predators, efficient use of food resources, and a wider range of learning opportunities for the young. (Corbis)

the predator. A large group can also mount a more effective defense against a predator. Large groups of baboons use both of these tactics.

The hamadryas baboon (*Papio hamadryas*), on the other hand, lives in deserts in which the food supply is not only scattered but also often found in small patches. The hamadryas baboon's social structure contrasts with that of the olive baboon, perhaps because the small patches do not supply enough food to support large groups. A single adult male, one or a few adult females, and their young make up the basic group of fewer than twenty individuals. Several of these family groups travel together under certain conditions, forming a band of up to sixty animals. Within the band, however, the family groups remain intact. The male of each group herds his females, punishing them if they do not follow him. The bands are probably formed in defense against predators. They break up into family units if predators are absent. At night, hamadryas baboons sleep on cliffs, where they are less accessible to predators. Because suitable cliffs are limited, many family groups gather at these sites. Hundreds of animals may be in the sleeping troop, probably affording further protection against predators.

Though there are exceptions, forest primates consistently live in smaller groups. In many species, fewer than twenty individuals make up the social group at all times. These consist of one or a few mature males, one or a few mature females, and their offspring. The groups are more evenly distributed throughout their habitat than are groups of savanna or desert primates. In forests, the food supply is more abundant and more evenly distributed. Escape from predators is also more readily accomplished—by climbing trees or hiding in the dense cover. Under these conditions, the advantages of large groups are minimal and their disadvantages become apparent. For example, in small groups the competition for mates and food is less.

Ungulate Social Organization

The ungulates have all levels of social organization. African antelope demonstrate social orga-

nizations that, in some ways, parallel those of the primates. Forest antelope such as the dik-dik (*Madoqua*) and duiker (*Cephalophus*) are solitary or form small family groups, and they are evenly spaced through their environment. Many hold permanent territories containing the needs of the individual or group. They escape predators by hiding and are browsers, feeding on the leaves and twigs of trees.

Many grassland and savanna antelope, such as wildebeest (*Connochaetes*), on the other hand, occur in large herds. They outrun or present a group defense to predators and are grazers, eating the abundant grasses of their habitat. In many cases, they are also migratory, following the rains about the grasslands to find sufficient food. The social unit is a group of related females and their young. Males leave the group of females and young as they mature. They join a bachelor herd until fully mature, at which time they become solitary, and some establish territories. The large migratory herds are composed of many female/young groups, bachelor herds, and mature males. The social units are maintained in the herd. Though it may seem strange to speak of solitary males in a herd of thousands, that is their social condition. The male territories are permanent in areas that have a reliable food supply year-round, but they cannot be in regions in which the species is migratory. Under these conditions, the males set up temporary breeding territories wherever the herd is located during the breeding season.

There are parallels with primate social patterns. Large groups are formed in grasslands, and these roam widely in search of suitable food. The groups are effective as protection against predators in habitats with few hiding places. Smaller groups are found in forests, where food is more evenly dispersed and places to hide from predators are more readily found.

Rodent Social Organization

Rodents also have all kinds of social organizations. The best known, and one of the most complex, is the social system of the black-tailed prairie dog (*Cynomys ludovicianus*). The coterie is the family unit in this case, and it consists of an adult male, several adult females, and their young. Members maintain a group territory defended against members of other coteries. Coterie members maintain and share a burrow system. Elaborate greeting rituals have developed to allow the prairie dogs of a coterie to recognize one another. Hundreds of these coteries occur together in a town. The members of these towns keep the vegetation clipped—as a result, predators can be seen from a distance. Prairie dogs warn one another with a "bark" when they observe a predator, and the burrow system affords a refuge from most predators.

The only vertebrate known to be eusocial is the naked mole rat. It occurs in hot, dry regions of Africa. The colony has a single reproductive female, a group of workers, and a group of males whose only function is to breed with the reproductive female. The workers cooperate in an energetically efficient burrowing chain when enlarging the burrow system. In this way, they are able to extend the burrow system quickly during the brief wet season. Digging is very difficult at other times of the year. The entire social system is thought to be an adaptation to a harsh environment and a sparse food supply.

Carnivore Social Organization

Most carnivores are not particularly social, but some do have elaborate social organizations. Many of these are based on the efficiency of group hunting in the pursuit of large prey or on the ability of a group to defend a large food supply from scavengers. The gray wolf (*Canis lupus*) and African hunting dog (*Lycaon pictus*) are examples. In both cases, the social group, or pack, consists of a male and female pair and their offspring of several years. There are exceptions, solitary carnivores and carnivores that form temporary family units during the breeding season, such as the red fox (*Vulpes vulpes*), which hunt prey smaller than themselves. The coyote (*Canis latrans*) can switch social systems to use the food available most efficiently. It forms packs similar to those of the gray wolf when its main prey is large or when it can

scavenge large animals and is solitary when the primary available prey is small.

These examples and many others show that the social groups of mammals are based on the family group. The particular social organization employed by a species is determined by the ecological situation in which it occurs. The specific aspects of the environment that seem to be most important include food abundance, food distribution, food type, and protection from predators.

Fieldwork and Laboratory Studies

Observation has been a very important method of studying mammal societies. One of the reasons that primate and ungulate societies are so well known is that they are large and active during the day, and so are easily observable. The observer must take great pains to be inconspicuous or, in some cases, to become a part of the subject's environment. Small mammals (and sometimes larger mammals) have been kept in enclosures and observed to learn more about their social lives. The observer maps movements, records activities and interactions, and analyzes the data that result.

Simple observation is enhanced by manipulating the subjects in various ways. Individual animals can be marked, or in some cases they can be identified by natural color patterns, scars, or other marks. These marked individuals can be followed, and their behavior and interactions with other individuals observed. Radios and radioactive tracer elements are sometimes implanted in individuals, and these individuals are followed in the field. Much can be learned about a species' social behavior by following the locations of such tagged animals. In addition, they are more readily locatable for direct observation.

Small mammal species that are not readily observable are trapped live, marked, released, and recaptured. Mutually exclusive use of certain areas, areas used in common, and patterns of multiple captures in individual traps are some types of information from trapping that can be interpreted in terms of social behavior. Experiments are sometimes carried out in the natural context. A group or a specific individual is presented with an artificial situation, and any reactions to it are recorded, often on film or videotape.

Laboratory studies are also used to supplement the field observations. Psychological and physiological capabilities of organisms can best be studied in the controlled confines of a laboratory experiment. These data, however, must always be put back in the context of the field observations to make a meaningful contribution to the understanding of the species' social behavior.

Computer simulations and mathematical models have been used to explore the possible reactions of social systems to various environmental pressures. As with laboratory results, it is important to test predictions generated in these ways against the social system in nature before assuming their validity. Comparative studies of all the above types are of great importance. Related species, or different populations of the same species, that occur in different regions are studied and compared; these studies are tantamount to reading the data from a natural experiment.

Social Organization and Food

Mammalian societies are always organized around one or more females and their offspring. Males may also be part of the group, or they may form separate groups. The size and structure of the group are determined by the ecological setting in which it evolves. The particular ecological factors that seem to be of greatest importance in this determination are food supply, the distribution of the food, and predation (including the hiding places and escape routes available in the habitat).

Large groups occur when food is scattered in a patchy distribution. These groups are largest when the patches contain abundant food. Many organisms are more likely to find the scattered patches than is a single individual. As long as the patches have enough food for all members of the group, it is to each member's advantage to search with the group. On the other hand, if food is evenly dispersed in small units throughout the environment, the advantage of a group search is lost. Each individual will be better off searching for itself, and some strategy involving a very small social group or even solitary existence would be advantageous.

A somewhat similar argument follows for predators. If large prey are taken, a group of predators should be able to subdue the prey and protect its remains from scavengers more efficiently. If small prey are taken, solitary predators have the advantage, since the prey is easily dispatched and the predator will have it to itself. Many other factors are involved in determining the final form of a species' social organization, but the family unit and environmental context are fundamental in determination of all mammalian social structures.

Conservation of the mammal species that still exist on the earth requires knowledge of their social organization. Understanding that mammalian social organizations are responses to the environmental context in which they have evolved emphasizes the need to conserve entire ecosystems, not only the individual species that exist within them.

—Carl W. Hoagstrom

See also: Altruism; Communication; Communities; Competition; Ecological niches; Ethology; Groups; Herds; Insect societies; Mammals; Mating; Packs; Population analysis; Primates; Territoriality and aggression.

Bibliography

Dunbar, Robin I. M. *Primate Social Systems*. Ithaca, N.Y.: Cornell University Press, 1988. One view of the social systems of the most social of all mammal groups. Assumes an elementary understanding of probability theory and uses mathematical models, but much can be followed without a complete understanding of the mathematics. Index. Extensive bibliography.

Eisenberg, John F., and Devra G. Kleiman, eds. *Advances in the Study of Mammalian Behavior*. Special Publication 7. Shippensburg, Pa.: American Society of Mammalogists, 1983. A group of papers on behavior; several deal with social behavior. This is aimed at the professional or student mammalogist, but many papers are accessible to a wider audience. Includes several case studies of social systems (duck-billed platypus, spiney anteaters, jackals, mongoose, seals, and ground squirrels). The last paper deals with the evolution of social structure in mammals. Brief index. Elaborate lists of references.

Gittleman, John L., ed. *Carnivore Behavior, Ecology, and Evolution*. Ithaca, N.Y.: Cornell University Press, 1989. A group of papers on carnivores, including a large section on behavior. Aimed at the professional or student zoologist. Chapter 5 deals with canid social systems and chapter 7 with comparative carnivore social systems. Index. Each paper has an extensive list of references.

Immelmann, Klaus, ed. *Grzimek's Encyclopedia of Ethology*. New York: Van Nostrand Reinhold, 1977. A collection of articles on ethology (behavior) aimed at the informed layman. Chapter 31 deals with the evolution of social behavior in animals, and chapters 34 and 35 deal with group behavior in ungulates and primates, respectively. Several other chapters have relevant information. Brief index, dictionary of ethological terms, and bibliography.

Macdonald, David W. *European Mammals: Evolution and Behavior*. London: HarperCollins, 1995. An in-depth look at mammalian adaptations to life on land, in the sea, and in the air, and the effect of these evolutionary adaptations on mammalian behavior.

Nowak, Ronald M., and John L. Paradiso. *Walker's Mammals of the World*. 6th ed. 2 vols. Baltimore: The Johns Hopkins University Press, 1999. Outlines the distribution and characteristics of most mammals. In many cases, a brief but informative outline of the social system of the species is given. Intended for the general public, it is also useful to the professional. Index. Extensive list of references.

Rosenblatt, Jay S., and Charles T. Snowdon, eds. *Parental Care: Evolution, Mechanisms, and Adaptive Significance*. Advances in the Study of Behavior 25. San Diego, Calif.: Academic Press, 1996. Covers parenting in species from invertebrates through primates, focusing on physiology and behavior and the links between embryology, physiology, and parental activity.

Slater, P. J. B. *An Introduction to Ethology*. Reprint. London: Cambridge University Press, 1990. A brief introduction to animal behavior intended for the undergraduate student. Chapter 9 covers social organization in general and uses some mammalian examples. Brief index and bibliography.

Vaughan, Terry A. *Mammalogy*. 4th ed. Philadelphia: Saunders College Publishing, 2000. A text on mammals that is intended for upper-level college students but is also understandable to a much wider audience. Chapter 20 deals with behavior, and a large part of it covers social behavior. Aspects of mammalian sociobiology are covered in several other chapters as well. Index. Extensive bibliography.

Wrangham, Richard W., W. C. McGrew, Frans B. M. De Waal, and Paul G. Heltne, eds. *Chimpanzee Cultures*. Cambridge, Mass.: Harvard University Press, 1994. A collection of essays on the extent to which chimpanzees exhibit culture, in the sense of socially transmitted adjustable behavior.

MAMMALS

Type of animal science: Classification
Fields of study: Anatomy, systematics (taxonomy), zoology

Mammals are four-legged animals with backbones; they have a number of unique characteristics, including hair, constant warm body temperatures, mammary glands, and specialized teeth that are replaced only once. Mammals originated more than two hundred million years ago from mammal-like reptiles, and they eventually evolved into tree-dwelling, flying, burrowing, and aquatic forms.

Principal Terms

GENUS (pl. GENERA): a group of closely related species; for example, *Felis* is the genus of cats, and it includes the species *Felis catus* (the domestic cat) and *Felis couguar* (the cougar or mountain lion)

MAMMARY GLANDS: the milk glands that female mammals use to nurse their young

MARSUPIAL: a mammal that gives birth to a premature embryo and then lets it finish its development in a pouch

MONOTREME: a primitive mammal, such as the platypus and spiny anteater, which lays eggs and has other archaic features

ORDER: a group of closely related genera; in mammals, orders are the well-recognized major groups, such as the rodents, bats, whales, and carnivores

PLACENTALS: mammals that carry the embryo in the mother until it is born in a well-developed state; it is nourished in the womb by a membrane (the placenta)

VERTEBRATE: an animal with a backbone; this includes fish, amphibians, reptiles, birds, and mammals

Mammals are a group of vertebrates (animals with backbones) that have been the dominant animals on land and in the sea since the dinosaurs died 65 million years ago. Indeed, this period of time, the Cenozoic era, is often called "the age of mammals." About 4,170 species of mammals are alive today, but at least five times that many are now extinct. About 1,010 genera of mammals are living, but according to a 1945 tabulation, there were an additional two thousand extinct genera, and that number has greatly increased since 1945 as taxonomical research has progressed. Mammals have been very successful in occupying a great variety of terrestrial and aquatic ecological niches. These mammals include terrestrial meat-eaters and plant-eaters, tree-dwellers, burrowing forms, and aquatic forms. The largest mammals today are elephants, but the extinct hornless rhinoceros Paraceratherium was much larger, reaching six meters at the shoulder and weighing about twenty thousand kilograms. The largest mammals, however, are whales, which can weigh up to 150,000 kilograms in the case of the blue whale. That is larger than even the largest dinosaurs.

The Features of Mammals

Living mammals are easily distinguished from all other vertebrates by a number of unique evolutionary specializations. Unlike other vertebrates, mammals all have hair, have mammary glands to nurse their young, and bear live young (except for the most primitive egg-laying mammals, the platypus and the spiny anteater). Mammals maintain a constant, relatively high body temperature. They have a four-chambered heart and a very efficient digestive and respiratory system. Mammals grow rapidly as juveniles and then stop growing when they reach adult size, unlike other animals, which grow continuously throughout their lives.

In addition to these features, mammals have a number of features in their skeletons that make them easy to distinguish from birds or reptiles. Their limbs are designed for efficient walking or running and so are aligned straight under the body (rather than sprawling, as in reptiles). Their ribs are locked into a solid rib cage, so they do not use rib muscles to pump their lungs; instead, they have a diaphragm in the chest cavity to aid in breathing. Their skulls are highly modified, with fewer bones than in birds or reptiles. Some of these modifications include a single nasal opening, a bony palate that separates the breathing passages from the mouth, a large opening on the side of the skull for jaw muscles, and a pair of joints around the spinal column to hold up the head. They have only a single bone in their jaw, and the bones that form extra jaw bones in the reptile have been modified to form the three sound-conducting bones of the ear: the hammer, anvil, and stirrup. Finally, unlike the simple, peglike teeth of reptiles, mammal teeth are highly specialized: There are nipping incisors in front, big stabbing canines for grasping prey, and grinding teeth (molars and premolars) in the back of the jaw for processing food. These teeth are replaced only once (baby teeth are replaced by adult teeth), unlike the continuous replacement found in other animals.

Monotreme and Marsupial Mammals

Living mammals can be divided into three major groups, based on their mode of reproduction. These are the monotremes (egg-laying mammals), marsupials (pouched mammals), and placentals (mammals that carry their young to full term). The monotremes include the platypus and the spiny anteater of Australia and New Guinea. They lay a leathery egg much like that of reptiles, although the young is born after ten days and is then carried in a pouch. The females have mammary glands but no nipples, so the young must lap up the milk as it oozes from their skin. Monotremes have a number of other primitive reptilian features, such as a variable body temperature and a venomous claw in male platypuses. The platypus has a flexi-

ble, ducklike "bill," webbed feet, and a beaverlike tail, which it uses to swim in streams and catch freshwater insects and crayfish. The spiny anteater is covered with thick spines and lives on ants and termites, which it catches with its long, sticky tongue.

The more advanced mammals bear live young. One group, the pouched mammals, or marsupials, give birth to a premature, partially developed embryo. The embryo then climbs into the mother's pouch and fastens onto a nipple, where it finishes its development. Although this sequence makes the young more vulnerable than a placental embryo, which is always carried in the mother, there are advantages. If conditions are bad, a marsupial mother can abort the baby without losing her own life and survive to breed again. A female marsupial can also carry one baby in the pouch and become pregnant with another, allowing a higher rate of reproduction.

The most familiar marsupials are the kangaroo, koala, Tasmanian devil, and opossum, although there have been many other types of marsupials in the past, and many are still alive today in Australia and South America. Where marsupials lived in isolation with no competition from placental mammals, they evolved into many different body forms, which converge on the body forms of their ecological equivalents in the placentals. In Australia today, there are marsupial equivalents of cats, wolves, mice, flying squirrels, rabbits, moles, tapirs, and monkeys. Among extinct marsupials of South America, there were the equivalents of lions and of saber-toothed cats. As similar as these animals look to their placental equivalents in their external body form, they are not related to true cats, wolves, or the rest, since they are all pouched mammals.

Placental Mammals

Unlike marsupials, placentals must carry the embryo in their womb through their full development. To allow this, the embryo is nourished by an extra membrane surrounding it in the womb. This membrane, the placenta, is shed when the baby is born and is part of the "afterbirth." This mode of

Mammals come in many shapes, colors, and sizes. (PhotoDisc)

reproduction makes the placental embryo less vulnerable than a marsupial, but it means that the mother is more vulnerable, since she must carry a larger embryo for a longer time.

After the extinction of the dinosaurs, about 65 million years ago, the earth was open for a new group of large animals to evolve and take over the vacant ecological niches. Placental mammals underwent a tremendous diversification, until they occupied many ecological niches, and some reached the size of sheep. Most, however, were no larger than a cat. The placentals soon diversified into the edentates (anteaters, sloths, armadillos, and their relatives) and the rest of the mammalian orders (groups of genera). Like marsupials, edentates had their greatest success in isolation in South America, although the armadillo is successfully spreading northward. Edentates have a very primitive womb and a slow, variable metabolism compared to other placentals. Although the name "edentate" means toothless, only the anteaters are actually toothless; sloths and armadillos have simple, peglike teeth. Anteaters and armadillos eat ants and termites, and sloths hang upside down from branches, slowly munching leaves.

The rest of the placental mammals have occupied a variety of niches. One group, the insectivores, includes the shrews, moles, and hedgehogs. All these animals are small in size and have sharp teeth for eating insects and worms. The smallest shrews are only three centimeters long and weigh only two grams. They must eat almost continuously in order to make up for their small body size and rapid heat loss. They are so active and aggressive that they will attack animals many times their size. Moles live exclusively underground; they are nearly blind and have well-developed digging claws. Hedgehogs are covered with thick spines and roll up in a ball when they are threatened.

Carnivores and Archontans

In addition to the insectivores, the primary mammalian predators are the carnivorans. They include the feliforms (cats, hyenas, and mongoose) and the caniforms (dogs, bears, seals, sea lions, walruses, pandas, raccoons, weasels, and their relatives). Carnivora (except the panda) all live by killing prey and eating the meat. For this purpose, they have sharp cutting and slicing teeth and enlarged front canine teeth for stabbing. Most have sharp claws, and cats have claws that are retractable. Some carnivora, such as bears, eat fruit, berries, insects, fish, and almost anything else that is available. The pinnipeds, another group of carnivorans related to the bears, have become secondarily aquatic. These include the seals, sea lions, and walruses. Their aquatic specializations include a streamlined body for swimming, with hands and feet developed into flippers.

Another group of placentals, the archontans, have taken to the trees and the air. These include the primates (lemurs, monkeys, apes, and humans), the bats, the elephant shrews, and the colugos, or "flying lemurs." Primates developed front-facing eyes with binocular color vision and agile, grasping hands and feet (with nails rather than claws) for life in the trees. Most have long tails, and New World monkeys can grasp with their tails. Apes and humans, however, have lost their tails. Most primates eat fruit, leaves, or seeds, and they have complex social behavior and well-developed brains.

Bats, on the other hand, have developed a membrane between their fingers that allows them to become excellent fliers. Most bats catch insects in flight by means of their sonar, and they fly at night, sleeping in caves during the day. Fruit bats, however, are much bigger, living in trees and eating fruit during the day.

Glires and Ungulates

The most successful group of mammals is certainly the Glires, or rodents and rabbits. There are more than 1,700 species of rodents alone, or about 40 percent of the mammals, and in numbers and rate of reproduction, they also are far more abundant than any other group of mammals. Rodents (including rats, mice, hamsters, gophers, squirrels, chipmunks, beavers, porcupines, guinea pigs, chinchillas, capybaras, and hundreds of less familiar forms) have developed chisellike incisors (front teeth), which are used for gnawing. They have adapted to a tremendous variety of ecological niches, including sheep-sized browsers (capybaras), terrestrial fruit-, seed-, and insect-eaters, tree-dwellers, and a variety of burrowing forms (such as gophers, ground squirrels, prairie dogs, and mole rats). The rabbits, hares, and pikas form another group that is distantly related to the rodents. Like rodents, they have chisel-like gnawing incisors, but there are two pairs instead of the single pair found in rodents.

Most of the large, plant-eating mammals are ungulates, or hoofed mammals. These include the even-toed artiodactyls (pigs, hippos, camels, deer, antelope, goats, sheep, and cattle), the odd-toed perissodactyls (horses, rhinos, tapirs, hyraxes, and their extinct relatives), the tethytheres (ele-phants, manatees, and sea cows), and the whales, along with a number of extinct archaic groups. Most ungulates have developed hooves, or hard protection on their toes for better running, and many have elongate limbs for fast running. Since they are nearly all herbivorous, most ungulates have developed a complex stomach system to digest large quantities of low-quality, relatively indigestible plant material. Ungulates have also modified their teeth, so that they have larger grinding teeth, which grow almost continuously. That allows them to chew tough, gritty vegetation without becoming toothless. Whales, dolphins, and porpoises are completely aquatic, having lost their hind limbs and developed flippers for front feet. Although they do not look like other hoofed mammals, they had terrestrial ancestors that looked like large bears and were similar to the most primitive ungulates.

Field and Laboratory Studies

Mammalogists use a wide variety of techniques. Traditionally, most studies of the behavior of mammals require going into the wild and observing their ecology. These are the types of studies that are most familiar from nature programs on television. Individual mammals can be studied in detail; in addition, scientists can observe and film their behavior. Often, they are trapped, labeled with a radiocollar or some permanent marking, and freed to be followed or recaptured later. To learn more, however, scientists must collect specimens and study them in the laboratory. There, the anatomy of the specimen can be studied in detail, or the specimen can be kept alive and its behavior and physiology examined much better than in the wild. Mammalogists also have begun to study the details of the biochemistry and molecular biology of mammals. Such studies allow the analysis of the genetic diversity, the evolutionary relationships, and the detailed molecular basis for many of the properties of mammals that were poorly understood before.

—Donald R. Prothero

See also: Aardvarks; American pronghorns; Antelope; Armadillos, anteaters, and sloths; Baboons; Bats; Bears; Beavers; Camels; Cats; Cattle, buffalo, and bison; Cheetahs; Chimpanzees; Deer; Dogs, wolves, and coyotes; Dolphins, porpoises, and toothed whales; Donkeys and mules; Elephant seals; Elephants; Elk; Foxes; Fur and hair; Giraffes; Goats; Gophers; Gorillas; Grizzly bears; Hippopotamuses; Hominids; Hormones in mammals; Horns and antlers; Horses and zebras; Hyenas; Hyraxes; Jaguars; Kangaroos; Koalas; Lemurs; Leopards; Lions; Mammalian social systems; Mammoths; Manatees; Marsupials; Meerkats; Mice and rats; Moles; Monkeys; Monotremes; Moose; Mountain lions; Neanderthals; Orangutans; Otters; Pandas; Pigs and hogs; Placental mammals; Platypuses; Polar bears; Porcupines; Primates; Rabbits, hares, and pikas; Raccoons and related mammals; Reindeer; Reproductive system of female mammals; Reproductive system of male mammals; Rhinoceroses; Rodents; Ruminants; Seals and walruses; Sheep; Shrews; Skunks; Squirrels; Tasmanian devils; Tigers; Ungulates; Vertebrates; Warm-blooded animals; Weasels and related mammals; Whales, baleen.

Bibliography

Anderson, Sydney, and J. Knox Jones, Jr., eds. *Recent Mammals of the World: A Synopsis of Families*. New York: Ronald Press, 1967. A detailed review of the mammalian families and genera living today, emphasizing their taxonomy.

Benton, M. J., ed. *Mammals*. Vol. 2 in *The Phylogeny and Classification of Tetrapods*. London: Clarendon Press, 1988. A review of the relationships and evolutionary history of the major groups of mammals, utilizing the techniques of both morphological and molecular analysis. Many of the chapters are classics, and they may revolutionize the study of mammals.

Eisenberg, John F. *The Mammalian Radiations: An Analysis of Trends in Evolution, Adaptation, and Behavior*. Chicago: University of Chicago Press, 1983. A highly original book that not only reviews the mammals but also points out many common patterns in their ecology and evolution. The innovative use of quantitative analysis and plotting characteristics on multiple axes was revolutionary in its impact on mammalogy.

Grzimek's Encyclopedia of Mammals. 4 vols. New York: McGraw-Hill, 1990. Exhaustive treatment of mammalian life, organized by classification. Emphasizes study of mammals in their natural environments. Numerous charts and illustrations. Includes common names of animals in English, French, and German, as well as Latin taxonomy.

Macdonald, David W., ed. *The Encyclopedia of Mammals*. Reprint. Totowa, N.J.: Barnes & Noble Books, 1999. An excellent, colorful review of the living mammals, with some very good sections on extinct mammals as well.

Nowak, Ronald M. *Walker's Mammals of the World*. 6th ed. Baltimore: The Johns Hopkins University Press, 1999. The standard guide and reference on all the living species of mammals, with detailed information on their taxonomy, geography, behavior, and ecology. The first place to turn for specific information about a particular mammal.

Savage, R. J., and M. R. Long. *Mammal Evolution: An Illustrated Guide*. New York: Facts on File, 1986. A beautifully illustrated account of mammal evolution for the general reader. It groups the mammals by their ecology, rather than by their normal classification, however, and contains some inaccuracies.

Vaughn, T. A. *Mammalogy*. Philadelphia: Saunders College Publishing, 1978. One of the best single college textbooks on the subject of living mammals, with excellent sections on how mammals are studied in the field and in the laboratory.

MAMMOTHS

Type of animal science: Classification
Fields of study: Anatomy, genetics, paleontology, zoology

Mammoths were prehistoric herbivorous mammals that became extinct approximately 10,000 to 3,800 years ago.

<table>
<tr><td>

Principal Terms

IVORY: a white or honey-colored, bony substance

PATHOGENS: bacterium or viruses which cause diseases

PERMAFROST: soil layer below the earth's surface which never thaws

PLEISTOCENE: epoch occurring from 1.7 million to 10,000 years ago

TUNDRA: vast, mostly treeless area coated by ice and snow

</td></tr>
</table>

Mammoths lived during the Ice Ages of the Pleistocene epoch. The oldest recovered mammoth fossils, found in Africa, are four million years old. Mammoths migrated from Africa, and several species of varying sizes and appearances evolved on other continents. Some mammoths crossed the Bering land bridge. Most mammoths favored tundra habitats formed when Ice Age glaciers covered parts of the northern hemisphere.

Anatomy

Most mammoths were large mammals, standing 3.5 to 4.3 meters (12 to 14 feet) tall when measured from the ground to the top of their shoulders. They weighed an average of seven metric tons (eight tons). Other mammoths were dwarves that only stood 1.8 meters (6 feet) tall. These small mammoths lived on islands, including Siberia's Wrangel Island and California's Channel Islands.

Woolly mammoths grew long, thick brown fur over a short undercoat, which protected them in cold climates. Fat layers, which were 7.5 centimeters (3 inches) thick, underneath their skin and in their shoulder hump and prominent brow, insulated their bodies. Each mammoth had two ivory tusks, which extended from their upper jaw. Mammoths' tusks were as long as five meters (sixteen feet) and curved toward and crossed over their trunk. During their lifespan, they had a total of twenty-four molars, which they shed. Because of their cold habitat, mammoths' ears were small, to retain heat.

Life Cycle

Mammoths roamed in herds, sometimes covering great distances in their search for vegetation, such as sedge and Arctic sagebrush, to eat. They spent an average of twenty hours a day finding and eating food. Chewing plants smoothed their teeth. When frozen land inhibited plant growth, the mammoths' stored fat provided essential nutrients and vitamins necessary to live.

Mammoths acted protectively toward immature, elderly, and infirm herd members. Males often fought over females during breeding seasons. Gestation lasted twenty-two months. Females usually produced one calf. Mammoths were unique because their mammary glands were near their front legs, unlike other mammals. Young mammoths remained with their mothers for several years, gradually becoming independent. Mammoths reached maturity by the age of twelve years.

Because of their size, adult mammoths had few predators. They moved slowly, which made them more vulnerable. The saber-tooth tiger killed mammoths for their flesh, and prehistoric humans hunted mammoths for their meat, hides, bones, and tusks. Human hunters devised strategies to capture mammoths in traps that confined their legs. Prehistoric people used mammoth tusks and

In 1999, a nearly perfectly preserved wooly mammoth was discovered in the Taimyr Peninsula of Siberia. Here, a scientist counts the enamel strips on the mammoth's upper molar to determine its age at death: forty-seven years old. (AP/Wide World Photos)

bones to build shelters. Many mammoths died when they fell into naturally occurring sinkholes or were mired by quicksand and tarpits. Sometimes, male mammoths' tusks became entangled while fighting, and they starved to death.

Although mammoths were abundant during the Ice Ages, they became extinct in the period between 13,000 and 2000 B.C.E. Paleontologists estimate that the last mammoths died circa 1800 B.C.E. Researchers hypothesize that mammoths' extinction occurred because they were unable to adapt to the altered environment, specifically the warmer climate and different edible plants that existed after the final Ice Age, or that humans hunted them into extinction. Evidence exists for both theories. Mammoths migrated north as glaciers began to recede. When the final Ice Age glaciers disappeared about 9500 B.C.E., many placen-

tal mammals became extinct. Fossils suggest that some mammoths survived on an isolated Siberian Arctic island until 1500 B.C.E.

Recovering Specimens

Humans have found mammoth fossils on land and in bodies of water, such as the North Sea, which used to be dry land where mammoths roamed. The Siberian permafrost has preserved carcasses that have not decomposed or become too dry for examination. A baby mammoth that scientists called Dima was located in 1977 and had intact red blood cells. Many mammoth fossils have been located during the digging of the Eastside Reservoir, near Los Angeles, California, including a female found in April, 1999, which is considered to be the best preserved mammoth specimen.

Mammoth Facts

Classification:
Kingdom: Animalia
Subkingdom: Metazoa
Phylum: Vertebrata
Class: Mammalia
Subclass: Eutheria
Order: Proboscidea
Suborder: Elephantoidea (elephant-like)
Family: Elephantidae
Genus and species: Mammathus africanavus (North African), *M. columbi* (Columbian), *M. exilis* (dwarf), *M. imperator* (North American), *M. jeffersonii* (southern North America), *M. meridionalis* (European), *M. primigenius* (wooly), *M. subplanifrons* (southeast Africa), *M. trogontherii* (steppe mammoth)
Geographical location: Africa, Northern Europe, North America, Eurasia, Siberia
Habitat: Tundra, grasslands
Gestational period: Twenty-two months
Life span: Sixty years
Special anatomy: Curved tusks, fatty hump, protruding brow

In 2000, paleontologists recovered what they believed to be the first completely frozen mammoth on Siberia's Taimyr Peninsula. Named the Jarkov mammoth, it was airlifted to a cold cave research facility, encased in a block of permafrost that included plants, insects, and pathogens, which scientists hoped would provide insights to mammoths' environments. They planned to test a hypothesis that mammoths became extinct due to an epidemic. Initially, scientists hoped to secure enough DNA to clone the mammoth, but it turned out that radar readings had incorrectly shown more of the carcass remaining than actually existed.

—Elizabeth D. Schafer

See also: Elephants; Extinction; Prehistoric animals; Teeth, fangs, and tusks.

Bibliography

Agenbroad, Larry D., and Jim I. Mead, eds. *The Hot Springs Mammoth Site: A Decade of Field and Laboratory Research in Paleontology, Geology, and Paleoecology.* Hot Springs, S.Dak.: Mammoth Site of Hot Springs, South Dakota, Inc., 1994. An essay collection discussing mammoth specimens in addition to fossils of predators and other organisms found at this South Dakota site. Includes photographs and diagrams and information about the excavation history and geology of the area.

Agenbroad, Larry D., Jim I. Mead, and Lisa W. Nelson, eds. *Megafauna and Man: Discovery of America's Heartland.* Flagstaff: Northern Arizona University, 1990. Scientific papers from a 1989 symposium which featured scholars researching North American mammoths.

Frison, George C., and Lawrence C. Todd. *The Colby Mammoth Site: Taphonomy and Archaeology of a Clovis Kill in Northern Wyoming.* Albuquerque: University of New Mexico Press, 1986. Describes the interactions of Paleo-Indians and mammoths.

Haynes, Gary. *Mammoths, Mastodonts, and Elephants: Biology, Behavior, and the Fossil Record.* New York: Cambridge University Press, 1991. Scientific discussion of mammoths and their modern relatives.

Lister, Adrian, and Paul Bahn. *Mammoths.* New York: Macmillan, 1994. Thorough and well-illustrated account of mammoths before and after extinction and their role in ecosystems and human cultures. Discusses computer modeling and scientific investigations concerning mammoths. Useful for the general reader and includes a glossary, list of museums and mammoth sites, and maps.

MANATEES

Type of animal science: Classification
Fields of study: Conservation biology, ethology, evolutionary science, marine biology, physiology, population biology, wildlife ecology, zoology

Manatees are aquatic, herbivorous animals of the order Sirenia that inhabit warm, shallow waterways. Decreasing habitat and other challenges from humans have caused all four of its species to be threatened or endangered.

Principal Terms

METAZOA: organisms which are multicellular

MOLARS: flat, stout teeth used for grinding food

NOTOCHORD: longitudinal, flexible rod located between the gut and nerve cord

PLACENTA: structure that connects a fetus to the mother's womb; indicative of internal gestation of young

SIRENIAN: refers to animals in the family Sirenia, the only completely aquatic herbivorous mammals; the two families, those of the dugongs and manatees, contain four species

VIBRISSAE: whiskers or bristles, used to provide positional information to an animal

The Florida manatee (*Trichechus manatus latirostris*) is found primarily in the Florida peninsula, but can occasionally be sighted as far north as Virginia and west to Mississippi. The Antillean manatee (*Trichechus manatus manatus*) shows a patchy distribution, including northeastern South America, southern Texas, Mexico, and the Caribbean. *Trichechus senegalensis*, the West African manatee, can be found on the west coast of Africa from Senegal to Angola. The Amazonian manatee, *Trichechus inunguis*, is found throughout the Amazon river drainage basin.

The species are distinguished by their geographical distributions, and while they are physically very much alike, some differences exist. Manatees all exhibit a streamlined body, full around the middle with no visible neck, tapering to a paddle-shaped tail used for propulsion. They have two small pectoral flippers on their upper bodies that are used for steering and bringing food to the mouth. The lips are long and flexible, and help in funneling plants into their mouths. Manatees are grayish-brown in color and have sparse, bristly hair scattered thinly over their torsos. They have molars at the back of their mouths, and unlike most other mammals, as the front molars wear down they are continually replaced by new teeth from the back of the mouths. Adult West Indian and West African manatees average about ten feet in length and weigh approximately 800 to 1200 pounds. Amazonian manatees are smaller, shorter, and more slender, averaging about eight feet in length and less than eight hundred pounds.

Adapted for the Aquatic Life

Streamlined bodies and flippers are ideal for the aquatic life of a manatee. They are agile and typically cruise in search of food at speeds of two to six miles per hour, but can swim as fast as fifteen miles per hour. While they have been sighted diving to depths of about thirty-three feet, they primarily forage for food no deeper than ten feet. Manatees can stay underwater for up to twenty minutes, but typically surface every two to three minutes for air.

Manatees spend most of their day foraging for aquatic plants and will consume between 4 and 9 percent of their body weight in wet vegetation daily. The animals are nonaggressive and non-

Manatees have streamlined bodies with no visible neck, which increases their efficiency in swimming. (Digital Stock)

territorial and spend the rest of their time traveling, investigating objects, and socializing. Manatees are considered semisocial, and the typical social unit is a female and her calf; however, congregations of up to two hundred individuals can be found near warm water sources such as power plant outfalls and hot springs, especially during winter months.

Male manatees are sexually mature at about nine to ten years of age, and females at about seven to eight years. Mating takes place in the water, and the female is accompanied by, and mates with, several males. Gestation takes about twelve months; the newborn calf measures about four feet and weighs sixty to seventy pounds. Calves remain close to their mothers for up to two years. The calf is dependent on its mother for nutrition, and for learning about feeding and resting areas, migratory routes, and warm water refuges. It is believed that manatees can live to sixty years in the wild; however, habitat destruction and injury by boats and fishing lines are seriously endangering manatees worldwide.

—*Karen E. Kalumuck*

See also: Fins and flippers; Lakes and rivers; Marine animals; Marine biology; Seals and walruses.

Bibliography

Dold, Catherine. "Hearing of Manatees May Prove to Be Key to Protecting Species." In *The Science Times Book of Mammals*, edited by Nicholas Wade. New York: Lyons Press, 1999. This article, originally from *The New York Times*, discusses the role of manatee hearing in their accidents with motor boats, and potential means to remedy the situation.

Manatee Facts

Classification:
Kingdom: Animalia
Subkingdom: Bilateria
Phylum: Chordata
Subphylum: Vertebra
Class: Mammalia
Subclass: Eutheria
Order: Sirenia
Family: Trichechidae (manatees)
Genus and species: Trichechus manatus (West Indian manatee, with subspecies *latirostris*, the Florida manatee, and *manatus*, the Antillean manatee); *T. senegalensis* (West African manatee); *T. inunguis* (Amazonian manatee)
Geographical location: Florida, Caribbean, and northern South America; West Africa; Amazon River basin
Habitat: Warm tropical and subtropical water, including shallow coastal waters, estuaries, and rivers
Gestational period: Approximately twelve months
Life span: Twenty-eight years in the wild; fifty to sixty years in captivity
Special anatomy: Flat, paddle-shaped tail; two pectoral flippers; nasal openings at the top of the snout; large, flexible upper lip equipped with vibrissae attached separately to nerve endings and each with its own blood supply

Macdonald, David, ed. *The Encyclopedia of Mammals.* 2d ed. New York: Facts on File, 1995. This lavishly illustrated compendium of mammals gives detailed, accessible information on hundreds of wild mammals, organized by phylogenetic order. Of particular interest are the introductory chapter on mammals and the chapter on sea cows and manatees.

Ohara, Rei, and Akemi Hotta. *Manatee.* San Francisco: Chronicle, 1998. This book features superb and unusual underwater photographs of manatees, illuminated by a simple yet informative text.

Reynolds, John, and Daniel Odell. *Manatees and Dugongs.* New York: Facts on File, 1991. This authoritative book written by two leading marine scientists is a thorough, clear, engaging, and beautifully illustrated treatise on the lives of sirenians.

Wlodarski, Loren, et al. *Siren's Song: The Story of Manatees.* San Diego, Calif.: Sea World, 1998. A concise introduction to the biology of manatees and current conservation efforts focused on the Florida manatee.

MARINE ANIMALS

Type of animal science: Ecology
Fields of study: Behavior, marine biology, oceanography, physiology

The world's oceans contain the largest and most varied array of life forms on earth. The marine environment is divided into coastal, open water, deep-sea, and bottom zones, and the lives of animals living in each of these regions are dictated by the physical conditions present in these zones. Due to the inaccessibility and hostility to human life of most of the ocean environment, much remains to be learned about the biology and ecology of most of the animals living in the marine world.

Principal Terms

BENTHOS: organisms living upon, or below, the surface of the substrate that forms the ocean floor

ESTUARY: the region where freshwater rivers empty into and mix with the marine environment of oceans or seas

INTERTIDAL ZONE: the portion of the marine environment located between low and high tide marks

NEKTON: larger marine animals that have sufficient powers of locomotion to move independently of water currents

NERITIC ZONE: the shallow water areas that extend over the continental shelves up to the low tide mark

PELAGIC ZONE: portions of the marine environment that are located away from the shorelines; the open ocean environment

PLANKTON: small animals and plants that drift with the water currents in the marine environment

---

Approximately 71 percent of earth's surface is covered by salt water, and the marine environments contained therein constitute the largest and most diverse array of life on the planet. Life originated in the oceans, and the salt water that comprises the largest constituent of the tissues of all living organisms is a vestigial reminder of the aquatic origins of life.

Marine Zones

The marine environment can be divided broadly into different zones, each of which supports numerous habitats. The coastal area between the high and low tide boundaries is known as the intertidal zone; beyond this is the neritic zone, relatively shallow water that extends over the continental shelves. The much deeper water that extends past the boundaries of the continental shelves is known as the oceanic zone. Open water of any depth away from the coastline is also known as the pelagic zone. The benthic zone is composed of the sediments occurring at the sea floor. Areas in which freshwater rivers empty into the saltwater oceans produce a continually mixed brackish water region known as an estuary. Estuarine zones often also include extensive wetland areas such as mud flats or salt marshes.

Zones in the marine environment are distributed vertically as well as horizontally. Life in the ocean, as on land, is ultimately supported by sunlight in most cases, used by photosynthetic plants as an energy source. Sunlight can only penetrate water to a limited depth, generally between one hundred and two hundred meters; this region is known as the photic or epipelagic zone. Below two hundred meters, there may be sufficient sunlight penetrating to permit vision, but not enough to support photosynthesis; this transitional region may extend to depths of one thousand meters and is known as the disphotic or mesopelagic zone. Below this depth, in the aphotic zone, sunlight cannot penetrate and the environment is perpetually dark, with the exception of small amounts of light produced by photoluminescent invertebrate and vertebrate animals. This aphotic zone is typically divided into the bathypelagic zone, between seven hundred and one thousand meters as the upper range and two thousand to four thousand meters as the lower range, where the water temperature is between 4 and 10 degrees Celsius. Beneath the bathypelagic zone, overlying the great plains of the ocean basins, is the abyssalpelagic zone, with a lower boundary of approximately six thousand meters. Finally, the deepest waters of the oceanic trenches, which extend to depths of ten thousand meters, constitute the hadalpelagic zone. In each of these zones, the nature and variety of marine life present is dictated by the physical characteristics of the zone. However, these zones are not absolute, but rather merge gradually into each other, and organisms may move back and forth between zones.

Plankton

Marine life can be divided broadly into three major categories. Those small organisms that are either free-floating or weakly swimming and which thus drift with oceanic currents are referred to as plankton. Plankton can be further divided into phytoplankton, which are plantlike and capable of photosynthesis; zooplankton, which are animal-like; and bacterioplankton, which are bacteria and blue-green algae suspended in the water column. Larger organisms that can swim more powerfully and which can thus move independently of water movements are known collectively as the nekton. Finally, organisms that are restricted to living on or in the sediments of the seafloor bottom are referred to as the benthos.

The phytoplankton, which are necessarily restricted to the photic zone, are by far the largest contributors to photosynthesis in the oceans. The phytoplankton are therefore responsible for trapping most of the solar energy obtained by the ocean (the primary productivity), which can then be transferred to other organisms when the phytoplankton are themselves ingested. The phytoplankton are composed of numerous different types of photosynthetic organisms, including diatoms, which are each encased in a unique "pillbox" shell of transparent silica, and dinoflagellates. The very rapid growth of some species of dinoflagellates in some areas results in massive concentrations or blooms that are sometimes referred to as red tides. Chemicals that are produced by red tide dinoflagellates often prove toxic to other marine organisms and can result in massive die-offs of marine life. Smaller photosynthetic plankton forms comprise the nanoplankton and also play an important role in the photosynthetic harnessing of energy in the oceans.

The zooplankton are an extremely diverse group of small animal organisms. Unlike the phytoplankton, which can make their own complex organic compounds via photosynthesis, the phytoplankton must ingest or absorb organic compounds produced by other organisms. This is accomplished by either preying upon other planktonic organisms or by feeding on the decaying remains of dead organisms. A number of zooplankton species also exist as parasites during some portion of their life cycles, living in or upon the bodies of nekton species. The largest group of zooplankton are members of the subphylum Crustacea, especially the copepods. These organisms typically possess a jointed exoskeleton, or shell, made of chitin, large antennae, and a number of jointed appendages. Space precludes a definitive listing of all of the zooplanktonic organisms; however, virtually all of the other groups of aquatic invertebrates are represented in the bewildering variety of the zooplankton, either in larval or adult forms. Even fishes, normally a part of the nekton, contribute to the zooplankton, both as eggs and as larval forms.

The bacterioplankton are found in all of the world's oceans. Some of these, the blue-green algae (cyanobacteria), play an important role in the photosynthetic productivity of the ocean. Bacterioplankton are usually found in greatest concentrations in surface waters, often in association with organic fragments known as particulate organic carbon, or marine snow. Bacterioplankton play an important role in renewing nutrients in the photic zones of the ocean; such renewal is important in maintaining the photosynthetic activity of the phytoplankton, upon which the rest of marine life is in turn dependent.

One of the principal problems facing plankton is maintaining their position in the water column. Since these organisms are slightly denser than the surrounding seawater, they tend to sink. Clearly this is a disadvantage, particularly since plankton typically have very limited mobility. This is especially true for the photosynthetic phytoplankton, which must remain within the photic zone in order to carry on photosynthesis. A number of strategies have evolved among planktonic species to oppose this tendency to sink. Long, spindly extensions of the body provide resistance to the flow of water. Inclusions of oils or fats (which are less dense than water) within the body provide positive buoyancy by decreasing the overall density of the plankton. Finally, some species, such as the Portuguese man-o'-war, generate balloonlike gas bladders, which provide enough buoyancy to keep them at the very surface of the epipelagic zone.

Nekton

The nekton is composed of those larger animals that have developed locomotion to a sufficient degree that they can move independently of the ocean's water movements. Whereas the plankton are principally invertebrates, most of the nekton are vertebrates. The majority of the nekton are fishes, although reptile, bird, and mammalian species are also constituent parts. The oceanic nekton are those species which are found in the epipelagic zone of the open ocean. These include a wide variety of sharks, rays, bony fishes, sea birds, marine mammals, and a few species of reptiles. Some members of the oceanic nekton, such as blue sharks, oceanic whitetip sharks, tuna, flying fish, and swordfish, spend their entire lives in the pelagic environment; these are said to be holoepipelagic. Others, the meroepipelagic nek-

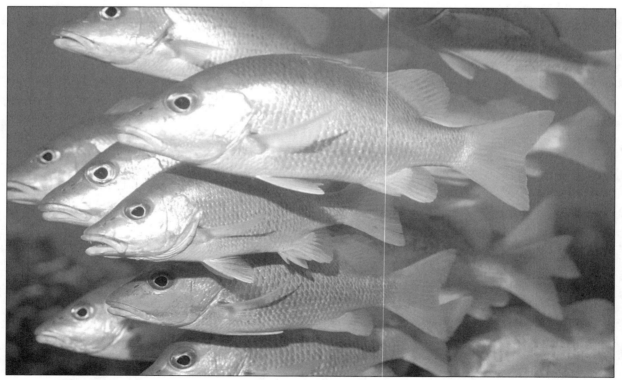

The nekton is made up of animals with independent locomotion, such as fish. (Digital Stock)

Hydrothermal Vents

One of the most exciting discoveries made by deep-sea biologists occurred as recently as 1977. Scientists aboard the deep submersible *Alvin* happened upon a previously unknown hydrothermal vent community while studying the Galápagos Rift Zone, 2,700 meters below the sea surface. These communities were composed of high densities of organisms never before seen by humans, in an otherwise barren landscape of sparse benthic life. Since their discovery, over four hundred species have been recovered from hydrothermal vents, including one new class, three new orders, and twenty-two new families of animals. Among the more notable animals are giant clams, mussels, and vestimentiferan tubeworms. Vestimentiferan tube worms are about two meters long and lack a digestive tract.

These dark ecosystems obviously cannot be supported by photosynthetic organisms, but instead, chemoautotrophic bacteria that derive energy from sulfide are the primary producers that feed this unique food web. Vestimentiferan tube worms have these chemoautotrophic bacteria living symbiotically within their tissues. These bacteria directly provide the tube worms with the fixed carbon they need to survive, which is why the worms do not have digestive tracts. They do not need to eat. Since the discovery of hydrothermal vent communities in the Galápagos rift, similar communities have been discovered along rifts in the Atlantic, Indian, and western Pacific Oceans.

ton, only spend a portion of their lives in the epipelagic zone, returning to coastal areas to mate, as with herring and dolphins, or returning to freshwater, as with salmon and sturgeon. Sea birds are a special case: Although they spend much of their time flying over the epipelagic zone and nest on land, they feed in the epipelagic zone and some species may dive as deep as one hundred meters in search of prey. Some members of the nekton enter the epipelagic only at certain times in their life cycle. Eels of the family Anguillidae spend most of their lives in freshwater but return to the epipelagic zone to spawn. Additionally, at night many species of deep-water fishes migrate up into the epipelagic to feed before returning to deeper waters during the daylight hours.

The pelagic environment, unlike the terrestrial one, is profoundly three-dimensional. Nektonic animals can move both horizontally and vertically within the water column. Furthermore, since most of the pelagic environment is essentially bottomless, since there is no apparent or visible ground or substrate, the environment is basically uniform and featureless. These characteristics play an important role in the evolution of the behavior of nektonic animals. Fishes suspended in an essentially transparent and featureless medium have no shelter in which to hide from predators, nor are there any apparent landmarks to serve as directional cues for animals moving horizontally from place to place. Life in the open ocean has therefore favored adaptations for great mobility and speed with which to move across large distances and escape from predators, as well as camouflage and cryptic coloration designed to deceive potential predators or prey.

As is the case for plankton, most nektonic animals are denser than the surrounding seawater, and maintaining position in the water column is of the first importance. Most fishes possess a swim bladder, a gas-filled membranous sac within their body that opposes the tendency to sink and provides the fish with neutral buoyancy. Sharks and rays lack a swim bladder, but accumulate large concentrations of fats and oils in their liver, which also help counter the tendency to sink. Large, fast-swimming species of shark, tuna, and many billfish also rely on the generation of hydrodynamic lift to maintain vertical position in the water column. The tail and body of these fishes generate forward thrust, moving the animal through the water, and the fins, notably the pectoral fins, generate lift from the water flowing over them in a manner similar to that of an airplane's wing. Thus these animals fly through the water, but are in turn required to move continuously in order to generate lift.

All members of the nekton are carnivores, feeding on other nektonic species or upon plankton, particularly the larger zooplankton. In general, the size of the prey consumed by nekton is directly related to the size of the predator, with larger species consuming larger prey. However, the organisms that feed upon plankton, the planktivores, include a wide variety of fish species such as herring, salmon, and the whale shark, the largest extant fish species. They also include the largest marine animals of all, the baleen whales. The case of large animals feeding upon very small plankton directly addresses the need of all animals to meet their energy requirements. For all animals, the amount of energy obtained from food consumed must necessarily exceed the energy expended in acquiring the prey. Very large animals, such as whales and whale sharks, require a great deal of energy to move their bodies through the aquatic environment, but because of their great size they are necessarily less agile than smaller forms. The amount of energy required to chase and catch these smaller animals would generally exceed the energy derived from ingesting them. Plankton, however, are relatively easy to obtain due to their very limited mobility. However, because of their small size, vast quantities of plankton must be ingested in order to meet the metabolic requirements of large marine animals. Some very large species that are not planktivores solve the energy problem by evolving behaviors for acquiring specialized diets that yield higher energy. White sharks, for example, feed on fish when young, but as they age and increase in size, marine mammals, notably seals and sea lions (pinnipeds), become a major part of their diet. Marine mammals all possess blubber, an energy-rich substance that yields much more energy than fish. Similarly, sperm whales, the largest hunting carnivores on the planet, have a diet that consists in large part of giant squid, which are hunted in the ocean depths largely using the whale's acoustic echolocation sense. Orcas (killer whales) effectively use pack hunting techniques to hunt larger whales and other marine mammals.

The deeper regions of the ocean are dominated by different types of nekton. However, we know even less about their ecology due to their relative inaccessibility. The disphotic or mesopelagic zone contains many animal species that migrate vertically into surface waters at night to feed upon the plankton there. Many of these organisms possess large, well-developed eyes and also possess light organs containing symbiotic luminescent bacteria. The majority of the fish species in this group are colored black and the invertebrates are largely red (red light penetrates water less effectively than do longer wavelengths, and these animals appear dark-colored at depth). Beneath this zone, in the bathypelagic and abyssalpelagic zones, there are many fewer organisms and much less diversity than in the shallower levels. Animals in this region are typically colorless and possess small eyes and luminescent organs. Because organisms in these deep regions are few and far between, many species have become specialized in order to maximize their advantages. Thus, deep-sea fish are characterized by large teeth and remarkably hinged jaws that allow them to consume prey much larger than might be expected from their size. Similarly, since encounters with potential mates are presumably scarce, a number of unique reproductive strategies have evolved. In the anglerfish (Ceratius), all of the large individuals are female and the comparatively tiny males are parasitic, permanently attaching themselves to the female. Much, however, still remains to be learned of the ecology of these deep-sea organisms.

Benthos

The benthos of the world's oceans consists of animals that live on the solid substrate of the water column, the ocean floor. Scientists typically divide benthic organisms into two categories, the epifauna, which live on the surface of the bottom at the sediment-water interface, and the infauna, those organisms living within the sediments. In shallow water benthic communities, members of virtually every major animal group are represented. Ecologists generally differentiate between soft bottom benthic communities (sand, silt, and mud, which comprise the majority of the benthic zone) and rocky bottom communities, which are less common proportionately. Soft bottom communities have an extensive diversity of burrowing infauna, such as polychaete worms, and mollusks, such as clams. Rocky bottom communities possess a larger proportion of epifauna, such as crustaceans and echinoderms (starfish, sea urchins, and brittle stars), living on the surface of what is essentially a two-dimensional environment. Vertical faces of the hard bottom environment, such as canyon walls or coral reefs, are often home to a wide variety of animals occupying various crannies and caves. In some parts of the world, kelp plants that are anchored to the substrate and which extend to the water surface dominate the rocky bottom substrate. In these kelp forests, large kelp plants (actually a species of brown algae) form a forestlike canopy that plays host to a wide and complex array of animals extending throughout the water column. On the deep ocean floor, the benthos is composed of representatives of virtually every major animal group: crustaceans such as amphipods, segmented polychaete worms, sea cucumbers, and brittle stars. Less common are starfish, sea lilies, anemones, and sea fans. The fishes of the deep benthos include rat tails and a number of eel species.

Estuaries, where freshwater rivers empty into marine environments, are typified by large, cyclic changes in temperature and salinity. Although estuaries have played an important role in human history as the sites of major ports, the variety and number of estuarine species tend to show less diversity of animal species due to the difficulty in adapting to the large swings in environmental conditions.

Animal life in the sea, like that on land, shows an astonishing variety of forms and behaviors, the result of natural selection. The inaccessibility and hostility of much of the world's oceans to human exploration and observation leaves much yet to be learned about the biology of marine life. Much remains to be achieved in order to obtain a useful body of knowledge concerning life in the sea.

—John G. New

See also: Bioluminescence; Clams and oysters; Coral; Crabs and lobsters; Crustaceans; Deep-sea animals; Dolphins, porpoises, and toothed whales; Echinoderms; Eels; Elephant seals; Fins and flippers; Fish; Horseshoe crabs; Ichthyosaurs; Jellyfish; Lakes and rivers; Lampreys and hagfish; Lungfish; Manatees; Marine biology; Mollusks; Octopuses and squid; Otters; Penguins; Protozoa; Reefs; Reptiles; Seals and walruses; Salmon and trout; Seahorses; Seals and walruses; Sharks and rays; Snakes; Sponges; Starfish; Swamps and marshes; Tentacles; Tidepools and beaches; Turtles and tortoises; Whale sharks; Whales, baleen; White sharks; Zooplankton.

Bibliography

Niesen, T. M. *The Marine Biology Coloring Book.* 2d ed. New York: HarperResource, 2000. A fun and very useful way to learn modern concepts in marine biology, excellent for both adults and children. Illustrations by W. Kapit and C. J. Simmons.

Nybakken, J. W. *Marine Biology: An Ecological Approach.* 5th ed. San Francisco: Benjamin/ Cummings, 2001. A useful, college-level text dealing with general concepts of marine biology and ecology.

Robison, B. H., and J. Connor. *The Deep Sea.* Monterey Bay, Calif.: Monterey Bay Aquarium Press, 1999. A fascinating and beautifully illustrated look at the often bizarre creatures that inhabit the ocean's depths.

Safina, C. *Song for the Blue Ocean: Encounters Along the World's Coasts and Beneath the Seas.* New York: Henry Holt, 1998. A fascinating, readable, and important work relating modern ecological concepts of marine biology and the impact of human activity on the state of the world's oceans.

MARINE BIOLOGY

Type of animal science: Fields of study
Fields of study: Conservation biology, ecology, environmental science, invertebrate biology, marine biology, oceanography, zoology

Marine biology is the study of the plants, animals, fungi, protists, and microbes that live in the oceans. Because humans are not well adapted to marine environments, it is difficult to observe and study marine organisms in their natural surroundings. However, with modern technology such as scuba, submarines, and remote cameras, marine biologists are gaining considerable information about marine organisms.

Principal Terms

AUTOTROPHS: primary producers; organisms that are self-feeding; includes photosynthetic and chemoautotrophic organisms

BENTHIC: the area of the ocean floor; organisms associated with the sea bottom

EPIFAUNA: animals that live on the sea floor

HETEROTROPH: consumers; organisms that must acquire energy by consuming organic material

INFAUNA: animals that live in the sea floor

INVERTEBRATES: animals that lack backbones

LITTORAL: the area in the intertidal zone; organisms that live in the intertidal

NEKTON: organisms that are strong swimmers and can move against ocean currents

PELAGIC: the area of open water in the oceans; organisms that occur in the water column

PLANKTON: organisms that drift in the ocean currents because they have limited power of locomotion

---

The earliest known life forms were marine. The oceans cover 71 percent of earth's surface and have an average depth of 3,800 meters, which means that they represent 99 percent of the living space on the planet. Marine biology is the study of all the organisms that occupy this space. Despite having 99 percent of the planet's living space at their disposal, only 250,000 of approximately 1.8 million described living species (14 percent) are marine. While the oceans may lack diversity at the level of species, they are home to members of thirty-one of thirty-four animal phyla, about twice the number of phyla that are found on land or in freshwater.

Because of its vastness and humans' inability to easily visit deep waters, the oceans remain the least studied habitats on earth. For example, scientists know that giant squid eighteen meters in length, larger than any other invertebrate, exist because they have been found in the stomachs of sperm whales and washed up on beaches. However, no human has ever seen one of these creatures living in their natural habitats. If this sheet of paper represented the area of the ocean, the area directly observed by people would be smaller than the period at the end of this sentence. Despite the small amount of study in marine biology, research in these systems has contributed greatly to the understanding of living systems.

Oceanographers divide the ocean into zones that have distinct physical characteristics. These physical conditions, in turn, select for different organisms that are adapted to the set of conditions. The benthic realm refers to the sea floor and extends from the high intertidal zone, where ocean meets terrestrial land, to depths of eleven kilometers at oceanic trenches, the deepest parts of the ocean where the ocean floor slowly sinks back into the interior of the earth. Organisms that live in, on, or near the ocean floor are appropriately called benthic organisms, and represent 98 percent of all marine creatures. The pelagic realm of the ocean refers to the open ocean, basically the space between the benthos and the sea surface. Pelagic organisms, representing the remaining 2 percent of marine species, live in the water column. The pelagic zone can be divided into the photic and aphotic zones, a distinction that is especially important for photosynthetic organisms. The photic zone is the shallow part of the ocean that receives enough sunlight to support photosynthesis, which is about two hundred meters deep in the clearest waters, and as shallow as three meters in turbid coastal waters. The aphotic zone is where there is not enough light to support photosynthesis, and extends from the bottom of the photic zone to the ocean floor.

Organisms of the Open Ocean
Pelagic creatures can be further categorized into plankton, which drift around in the ocean currents, and nekton, which are capable of swimming against currents. Photosynthetic plankton live in the photic zone and are called phytoplankton. These algae form the base of the food web in the open ocean and are eaten by zooplankton (heterotrophic plankton including protists and small crustaceans such as copepods). Zooplankton are eaten by small fishes, which are eaten by larger fishes, which are eaten by sharks, birds of prey, and people. The open ocean food web is one of the longest food webs known, partly because it starts with the smallest photosynthetic organisms.

The largest animal migrations on earth occur every day in the open oceans, in a process called diel vertical migration. Zooplankton, midwater fish, squid, and krill migrate to shallow waters at night and then return to the dark depths during the day. The main reason for this daily migration is probably to feed, and to avoid being eaten. Food densities are highest in the shallow productive waters, so these predators move to the shallows to feed at night. Because they would be susceptible to their own visually oriented predators in the well-lit shallows, they return to the dark depths during the day to avoid being eaten.

In the aphotic zone, there are only heterotrophic organisms that are supported mostly by organic material that rains down from the lit environments above. These animals live in darkness, with the exception of light produced by the animals themselves, called bioluminescence. It is common for sea creatures (especially ones at intermediate depths) to house luminescent bacteria within their tissues, which are able to produce light for communication, as a lure to attract prey, or to light their bottom surface to conceal their silhouette against the dimly lit background from above. Anglerfish are deep-sea predators that attract prey near their mouths by dangling a bioluminescent lure in front of their head.

The density of organisms in the deep sea is low. Because of this low density, a long period of time can pass between meals, or between encounters with the opposite sex. To deal with the problem of infrequent meals, deep-sea creatures are often gigantic compared to shallow-water relatives. Large size allows for storage of food reserves that sustain the animals between meals. Predatory fish also have large mouths and stomachs that allow them to take full advantage of any meal, regardless of size. To overcome the problem of rare mates, the miniature males of some anglerfish have the unusual adaptation of attaching themselves to the female, where they live the rest of their lives as parasitic sperm producers.

Organisms of the Bottom
Most marine species are found at the ocean floor. They occur in such familiar marine habitats as mud flats, sand flats, beaches, coral reefs, kelp forests, and the rocky intertidal. The main primary producers in benthic habitats are macroscopic seaweeds that grow attached to the bottom or microscopic algae that grow within the tissues of animals such as corals, sponges, and bryozoans. Benthic animals include mobile creatures such as fish, crabs, shrimp, snails, urchins, sea stars, and slugs. Additionally, there are numerous animals that, unlike familiar terrestrial animals, never move around as adults. These sessile animals include

Oases on the Deep Ocean Floor

Although animal life exists on the ocean floor, much of that zone is barren, with relatively few animals distributed at large intervals. One of the most exciting discoveries in marine biology during the last quarter of the twentieth century occurred in 1977 with the discovery of abundant animal life in the regions immediately surrounding four hot water geysers on the ocean floor. These hydrothermal vents, located in the eastern Pacific Ocean at a depth of approximately 2,700 meters, boasted a spectacular variety of species all located within a few meters of the vent. The geyser is caused when water is heated by Earth's mantle and rises rapidly through the cooler waters above. Water in the geyser reaches temperatures of 8 to 16 degrees Celsius, as opposed to the surrounding water, which is about 2 degrees Celsius. This warm water is also rich in dissolved hydrogen sulfide gas, which serves as a source of energy for bacteria at the vent. The bacteria, in turn, act as food for the various animal species in the area of the vent. Similar hydrothermal vents, each with a population of animals surrounding it, have since been discovered at other sites in the Atlantic and Pacific Oceans.

The largest animals located at these vents include large clams and mussels and several species of tube (vestimentiferan) worms. Other species include crabs, snails, segmented worms and spaghetti worms, so called because of their appearance. Many of these species had not been previously observed or described. All vent communities appear to have similar types of organisms in association, but not always the same species. Current estimates indicate that the life span of the vents is quite short (probably in the tens of years) and that when the vent dies, so do the animal communities surrounding it. How these vents are located and colonized by organisms living on the deep ocean floor is just one of many interesting questions concerning these deep-sea oases that remain to be answered.

barnacles, sponges, oysters, mussels, corals, gorgonians, chrinoids, hydroids, and bryozoans.

The commonness of sessile animals in the marine benthos suggests that it is a successful way of life. These animals' lifestyle combines facets of plant and animal lifestyles. Sessile invertebrates are plantlike in that they obtain some of their energy from sunlight (the animals themselves do not photosynthesize, but they house photosynthetic symbionts), they are anchored in place, and they grow in a modular fashion just as the branches of a tree do. They are animal-like in that they capture and digest prey and they undergo embryonic development, often involving metamorphosis. In fact, nearly all benthic animals start life in the pelagic realm, drifting around as planktonic larva, dispersing to new habitats as they develop and feed. After a few hours to weeks of pelagic living, they sink to the ocean floor to complete life as adults.

Being stuck in one place presents special challenges for sessile animals, including food acquisition, predator avoidance, and mating. Sessile animals feed by having symbiotic algae and by filtering organic particles from passing water currents. Like plants, sessile animals use structural and chemical defenses against predators, and have tremendous regenerative abilities to recover from partial predation events. Most benthic animals mate via external fertilization: Sperm and eggs are spawned into the water column and fertilization occurs outside the body of the female. Amazingly, sessile barnacles must copulate to achieve internal fertilization. These animals increase their reproductive success by being hermaphroditic, thus assuring that any neighbor is a potential mate; being gregarious to assure a high density of mates; and by having a penis long enough to deliver sperm to an individual seven shell lengths away.

Trophic Cascades and Keystone Predators

Marine organisms live in environments that are foreign to humans, and they have lifestyles that are unique to them and different from terrestrial creatures. However, the study of marine organisms has led to advances in ecological theory that have proven to be useful in understanding terrestrial communities as well. The concept of keystone species, where a relatively rare species has a disproportionately large effect on the community structure, was discovered from research conducted on rocky intertidal habitats. Professor Robert Paine removed starfish from rocks off the coast of Washington state and observed that mussels soon crowded out seaweeds and barnacles, resulting in about a 50 percent reduction in species richness. He concluded that starfish are keystone predators because their predation prevented mussels from excluding less-competitive species from the habitat.

Not far offshore from the rocky intertidal in kelp forests, another keystone species was shown to influence the community by a trophic cascade (where the effects from a top predator "cascade" down to lower trophic levels). Sea otters, by preying on sea urchins, protect the kelps that make kelp forests, an important habitat that many marine species rely on. In the 1800's, hunters greatly reduced the number of otters by harvesting their thick furs. As a result, sea urchin populations exploded because they were relieved from predation, resulting in a decimation of kelps by the herbivorous urchins. Once otters received government protection, their numbers increased, urchins decreased, and kelp forests returned, at least in some areas. Interestingly, Aleutian killer whales began preying on otters in the 1980's, causing kelp forests to begin disappearing again. The reason orcas began eating otters is probably due to concurrent declines in seal and sea lion numbers, the normal prey of these killer whales. The ultimate causes of these altered food webs is uncertain, but it almost certainly is the result of human activity. Even though humans are poorly adapted for a marine existence, the evidence is mounting that humans are altering marine ecosystems in complex, novel, and unpredictable ways. The science of marine ecology is best equipped to study these effects and offer information to protect and manage ocean life.

—Greg Cronin

See also: Bioluminescence; Clams and oysters; Coral; Crabs and lobsters; Dolphins, porpoises, and toothed whales; Echinoderms; Eels; Fins and flippers; Fish; Horseshoe crabs; Jellyfish; Lakes and rivers; Lampreys and hagfish; Lungfish; Manatees; Marine animals; Mollusks; Protozoa; Reefs; Reptiles; Salmon and trout; Seals and walruses; Sharks and rays; Snakes; Sponges; Starfish; Swamps and marshes; Tentacles; Tidepools and beaches; Turtles and tortoises; Whale sharks; Whales, baleen; White sharks; Zooplankton.

The interrelationship between orcas (pictured), seals, sea otters, and kelp forests can cause a trophic cascade when one element of the food chain is disturbed. (Corbis)

The Problem of Ocean Navigation

The epipelagic environment is essentially a featureless one, without many of the landmarks or other indicators that might serve as directional indicators for migratory ocean species. Yet many animal species are known to make long and highly accurate migrations through the apparently trackless regions of the world's oceans.

The best-known example of oceanic migration is probably in salmon, which as young animals move from the freshwater streams in which they hatched out to the open ocean in which they mature. Upon reaching sexual maturity, the salmon return to their natal freshwater streams, where they spawn and die. The later portions of this return migration, when the animals are moving up the river system and locating the precise stream in which they were born, has been the most intensely studied, and the sense of smell appears to play an important role. However, the salmon are also faced with the difficult navigational task of locating the mouth of their home river from the vast expanse of the open ocean. How they accomplish this task is completely unknown.

Other astonishing examples of animals making precise migratory movements are known. Green sea turtles (*Chelonia mydas*) from the Caribbean and North and South America are known to accurately navigate to Ascension Island in the middle of the Atlantic Ocean to lay their eggs. American and European eels leave their freshwater homes in streams and ponds upon reaching sexual maturity and migrate out to sea. Upon reaching the marine environment, these animals transform from freshwater eels to marine deep-sea eels of strikingly different appearance, and navigate to a region in the North Atlantic known as the Sargasso Sea, where they spawn and die.

The cues that are used by these animals in their migrations are unknown. Although it is tempting to call this instinctive behavior, instinct is merely a label for mechanisms that are unlearned and innate. These animals are clearly making choices of which direction to proceed in order to reach their goal, but the manner in which those choices are made remain mysterious and fascinating.

Bibliography

Cousteau, Jacques-Yves. *The Ocean World*. New York: Abradale Press, 1985. Virtually an encyclopedia of the undersea world, this comprehensive volume covers several aspects of marine biology by the famous explorer who first strapped on a scuba cylinder to enter the aquatic world. It contains over four hundred photographs, maps, and diagrams.

Earle, Sylvia A. *Sea Change: A Message of the Oceans*. New York: G. P. Putnam's Sons, 1995. A fascinating autobiographical book by an accomplished marine biologist and deep-sea explorer who shares a passion for exploration of the sea and warns of looming dangers to these vast systems.

National Geographic Society. *Ocean Drifters*. Washington, D.C.: Author, 1993. This videotape follows a loggerhead sea turtle into the open ocean and examines the wonderful creatures it encounters. Incredible cinematography.

Nybakken, James Willard. *Marine Biology: An Ecological Approach*. 5th ed. San Francisco: Benjamin/Cummings, 2001. An undergraduate textbook that covers marine organisms, their oceanic environment, and their interactions with their biotic and abiotic environment.

Sumich, James L. *An Introduction to the Biology of Marine Life*. 7th ed. Boston: WCB/McGraw-Hill, 1999. A comprehensive, colorful introductory textbook that is geared toward high school and undergraduate students.

MARK, RELEASE, AND RECAPTURE METHODS

Type of animal science: Ecology
Fields of study: Conservation biology, genetics, population biology, wildlife ecology

The number of animals present within an area can be determined by the ratio of marked to unmarked animals. Knowledge of these numbers provides the biologist with information on the impact of environmental events. It is often necessary to know animal abundance to prevent exploitation, allow effective management of the population, and project future animal numbers.

Principal Terms

DENSITY: the number of animals present per unit of area being sampled

EMIGRATION: the movement of animals out of an area; one-way movement from a habitat type

HABITAT: the physical environment, usually that of soil and vegetation as well as space, in which an animal lives

HOME RANGE: the space or area that an animal uses in its life activities

IMMIGRATION: the movement of animals into an area; a one-way movement into a habitat type

MARKED: an individual animal that is identifiable by marks that may be either human-made, such as metal bands or tags, or natural, such as the pattern of a giraffe

N: a standard abbreviation for the size of an actual population; if capped with a ^ ($\hat{n}$) it is an estimated value

POPULATION: a group of animals of the same species occupying the same physical space at the same time

RECAPTURED: a previously marked animal that is either seen, trapped, or collected again after its initial marking

SAMPLING: the process of collecting data, usually in such a manner that a statistically valid set of data can be acquired

There are four different ways to determine the number of animals within a habitat or population: counting the total number of animals present, sampling part of the area (a quadrat) and extrapolating to find a total, sampling along a line-transect and measuring the distance and angle to where the animal being counted was first seen, and using the mark-recapture approach. Different types of animals require different techniques for population estimation. The mark-recapture approach is often used for groups of animals whose populations are too large or too secretive for other methods. These are usually vertebrates, although mark-recapture procedures have also been used for invertebrates, such as grasshoppers.

All mark-recapture calculations are based on how many individual animals are marked (denoted M) in the population being studied, how many animals are captured during sampling (n), and how many of the captured animals have been previously marked (m). The estimated population is commonly indicated $\hat{n}$. The basic mathematical relationship of these data is $\hat{n}/M = n/m$, when $\hat{n} = Mn/m$. An estimate of population density can be obtained by dividing the area being sampled by the estimate of N.

Marking

Animals may be either temporarily or permanently marked. Temporary marking may be daubing paint on an animal's body, clipping some hair off a mouse's back, or pulling off a few scales from a snake's belly. If each animal is given a unique mark, such as a number or symbol, then it is possible to determine how long the particular animal lives, its home range, and patterns of movement, such as immigration and emigration rates. Some mark-recapture calculations require that the number of times an individual animal is captured be known; "recaptured" must be separated individually for these calculations.

There are pitfalls in this census method that should not be overlooked. Several conditions must hold if mark-recapture population estimates are to be valid. The marked animals must neither lose nor gain marks. Care must be taken if natural marks, such as missing toes on a mouse, are used; additional mice losing toes would lead to an error

in population estimation. Marked animals must be as subject to sampling as unmarked ones. Because of the excitement of being captured, many animals will not return to a live-trap a second time, leading to an overestimation of animal abundance. If an animal becomes easily caught and returns frequently to the trap, such as kangaroo rats often do in the desert, then this trap-happy animal produces an underestimate of population size. The marked animals must also suffer the same natural mortality as unmarked ones, and the stress of being captured and marked may cause a higher mortality rate in the animals that are marked. If this occurs, the population estimate will be too high.

The marked animals must become randomly

A veterinarian prepares to shoot a tranquilizing dart into a bull elephant so that it can be fitted with a collar containing a Global Positioning System (GPS) beacon. The GPS will allow the elephant's movements to be tracked. (AP/ Wide World Photos)

mixed with the unmarked ones in the population, or the distribution of sampling effort must be proportional to the number of animals in different parts of the habitat being studied. If the animals are "clumped," population estimates will be either too high or too low, depending on whether the clumps of marked animals are included in the sample. Marked animals must be recognized and reported on recovery. Technicians working with the animals must be able to recognize marked animals and/or read the individual numbers per animal correctly. If marked animals are not recognized, population estimates are too high. There can only be a negligible amount of recruitment or loss to the population being sampled during the sampling period; emigration, if occurring, should be balanced by immigration. A short time between marking the animals and collecting the additional samples for the population estimate is necessary, or the ratio of N to M changes from that existing when n:m was established.

Even under the ideal conditions above, it is apparent, according to the laws of chance, that the ratio of marked to unmarked animals in the sample will not always be the same as that of marked to unmarked animals in the population; in fact, the two ratios may seldom be the same. Possibilities for sampling error can be decreased by enlarging the size of the sample. As the sample size approaches the population in size, chances for error become smaller. When the point is reached where the sample includes the entire population, there can be no error in estimation. In general, at least 50 percent of the population should be marked, and the number of marked animals in the sample should be 1.5 times the number of unmarked animals in the sample. In actual practice, it is difficult (if not impossible) to meet all these requirements. Consequently, it is often best that mark-recapture population estimates be used as measures of trends in major population fluctuations from year to year.

Applications of the Method

There are many different formulas for utilizing the mark-recapture data to produce estimates of population size for any animals that can be marked and recaptured or observed later. The first use of the ratio of marked to unmarked animals for population estimation was for fish and ducks; the technique is usually called the Lincoln-Petersen method or index. Its formula is $N = Mn/m$, where N is the total estimated population, n is the number of animals sampled or captured, M represents the number of animals marked in the population before sample size n is drawn, and m equals the number of previously marked animals recovered in sample size n.

An example of how the calculations for the Lincoln-Petersen index would be made is shown by the following information. If 375 quail were banded and later, in a sample of 545, there were 85 previously banded birds recovered, therefore $N = Mn/m$, or $375 \times 545/85 = 2,404$ quail estimated to be present in the population being sampled.

When, as is the case with ring-necked pheasants, there is a variation in the capture of the sexes, caused perhaps by capturing technique, the formula can be applied to both sexes, or even to age classes, to arrive at a better estimate of the total population. For example, if 500 males and 750 females were banded before hunting season and then 360 males and 150 females were recovered after the harvest, with 150 banded males found in the 360 males checked and 50 banded females recovered in the 150 females checked, a population estimate can be made. The estimate would be $500 \times 360/150$, equaling 1,200 males in the population. The female population estimate would be $750 \times 150/50$, equaling 2,250 females in the population. The total population of pheasants would be estimated to be 1,200 males and 2,250 females, equaling a total of 3,450 pheasants.

The Lincoln-Petersen index differs from other mark-recapture calculations in that only two periods, the initial period when animals are marked and the second period when the sample (n) is collected, are used. If several capture periods are used, sequential formulas, such as the Schnabel method, must be utilized. In each sample taken, all unmarked animals are marked and returned to the population; marking and recapture are done concurrently. The sequential approach makes allowances for the increasing number of marked animals in the population (M). M usually increases with time, but it may decrease with known mortality or removal of marked animals from the population. All the assumptions for the Lincoln-Petersen index should also be met for the Schnabel method to produce accurate population estimates.

The Schnabel method formula for multiple sampling periods is $N = \Sigma(C_T M_T)/R_T$. Each line of the Schnabel method calculation corresponds to a line in the Lincoln-Petersen index calculation. C represents the number captured during sampling time one, M is the total marked, and R is the number of recaptures. The subscript T is the sample time.

An example of the Schnabel calculation can be demonstrated with the following data. For four days of trapping, the following data were obtained: day one, five animals captured, no recaptures; day two, ten animals captured, five previously marked animals in the population at the start of the second day of trapping, three previously marked animals in the day two sample; day three, fifteen animals captured, fifteen previously marked animals in the population at the start of day three of trapping, three previously marked animals in day three of trapping; day four, ten animals captured, twenty previously marked animals in the population at the start of day four, and four previously marked animals among the animals captured on day four. For the four days of trapping, then, ten animals were recaptured. From these data, the Schnabel method estimate of population density would be $N = (5 \times 0) + (10 \times 5) + (15 \times 15) + (10 \times 20)/(0 + 3 + 3 + 4)$ with $N = 50.5$.

Another approach to the estimation of animal numbers has been developed that differs from the usual approach to mark-recapture population estimation calculations. The Eberhardt method is based not on the ratio of marked to unmarked animals but on the number of times an individual is recaptured during the recovery operations; the assumption is that this recapture frequency is related to the total population size. The relationship is believed to be a hypergeometric one. Eberhardt's formula is $N = n/1 - (n/t)$, where n = number of individuals handled in recovery operations and t = the total number of captures of individuals. In this tally, t will always be greater than n unless all animals are captured only once. To use these modified mark-recapture data, individual animals must be recognizable. This calculation has been used for a number of mice studies. For example, if twenty different kangaroo rats in the Mojave Desert were captured thirty-five times, the estimated population would be $N = 20/1 - (20/35)$; $N = 46.67$ animals.

These represent only a few of the mark-recapture formulas available for the estimation of population size and animal density. Many other, more complicated, formulas for calculation of population estimates based on mark-recapture ratios (such as the Schumacher-Eschymeyer, DeLury, and Jolly procedures) exist, but the simplest formulas often provide the best, most usable estimates of animal numbers.

Uses of the Method

The use of mark-recapture procedures allows the biologist to determine numbers of animals present in a given area. Without these numbers, the future of these animals cannot be predicted. This knowledge allows appropriate management strategies to be developed, either protecting them or providing needed control activities, such as spraying insecticides on agricultural crops before severe economic damage to the crops results. Information about animal populations is essential for determining the effects of environmental changes or human activities, such as construction, on animal communities.

The populations of different areas may be compared; population numbers between seasons or years may also be studied. The fact that certain areas have high numbers of individuals implies that these areas have good conditions for them. Wildlife managers need to know why these areas have higher numbers so that they can improve the relevant conditions in other areas. Learning this would not be possible without knowing population sizes on these respective sites. The success of management work can be judged by changes in population size.

Mark and recapture techniques are applicable to more population situations than are the other options for population estimation. The wide variety of methods available for marking animals often allows previously marked animals to be identified without actually being handled. This minimizes the stress on the marked animal by reducing human contact.

If information on how long an animal lives in the wild is needed, individual marking from population work can also serve this purpose. From their recapture points, the area used by individuals on a daily, seasonal, and yearly basis, known as the home range, can be determined. The degree of movement of individuals within the population can also be estimated. These data are economical to obtain, because the marking and capture of the animals for the population estimate also funds the cost of obtaining home range, movement, and longevity data. The rate of exploitation of the population and the rate of recruitment of new members into it can also be calculated from the ratio of marked to unmarked animals collected during the population studies. The biology of organisms in the wild cannot be adequately studied without accurate estimates of their population levels and fluctuations being known. Mark and recapture procedures are among the important scientific tools for the collection of this information.

—*David L. Chesemore*

See also: Demographics; Ecology; Extinction; Population analysis; Population fluctuations; Population growth; Wildlife management.

Bibliography

Blower, J. G., L. M. Cook, and J. A. Bishop. *Estimating the Size of Animal Populations*. Winchester, Mass.: Allen & Unwin, 1981. An excellent introduction to mark-recapture calculations with many easy-to-follow examples. Chapter 4, devoted to mark-recapture, includes data recording, estimation of survival rates, and Jolly's procedure.

Davis, D. E. *Handbook of Census Methods for Terrestrial Vertebrates*. Boca Raton, Fla.: CRC Press, 1982. An encyclopedia of techniques for estimating the population of vertebrates living on land. Each species or group of animals has an entire chapter devoted to estimation of their numbers. Excellent literature review.

Emery, Lee, and Richard S. Wydoski. *Marking and Tagging of Aquatic Animals: An Indexed Bibliography*. Washington, D.C.: U.S. Department of the Interior, Fish and Wildlife Service, 1987. Reviews many of the available references on marking aquatic invertebrates and vertebrates. An extremely good reference work in which to find more information on marking and tagging animals.

Krebs, C. J. *Ecological Methodology*. 2d ed. New York: Harper & Row, 1999. An excellent book treating mark-recapture work and all other ecological techniques. Written for introductory work in population ecology as well as advanced studies. Has an excellent bibliography.

Kunz, T. H. *Ecological and Behavioral Methods for the Study of Bats*. Washington, D.C.: Smithsonian Institution Press, 1988. Kunz provides an excellent review of population estimation techniques specifically for bats. Techniques for studying this fascinating group of animals are clearly discussed.

Otis, D. L., K. P. Burnham, G. C. White, and D. R. Anderson. *Statistical Inference from Capture Data on Closed Animal Populations*. Washington, D.C.: Wildlife Society, 1978. A complex discussion, with sophisticated mathematics, of mark-recapture calculations. Excellent examples, with step-by-step calculations. A computer program is also available for use with these formulas.

Scheiner, Samuel M., and Jessica Gurevitch, eds. *Design and Analysis of Ecological Experiments*. 2d ed. New York: Chapman and Hall, 2001. A collection of essays discussing techniques used in mark, release, and recapture experiments.

Schemnitz, Sanford D., ed. *Wildlife Management Techniques Manual*. 4th ed. Washington, D.C.: Wildlife Society, 1980. An encyclopedia of techniques for studying wild animals. Covers capturing and marking animals, estimating the numbers of wildlife populations, and how to determine the vital statistics of animal populations. Almost three thousand references in the literature cited section.

Seber, G. A. F. *Estimation of Animal Abundance and Related Parameters*. 2d ed. Reprint. New York: Macmillan, 1994. This is for readers with a strong mathematical background, as it is a technical book. It provides an excellent treatment of mark-recapture methods, but it is very advanced reading.

Skalski, John R., and Douglas S. Robson. *Techniques for Wildlife Investigations: Design and Analysis of Capture Data*. San Diego, Calif.: Academic Press, 1992. A guide for designing mark, release, and recapture experiments, written for professionals.

MARSUPIALS

Types of animal science: Anatomy, classification, ecology, evolution, geography, reproduction
Fields of study: Anatomy, ecology, physiology, zoology

Marsupials are a primitive group of mammals that separated early from the dominant group of mammals today, the placentals. Their present strongholds are in Australia, New Guinea, Tasmania, Mexico, Central America, and most of South America.

Principal Terms

ADAPTIVE RADIATION: the rapid evolution of new animal forms to fill unoccupied ecological niches

CONVERGENT EVOLUTION: the process by which unrelated animals tend to resemble one another as a result of adaptations to similar environments

MARSUPIUM: the abdominal pouch possessed by most marsupials, in which immature young attach to nipples to complete their development

MONOTREME: the egg-laying mammals, including the platypus and spiny anteaters

OMNIVOROUS: description of a diet utilizing a great variety of food material

---

Marsupials are pouched animals that form a distinctive group within the class Mammalia. They possess the diagnostic features of typical mammals, including high and stable body temperature, furry pelt, simple lower jaw, and mammary glands. However, there are other features that distinguish them from what are considered to be typical mammalian features.

The kangaroo is the most commonly known marsupial, but a vast array of marsupials exist. Most marsupials are crepuscular or nocturnal, so most zoo visitors are unable to observe them. Most marsupials are found in Australia and New Zealand. Outside of Australia, it is rare to see marsupials in zoos. Australian authorities impose strict export sanctions to protect their numerous endangered species. The only naturally occurring marsupial found in the United States is the opossum, *Didelphis marsupialis*. The opossums of North and South America are the most diverse of three families of extant marsupials outside of Australia.

There are three families of marsupials, Didelphidae, Microbiotheriidae, and Caenolestidae, that inhabit South and Central America. One species of didelphid, the Virginia opossum, extends across North America and beyond the Canadian border. The American marsupials alive today are mostly small, ranging from mouse to rabbit size. These are generally either carnivorous or omnivorous, living in forests and feeding on insects.

Marsupial Development

Marsupials are an example of adaptive radiation. This adaptation to their varied habitats has led to their enormous diversity of forms and niches. They are also an example of convergent evolution, as indicated by the similarities between marsupials and placentals in the rest of the world. The marsupial gliders resemble the flying squirrels and lemurs, the Tasmanian tiger or thylacine was doglike, and marsupial moles resemble eutherian moles. There are many physiological similarities as well. Wombats process grasses and sedges as horses do and numbats feed on termites as anteaters do.

With few exceptions, marsupials are not conspicuous in coloration or any external physical attributes. The greatest majority of them are small, ranging in size between that of a mouse and of a small rabbit. They developed from small carnivores into herbivores the size of hippopotamuses. The larger marsupials died out only several thousand years ago.

For the most part, marsupials have remained curiosities for the general public. Humans have not traditionally exploited marsupials. They have never been kept as pets, the meat of larger kangaroos is mostly used for dog and cat food, and the furs of only a few marsupials have commercial value.

Marsupials include 18 families, 76 genera, and over 266 species, but these divisions and categorizations are currently being debated. Marsupials are the only order in the subclass Metatheria. There is no other group within the higher mammals that contains such a diversity of higher species, genera, and families as the marsupials.

There are marsupials that spring about on their hind legs, as well as climb, glide, burrow and even swim, and they range in adult size from 147 pounds to only 0.1 ounce. They are found in habitats as diverse as freshwater, alpine areas, hot deserts, and tropical rain forests. Their diet ranges from purely insects to vertebrates, fungi, underground plant roots, bulbs, rhizomes and tubers, plant exudates such as saps and gums, seeds, pollen, terrestrial grasses, herbs and shrubs, and tree foliage. Because of this vast diversity it is impossible to categorize marsupials with a simple description. Instead, the physiology of marsupials must be used to categorize them.

Physiology

There are three types of reproductive patterns in mammals. There are monotremes, which are egg-laying mammals, such as the duck-billed platypus. There are placentals, whose embryo develops inside the uterus, and the placenta formed in the uterus provides nutrients to the developing embryo. In placentals, the offspring are born completely developed, as in humans. Finally, there are marsupials, which functionally fall in between monotremes and placentals.

Marsupial Facts

Classification:
Kingdom: Animalia
Subkingdom: Bilateria
Phylum: Chordata
Subphylum: Vertebrata
Class: Mammalia
Subclass: Theria
Infraclass: Metatheria
Order: Marsupialia
Suborders: Polyprotodonta and Diprotodonta
Families: Didelphidae (American opossums, eleven genera, seventy-five species); Microbiotheriidae (Monito del montes); Caenolestidae (shrew or rat opossums, three genera, seven species); Dasyuridae (quolls, dunnarts, and marsupial mice, eighteen genera, fifty-two species); Myrmecodoiidae (numbats); Thylacinidae (thylacines); Notoryctidae (marsupial mole); Peramelidae (bandicoots, seven genera, seventeen species); Phalangeridae (cuscuses and brishtails, three genera, fourteen species); Burramyidae (pygmy opossums, four genera, seven species); Pseudocheiridae (ringtail opossums, two genera, sixteen species); Petauridae (gliders, three genera, seven species); Macropodidae (kangaroos and wallabies, eleven genera, fifty species); Potoridae (rat kangaroos, five genera, ten species); Phascolarctidae (koalas); Vombatidae (wombats, two genera, three species); Tarsipedidae (honeypossums)

Geographical location: Australian region; North, Central, and South America

Habitat: Varied depending on species, but includes major terrestrial habitats and some arboreal habitats

Gestational period: Characteristically short, as the newborn completes its development attached to a nipple inside the marsupium

Life span: Varies by species

Special anatomy: The marsupium, or abdominal pouch, is characteristic, although some forms, such as the murine opossum of Central and South America, lack one

Marsupials are often thought of as pouched mammals. Their embryo develops inside the uterus but, unlike placental mammals, the marsupial is born very early in its development. It completes its embryonic development outside the mother's body, attached to teats of abdominal mammary glands, which are often but not always enclosed in a pouch called the marsupium. The helpless embryonic form has forelimbs that are strong enough to climb from the birth canal to the mother's nipples, where it grabs on and nurses for weeks or months depending on the species.

When the young are born, their eyes and ears are closed, hind limbs and tails are stumps, and they are completely hairless. Their olfactory senses

Kangaroos are the archetypal marsupials, carrying their young in an abdominal pouch, or marsupium, for up to eleven months. (Digital Stock)

are greatly developed, as are their tactile senses, allowing them to navigate their way to the marsupium.

The marsupium is formed in diverse ways, ranging from the "primal pouch" (the annular skin creasing around each teat), to common marsupial walls surrounding all teats, and finally to a closed marsupium, which can be opened to the front or to the rear. Marsupials are usually woolly, with shortened forelimbs and elongated hind limbs. In kangaroos, these physical features allow locomotion in a hopping movement only. However, at an equivalent speed, allowing for the differences in weight, a hopping kangaroo uses less energy than a running horse or dog.

In several families, second and third toes of the hind foot function as grooming claws and the first toe is always clawless, except in the shrew opossum. Vision is usually poorly developed and olfactory, tactile, and auditory senses are well developed.

The gestation period is eight to forty-two days, after which the young is carried in the marsupium for between thirty days and seven months. Litter sizes range from one to twelve per birth. The young are weaned anywhere between six weeks and one year. The relationship between mother and offspring is long lasting in many species. Sexual maturity is reached between ten months and four years, depending on the species. The longer range is associated with the male koala.

Behavior

Marsupials range from pure carnivores to pure herbivores, with all the intermediate stages in between. They are usually nocturnal and crepuscular. Some species are solitary, while others live in family groups.

In all mammals, because of the milk produced by the mother, male assistance in feeding the young is less important than in birds, for example. In many marsupials, the role of the male is further reduced because the pouch takes over the functions of carrying and protecting the young and keeping it warm. A female's need for assistance in rearing young does not appear to be an important

The Tasmanian Tiger

One unique marsupial is the Tasmanian tiger, or thylacine, a carnivorous marsupial now believed to be extinct. This mammal has attracted special attention because of its striking similarity to the dog. Also unique to this animal is its tigerlike dorsal stripes, thirteen to nineteen dark, transverse stripes that were distinct over the light gray to yellow brown coat. Reports of size vary, but it was similar in size to the American wolf. Reliable biological and behavioral information on this animal is very rare.

With the arrival of sheep ranchers, the Tasmanian tiger's numbers declined rapidly. Due to a government bounty program, between 1888 and 1905 more than two thousand Tasmanian tigers were officially killed, with the actual numbers likely much higher. What was believed to be the last wild Tasmanian tiger was shot in 1930. Tasmanian tigers never reproduced in captivity. Ironically, the Tasmanian tiger embellishes the Tasmanian coat of arms, designated by King George of England in 1917.

factor promoting the formation of long-lasting male-female pairs or larger social groups. The majority of marsupial species mate promiscuously. There are few examples of long-lasting bonds and they do not live in groups. Some species form monogamous pairs and harems. It is hypothesized that the lack of frequent examples of this sort is due to the lack of external pressures.

Evolution

Marsupial evolutionary development is not yet clearly understood. Fossil records suggest that they may have evolved simultaneously with the placental animals about 100 million years ago, in the Cretaceous period. The oldest geological finds come from the recent Upper Cretaceous of North America, about seventy-five million years ago. Although there was some development of marsupials in North America, they later declined as placentals increased in diversity. In contrast, South America has a considerable diversity of marsupial fossil forms, indicating their persistence for more than sixty million years. Seven families of living and fossil marsupials are known from South America. About two to five million years ago, a land connection between the two Americas

was established again, and more placental animals reached South America, including carnivores such as the jaguar. In the face of such competition, the large carnivorous marsupials disappeared, but the small omnivores have persisted successfully to the present day. Some of them moved north to colonize in North America.

The earliest marsupials found in Australia are dated from twenty-three million years ago. Most modern families and forms were clearly established by that time. There is no clear evidence to establish whether marsupials originated in North America, South America, or Australia. The lack of fossil records of marsupials in Asia or Africa makes the most likely route of migration from South America to Australia via Antarctica. At that time, all three southern continents were united in the land mass known as Gondwanaland. This mass of land began breaking up 135 million years ago, with South America and Antarctica still being connected until about 30 million years ago. One land mammal fossil has been found in Antarctica which is a marsupial dated to be forty million years old. The Australian plate then gradually drifted northward for another thirty million years before reaching its current latitude. This long isolation allowed the extensive development of the marsupials in Australia in the absence of competition from other placentals.

As marsupials evolved in Australia, so did the placentals in the rest of the world, filling the same ecological niches. In many cases, they adopted similar morphological solutions to ecological problems. One example is the convergent evolution of the carnivorous Tasmanian devil, a marsupial, and placental wolves of other continents. The marsupial mole is very similar in form to the placental mole. The marsupial sugar glider and the two flying squirrels of North America are also very similar.

Habitat

The arrival of European settlers and the influx of new species—sheep, cattle, rabbits, foxes, cats, dogs, donkeys, and camels—have caused a large-scale modification of the marsupial's habitat in Australia. The first major change was in the late Pleistocene, with the extinction of whole families of large terrestrial marsupials. Included in this extinction was Diprotodon, the largest browsing kangaroo. It is likely that the climatic fluctuations increased aridity and reduced the available favorable habitat. Many of the species were already under stress when man arrived.

Approximately nine species have become extinct in Australia and fifteen to twenty have suffered gross reduction in range. The most affected have been small kangaroos, bandicoots, and large carnivores such as the thylacine and native cats.

Not all the environmental changes have been unfavorable for marsupials. Many of the larger herbivores have fared well with the advent of ranching and available grazing land and watering holes already set up for stock animals. As these marsupials become competition for sheep and cattle, Australian authorities have developed programs to keep their population controlled by allowing a certain number to be shot. Most species of marsupials have little or no importance as pests and their continued existence depends largely on the maintenance of sufficient habitat to support secure populations. The control of feral foxes and cats is very important to keep predation limited.

Marsupials outside Australia appear to have suffered no ill effects due to the destruction of habitat in North and South America.

—*Donald J. Nash*

See also: Embryology; Fauna: Australia; Fertilization; Kangaroos; Koalas; Opossums; Pregnancy and prenatal development; Reproduction; Reproductive strategies; Reproductive system of female mammals; Tasmanian devils.

Bibliography

Dawson, Terence. *Kangaroos: The Biology of the Largest Marsupials.* Ithaca, N.Y.: Comstock, 1995. A very comprehensive book covering all aspects of the most familiar marsupial, the kangaroo. Easy-to-understand language covers the social organization, activities, habitat, and general characteristics of kangaroo life. Also covers kangaroos' history of coexistence with humans.

Hume, Ian D. *Marsupial Nutrition.* New York: Cambridge University Press, 1999. Discusses all food resources utilized by marsupials, as well as their digestion and metabolism. Provides nutrition and digestion information applicable to all mammals.

Lee, Anthony K., and Andrew Cockburn. *Evolutionary Ecology of Marsupials.* New York: Cambridge University Press, 1985. A good discussion of the rich diversity of marsupials and the various coping techniques each has developed throughout evolution.

Paddle, Robert. *The Last Tasmanian Tiger: The History and Extinction of the Thylacine.* New York: Cambridge University Press, 2000. An insightful and up-to-date examination of the history of one of Australia's most enduring folkloric beasts—the thylacine, otherwise known as the Tasmanian tiger.

Saunders, Norman, and Lyn Hinds, eds. *Marsupial Biology: Recent Research and New Perspectives.* Sydney, Australia: New South Wales University Press, 1997. Twenty-two articles written by leading experts in the field. Articles are divided by subject matter: reproduction and development, genetics, ecology, pathology and homeostasis, and developmental neurobiology. A good reference text for graduate students.

Tynedale-Biscoe, Hugh, and Marilyn B. Renfree. *Reproductive Physiology of Marsupials.* New York: Cambridge University Press, 1987. An advanced discussion of reproductive characteristics and behaviors of marsupials.

MATING

Type of animal science: Behavior
Fields of study: Ethology, invertebrate biology, zoology

Mating is the process by which individuals of a species accomplish the fertilization of ova, or eggs. Mating systems vary widely in the animal kingdom, but every system must bring together two animals, usually of different sexes of the same species.

Principal Terms

ASEXUAL REPRODUCTION: reproduction without the union of male and female sex cells

DIOECIOUS: having two separate sexes, namely male and female

ESTRUS: the period of the sexual cycle during which a female is sexually receptive

GAMETE: a sex cell, either male or female

HERMAPHRODITE: an animal with both male and female sex organs

MONOGAMY: a mating system in which one male and one female comprise the main breeding unit

POLYGAMY: a mating system in which a single adult of one sex mates with several members of the opposite sex.

As one observes the diverse and elaborate movements and behaviors seen in courtship displayed by all types of animals, it is easy to lose sight of what courtship is meant to accomplish. Whether it is relatively simple or highly complex, the ultimate goal of courtship is to bring together two animals of different sexes of the same species to bring about successful mating and reproduction. Although they may not always be as obvious, mating patterns among animals also are quite diverse.

Reproduction, one of nature's most fundamental and essential functions, is the process by which all living organisms produce offspring. The need to propagate is a cardinal necessity for the preser-vation of the species. Each living organism has its own unique way of accomplishing this requirement. The higher the animal, the more intricate the process. In one-celled organisms reproduction is usually asexual, where only one living entity is required for procreation. The one-celled animal simply splits in two, losing its original identity and thereby creating two new organisms that are characteristically exact duplicates of the parent. The process is known as fission. Some single-celled entities reproduce sexually, where two similar organisms fuse, exchange nuclear materials, and then break apart, after which each organism reproduces by fission. This form of unicellular reproduction is known as conjugation. Sometimes after conjugation, the participating organisms do not reproduce. It appears that the process is merely to revitalize the organisms. This is the most primitive method of sexual reproduction. The new organisms that are produced are from two distinct parents, having definite genetic characteristics of their own.

The procreation process in most multicellular animals involves a more complex form of sexual reproduction. Here, unique and differentiated male and female reproductive cells called gametes unite to form a single cell known as the zygote. The zygote undergoes successive divisions to form a new multicellular organism, where half the genes in the zygote come from one parent and half from the other, creating a singularly different living creature.

Since most species of animals have sexual reproduction, it must offer some advantages. Sexual reproduction results in maintaining high levels of genetic variability within a population. Genetic variability produces variability in behavior, structure, and physiology and provides a species with greater flexibility in meeting changing environmental conditions. There must also be some long-term evolutionary advantages.

A dictionary definition of mating encompasses the idea of individuals coming together to form a pair, with the implication of doing so to produce offspring. If an emphasis is placed on the latter part of the definition, then it is seen that mating may involve all degrees and types of relationships and interactions among animals. In animals that have sexual reproduction, the union of male and female sperm and eggs, also known as gametes or sex cells, may be accomplished by either external or internal fertilization. Simple or elaborate courtship rituals ready animals for mating and fertilization. Several factors may influence reproduction. Most animals have a distinct time of the year, the breeding season, when reproduction is possible. Depending on the species, the breeding season may be the spring, summer, fall, or winter. In other species, including the human and some primates, breeding may occur throughout the year. The breeding season starts with the onset of courtship and is finished when the last offspring are weaned. The same species may have different breeding seasons depending on where they live. In temperate regions of the world, there may be much variation in the breeding in different years. Any factor that affects the health of animals may modify their breeding season, including food availability, temperature, light, and population density. In general, the breeding season usually is coordinated to maximize the likelihood of survival of the young.

Mating Systems

The means by which males and females are brought together in courtship and, ultimately, copulation is achieved is by some type of mating

Sexual reproduction, the form of mating that involves the mixture of genetic material from a male and a female organism, increases a species' genetic flexibility in coping with environmental changes. (Corbis)

Sexual Reproduction—With and Without Copulation

When one thinks of mating and sexual reproduction, one usually thinks of the act of copulation and the physical coupling of the male and female. However, sexual reproduction does not always involve actual copulation and there is much variation among animals. Sexual reproduction which involves the union of male and female sex cells may be achieved by means of internal or external fertilization. In most animals, the male and female sex cells are of unequal size, and typically, a small gamete fertilizes a large gamete. In certain flagellates, however, the two gametes are identical but still function to produce a new individual.

External fertilization is accomplished in a number of invertebrates as well as in some vertebrates. Among many aquatic animals, large numbers of sperm and eggs may be simply expelled into the water, and although the system is inefficient, the large number of sex cells assures that some fertilization will be effected. Among the vertebrates, fertilization is internal in some fishes and external in others. Most frogs and toads have external fertilization: As eggs are released by the female, they are fertilized by the males.

Even in those vertebrates that have internal fertilization, the copulatory organ may not necessarily be a penis. Among fishes, there are modified fins and claspers which serve to transfer sperm for internal fertilization. In most species of birds there is no penis, and during copulation sperm from the male is transferred from his cloaca (a common cavity into which the reproduction, urinary, and intestinal canals open) to the female's cloaca by direct contact, in what biologists refer to as a "cloacal kiss." All mammalian males have a copulatory organ, although there is much variation in the size and structure of the penis. In the egg-laying mammals, the penis is attached to the wall of the cloaca and its canals are used only for the transfer of sperm to the cloaca of the female.

In many mammals, but not all, the penis contains a bone, the os penis or baculum. A baculum is found in most carnivores and aquatic mammals, as well as in some rodents and a variety of other mammals. Other interesting structural variations in the penis include the corkscrew-shaped penis of the pig and the forked tip of the penis of the opossum.

Regardless of the way in which sexual reproduction is accomplished, it succeeds in producing new genetic combinations, and this increased genetic variability appears to be a major advantage to species.

system or pattern. Systems vary widely throughout the animal kingdom, but there are several general groupings. (A reading of the literature in cultural anthropology shows that virtually all of these mating systems can be found in different human cultures, as well.) Mating systems typically are classified under three headings: promiscuity, monogamy and polygamy. It also is useful to delineate subcategories within some of these.

Promiscuity, as it defines a mating system, means that no pair bond is formed between individuals. A technical term for this system is polybranchygamy. which literally means "many brief matings." In this system, any male may mate with any female and no one individual has exclusive rights over any individuals of the opposite sex. Males and females may copulate with from one to many of the opposite sex. Promiscuity is found more often in males than in females, and this observation is probably related to the fact that in such species males have a far lesser investment in offspring than do females. The system is found in a small percentage of bird species and many species of mammals. Promiscuity is demonstrated well in those species that form leks. A lek is an area or territory used for communal courtship displays and mating by certain species. Leks have been observed among grouse, some African antelope, a species of bat, and some insects. Leks are used solely for mating. Females are attracted to the leks by elaborate courtship displays by the males, and mate with one or a small number of the males. Usually, only a select few of the males perform most of the matings. In one study of grouse, less than 10 percent of the males carried out more than 75 percent of all copulations. The African antelope, the Uganda kob, is an example of a mammal that utilizes a lek. Here, also, a small percentage of males breed most of the adult females. The system is very effective, as nearly all of the females produce offspring.

A second major type of mating system is monogamy. It is a system in which a pair-bond is formed between one female and one male. The pair-bond may exist for only one breeding season (annual monogamy) or it may persist for one or more breeding seasons (perennial monogamy). Monogamy is very common in birds, with a large majority of species showing it. Swans and eagles are examples of species showing perennial monogamy, and sparrows and warblers are examples of species showing annual monogamy. Although it has been thought that many birds practice true monogamy, evidence has accumulated that in monogamous birds there are frequent matings of the males with females outside of the primary pair-bond relationship. These matings are known as extra-pair-bond copulations, and offer advantages to the males if they result in successful fertilization of additional eggs. Monogamy seems to occur when both the male and female have nearly identical roles in the rearing of the young.

Monogamy is far less common in mammals than it is in birds, but there are some good examples. If monogamy is likely to occur where there is equal parental investment by both males and females, it might be predicted there is less need for male mammals to practice monogamy, since only the females nourish the young with milk. Among mammals, monogamy can be observed in gibbons, foxes, wolves, beaver, red fox, and even among some small rodents. Why should there be monogamy among any mammals? It would be advantageous for males to be monogamous if by doing so they increased the likelihood of survival of their offspring. Males could do so if they helped to feed the young and helped to defend the territory against predators. Monogamy can also be looked at from the perspective of the female, a so-called female-enforced monogamy, which may

take place if females can gain sole benefit from the male's efforts without having to share them with other females.

The third major type of mating system is polygamy. In polygamy, an individual of one sex forms a pair-bond with several members of the opposite sex. The two major subtypes of polygamy are polyandry, where one female mates with more than one male, and polygyny, where one male mates with more than one female. Within both polygyny and polyandry there may be either a serial type or a simultaneous type. In the former case, one male or one female bonds with several members of the opposite sex but only one at a time. In the latter case, one male or one female bonds with several individuals of the opposite sex at the some time. Simultaneous polygyny is often referred to as harem polygyny. Although it appears to resemble promiscuity, it differs from promiscuity in that a pair-bond is formed even though it may be temporary. In altricial birds, in which the young are born in a helpless state, it is often observed that polygyny is present in habitats where food is unevenly distributed and one male is able to provide food for more than one female. Polygyny also may result in cases where there is a lack of availability of suitable territories for breeding and where there may be a pressure of heavy predation.

Hermaphroditism

Although most species of animals are dioecious, meaning that they have two separate sexes, male and female, some individuals and even whole species are monoecious and have both sexes in one individual. Monoecious individuals are also called hermaphrodites. Although hermaphroditism is unusual in humans and most mammals, in some species it is the rule. There are many invertebrates in which the same individual produces both eggs and sperm. Some examples of hermaphroditic species include garden snails, free-living flatworms, the common earthworm, and some fish. It might seem that the hermaphroditic condition is a beneficial and efficient one, since any single individual is capable of producing and delivering both eggs and sperm. However, there is at least one major drawback to the system, the possibility of self-fertilization. Self-fertilization does not lead to as much genetic variability in the offspring compared to what might be expected if there was cross-fertilization. In species in which hermaphroditism is the rule, there are a number of processes at work which make self-fertilization unlikely, if not impossible. Some animals produce eggs at one time and sperm at a different time, making it impossible to self-fertilize. In animals in which the sex organs mature at different times, a condition known as protandry, an individual alternates between being different sexes, functioning as a female or male first and then becoming the other sex at a later time. In some species, including the earthworm, two individuals come together to engage in a mutual copulation. The two worms are held together by a secretion produced by both of them. The togetherness allows sufficient time for sperm from one worm to travel to the other worm and fertilize that worm's eggs and vice versa. Although in most animals the act of copulation is quite short, in the earthworm it may last three hours.

It would be remiss not to mention briefly an ex-ample of parthenogenesis in invertebrates. There are a few species of fish and lizards in which only females are known and in which the offspring are produced from eggs without fertilization by sperm. The Amazon moly, a fish, goes through the acts of courtship and mating with a male, but it is with males of other species, and sperm do not fertilize any of her eggs. The eggs develop parthenogenetically and produce another generation, apparently only of females.

Even a brief discussion of mating in animals reveals the complex and diverse methods that are used to produce another generation of the rich diversity of animal life on earth.

—*Donald J. Nash*

See also: Asexual reproduction; Breeding programs; Cleavage, gastrulation, and neurulation; Cloning of extinct or endangered species; Copulation; Courtship; Determination and differentiation; Development: Evolutionary perspective; Estrus; Fertilization; Gametogenesis; Hermaphrodites; Hydrostatic skeletons; Parthenogenesis; Pregnancy and prenatal development; Reproduction; Reproductive strategies; Reproductive system of female mammals; Reproductive system of male mammals; Sexual development.

Bibliography

Choe, Jae, and Bernard Crespi. *The Evolution of Mating Systems in Insects and Arachnids.* New York: Cambridge University Press, 1997. This book investigates sexual competition in insects and arachnids, and the ways that the sexes pursue, persuade, manipulate, control and help each other.

Daly, Martin, and Margo Wilson. *Sex, Evolution, and Behavior: Adaptations for Reproduction.* 2d ed. North Scituate, Mass.: Wadsworth, 1990. This book provides elementary-level discussion of theory relating to evolutionary and adaptive aspects of reproductive behavior.

Jacobs, Merle. *Mr. Darwin Misread Miss Peacock's Mind: A New Look at Mate Selection in Light of Lessons from Nature.* Berkeley, Calif.: Nature Books, 1999. Challenging long-held assumptions about mate selection, the author contends that female attraction is motivated by a food stimulus rather than aesthetic attraction.

Wallen, Kim, and Jill Schneider. *Reproduction in Context: Social and Environmental Influences on Reproduction.* Cambridge, Mass.: MIT Press, 2000. This book explores the neuroendocrine bases of reproduction in relation to environment and social contexts by providing accounts of reproductive behaviors in animals ranging from turtles to humans.

MEERKATS

Type of animal science: Classification
Fields of study: Ethology, population biology, wildlife ecology, zoology

Meerkats are close relatives of mongooses and are particularly noted for their close, cooperative societies.

Principal Terms

CARNASSIAL TEETH: teeth designed to sheer flesh; the characteristic that unites all members of the order Carnivora
DENTITION: referring to the teeth
DIURNAL: active during the day
METAZOA: organisms which are multi-cellular
NOTOCHORD: longitudinal, flexible rod located between the gut and nerve cord
PLACENTA: structure that connects a fetus to the mother's womb; indicative of internal gestation of young

---

Meerkats, also called suricates, have the long body and short legs characteristic of most mongooses. The body length is from ten to thirteen inches, and the tail adds an additional seven to nine inches in length. The short fur of the meerkat is grayish-brown to light gray in color, featuring dark stripes across the back. The eyes and nose are dark and form a contrast with the light-colored head and throat. The belly fur of the meerkat is rather thin, which helps to regulate the animal's body temperature. To warm up, the animal will bask in the sun while sitting up or lie belly-down on warm earth, and to relieve itself from the hot desert sun, a meerkat will lie belly-down in a cool, shaded area or inside its burrow.

Meerkats are desert creatures, primarily inhabiting the Kalahari and Namib deserts and other dry, open areas of southern Africa. One of the most notable characteristics of meerkats is their extreme gregariousness. They typically live in family units of up to thirty individuals. The colony of animals occupies a home territory, and digs several systems of burrows with multiple entrances connected by tunnels and a number of distinct chambers. The colony moves several times during a year as food becomes depleted, and establishes a new system of burrows or occupies those left from a previous occupation by the group.

Group Life

Because life in the open desert is harsh and predators, such as jackals, hyenas, and birds of prey, are plentiful, the meerkat group social structure ensures that there are many individuals to act as sentries while the group is foraging for food. The sentry takes its place on a raised area such as a sandbank, bush, or tree, and constantly scans the sky and the horizon for potential predators. The keen eyes and ears of meerkats aid in their constant vigil against danger. Should a predator be spotted, the sentry barks a warning and all members rush to the protection of the burrows. When larger animals, such as hyenas, have threatened the safety of the meerkats inside the burrow, the entire meerkat colony has been seen to band together, stand on their hind legs, and confront the intruder with barking threats, successfully repelling the threatening advances of the much larger animal.

The diet of meerkats consists largely of whatever is available in the harsh desert habitat. Insects, spiders, centipedes, scorpions, lizards, snakes, small mammals, birds and their eggs, roots, tubers, and other plant matter are all staples of the meerkat diet. Each individual takes its turn

The meerkat's distinctive upright pose allows it to search the environment for possible predators. (Digital Stock)

Meerkat Facts

Classification:
Kingdom: Animalia
Subkingdom: Bilateria
Phylum: Chordata
Subphylum: Vertebra
Class: Mammalia
Subclass: Eutheria
Order: Carnivora
Family: Viverridae
Genus and species: Suricata suricatta (gray meerkat)
Geographical location: Southwestern Angola to southern Africa
Habitat: Desert, open country, savanna, and bush
Gestational period: Approximately eleven weeks
Life span: About ten years in the wild, up to seventeen in captivity
Special anatomy: Long, stiff tails that are used to stand upright; keen eyes located at the front of their heads provide good depth of vision

---

at sentry duty while the others forage, and the sentry is changed about every thirty minutes. Meerkats are diurnal creatures, spending much of their time foraging for food during daylight hours, except for the hottest portion of the day when they rest in the shade.

Mating generally occurs between October and April, and females give birth to two to five young after a gestational period of about seventy-seven days. Mothers give birth in a specific chamber in the burrow and young are born blind, but their eyes open in twelve to fourteen days. Meerkat young begin eating solid food at about three or four weeks of age but typically nurse for eight to twelve weeks. In meerkat society, there are many individuals who "babysit" young while the mothers forage for food. Meerkat young are sexually mature about one year after birth.

—*Karen E. Kalumuck*

See also: Altruism; Communities; Deserts; Fauna: Africa; Groups; Mammalian social systems.

Bibliography

Macdonald, David, ed. *The Encyclopedia of Mammals.* 2d ed. New York: Facts on File, 1995. This lavishly illustrated compendium of mammals gives detailed, accessible information on hundreds of wild mammals, organized by phylogenetic order. Of particular interest are the introductory chapter on mammals, and the chapter on mongooses, including the meerkat.

National Geographic Society. *National Geographic Book of Mammals.* Washington, D.C.: Author, 1998. This excellent book for children and adults alike features detailed essays and marvelous photographs of over five hundred animals in their wild habitats.

Ricciuti, Edward and Jenny Tesar. *What on Earth Is a Meerkat?* Woodbridge, Conn.: Blackbirch, 1997. This superbly illustrated, engaging book designed for elementary children is a useful resource for adults seeking to learn more about meerkats.

Weaver, Robyn. *Meerkats.* Mankato, Minn.: Bridgestone Books, 1999. This informative book provides an excellent means for parents to introduce young children to the fascinating world of meerkats.

Whitfield, Philip, ed. *The Simon and Schuster Encyclopedia of Animals.* New York: Simon & Schuster, 1998. This book, a who's who of the world's creatures, gives concise information on the meerkat and related creatures; its organization helps the reader compare animals across groups.

METABOLIC RATES

Type of animal science: Physiology
Field of study: Biochemistry

Metabolic rate is a measure of the amount of energy used by an organism in a period of time (calories per hour). By studying factors that affect metabolic rates, scientists can calculate food requirements for animals in particular conditions and explain energy adaptations to different habitats.

Principal Terms

ADENOSINE TRIPHOSPHATE (ATP): the primary energy storage molecule in cells; links energy-producing reactions with energy-requiring reactions

ANABOLISM: a series of chemical reactions that builds complex molecules from simpler molecules using energy from ATP

BASAL METABOLIC RATE (BMR): the rate of metabolism measured when the animal is resting and has had no meals for twelve hours; used to compare different species

CATABOLISM: a series of chemical reactions that break down complex molecules into simple components, usually yielding energy

ELECTRON TRANSPORT CHAIN: a series of electron carrier molecules found in the membrane of mitochondria; oxygen is used and ATP is made at this site

HORMONE: a chemical messenger molecule within organisms; acts as a regulator of cell activities

MITOCHONDRION (pl. MITOCHONDRIA): a cell organelle found in plants, fungi, animals, and protists; the site of most aerobic metabolism

SPECIFIC METABOLIC RATE: the rate of metabolism per unit body mass (calories per gram per hour)

All living things require energy to sustain life as well as to carry out normal activities, develop, and grow. All the chemical reactions that allow energy acquisition and use are collectively called metabolism. Animals usually obtain energy by breaking down food in the presence of oxygen, so the amount of oxygen they use can be considered a measure of their rate of metabolism. Oxygen consumption is not always a good indicator of rate of energy conversion and use. Some organisms, such as fish, that live in oxygen-poor environments gain energy without using oxygen through anaerobic metabolism. Since all energy used by living things is eventually converted to heat, metabolic rate can be defined as the total amount of heat produced by an organism in a certain time period.

Rates of metabolism measured on whole organisms can be used to study the effects of factors such as temperature, size, age, and sex on rates of energy use. Studies of metabolic rates of different species of organisms are used to determine food requirements and energy adaptations in different environments. Metabolic rate can also be measured on isolated tissues, cells, or cell organelles in order to study the different biochemical reactions that occur in tissues and cells.

Factors Affecting Metabolic Rate

One of the most important factors that affects metabolic rate is the temperature of the organism, since within limits all chemical reactions of metabolism proceed faster at higher temperatures. The internal temperature of most invertebrate animals, fish, and amphibians is the same as the temperature of the environment in which they live. Such organisms are called poikilotherms. In poikilothermic organisms, metabolic rate increases as the environmental temperature increases. Such organisms move slowly and grow slowly when the temperature is cold, since their metabolic rate is very low at cold temperatures. To compare the metabolic rates of different poikilotherms, one must measure their rate of metabolism under standard conditions. Standard metabolism is usually defined as the rate of energy use when the animal is resting quietly, twelve hours after the last meal, and is at a temperature of 30 degrees Celsius; however, for small invertebrates, protists, and bacteria, only temperature is usually controlled. Most reptiles, birds, and mammals can maintain their body temperature at a constant level even when the environmental temperature changes greatly. Such organisms are called homeotherms. Birds and mammals can maintain their body temperature through internal heat production (endothermic homeothermy), while reptiles must acquire the necessary heat from their environment by changing their behavior, body posture, or coloration (ectothermic homeothermy). Most endotherms can maintain a constant body temperature over a range of temperatures (thermal neutral zone) without affecting their rate of metabolism. At temperatures outside the thermal neutral zone, metabolic rate increases to maintain constant body temperature. At colder temperatures, increased muscular activity and shivering require increased metabolic rate. Sweating and panting can increase the rate of metabolism at high temperatures. To compare the metabolic rates of endothermic homeotherms, scientists measure the basal metabolic rate (BMR), which is also referred to as the energy cost of living.

Body size (mass in grams or kilograms) is another major factor that affects basal or standard metabolic rate. A 3,800-kilogram elephant has a metabolic rate of about 1,340 kilocalories per hour, while a 2.5-kilogram cat has a rate of 8.5 kilocalories per hour, which means an elephant needs about 150 times as much food as a cat each day. A different picture emerges if one looks at energy use per kilogram of mass. For each kilogram of mass, the cat actually uses ten times the energy of a kilogram of elephant. Metabolism per unit of mass (specific metabolism) decreases as body size increases for all organisms. Since small organisms or cells have a larger surface area relative to their total volume than large ones, they can lose more heat from the surface. For two organisms of the same mass, the taller or thinner organism will have a larger surface area and higher BMR than a shorter, fatter one. More oxygen, food, and waste products can diffuse across the larger surface area; thus cell size in single-celled organisms and in different types of cells in multicellular organisms is limited by rate of energy metabolism. Small animals move faster, breathe faster, and their hearts pump faster. A mouse has a heart rate of six hundred beats per minute, while an elephant's heart rate is thirty beats per minute. Even the length of life appears to be related to the faster metabolic rate of these small creatures. Mice live only two to three years, while an elephant can live sixty years or longer.

Age and sex also influence basal metabolism. Young animals that are growing rapidly have a higher BMR than adults. As adults age, the proportion of skeletal muscle decreases and the BMR declines. Muscle tissue is metabolically very active even at rest, contributing to the higher BMR in males as opposed to females, since males have a higher proportion of body mass that is muscle. Physical or emotional stress can increase metabolic rates by increasing the catabolism of fats through the action of the hormones epinephrine and norepinephrine.

Skeletal muscle activity causes rapid short-term increases in metabolic rate. In humans, for example, a few minutes of vigorous exercise causes a twentyfold increase in the rate of metabolism, and the metabolic rate remains high for several hours. Walking, swimming, running, and flying require more energy than sitting still; however, each of these activities influences metabolic rate differently. Water is denser and has higher viscosity and resistance to movement compared to air, so more energy must be expended to swim than to walk at a given speed. Running also increases energy use, and the faster one runs, the more energy is required. Large animals, however, increase their rate of metabolism less per kilogram of mass than do small animals, so there is a metabolic advantage to large body size. Intriguingly, for the same size animal, flying is less energy-expensive than running.

Measuring Metabolism

Rate of metabolism can be measured as the amount of heat produced by an organism in a time period. The traditional unit of heat is the calorie; a kilocalorie is one thousand calories. The two terms are frequently confused in popular literature. In the international system of units, heat is measured in joules, and 1 calorie is equal to 4.184 joules.

Metabolic rate can be determined from the energy budget of an animal. If the total energy excreted in urine and feces is subtracted from the total energy in food eaten during a period of time, the result would be a measure of metabolic rate. The energy content of food and waste products can be determined by burning these materials in a calorimeter. The amount of heat produced is used to raise the temperature of a known amount of water. This method assumes that the organism is not growing or changing the amount of fat stored during the measurement period. It is also difficult to control metabolic activity of gut microorganisms. Although this technique is cumbersome, it may be the best way to assess energy metabolism in a normally active state for animals in their natural habitat.

More controlled measures of basal and standard metabolism can be made by isolating the animal in a calorimeter and directly measuring heat produced. This method is more accurate than the energy budget approach but still assumes that no new molecules are being produced and no activity or work is being performed. This technique is most useful for birds and small mammals that have relatively high rates of metabolism. Normal behavior and function may be altered by the confined conditions.

Indirect calorimetry is the most often used method in assessing metabolic rate in whole organisms, isolated cells, and cell components. Some factor related to energy use, such as oxygen consumed or carbon dioxide produced, is measured as an index of energy use. For aerobic metabolism, the amount of heat produced is related to oxygen use by the organism. Respirometry is the method used to monitor the oxygen used and carbon dioxide produced by an organism in a closed chamber. Oxygen consumption can be measured by absorbing carbon dioxide with soda lime and measuring the change in gas pressure in the closed system by a manometer. Oxygen electrodes can also be used to measure the decrease in oxygen concentration in water within the chamber if the animal is a water-dweller. For air-breathing animals, oxygen in the gas phase can be measured by a mass spectrometer. Carbon dioxide can also be measured by an infrared unit. In such closed systems, the gases can be monitored in the air as it enters and leaves the chamber. Respirometry can also be accomplished in open systems if the animals are fitted with breathing masks and the respired air is collected and analyzed. Oxygen consumption is a good index of metabolic rate for most animals at rest, since most of their metabolism is aerobic. Animals that live in oxygen-poor environments, such as internal parasites and mud-dwelling invertebrates, and all animals under extreme exercise often metabolize anaerobically.

To translate the amount of oxygen used into heat produced, one must know the proportion of fat, carbohydrate, and protein in the diet, since the amount of heat produced for each liter of oxygen consumed differs. In practice, it is usually assumed that only carbohydrates are being used.

Assimilating and digesting food cause large increases in metabolic rate. This increase reaches a maximum about three hours after a meal and remains above basal level for several hours in birds and mammals and up to several days in poikilotherms. Foods differ in the amount of increase in metabolic rate. Proteins, for example, cause about three times the increase in rate compared to carbohydrates or fats. The increase partially results from increased activity in cells of the digestive tract and partly from higher activity of liver and muscle cells preparing these foods for storage. Basal metabolism must then be measured after a twelve-hour fast to minimize this effect.

Since body temperature, body size, and activity affect metabolic rate of organisms, one can see that available food supplies, oxygen levels, and environmental temperatures limit the physiology of energy metabolism in different habitats. Scientists study metabolic rates in whole organisms to explain their food habits and their distributions in different habitats and to calculate energy requirements for raising animals under different conditions.

—Patricia A. Marsteller

See also: Carnivores; Cold-blooded animals; Digestion; Hibernation; Ingestion; Omnivores; Thermoregulation; Warm-blooded animals.

Bibliography

Peters, Robert H. *The Ecological Implications of Body Size.* New York: Cambridge University Press, 1983. A clear, readable book that provides an advanced treatment of metabolic rates in different organisms. Much original data and extensive bibliographies. For the advanced student.

Randall, David, Warren Burggren, Kathleen French, and Russell Fernald. *Animal Physiology: Mechanisms and Adaptations.* 4th ed. New York: W. H. Freeman, 1997. An intermediate-level college text. The chapter on animal energetics and temperature relations provides extensive treatment of methods of measuring metabolic rates and on the effects of size and thermoregulation on metabolism. Suggested readings provided at the end of each chapter permit interested students to investigate these topics further.

Salway, J. G. *Metabolism at a Glance.* 2d ed. Malden, Mass.: Blackwell Scientific Publications, 1999. A complete review of metabolic pathways, focusing on human systems. Written for medical students.

Schmidt-Nielsen, Knut. *Animal Physiology: Adaptation and Environment.* 5th ed. New York: Cambridge University Press, 1997. An intermediate-level college text. Clear, interesting stories of animal adaptations in metabolism make this text useful for advanced high school students. Extensive bibliographies.

_____. *How Animals Work.* New York: Cambridge University Press, 1972. Very readable book addressing intermediate-level college students. Clear examples and superior writing style capture the interest of high school students. Less extensive treatment than his *Animal Physiology.*

Solomon, Eldra, et al. *Biology.* 3d ed. Fort Worth, Tex.: Saunders College Publishing, 1993. An introductory college text with clear writing style that makes it useful for most high school students. Chapter 41, "Nutrition and Metabolism," and chapter 6, "The Energy of Life," are useful. Annotated recommended readings.

METAMORPHOSIS

Types of animal science: Development, physiology
Fields of study: Biochemistry, entomology, invertebrate biology, zoology

Metamorphosis is the major change in body form that occurs after embryonic development is complete in many types of animals, including insects.

Principal Terms

CORPORA ALLATA: a gland in insects that synthesizes and secretes juvenile hormone (JH)

ECDYSONE: a hormone that triggers both molting and metamorphosis in insects as well as in many other species of animals

IMAGINAL DISK: flat sheets of cells within an insect larva; these cells will change shape during metamorphosis and form the external structures of the adult

JUVENILE HORMONE (JH): a species-specific hormone which controls whether a molt will produce a larger larva or initiate metamorphosis

LARVA: the reproductively immature feeding stage in the development of many species of animals, including those insects which undergo complete metamorphosis

NYMPH: the sexually immature feeding stage in the development of those insects which undergo incomplete metamorphosis

PROTHORACIC GLAND: the gland where ecdysone is made in insects

PROTHORACICOTROPIC HORMONE (PTTH): a hormone made in the brain of insects which stimulates the prothoracic gland to make ecdysone

PUPA: the stage of insect development during which metamorphosis occurs

Unlike mammals, most animals become adults by first going through a distinctly different immature, or larval, stage of development. Larvae look nothing like the adults they will form. They are specialized primarily for growth and feeding and are unable to reproduce. At a certain time in their growth, often in response to signals from hormones or the environment, the larva undergoes a second spurt of development called metamorphosis.

The word "metamorphosis" comes from the Greek word for transformation and is today defined as a major change in body form that occurs after embryonic development is completed. Metamorphosis is very widespread in the animal kingdom but it has been studied most thoroughly in three different groups of organisms: the arthropods (crabs, lobsters, and insects), the amphibians (frogs and salamanders), and the echinoderms (sea urchins and starfish).

Metamorphosis affects the ways animals eat, breathe, and move; it can also change the nature of the environment in which they live. In echinoderms, the larval stage is microscopic and free swimming, while the adults are large and move very little if at all. The changes in body form thus coincide with significantly different ways of feeding and moving.

Amphibian Metamorphosis

In amphibians, metamorphosis prepares the organism for its transition from living in water to living on land. Internally, the digestive system changes to accommodate the new diet as the animal switches from consuming plants to animals. Sense organs like the eyes and ears also change to adapt to functioning predominantly in air. Finally, many of the chemical reactions which occur in the individual cells of the frog also change at this time.

Metamorphosis in amphibians is controlled by a pair of hormones. Prolactin, a protein secreted by the anterior pituitary gland, controls the rate of growth of the tadpole and suppresses metamorphosis. Thyroxine is a modified amino acid made in the thyroid gland of the tadpole and causes metamorphosis to begin. After the tadpole has grown to a certain minimal size, the thyroid gland is stimulated by environmental conditions to produce large quantities of thyroxine, which reverses the suppression exerted by prolactin and begins metamorphosis. The hormones pass through the tadpole's circulation and instruct different tissues to activate and deactivate different sets of genes that cause some tissues to degenerate, others to change, and others to grow.

These same hormones, prolactin and thyroxine, are produced by other vertebrates; nature often uses the same molecules to produce very different results in different animals. In humans, thyroxine regulates the rate of metabolism, while prolactin is crucial for milk production in nursing women. In fish, however, prolactin is crucial for keeping the cell's content of salt in balance.

Insect Metamorphosis

The hormonal control of metamorphosis in one group of animals, the insects, has yielded basic information about how genes and hormones interact to guide development in all animals. This process is understood in greater detail in insects than in any other group of animals.

Insects, like all arthropods, have a rigid exoskeleton, called a cuticle, that supports their body mass and allows the attachment of muscles for movement. Because it is external and fixed in size, however, the cuticle must be shed periodically for growth to occur. The process of insects shedding their cuticle and producing a new, larger one is called molting, and the number of molts normally occurring is regulated by the insect's genes. Many insects have developed complex behaviors to ensure that they molt in a secluded location and avoid predation. Without their hard, external covering, they are relatively defenseless.

About 10 percent of all insect species, such as the grasshoppers and true bugs, go through a process of incomplete metamorphosis. Here, the egg hatches into a juvenile form called a nymph that resembles the adult but is much smaller and is not capable of reproduction. At each molt, the nymph sheds its cuticle and grows before the newly produced cuticle hardens. When the signal for metamorphosis arrives, the molt produces an adult, often with wings, but, more important, fully capable of reproduction. Male and female adults can then mate and lay a new generation of eggs.

Complete metamorphosis, on the other hand, is a very different process that is undergone by 90 percent of all species of insects, including the ants, bees, flies, butterflies, and moths. Here, the larval form looks nothing like the adult. For example, the larval form of a butterfly is a caterpillar and the larval form of a fly is a wormlike maggot. Each

When an insect nymph undergoes metamorphosis within a chrysalis, its tissues completely break down into a kind of sludge, which is reformed under the directions of cellular imaginal discs. (Digital Stock)

larva undergoes a series of molts so as to be able to grow. The larva of the cecropia moth increases in mass by five thousand times during its larval development. When the trigger for metamorphosis occurs, the insect undergoes a radical change. The larva stops feeding and moving, anchors itself to a twig, leaf, or rock, and either spins a cocoon or encloses itself in its own hardened cuticle.

Silkworm Metamorphosis

The silkworm, *Bombyx mori*, undergoes complete metamorphosis, and its development begins when the egg hatches and includes five distinct larval stages. Each stage is larger than the one preceding as the larva eats, voraciously consuming several times its own body weight in food each day. At the end of the fifth larval stage, the molting event that follows is very different from the previous larval molts as the caterpillar spins a cocoon made out of silk (which is the commercially valuable product of this organism) and becomes a pupa. The next molt occurs within the pupal case when metamorphosis begins. When the adult has formed, the cocoon breaks open and the adult emerges to begin reproduction.

Metamorphosis involves the complete replacement of one body form with another. Inside the pupal case, the larval tissues break down and their molecules are reutilized in the construction of the cells and tissues of a very different looking animal, the adult. Certain groups of cells, called imaginal disks, along with the larval brain are generally the only tissues that are not broken down in this process. Opening a cocoon at this time shows that it is filled with a white, milky sap and little else as the caterpillar has been completely broken down.

The imaginal disks, round, flat sheets of cells, begin to evert or "telescope" and form the external structures characteristic of the adult cuticle. There is an imaginal disk for each eye, for each antenna, for the two or four wings, and for each of the six legs. These structures attach and the adult insect is constructed. In the well-studied fruit fly, *Drosophila*, this process takes about a week, while it can take months for other insects.

Hormonal Regulation of Metamorphosis

In insects, three very different hormones combine to regulate the timing of both molting and metamorphosis. Each of these three hormones is produced in a different tissue, and each has a different chemical structure and mode of function. The signal to molt originates in a small group of cells within the caterpillar's brain in response to neural or environmental signals. The hormone produced there, prothoracicotropic hormone (PTTH), is a small protein that passes through the insect's hemolymph (blood) to all parts of the body. As is true of all hormones, only certain target tissues are genetically programmed to respond to the production of this hormone. In this case, the prothoracic gland responds by producing a second hormone called ecdysone, the molting hormone. Ecdysone is a steroid, a chemical derivative of the cholesterol that the insect requires in its diet. Ecdysone is not actually the molting hormone itself, but must first undergo some minor chemical changes before it becomes active. More important, ecdysone and its derivatives are the molting hormone not only for all insects, but for all arthropods and many other animals. This chemical signal thus evolved before divergence of these organisms from a common ancestor.

Different tissues respond differently to ecdysone, but the hormone's major effect is to trigger molting by causing the hardening of the cuticle and the separation of the living cells beneath the cuticle from it. The cuticle then dries and cracks and the larva can then emerge from its old skin and grow. If PTTH production stops, the amount of ecdysone released from the prothoracic gland also falls. This happens normally in the period between molts, but can also occur during a molt. In cecropia moths, the level of PTTH drops during the pupal stage and the subsequent drop in ecdysone production causes diapause, a programmed pause in development. The pupa will remain in diapause until an environmental signal, consisting of a minimum of two weeks in the cold followed by a normal spring warming, triggers the resumption of PTTH secretion and the completion of metamorphosis.

Ecdysone acts specifically on certain groups of insect genes and not others. Many genes required for larval functions are turned off in response to ecdysone, while those genes required for molting and metamorphosis are turned on. This effect can be seen when imaginal disks are placed in a culture dish along with physiological levels of ecdysone. The disks stop growing and begin metamorphosis. Such development can even produce normal-looking legs floating free in a culture dish. The changes in shape of the disk cells can be directly attributed to the function of genes turned on by ecdysone.

The hormone ecdysone therefore not only causes metamorphosis, but also triggers certain simple molts. The third hormone involved in the control of metamorphosis is responsible for this choice, the choice to molt and grow or to undergo metamorphosis to the adult form. This compound is called juvenile hormone (JH). JH is produced in a gland called the corpora allata, and the hormone has yet a third chemical structure, a derivative of a class of molecules called terpenes. Unlike ecdysone, however, the JH of each species has a different chemical structure.

When a molt is triggered by the production of ecdysone, the type of molt that occurs will be determined by the level of JH present. During larval development, the amount of JH in the hemolymph is high and the tissues respond to ecdysone by undergoing a molt to become a larger larva. Later in development, the corpora allata stops producing JH and any new molt triggered by ecdysone causes the organism to proceed to the pupal stage and begin metamorphosis. The interaction of ecdysone and JH thus governs which type of molt will occur.

Although these hormones are crucial for controlling molting and metamorphosis, they play other roles as well. For example, adult female insects produce large quantities of ecdysone and adult males do not because ecdysone is important to egg production.

Laboratory Studies of Metamorphosis
Various parts of the elaborate interplay of the three hormones that regulate metamorphosis in insects

were discovered in a number of laboratories. The role of ecdysone was discovered first by Carroll Williams and Vincent Wigglesworth, who found that the prothoracic gland of the larva was responsible for producing a substance that triggered metamorphosis. They tied a fine string around the middle of a cecropia larva and observed that only the front half underwent metamorphosis. The signal was thus produced somewhere in the front half and could not reach the rear of the insect. The prothoracic glands were later discovered to be the source of the molting signal when glands transplanted to the rear half of a ligated larva caused metamorphosis to occur there as well.

Ecdysone was painstakingly purified and chemically identified by Peter Karlson's group in Germany. In a monumental effort, they extracted twenty-five milligrams of pure ecdysone from a ton of silkworms. Much of the difficulty encountered in this isolation stemmed from the lack of an easy method for identifying the presence of ecdysone in the extract they were producing.

A tedious and relatively insensitive bioassay was used. In this assay, the extract to be tested during the isolation was dissolved in an organic solvent and painted on the cuticle of an insect. If the extract contained ecdysone, the larva would begin to molt. Although time-consuming and not very reproducible, the bioassay allowed the purification of ecdysone from insects.

The role of the corpora allata and JH was found by a similar set of experiments. Wigglesworth found that if he decapitated a fourth-stage Rhodnius nymph and sealed the body with wax to that of a decapitated fifth-stage nymph, the insects could survive for a long period of time. This technique is called parabiosis. The mature nymph never underwent metamorphosis but rather molted in response to its own ecdysone or to ecdysone painted on its cuticle to form a larger nymph. Once again, tissue transplantation experiments showed that the hormone coming from the immature nymph that prevented metamorphosis was produced by the corpora allata. Removing the corpora allata from a larva or nymph caused either death or premature metamorphosis to the adult stage.

The role played by PTTH was also elucidated by a similar set of surgical experiments. Removing the brain of a larva prevented the production of ecdysone from the prothoracic gland so long as the removal occurred before the brain released the PTTH. After the PTTH signal was released, removing the brain had no effect on the subsequent production of ecdysone for that molt.

Today, the presence and amount of the three hormones are measured by easy, sensitive, and highly reproducible immunological techniques. In these procedures, antibodies, proteins having the chemical ability to bind tightly to a specific insect hormone, are produced. If an extract contains one of these hormones, adding its specific antibody will cause the hormone to bind to the antibody, and the amount of bound antibody can be accurately measured in a number of different ways.

These immunological procedures have allowed researchers to monitor the changing levels of the three hormones during the development of any insect and to correlate these changes with the progress of metamorphosis. Coupled with new information about the ability of hormones to turn genes on and off directly, these procedures have advanced the understanding of the mechanisms of control of metamorphosis at the molecular level.

—*Joseph G. Pelliccia*

See also: Amphibians; Determination and differentiation; Endocrine systems of invertebrates; Insects; Morphogenesis; Physiology.

Bibliography

Danks, H. V., ed. *Insect Life-Cycle Polymorphism: Theory, Evolution, and Ecological Consequences for Seasonality and Diapause Control.* Boston: Kluwer, 1994. Written for biologists and entomologists. Reviews recent work on the effects of genetic polymorphism in structuring insects' seasonal life cycle.

DePomerai, David. *From Gene to Animal.* 2d ed. New York: Cambridge University Press, 1990. An intermediate-level college text that emphasizes the role various genes and molecules play in the control of metamorphosis. Chapter 7, "Insect Development," discusses the anatomical, genetic, and biochemical changes which occur in metamorphosis. Detailed bibliography.

Gilbert, Lawrence I., and Earl Frieden, eds. *Metamorphosis: A Problem in Developmental Biology.* 2d ed. New York: Plenum, 1981. An advanced college-level discussion of metamorphosis controls in all kinds of animals. The chapter on hormonal control of insect metamorphosis is particularly good. Very complete bibliography.

Gilbert, Scott F. *Developmental Biology.* 6th ed. Sunderland, Mass.: Sinauer Associates, 2000. An intermediate-level college textbook that has become a standard in the teaching of developmental biology. Metamorphosis in insects, amphibians, and echinoderms is covered in good detail. Gilbert provides an excellent overview of the anatomical, genetic, and biochemical events of these complex processes. Excellent bibliography.

Koolman, Jan, ed. *Ecdysone: From Chemistry to Mode of Action.* New York: G. Thieme Verlag, 1989. A collection of seventy-seven review essays by experts in the field of ecdysone biology, serving as a good introduction to research on the role of these steroids in insect metamorphosis, as well as their effects on other animals.

Raven, Peter H., and George B. Johnson. *Biology.* 6th ed. St. Louis: Times Mirror/Mosby, 2001. A good, introductory college-level text that is also suitable for high school students. Metamorphosis in insects and other animals is clearly discussed with very good diagrams (pages 836-838) that show insect development and the roles played by insect hormones. Glossary.

MICE AND RATS

Types of animal science: Behavior, classification, ecology
Fields of study: Ecology, physiology

House mice and Norway rats are found worldwide and almost always are found related to humans and human activities. No other mammals are greater pests than are mice and rats.

Principal Terms

ALTRICIAL: born in a helpless state and completely dependent on the parent(s)

COMMENSAL: living in close association with humans; the word means "sharing the table"

OMNIVOROUS: eating all kinds of food, both plant and animal

RODENTS: the gnawing mammals characterized by specialization of the incisor teeth into gnawing teeth

---

Although there are many types of mouse and rat, the words most often are used to refer to the house mouse, *Mus musculus*, and the Norway rat, *Rattus norvegicus*. These two species belong to a family of rodents known as the Muridae, the Old World mice and rats, and there are over 450 species within this group. Both species underwent their early evolution in the wilds of Asia. As humans appeared and began to settle in farms and villages, mice and rats became associated with them. As humans migrated to other parts of the world and as commercial exchanges took place around the world, mice and rats went along for the ride and became established in the Old World and the New World, and today may be found almost any place on earth where humans are found.

At birth, mice and rats are unimposing animals, as the pups are born in an altricial state and are naked, sightless, and helpless. They develop rapidly, however, and by three weeks of age are weaned from their mother. By 1.5 to 2 months of age for the mouse, and 2 to 3 months of age for the rat, they are sexually mature. Their powers of reproduction are phenomenal and many litters, some containing more than a dozen young, can be produced in a single year. A single female mouse may produce over one hundred young in one year.

As adults, mice and rats still are not imposing. They have a long, scaly, scantily haired tail, and are a grayish brown color, which is somewhat paler in the belly. Although they have poor sight, their senses of smell, hearing, and taste are all excellent.

The Good and Bad of Mice and Rats

All of the other mammalian species combined, as well as all other animals, do not cause as much damage and destruction to humans as do mice and rats. The two species have been able to adjust to living with and near humans and their habitations. Their diets are omnivorous and include all types of foods, grains, and grain products. They can climb, burrow, and swim, and can invade nearly all buildings, houses, barns, warehouses, and other structures. As a result, they cause billions of damage each year around the world. Human food is eaten or destroyed by contamination with urine and feces. The urine may contain bacteria, causing diseases such as leptospeosis. Food poisoning may be caused by salmonellosis in their feces. Many children and some adults are bitten each year by rats, especially in rundown urban areas. Even those mice and rats that live under semi-wild conditions may affect humans in their effect on other wildlife. Ground-nesting birds and the young of other animals are susceptible to predation by mice and rats. The costs of these pests to humans is tremendous.

Some consolation for having to deal with the extensive bad side of mice and rats may be found in their beneficial use as laboratory animals. Both species are used widely in medical and genetic research. The mouse has been used to elucidate many basic principles of genetics, and today, its many genetic variants serve as models for many human disorders, including genetic diseases and cancer. As studies define the human genome, similar studies are defining the mouse and rat genomes. The laboratory rat, although not as

Mouse and Rat Facts

Classification:
Kingdom: Animalia
Subkingdom: Bilateria
Phylum: Chordata
Subphylum: Vertebrata
Class: Mammalia
Subclass: Theria
Infraclass: Eutheria
Order: Rodentia
Suborder: Myomorpha
Family: Muridae, with fifteen subfamilies, 241 genera, and 1,082 species
Geographical location: Worldwide, through introduction by humans
Habitat: Occurs in highly varied habitats, usually in association with humans
Gestational period: Nineteen to twenty-two days, although this may be lengthened in nursing females due to delayed implantation
Life span: Around one year in the wild; laboratory rats and mice may live three to four years or longer
Special anatomy: The scantily haired tail helps to distinguish house mice and Norway rats from most other types of mice and rats

Small rodents, such as these deer mice, often sleep in piles to conserve body heat. (Rob and Ann Simpson/Photo Agora)

genetically well known as the mouse, has been used extensively in physiological and psychological research.

It is unlikely that humans will ever be able to eliminate completely mice and rats and control their destruction, but it also is unlikely that their importance to both basic and applied science will ever be diminished, at least in the foreseeable future.

—Donald J. Nash

See also: Beavers; Gophers; Porcupines; Rodents; Squirrels.

Bibliography

Barnell, S. A. *The Rat: A Study in Behavior.* Rev. ed. Chicago: University of Chicago Press, 1975. A classic book which presents all kinds of interesting material about behavior in rats.

Chapman, Joseph A., and George A. Feldhamer, eds. *Wild Mammals of North America: Biology, Management, and Economics.* Baltimore: The Johns Hopkins University Press, 1982. Rats and mice are covered in the section on exotic species. Useful information on the natural history of these species is included.

Crowcroft, Peter. *Mice All Over.* Chester Springs, Pa.: Dufour Editions, 1966. Out-of-print but well worth searching libraries. Describes fascinating accounts of the behavior and ecology of freely growing confined populations of mice.

Silver, Lee M. *Mouse Genetics: Concepts and Applications.* New York: Oxford University Press, 1995. A comprehensive but readable book on genetics of the house mouse and it includes useful information on the origin and spread of mice.

U.S. National Research Council. *Urban Pest Management.* Washington, D.C.: National Academy Press, 1980. This work discusses problems and solutions of control of pest rodents in cities.

MIGRATION

Type of animal science: Behavior
Fields of study: Biophysics, ethology, neurobiology, marine biology, ornithology

Migration is an important ecological process that results in the redistribution of animal populations from one habitat to another. Adaptive habitat changes are fundamentally important aspects of the life histories of many species.

Principal Terms

BIOLOGICAL CLOCK: an inherent sense of timing regulating certain types of behavioral activity such as migration

CLOCK SENSE: an inherent awareness of time or time intervals used, for example, to compensate for celestial movements in navigation

DISPERSAL: the spreading apart of individuals away from one another and away from a place; includes a directional component when passive animals are moved by winds or currents

HABITAT: a specific, recognizable geographical region in which a particular kind of organism lives

MIGRANT: a group or species of animal that moves from one habitat or geographical region to another

NAVIGATION: to follow or control the course of movement from the place of origin to a specific destination

NOMADS: migrants without a specific habitat; wanderers

ORIENTATION: an inherent sense of geographical location or place in time

POPULATION: a group of individuals of the same species geographically located in a given habitat at the same time

Migration is a general term employed by ecologists and ethologists to describe the nearly simultaneous movement of many individuals or entire populations of animals to or between different habitats. As defined, migrations do not include local excursions made by individuals or small groups of animals in search of food, to mark territorial boundaries, or to explore surrounding environments.

Nomads are migrants whose populations follow those of their primary food sources. Such animals (the American bison, for example) do not have fixed home ranges and wander in search of suitable forage. Some scientists view nomadic movements as a form of extended foraging behavior rather than as a special case of migration. In either context, the important point is that populations change habitats in response to changing conditions.

In contrast to migrations made by populations and excursions made by individuals, the spreading or movement of animals away from others is known as dispersal. Examples of dispersal include the drift of plankton in currents and the departure of subadult animals from the home range of their parents. In numerous species (sea turtles, rattlesnakes, and salmon, for example), dispersed members of a population may return to the place of origin after a variable interval of time.

Means and Reasons

Some migratory species can orient themselves—that is, they know where they are in time and space. Many birds and mammals, for example, have an inherent sense of the direction, distance, and location of distant habitats. Orientation and travel along unfamiliar routes from one place or habitat to another is called navigation. Navigators use environmental and sensory information to reach distant geographical locations, and many of them do so with a remarkably accurate sense of timing. Homing pigeons are perhaps the best-studied animal navigators. These birds are able not only to discover where they are when released but also to return to their home loft from distant geographical locations.

Much has been learned about how animals successfully navigate over long distances from the pioneering studies of Archie Carr. Carr proposed that green sea turtles successfully find their widely separated nesting and feeding beaches by means of an inherent clock sense, map sense, and compass sense. His investigations and those of many others continue to stimulate great interest in the physiology and ecology of navigating species and in the environmental cues to which they respond. Sensory biologists, biophysicists, and engineers have incorporated knowledge of how animals detect and use environmental information to develop new and more accurate navigational systems for human use.

Animals use a variety of cues to locate their positions and appropriate travel paths. Most species have been found to use more than one type of information (sequentially, alternatively, or simultaneously) to navigate. Among the animals known to navigate are birds (the best-studied group), lobsters, bees, tortoises, bats, marine and terrestrial mammals, fish, brittle starfishes, newts, toads, and insects. Included among the orientation guideposts that one or more of these groups may use are the positions of the sun and stars, magnetic fields, ultraviolet light, tidal fluctuations caused by the changing positions of the moon and sun, atmospheric pressure variations, infrasounds (very low frequency sounds), polarized light (on overcast days), environmental odors, shoreline configurations, water currents, and visual landmarks. Celestial cues also require a time sense, or an internal clock, to compensate for movements of the animal relative to changing positions of celestial objects in the sky. In addition to an absolute dependence on environmental cues, young or inexperienced members of some species may learn navigational routes from experienced individuals, such as their parents, or other experienced individuals in the population. Visual mapping remembered from exploratory excursions may also play a role in enhancing the navigational abilities of some birds, fish, mammals, and other animals.

The different categories of animal movements, however, are perhaps not so important as the reasons animals migrate and the important biological consequences of the phenomenon. As a general principle, migrations are adaptive behavioral responses to changes in ecological conditions. Populations benefit in some way by regularly or episodically moving from one habitat to another.

An example of the adaptive value of migratory behavior is illustrated by movement of a population from a habitat where food, water, space, nesting materials, or other resources have become scarce (often a seasonal phenomenon) to an area where resources are more abundant. Relocation to a new habitat (or to the same type of habitat in a different geographical area) may reduce intraspecific or interspecific competition, may reduce death rates, and may increase overall fitness in the population. These benefits may result in an increase in reproduction in the population. Reproductive success, then, is the significant benefit and the only biological criterion used to evaluate population fitness.

Programmed and Episodic Movements

While many factors are believed to initiate migratory events, most fall into one of two general categories. The first and largest category may be called programmed movements. Such migrations usually occur at predictable intervals and are important characteristics of a species' lifestyle or life cycle. Programmed migrations are not, in general, density-dependent. Movements are not caused by overcrowding or other stresses resulting from an excessive number of individuals in the population.

The lifestyle of a majority of drifting animals whose entire lives are spent in the water column, for example, includes a vertical migration from deep water during the day to surface waters at

Bird migrations are among the most noticeable animal migrations, with huge flocks passing through the skies on their way to a seasonal home. (Corbis)

with seasonal environmental change. The onset of migration in many vertebrates is evidenced by an increase in restlessness that seems, in human terms, to be anticipatory.

In addition to their daily vertical migrations (lifestyle movements), the life cycles of marine zooplankton involve migrations, and it is convenient to use them as examples. As discussed, most adult animal plankters are found at depth during the day and near the surface at night. In contrast, zooplankton eggs and larvae remain in surface waters both day and night. As the young stages grow, molt, and change their shapes and food sources, they begin to migrate vertically. The extent of vertical migrations gradually increases throughout the developmental period, and as adults, these animals assume the migratory patterns of their parents. Patterns of movement that change during growth and development are examples of ontogenetic, or life-cycle, migrations.

The second large category of migratory behavior includes episodic, density-dependent population movement. Such migrations are often associated with, or caused by, adverse environmental changes (effect) that may be caused by overlarge populations (cause). Local resources are adequate to support a limited number of individuals (called the "carrying capacity" of the environment), but once that number has been exceeded, the population must either move or perish. Unfortunately, migration to escape unfavorable conditions may be unsuccessful, as another suitable habitat may not be encountered. Migrations caused by overpopulation or environmental degradation are common. Pollution and habitat destruction by humankind's activities are increasingly the cause of degraded environments, and in such cases, it is

night. Thus, plankton exhibit a circadian rhythm (activity occurring during twenty-four-hour intervals) in their movements. An abundance of food at or near the surface, and escape from deepwater predators, are among the possible reasons for these migrations. Daily vertical movements of plankton are probably initiated by changes in light intensity at depth, and the animals follow light levels as they move toward the surface with the sinking sun. It is interesting to note that zooplankters living in polar waters during the winterlong night do not migrate.

Monarch butterflies and many large, vertebrate animals, such as herring, albatross, wildebeests, and temperate-latitude bats, migrate from one foraging area to another, or from breeding to foraging habitats, on a seasonal or annual basis. Annual migrations usually coincide with seasonal variation. Changes in day length, temperature, or the abundance of preferred food items associated with seasonal change may stimulate mass movements directly, or indirectly, through hormonal or other physiological changes that are correlated

reasonable to conclude that humans have reduced the carrying capacity of many animal habitats. Familiar examples of density-dependent migrations are those of lemmings, locusts, and humans.

Studying Migration

Methods used by scientists to study the mass movements of animals are quite varied and depend on the investigator's research interests and on the kinds of organisms being investigated. Environmental or physiological factors that initiate migrations may be of interest to sensory biologists and physiological ecologists; knowledge of variation in population distributions is important to biogeographers and wildlife biologists; and migrations in predator-prey relationships, competition, pollution, and life-history strategies are important aspects of classical ecological studies.

In addition to the specific aspect of migration being studied, the particular group of animals under investigation (moths, eels, elephants, snails) requires that different methods be used. Some of the approaches used in migration-related research illustrate how information and answers are obtained by scientists.

Arctic terns migrate from their breeding grounds in the Arctic to the Antarctic pack ice each year. The knowledge that these birds make a twenty-thousand-mile annual round-trip comes from the simplest and most practical method: direct observation of the birds (or their absence) at either end of the trip. Direct observation by ornithologists of the birds in flight can establish what route they take and whether they pause to rest or feed en route. Many birds have also been tracked using radar or by observations of their silhouettes passing in front of the moon at night. Birds are often banded (a loose ring containing coded information is placed on one leg) to determine the frequency of migration and how many round-trips an average individual makes during its lifetime. From this information, estimates of longevity, survivorship rates, and nesting or feeding site preferences can be made.

Factors that initiate migratory behavior in terns and in other birds can often be determined by ecologists able to relate environmental conditions (changes in temperature, day length, and the like) to the timing of migrations. Physiological ecologists study hormonal or other physiological changes that co-occur with environmental changes. Elevated testosterone levels, for example, may signal the onset of migratory behavior.

How Arctic terns orient and navigate along their migratory routes is usually studied by means of laboratory-conducted behavioral experiments. Birds are exposed to various combinations of stimuli (magnetic fields, planetarium-like celestial fields, light levels), and their orientation, activity levels, and physiological states are measured. Experiments involving surgical or chemical manipulation of known sensory systems are sometimes conducted to compare behavioral reactions to experimental stimuli. In such experiments, the birds (or other test animals) are rarely harmed.

Tags of several types are used to study migrations in a wide variety of animals, including birds, bees, starfishes, reptiles, mammals, fish, snails, and many others. Tags may be transmitting collars (located by direction-finding radio receivers); plastic or metal devices attached to ears, fins, or flippers; or even numbers, painted on the hard exoskeleton of bees and other insects. Additional types of tagging (or identifying) include radioactive implants and microchips that can be read by computerized digitizers; the use of brands and tattoos; and, of great interest, the use of biological tags. Parasites known to occur in only one population of migrants (nematode parasites of herring, for example) provide an interesting illustration of how the distribution of one species can be used to provide information about another.

The Importance of Migration

The causes, frequency, and extent of animal migration are so diverse that several definitions for the phenomenon have been proposed. None of these has been accepted by all scientists who study animal movements, however, and it is sometimes difficult to interpret what is meant when the term "migration" is used. Most researchers have adopted a broad compromise to include all but trivial population movements that involve some degree of habitat change.

It is important to recognize that few populations of animals are static; even sessile animals (such as oysters and barnacles) undergo developmental habitat changes, which are referred to as ontogenetic migrations. Aside from certain tropical and evergreen forest areas where migrations are relatively uncommon, a significant number of both aquatic and terrestrial species move from one habitat to another at some time during their lives. In the face of environmental change, including natural events such as seasonal variation and changes caused by resource limitations and environmental degradation, animals must either move, perish, or escape by means of drastic population reduction or by becoming inactive until conditions become more favorable (hibernation, arrested development and dormancy, and diapause in insects are examples of behavioral-ecological inactivity). Migration is the most common behavioral reaction to unfavorable environmental change exhibited by animals.

One cannot understand the biology of migrators until their distribution and habitats throughout life are known. The patterns of animal movements are fascinating, and it is useful to summarize some of the major differences between them. First, many species travel repeatedly during their lives between two habitats, on a daily basis (as plankton and chimney swifts do) or on an annual basis (as frogs and elks do). Second, some species migrate from one habitat (usually suitable for young stages) to another (usually the adult habitat) only once during their lives (for example, salmon, eels, damselflies, and most zooplankton, which live on the bottom as adults). Third, some species (many butterflies, for example) are born and mature in one geographical area (England, for example), migrate as adults to a distant geographical area (Spain, for example), and produce offspring that mature in the second area. These migrations take place between generations. In a fourth pattern, one may include the seasonal swarming of social arthropods such as termites, fire ants, and bees. A fifth but ill-defined pattern is discernible, exemplified by locust "plagues," irruptive emigration in lemmings and certain other rodents, and some mass migrations by humankind, as caused by war, famine, fear, politics, or disease. These are episodic and often, if not primarily, caused by severe population stress or catastrophic environmental change.

—Sneed B. Collard

See also: Birds; Competition; Demographics; Eels; Gene flow; Geese; Hormones and behavior; Instincts; Learning; Population fluctuations; Population growth; Reproduction; Reproductive strategies; Salmon and trout; Whales, baleen.

Bibliography

Able, Kenneth P., ed. *Gatherings of Angels: Migrating Birds and Their Ecology.* Ithaca, N.Y.: Comstock, 1999. A collection of essays by avian biologists on many aspects of a wide variety of bird migrations in the western hemisphere. Illustrations.

Aidley, David, ed. *Animal Migration.* New York: Cambridge University Press, 1981. An exceptionally well-written reference book covering all important aspects of animal migration. Contributing authors discuss major concepts in the ecology of migrations, as well as mechanisms of orientation and navigation in a variety of different animal groups. Emphasis is placed on birds, whales, fish, insects, and terrestrial mammals. Each chapter includes technical and general references. Suitable for college students and the general reader.

Begon, Michael, John Harper, and Colin Townsend. *Ecology: Individuals, Populations, and Communities.* 3d ed. Sunderland, Mass.: Sinauer Associates, 1996. One of the best modern textbooks in ecology. Presents a comprehensive overview of patterns of migrations using both familiar and exotic examples of different animal groups. Subject and taxonomic indexes, many references, outstanding illustrations. College or advanced high school level.

Dingle, Hugh. *Migration: The Biology of Life on the Move.* New York: Oxford University Press, 1996. Discusses a wide variety of animal movements, focusing on birds and insects. A good general reference. Numerous illustrations.

Eisner, Thomas, and Edward Wilson, eds. *Animal Behavior: Readings from "Scientific American."* San Francisco: W. H. Freeman, 1975. A collection of articles originally published in *Scientific American.* Includes many classic papers on ecological, physiological, and behavioral aspects of migration. Section 1 includes chapters on moths and ultrasound, the flight-control systems of locusts, and annual biological clocks. Habitat selection and homing behavior in pigeons and salmon are discussed in section 2. Clearly illustrated, very informative, and well written. Includes illustrations, a bibliography, and an index. College or general audience.

Newberry, Andrew, ed. *Life in the Sea: Readings from "Scientific American."* San Francisco: W. H. Freeman, 1982. A collection of articles published in *Scientific American.* Migrations of oceanic plankton are discussed in chapter 2. Chapter 16 focuses on migrations of the shad; biological clocks of the tidal zone are discussed with respect to migrations in chapter 19. As with all special-topic collections from *Scientific American,* each chapter is very informative and well illustrated. Includes a bibliography and an index. College or general audience.

Pyle, Robert Michael. *Chasing Monarchs: Migrating with Butterflies of Passage.* Boston: Houghton Mifflin, 1999. A well-known lepidopterist literally follows the migration of monarch butterflies from northern Washington state and Canada, south to their winter roosts in the mountains of Michoacan, in southwestern Mexico.

Rankin, Mary, ed. *Migration: Mechanisms and Adaptive Significance.* Port Aransas, Tex.: Marine Science Institute, University of Texas at Austin, 1985. A collection of scientific contributions presented at an international symposium on the subject of migrations. It is one of the most important and comprehensive monographs on the subject. Experimental and field approaches to the study of an extremely wide range of organisms are described, as are the evolutionary and ecological consequences and mechanisms of migration. The book was written by experts for other experts, but the book's organization and style make it an excellent source of further reading for high school and college students of biology. Includes many graphic illustrations, a general index, and topic bibliographies.

Reader's Digest Association. *The Wildlife Year.* Pleasantville, N.Y.: Author, 1993. Organized by month, covers stages in the yearly life cycles of 288 plants and animals, including seasonal migrations. Over four hundred photographs.

Schone, Hermann. *Spatial Orientation: The Spatial Control of Behavior in Animals and Man.* Translated by Camilla Strausfeld. Princeton, N.J.: Princeton University Press, 1984. Schone describes first the relationship between orientation, behavior, and ecology to provide the background for detailed discussions of types of orientation movements, sensory physiology, and the kinds of environmental information perceived and used by animals to direct their movements. An exhaustive treatment of the subject is presented, and the format is technical, but it should be understandable to advanced high school and college students with some background in biology.

MIMICRY

Types of animal science: Development, ecology, evolution
Fields of study: Ecology, evolution science, genetics, population biology

Mimicry is the process whereby one organism resembles another and, because of this resemblance, obtains an evolutionary advantage.

Principal Terms

ADAPTATION: a phenotype that allows those organisms that have it a competitive advantage over those that do not have it

CAMOUFLAGE: patterns, colors, and/or shapes that make it difficult to differentiate an organism from its surroundings

WARNING COLORATION: the bright colors seen on many dangerous and unpalatable organisms that warn predators to stay away

---

The broadest description of mimicry is when one organism, called the operator or dupe, cannot distinguish a second organism, called the mimic, from a third organism or a part of the environment, called the model. There are many different types of mimicry. Some mimics look like another organism; some smell like another organism; some may even feel like another organism. There are many ways that mimicking another organism could be helpful. Mimicry may help to hide an organism in plain sight or protect a harmless organism from predation when it mimics a harmful organism. It can even help predators sneak up on prey species when the predator mimics a harmless organism.

In the case of hiding in plain sight, the line between camouflage and mimicry is not sharply defined. Spots or stripes that help an organism blend with the surroundings are classified as camouflage, since those patterns allow the organism to remain hidden in many areas that have mixtures of sunlight and shadow, and the organism does not look like any particular model. As an organism's appearance begins to mimic another organism more and more closely, rather than being just a general pattern, it moves toward mimicry. As in all other areas of biology, there are arguments about where camouflage ends and mimicry begins. The stripes of a tiger and the spots on a fawn are certainly camouflage. The appearance of a stick insect is more ambiguous. Its body is very thin and elongated and is colored in shades of brown and gray. Is this mimicry of a twig or just very good camouflage? Many biologists disagree. The shapes and colors of many tropical insects, especially mantids, also fall into this gray area of either extremely good camouflage or simple mimicry.

Batesian and Müllerian Mimicry

In contrast to camouflage, which hides its bearers, many species of dangerous or unpalatable animals are brightly colored. This type of color pattern, which stands out against the background, is called warning coloration. Some examples are the black and white stripes of the skunk, the yellow and black stripes of bees and wasps, red, black, and yellow stripes of the coral snake, and the bright orange of the monarch butterfly. Several species of harmless insects have the same yellow and black pattern that is seen on wasps. In addition to mimicking the coloration of the more dangerous insects, some harmless flies even mimic the wasps' flying patterns or their buzzing sound. In each case, animals that have been stung by wasps or bees avoid both the stinging insects and their mimics. This mimicry of warning coloration

This eastern eyed click beetle has evolved coloration that mimics the appearance of the eyes of a much larger animal in order to scare away potential predators. (Rob and Ann Simpson/Photo Agora)

is called Batesian mimicry. Batesian mimicry is also seen in the mimicry of the bright red color of the unpalatable red eft stage of newts by palatable salamanders. Sometimes two or more dangerous or unpalatable organisms look very much alike. In this case, both are acting as models and as mimics. This mimicry is called Müllerian mimicry. Müllerian mimicry is seen in monarch and viceroy butterflies. Both butterflies have in their bodies many of the chemicals found in the plants they ate as larvae. These include many unpalatable chemicals and even toxic chemicals that cause birds to vomit. If a bird eats either a monarch or a viceroy that has these chemicals, the bird usually remembers and avoids preying on either species again—a classic Müllerian mimicry. Interestingly, not all monarchs or viceroys are unpalatable. It depends on the types and concentrations of chemicals in the particular plants on which they fed as larvae.

Birds that have eaten the palatable monarchs or viceroys do not reject either monarchs or viceroys when offered them as food, but birds that have eaten an unpalatable monarch or an unpalatable viceroy avoid both palatable and unpalatable members of both species. This represents both Batesian and Müllerian mimicry at work.

Aggressive Mimicry

Mimicry by predators is called aggressive mimicry. The reef fish, called the sea swallow, is a cleaner fish, and larger fish enter the sea swallow's territory to be cleaned of parasites. The saber-toothed blenny mimics the cleaner in both appearance and precleaning behavior, but when fish come to be cleaned, the blenny instead bites off a piece of their flesh to eat. Anglerfish have small extensions on their heads that resemble worms. They use the mimic worms to lure their prey close enough to be eaten. The alligator snapping turtle's tongue and the tips of the tails of moccasins, copperheads, and other pit vipers are also wormlike and are used as lures. Certain predatory female fireflies respond to the light flashes of males of a different species with the appropriate response of the female of that species. This lures the male closer, and when the unsuspecting male is close enough to mate, the female devours him. This mimicry is quite complex, because the predatory females are able to mimic the response signals of several different species.

There are many other instances of mimicry, but the world champion mimics may be octopuses. As predators, these animals show unbelievable aggressive mimicry of other reef organisms. Octopuses can take on the color, shape, and even texture of corals, algae, and other colonial reef dwellers. As a prey species, the octopus can use the same type of mimicry for camouflage, but can also be a Batesian mimic, taking on the color and shape of many of the reef's venomous denizens.

Since in each case, being a mimic helped the organism in some way, it is not hard to understand how mimicry may have evolved. In a population where some organisms were protected by being mimics, the protected mimics were the ones most likely to mate and leave their genes for the next generation while the unprotected organisms were less likely to breed.

—*Richard W. Cheney, Jr.*

See also: Adaptations and their mechanisms; Butterflies and moths; Camouflage; Coevolution; Defense mechanisms; Octopuses; Predation.

Bibliography

Brower, Lincoln P., ed. *Mimicry and the Evolutionary Process.* Chicago: University of Chicago Press, 1988. The proceedings of a symposium, offering a thorough look at the evolution of mimicry.

Ferrari, Marco. *Colors for Survival: Mimicry and Camouflage in Nature.* Charlottesville, Va.: Thomasson-Grant, 1993.

Owen, D. *Camouflage and Mimicry.* Chicago: University of Chicago Press, 1982. An interesting compilation of instances of mimicry.

Salvato, M. "Most Spectacular Batesian Mimicry." In *University of Florida Book of Insect Records,* edited by T. J. Walker. Gainesville: University of Florida Department of Entomolgy and Nematology, 1999. http://gnv.ifas.ufl.edu/~tjw/recbk.htm. An electronically published paper on the swallowtail butterfly.

Wickler, Wolfgang. *Mimicry in Plants and Animals.* New York: McGraw-Hill, 1968. An intriguing, small book on the subject of mimicry. Highly adaptive forms are beautifully illustrated in color; most examples are from the insect world.

MOLES

Types of animal science: Anatomy, classification, reproduction
Fields of study: Anatomy, reproduction, zoology

Moles spend most of their time in underground burrows, digging for food with specialized front paws. They perform a service to farmers by destroying grubs, caterpillars, and insects.

Principal Terms

BOLTRUN: mole burrow tunnel used as an emergency exit

GESTATION: time in which mammalian offspring develop in the uterus

MARSUPIAL: nonplacental mammal having a marsupium

MARSUPIUM: abdominal pouch containing mammary glands and sheltering offspring until they are fully developed

---

Moles inhabit Africa, Europe, Asia, North America, and Australia. They are voracious, continually burrowing in the ground for food, for a mole must eat its weight in food daily. This is accomplished by digging approximately twenty-five feet of burrow per hour. Mole burrows are close to ground surfaces and can cause surface ridges. A mole's home is also recognized by its large, central mound of earth. This mole hill is created from the earth dug up in the mole's search for food. There are twelve mole genera worldwide, five of which inhabit the United States.

Types of Moles

Moles have pointy snouts, rudimentary eyes, velvety fur, short legs, and powerful digging nails on their front legs. Moles are nearly or totally blind; however, their hearing is acute and they detect sounds from great distances. Common garden moles are six inches long with furless tails, and have huge forelegs whose broad, thick nails gouge out earth. The largest species in the United States, *Scapanus townsendii*, is nine inches long. Shrew moles, the smallest, are 3.5 inches long.

European moles inhabit grasslands and pastures in the British Isles, continental Europe, and Asia. They dig elaborate burrows that hold central chambers with round connected galleries. Passageways radiate in all directions from galleries (for example, there will be a boltrun exit in case of danger). Each burrow has a warmly lined nest

---

Mole Facts

Classification:
Kingdom: Animalia
Subkingdom: Bilateria
Phylum: Chordata
Subphylum: Vertebrata
Class: Mammalia
Order: Insectivora
Family: Talpidae, with twelve genera and twenty-nine species, including *Scalopus aquaticus* (common garden mole), *Scapanus townsendii* (largest U.S. mole), *Neurotrichus gobbsi* (shrew mole), *Talpa europaea* (European mole), *Condylura cristata* (star-nosed mole), *Notoryctes typhlops* (marsupial mole)

Geographical location: Europe, Asia, North America, and Australia

Habitat: Most inhabit grasslands and pastures, though some live in freshwater

Gestational period: Two to six weeks

Life span: Three to six years

Special anatomy: Huge forelegs with broad nails; leathery pads or tentacle rings covering nasal passages

---

Moles use their broad paws and strong forelimbs to dig extensive underground burrows. (Rob and Ann Simpson/ Photo Agora)

for the mole. Tunnels run from just below ground level to 2.5 feet deep and may be 170 feet long. European moles eat worms, beetle and fly larvae, and slugs. They are active day and night, alternating every four hours between digging or eating and resting, and they live alone except when mating.

Golden moles inhabit grassy forests, riverbanks, mountains, and semidesert areas of Africa. The smallest species is Grant's golden mole, 2.5 inches long, and the largest species is the giant golden mole, 9 inches long. Their golden fur repels water, keeping them dry while digging. When burrowing, leathery pads cover their noses, keeping dirt out. Burrowing allows foraging for insects, larvae, earthworms, crickets, slugs, snails, and spiders. Golden moles cannot see well and find food by touch. Sometimes they come to the

surface (for mating, as an example). They live alone and defend their homes when other moles—even others of their own species—intrude.

Star-nosed moles, of eastern Canada and the northeastern United States, are black-furred, five inches long (excluding the tail) and weigh three ounces. They live near water, swim well, and dig in soil along shorelines. They burrow day and night, foraging for earthworms, aquatic insects, fish, and mollusks. Their nose tips hold a twenty-two-tentacle, touch-sensitive star which is their main sense organ. Star-nosed moles are solitary.

The Australian central desert holds the only species of marsupial mole. Their bodies range from 2.5 to 8.5 inches long, and they weigh from 0.3 to 6 ounces. Like other moles, they have short legs and front paws holding digging claws. A bare patch of skin atop each mole's nose pushes aside dirt as it digs. They have yellow fur, small nose slits, and no functioning eyes.

Marsupial moles burrow in soil near rivers and grasslands in Australian desert. While burrowing, they eat moths and beetle larvae. They do not dig permanent burrows but travel widely, sleeping in temporary burrows. They are solitary and their females have marsupiums, to carry young.

Life Cycles of Moles

Life cycles are best known for the star-nosed, European, and golden moles. Male and female star-nosed moles mate between February and April. The female builds a nest of leaves and moss in her burrow and has two to seven furless, blind, helpless young after a six-week gestation. They develop quickly, leave the nest in three weeks, and can mate after ten months.

Male European moles enter the tunnels of females to mate. After a five-week gestation, three to seven blind, hairless young are born. The young grow fur, open their eyes, and leave the nest in five weeks. European moles live for three years.

Male golden moles wishing to mate can attract mates at any time of the year. Litters contain two to three hairless young. They are two inches long, blind, and weigh less than one ounce. After nursing, when they weigh about 1.5 ounces, the young leave the nest for good.

Moles are considered to be pests by farmers and home gardeners, who think they eat plant roots and kill crops. However, moles never eat plants and perform a real service by killing grubs, caterpillars, and insects. Nonetheless, farmers, gardeners, and people with lawns consider them nuisances and poison them or set traps in their tunnels because they "spoil gardens and fields."

—Sanford S. Singer

See also: Home building; Mammals; Marsupials; Shrews; Vision.

Bibliography

Bailey, Jill. *Discovering Shrews, Moles, and Voles.* New York: Bookwright Press, 1989. This book examines mole characteristics, eating habits, and behavior.

Garcia, Eulalia. *Moles: Champion Excavators.* Milwaukee: Gareth Stevens, 1997. This brief, juvenile book describes the physical characteristics and habits of moles.

Gorman, Martyn L., and R. David Stone. *The Natural History of Moles.* Ithaca, N.Y.: Comstock, 1990. This book covers the natural history of moles.

Mellanby, Kenneth. *The Mole.* New York: Taplinger, 1973. This very interesting illustrated book covers many mole topics, including habitats, diet, behavior, and reproduction.

MOLLUSKS

Type of animal science: Classification
Fields of study: Anatomy, invertebrate biology, physiology

Mollusks, the phylum Mollusca, have soft, boneless bodies. Most also have shells. There are fifty thousand species of mollusks, including snails, oysters, and clams.

Principal Terms

FOOT: bottom portion of gastropods, on which they walk; it subdivides into tentacles in cephalopods

HERMAPHRODITE: an organism having both male and female reproductive systems

MANTLE: the outermost living tissue of mollusks; it makes shells, mother-of-pearl, and pearls

RADULA: a tonguelike, toothed organ used to grind food or drill holes in shells of prey

SESSILE: an organism incapable of moving from its point of origin

---

Mollusks first appeared in early Cambrian times, 650 million years ago. They are the second largest animal phylum, Mollusca (from Latin *mollis*, "soft"). The only larger animal phylum is Arthropoda. According to current estimates, there are approximately 49,000 mollusk species. All mollusks have soft, boneless bodies. Most have shells, though some shells are poorly developed and are even absent in some cases. Mollusk shells are like coats of armor.

Mollusks form seven classes. Best known are gastropods, including approximately 38,000 species of snails and slugs; bivalves or pelecypods, approximately 8,000 species of clams and scallops; cephalopods, approximately 600 species of squid and octopuses; and polyplacophora, approximately 600 species of chitons. The other classes contain under 1 percent of mollusk species and are not discussed.

Mollusks are present in all of earth's habitats.

However, most of them, comprising the greatest species diversity, are in the oceans. There, the largest number are various species of gastropods (snails). All gastropods are univalves, possessing one shell. Many other mollusks are the familiar pelecypod—bivalve—mollusks (clams), which have two shells. Mollusk bodies look undeveloped and many lack apparent heads. However, all have well-developed nervous, circulatory, respiratory, and sensory systems. Except for cephalopods, mollusks are slow-moving, sluggish, or immobile (sessile). In many cases they spend their adult lives attached to rocks or dug into sand or mud, awaiting the approach of prey.

Some mollusks, mostly gastropods, occur in freshwater and on land. For example, snails and slugs are seen at the bottoms of ponds, rivers, and lakes, or under fallen trees and decaying logs. Some land snails even live in tree branches. Characteristic features of gastropods are a true head, a creeping surface (or foot), eyes, and tactile feelers. The foot gives gastropods their name, which means "foot on belly" in Latin. Most gastropods have a univalve shell, but some species have no shell. Others, such as garden slugs, have tiny, internal shells.

Physical Characteristics of Mollusks

The largest mollusk is the giant squid, which is also the largest invertebrate. It can be fifty-five feet long and weigh several tons. Most mollusks, however, range from 0.5 to 10 inches in length and only weigh up to a few pounds. One exception to the size rule is the tropical giant clam, maximum diameter 4.5 feet, maximum weight five hundred pounds.

Mollusk Facts

Classification:

Kingdom: Animalia

Phylum: Mollusca (soft-bodied)

Classes: Bivalva (two-shelled, no distinct head); Cephalopoda (no shell or internal shell, very mobile, tentacled, has head and eyes); Gastropoda (univalve shell, moves slowly, tentacled, has head); Polyplacophora (simple crawling mollusks or chitons); Scaphopoda (elongated, open-ended tusk shells, tentacled, but no head)

Orders: Protobranchia, Septibranchia, Filibranchia (mussels and oysters), Eulamellibranchia (clams), Tetrabranchia (Nautilids), Dibranchia (octopuses, squid, cuttlefish)

Geographical location: Every ocean and many bodies of freshwater on every continent except Antarctica

Habitat: Mostly salt water, although many live in freshwater and on the land

Life span: Some live for under a year, others for one to five years; giant squid and giant clams may live for twenty-five years or longer

Special anatomy: Foot (for locomotion), siphons (for jet propulsion), a shell of calcium carbonate and protein, gills or lungs, a tonguelike radula for drilling or grinding food

Mollusk shells thus function protectively, like the rib cages and pelvic bones of vertebrates. The shells of bivalves are divided into two valves, which are opened or tightly closed by strong muscles. These muscles must be cut in order to open the shell for examination. The gastropods possess one asymmetrical shell, often of spiral shape, into which they retreat from the world. In cephalopods the shell is internalized, but still yields organ protection.

The organs and their positions in mollusks, especially gastropods, are as follows. At the anus end of the body are two or more gills (lungs in land forms) for breathing. At the front end of the body are jaws and a tonguelike, toothed radula. It grinds food or procures it by drilling into shells of prey or ship timbers (in shipworm snails).

There are many mollusk variations. However, some features can be generalized in mollusk bodies, based on gastropods and bivalves. In motile forms, locomotion is due to crawling on flat, muscular, snail-like feet, though jet propulsion occurs in cephalopods. Also, the mollusk body has a head at one end and an anus at the other. Furthermore, much of the body is covered by a shell (internalized in cephalopods) of calcium carbonate and protein. Shells are made by mantle tissue. Inside a shell is a large part of the body, fragile parts of cardiovascular, digestive and excretory, nervous, reproductive, and respiratory systems.

Most mollusks, such as this swimming scallop, are distinguished by their calcified, bivalve shells, which provide armorlike protection for their soft, boneless bodies. (Digital Stock)

Around the jaws are tentacles, used for sensation in gastropods. Cephalopod tentacles are used to capture prey. Between the radula and the anus are the stomach and the gut. A heart, near the hind end of the body, receives blood and sends it into the body cavity. The mollusk nervous system—comparable to that in fish—is composed of nerves that surround the gut, a brain, and sense organs. Mollusks also have kidneys and gonads.

Many mollusks are herbivores, grazing on underwater plants. Some terrestrial gastropods, such as slugs, eat cultivated plants and are serious garden and agricultural pests. Scaphopoda feed on organic matter deposited on ocean and lake bottoms. Bivalves filter protozoa, eggs of sea animals, and diatoms out of the water in which they live. Many gastropods are hunter carnivores, eating slower-moving or sessile animals, such as other mollusks. Cephalopods are very active predators and prey on animals such as crabs.

The Life Cycles and Senses of Mollusks
Most mollusks have separate sexes and a few exhibit courtship behavior. With or without courtship, mollusks usually spawn sperm and many thousands of eggs into waters around them. There, fertilization and development of offspring occur. When eggs hatch, offspring undergo a larval stage. Larvae of most mollusk species are, at first, free-swimming, using for locomotion a ring of cilia. Most settle to lake, ocean, or stream bottoms as crawlers who mature into adults.

Many mature mollusks, such as oysters and mussels, are sessile, permanently affixed to rocks. Gastropods are capable of slow crawling on a muscular foot. Cephalopods, such as squid and octopuses, are an exception to the mobility rule for mollusks. They are very mobile at all stages of their lives. Adult cephalopods move very quickly by expelling jets of water from mantle cavities, using an organ called a siphon. Jets are expelled in the direction opposite to movement. Jet propulsion makes cephalopods the fastest-moving mollusks. It also suits them to a life of vigorous predation. In cephalopods, the foot differentiates to ten arms (squid) or eight arms (octopuses) that seize prey.

In some cases, fertilization in mollusks is internal, with protective coverings secreted around eggs. In others, fertilization and development are internal in females and offspring are delivered alive, as with live-bearing snails. Slow-moving snails and related mollusks are often hermaphro-

---

Oysters, Beds, and Pearls

Edible oysters, such as *Ostrea virginica*, are plentiful on European and North American coasts. Oysters were a major food of coastal Native American peoples. Colonists ate them too, and natural oyster supplies were severely depleted by the late 1800's. Cultivated oyster beds were then developed to enhance nature's yield. In the twenty-first century, many food oysters come from cultivated beds.

The beds begin with the millions of eggs spawned yearly by every female oyster. They are started by finding places holding many newly hatched oyster larvae. Bricks, flowerpots, old shells, and other objects are placed on shallow sea bottoms beneath them. Then, a week after hatching, larvae produce tiny shells and, weighed down, drop to the sea bed. They attach to the objects and remain attached for the rest of their lives. After attachment, oyster-laden objects are dredged up and moved to the new beds.

Some oysters are invaded by foreign bodies that lodge between mantle and shell. These irritants cause the mollusks to produce mother-of-pearl to seal them off. In time, an irritant encased inside many layers of mother-of-pearl becomes a pearl. Most pearls are irregularly shaped and much less valuable than a few perfectly spherical pearls used in pearl necklaces. Pearl color depends on oyster diet and bed temperature. The pearls sought most are white, rose, steel blue, or black. They arise naturally in small numbers, or are cultured by the insertion of mother-of-pearl beads into oysters.

dites (both male and female). This doubles the number of mates available to each such organism, and often self-fertilization occurs. Both options have great survival value. A few species reproduce by parthenogenesis, without fertilization. Also, mother mollusks may protect developing eggs, and some oysters raise their offspring within the mantle cavity.

The mollusks live for periods ranging from under a year to two years for many small to medium-sized varieties and four to five years for some snails and shellfish. Giant squid and giant clams reportedly can live for over twenty-five years. Such life spans occur only when organisms reach old age, an uncommon situation for wild animals.

Sensory ability varies in mollusks. Most mollusks see poorly or not at all. However, the cephalopods have eyes like vertebrates, complete with lenses and retinas. Squid eyes reportedly lack eyelids but otherwise look very much like those of humans. Although it is not ubiquitous in mollusks, some gastropods have well-developed abilities to smell and find food at considerable distances. Predators are often detected in this way, too. Conclusive studies of intelligence have not been carried out in mollusks. However, ability to learn from experience has been claimed for cephalopods.

Beneficial and Destructive Mollusks

Mollusks are very abundant. On the benefit side, they are important foods for bottom-feeding fish and whales. Numerous mollusks are also valuable foods for humans. Especially widely eaten are clams, mussels, oysters, octopuses, and scallops. More exotic food mollusks include land snails. Mollusks (especially oysters) also produce pearls, the only gems made by living organisms.

On the debit side, gastropod and cephalopod carnivores do serious damage. Gastropods prey on slow-moving or sessile organisms, wreaking havoc on clam and oyster beds that provide shellfish for human consumption. Also, cephalopods are active predators, diminishing the sea's yield of meats sought by humans.

Another mollusk problem is foreign invasion, exemplified by zebra mussels (*Dreissna polymorpha*). These pistachio-nut-sized European saltwater mussel invaders were released into the Great Lakes from the hulls of ocean-going vessels in the late 1980's. They flourish in the Arkansas, Hudson, Mississippi, Ohio, and Saint Lawrence rivers. Superabundant reproduction (a female lays a million eggs per year) yields myriad larvae that, as adults, clog power plant and factory water intakes, requiring costly cleanup. More problematically, they eat microbes in the water, damaging the lifestyles of indigenous wildlife that also use this food. It is estimated that the mussels cost industry and consumers over $500 million per year.

—*Sanford S. Singer*

See also: Clams and oysters; Locomotion; Marine animals; Octopuses and squid; Reproduction; Shells; Snails; Tentacles.

Bibliography

Hughes, Roger N. *A Functional Biology of Marine Gastropods.* Baltimore: The Johns Hopkins University Press, 1986. This text is clear, comprehensive, and contains a lot of useful information on gastropods and their biology, plus a complete bibliography.

Morton, John E. *Molluscs.* 5th ed. London: Hutchinson, 1979. This clear, thorough, solid, and illustrated work on mollusks touches many bases.

Pechenik, Jan A. *Biology of the Invertebrates.* 4th ed. Boston: McGraw-Hill, 2000. A good portion of this book is devoted to mollusks. It includes a bibliography and useful illustrations.

Yonge, C. M., and T. E. Thompson. *Living Marine Molluscs.* London: Collins, 1976. This classic work on mollusks includes good illustrations and a bibliography.

MOLTING AND SHEDDING

Type of animal science: Development
Fields of study: Developmental biology, invertebrate biology, physiology

Molting or shedding is a hormonally regulated process during which animals lose and then replace either their entire body surface covering or structures associated with the body surface such as hair or feathers.

Principal Terms

ARTHROPODS: animals with jointed exoskeletons, including insects, crayfish, and spiders

EXOSKELETON: a skeleton made up of proteins and minerals, found on the outer surface of an animal

HORMONES: chemical messengers produced by specialized organs and cells either in the endocrine system or the nervous system; they have specific effects on different cells of the body

INVERTEBRATE: animal without an internal skeleton made up of individual bones or vertebrae

METAMORPHOSIS: a change in the physical state of an animal, such as the transformation of a tadpole into a frog or the development of a moth from a caterpillar

STEROID HORMONE: a hormone that is made from cholesterol

As a normal part of growth and development, some species of invertebrate and vertebrate animals undergo a process commonly called molting or shedding. Scientists term this process ecdysis, which is derived from a Greek word meaning "to escape or slip out." It is common for animals to molt more frequently as larvae or juveniles and less often as they mature and become adults. Molting is also an integral part of metamorphosis. Immediately after a molt, animals may not be fully protected from the environment, and they may be more vulnerable to predators. For example, the new exoskeleton of arthropods may not be fully hardened, or some birds may lose flight feathers. During a molt, animals may shed and replace their entire body covering or structures associated with the body surface. Molting may involve the replacement of the skin, an exoskeleton, or a cuticle in its entirety. In temperate regions, it is not unusual in the summertime to see the old exoskeleton of a cicada still clinging to a tree. Feathers, fur, or hair, which are derived from skin cells, are often shed in a cyclic fashion rather than all at once. Animals may also lose body structures during a molt. Adult male deer shed their antlers at the end of the mating season and lemmings replace their claws. In most animals the interval between molts is called the intermolt, except insects, where it is referred to as an instar.

The Molting Process
Environmental factors such as stress, temperature, and light cycle can serve as stimuli to molting in different species of animals, but the actual sequence of events leading up to shedding and replacement is strongly influenced by hormones and by specific interactions between the nervous system and the endocrine system. Hormone production is the primary function of the endocrine system. However, regions of the brain also produce hormones, which in this case are called neurohormones. One class of hormones in particular, steroid hormones, is actively involved in the regulation of molting.

Before an animal molts, a new skin or exoskeleton will have formed beneath the old one. Often

Life Cycles and Molting

Animals in the phylum Arthropoda and class Insecta undergo many molts and possibly several metamorphoses as a normal part of their life cycle. By responding to appropriate environmental stimuli, the arthropod brain controls the processes of molting and metamorphosis. It does so by regulating the release of specific hormones. The amount of one hormone, commonly referred to as ecdysone, controls molting and growth. This hormone also stimulates the digestion and reabsorption of the old exoskeleton. Another, called juvenile hormone, controls metamorphosis. Some insects such as grasshoppers have a nymph stage where the juvenile undergoes considerable growth in body size as the nymph makes the transition to an adult. When an exoskeleton is shed, the new one will not immediately harden. This allows time for the skeleton to expand to the new body size. Greater than 80 percent of insects including moths and butterflies have a caterpillar or larval stage during which active growth occurs. The overlying cuticle, the exoskeleton, acts like a jacket to confine the caterpillar to a maximum size and therefore limit growth at each stage of development. Thus, the exoskeleton needs to be periodically shed. Animals often swallow air as one way to rupture the old skeleton and to resize the new one. After several molts in the caterpillar stage, the amount of juvenile hormone is decreased and the caterpillar transforms to a pupa. Finally, in the absence of juvenile hormone, a final molt will occur and the pupa undergoes metamorphosis into an adult. Once they become adults, insects no longer molt. However, the adults of other arthropods, such as crayfish and lobsters, continue to have periodic molts their entire life.

the old exoskeleton splits along the midline of the posterior surface and the animal then crawls out. Different animals have specific ways to stretch the new exoskeleton before it hardens. The best way to stretch the skeleton is by increasing body size. This might be accomplished by a growth stage or by artificially increasing size. A lobster, for example, might absorb enough water to increase body size by 20 percent, while other animals take up air.

Molting in Amphibians and Reptiles

Molting of an entire body surface such as the skin or exoskeleton requires the separation of the layer being shed from the underlying tissue that will become the new outer surface. The chemical and physical processes leading up to the separation will vary slightly between animal groups. Amphibians such as frogs, toads, and salamanders periodically shed the outermost surface layer of their skin. Prior to molting, mucus is secreted beneath this layer of cells. Since the mucus is secreted between the old and new surface layer, it may assist in the separation by creating a space between them. The mucus may also act as a lubricant and aid the animal in removal of the old skin. Frogs are also known to bloat themselves and increase movement to break out of the old skin. Animals will often eat the skin they have shed as a way to recycle nutrients.

Reptiles show a good deal of variability in the replacement of their scales. In fact, snakes may shed their skin several times a year. In the time immediately preceding ecdysis in snakes, a fragile zone develops. Old skin will now separate from the underlying, newly developed scales. During the molt the snake's eyes will appear cloudy, because the outer part of the eye, which is a part of the skin, is also detaching. The rattlesnake's rattle is formed from the part of the skin that remains attached to the tail at the end of each shedding cycle. Snakes shed their skin in one piece, but lizards do not. Turtles are different from both snakes and lizards. First, their scales do not overlap to form sheets and second, most turtles add new growth to existing scales and do not molt. The molting pattern in crocodiles and alligators, which shed and replace individual scales, more closely reflects events in birds than other reptiles.

Molting in Birds

Birds shed their feathers each year and it is not unusual for some species to molt several times during the year. During the molting season, a bird's behavior may be affected in several ways, including events associated with reproduction. Molting will be influenced by the time of year and the mating season. Many female chickens often stop laying eggs when they are molting. New feathers will develop in the same area of the skin as the old ones. As the new feathers grow from the follicle they may push the old feathers out of the skin or they may be pulled out. The primary purpose of molting is to replace worn feathers with new ones, and often a pattern of feather loss will be noted. Wing feathers closest to the body are lost first and the molt progresses outward along the wings.

It is common for male birds to molt so that they can replace their duller plumage with more colorful feathers associated with attracting a mate. This is called a prenuptial molt and generally occurs in late winter or early spring. A postnuptial molt is common in both males and females. The chicks will lose their original feathers (down) when they undergo a postnatal molt to juvenile feathers. If it is a species of bird where the males have a coloration pattern different from the females, the early juvenile coloration more closely resembles the female. As the juveniles mature they will undergo successive molts to adult plumage. The time frame for maturation can vary considerably. Eagles can take up to five years before a final molt into full adult feathers. Mature birds that depend upon flight as a method of escaping predation will not molt all of their feathers at one time. A complete molt would result in the loss of flight feathers on the wings as well as tail feathers which assist in stability and guidance during flight. Flight is less adversely affected if there is a symmetrical loss of feathers on the wings and if tail feathers are

This grasshopper has just emerged from a molted shell. (Kevin and Bethany Shank/Photo Agora)

shed in groups. Birds such as ducks and geese, which are able to spend extended periods of time in the water, are able to avoid most predators by swimming or hiding in tall grass. These birds do molt all of their flight feathers at one time. Male ducks are also somewhat unusual because, during the summer months, they molt from mating colors to plumage similar in color to females. This change is called an eclipse and makes it more difficult to tell males from females.

Molting in Mammals

When they lose hair, mammals, including humans, are molting. Like feathers, a hair grows outward from a follicle in the skin and, as new hairs grow, old hairs will be lost. Under normal conditions in humans, hair loss will be a gradual process over an individual's lifetime and it does not occur all at once. Molting in many mammals is directly influenced by length of day and interactions between the endocrine system and the nervous system. The number of molts per year varies, and many mammals molt twice a year, once in the spring and once in the fall. Foxes, however, molt once a year in the summer, and the snowshoe hare molts three times: summer, autumn, and winter. In general, it is common for the summer coat of mammals to be thinner than the winter one, and it is not unusual for the two coats to be different colors. Changing hair color provides one method of camouflage. White fur lacks pigment and blends well with snow and light surroundings. In contrast, darker shades blend better in the summer and fall. In mammals, like other animals, the transition from juvenile into adult can be associated with a molt. Deer fawns, for example, have very pronounced white spots, but when they become further developed they molt into the solid coloration of adults.

—*Robert W. Yost*

See also: Amphibians; Arthropods; Birds; Crustaceans; Feathers; Flight; Fur and hair; Hormones and behavior; Hormones in mammals; Insects; Reptiles; Shells; Skin; Snakes.

Bibliography

Gilbert, Scott F. *Developmental Biology*. 6th ed. Sunderland, Mass.: Sinauer Associates, 2000. An excellent text for more advanced students. Contains a full chapter on metamorphosis, regeneration, and aging.

Hickman, Cleveland P., Larry S. Roberts, and Allan Larson. *Biology of Animals*. 7th ed. New York: McGraw-Hill. 1998. A college freshman text with sections on ecdysis.

Lindsey, Donald. *Vertebrate Biology*. New York: McGraw-Hill, 2001. A comprehensive upper-level text with sections on molting. This text may be better for someone with a prior knowledge of biology.

Miller, Stephen A., and John B. Harley. *Zoology*. 4th ed. New York: McGraw-Hill, 1999. A general-audience textbook on animals, with sections on ecdysis.

Purves, William K., David Sadava, Gordon H. Orians, and H. Craig Heller. *Life: The Science of Biology*. 6th ed. New York: W. H. Freeman, 2000. College freshman biology book with a full chapter devoted to animals that molt.

MONKEYS

Type of animal science: Classification
Fields of study: Anatomy, wildlife ecology, zoology

Monkeys are a favorite in zoos around the world. Representing more than 140 species spread throughout the tropics, the term "monkey" is often misused as a synonym for any primate.

Principal Terms

ARBOREAL: living completely or primarily in the trees
OPPOSABLE: a thumb which can be turned so that its pad makes contact with the pad of each of the fingers (as in the human)
PREHENSILE: capable of grasping
QUADRUPEDAL: walking on all four feet
SEXUAL DIMORPHISM: the occurrence of anatomic or physiologic differences that distinguish males from females of a particular species

---

The term "monkey" is used to denote any higher primate (suborder Anthropoidea) that is not an ape. Thus, it includes both members of the New World monkeys (infraorder Platyrrhini) as well as the Old World monkeys (infraorder Catarrhini, superfamily Cercopithecoidea). Monkeys have little in common with each other except for the fact that most are quadrupedal, but this does not eliminate all other primates. It is unclear where the name "monkey" originated, although a common interpretation is that it relates to the medieval term "moneke," meaning manikin.

Old World (Catarrhine) Monkeys

The Old World monkeys are the largest and most diverse family of primates, covering about ninety-five species and ranging over most of Africa, Asia, and Indonesia. The name Catarrhine means "downward-nosed," referring to the fact that the nostrils are close together and point forward and down. Catarrhine monkeys include macaques, mangabeys, baboons, mandrills, velvet monkeys, guenons, colobuses, proboscis monkeys, and langurs. There are two subfamilies: the leaf-eating, arboreal Colobinae (examples include the colobus and the langur), and the omnivorous, often ground-dwelling Cercopithecinae (including the baboons, mandrills, macaques, and guenons). The Colobinae have a rather complex stomach and digestive system, whereas the Cercopithecinae have a simple stomach combined with cheek pouches in which food can be stored.

The macaques are the greatest in number among the Old World species, as well as the most widespread. The most northerly is the Japanese macaque, which can live in cold, snowy climates. Other macaques live in dry, almost desertlike conditions in the tropics.

Old World species are generally larger than New World species, and there is considerable sexual dimorphism. Most have bare buttock pads, which may be brightly colored. Their tails are seldom fully prehensile, and may be significantly reduced in size. Almost all are active during the day, with excellent vision, hearing, and sense of smell. They communicate almost entirely by sight and sound, displaying a wide range of calls. Many display a range of facial expressions, used for communication with their own species as well as with other species nearby. Most are fully arboreal, but baboons are ground feeders, and macaques live both on the ground and in the trees.

When more than one species of monkey dwells in the same locality, the various species generally occupy different vegetation levels in order to avoid competition. This behavior is known as arboreal stratification. Most authors recognize four layers of vegetation in the tropics: the ground layer, lower canopy, middle canopy, and upper canopy. For instance, in the African guenons (*Cercopithecus* spp.), DeBrazza's monkey lives at the ground level, the red-tailed monkey sleeps in the middle canopy but spends the day on the ground, the blue guenon lives in the upper canopy but forages in the middle, and the Diana monkey lives solely in the upper canopy.

New World (Platyrrhine) Monkeys

The New World monkeys are a highly successful and diversified group colonizing Central and South America. The term usually refers to the infraorder Platyrrhini, meaning "flat-nosed." As compared with the Catarrhine monkeys, the nostrils of the Platyrrhines are broadly separated and usually point to the sides. Members of the Platyrrhines include capuchins, howler monkeys,

sakis, woolly monkeys, squirrel monkeys, and uakaris, a total of about forty-five species.

New World monkeys have long, thin fingers on each hand, with flattened or curved nails. Although their thumbs are not opposable, as they are in the human, the big toe can be opposed against the other toes for gripping branches tightly. New World monkeys are excellent runners and jumpers, swinging and leaping through their densely wooded habitats. Their tails are fully prehensile; they can grasp objects at the tip and curl around a branch and support the full body weight of the animal. In almost all cases, the tail is at least as long as the head and body, and it acts as a balancing organ, often being held in a curled pattern.

None of the New World monkeys are ground dwellers, unlike the baboons and other Old World monkeys. None of them have cheek pouches, and sexual dimorphism is rarely seen. New World monkeys are gregarious and live in family-based

Most monkeys live in social groups that range over a defined territory in search of food. (PhotoDisc)

Monkey Facts

Classification:
Kingdom: Animalia
Phylum: Chordata
Subphylum: Vertebrata
Class: Mammalia
Order: Primate
Suborder: Anthropoidea
Families: Cercopithecidae (Old World monkeys, eight genera, forty-five species); Cebidae (New World, capuchin-like monkeys, eleven genera, and thirty species)
Geographical location: Africa and Asia (Catarrhines), Central and South America (Platyrrhines)
Habitat: Mostly forests, some grasslands
Gestational period: Old World monkeys, 5 to 6 months; New World monkeys, 4 to 7.5 months
Life span: Old World monkeys, twenty to thirty-one years; New World monkeys, twelve to twenty-five years
Special anatomy: Opposable thumbs, forward-facing eyes for binocular vision, large brain case

groups with much vocal and visual communication. They have highly developed olfactory organs that may also be used for communication. Males of many species contain a glandular patch on the sternum (breastbone) which they rub against tree branches to act as scent markers. Marking by means of urine and feces is also common. For instance, night monkeys coat their hands and feet with urine so that they leave a tell-tale scent wherever they go.

Families are well developed in most species of monkeys, although females do most of the caring for their offspring. Mothers usually carry their young on their backs until they are ready to move through the canopy on their own. Group size seems to depend primarily on the productivity and abundance of the foods typically eaten by the species. Species that live in small groups tend to feed on small, scattered, or scarce resources such as insects, small vine fruit, or newly emerged leaves of bamboo. Species that form large groups use abundant or clumped resources, such as fruits on large fig trees. Small family groups are typically one to three animals, while large groups may involve seven to twenty members.

—*Kerry L. Cheesman*

See also: Apes to hominids; Baboons; Cannibalism; Chimpanzees; Communication; Communities; Evolution: Animal life; Evolution: Historical perspective; Fauna: Africa; Gorillas; Groups; Hominids; *Homo sapiens* and human diversification; Human evolution analysis; Infanticide; Learning; Lemurs; Mammalian social systems; Orangutans; Primates.

Bibliography

Boitani, Luigi, and Stefania Bertoli. *Simon and Schuster's Guide to Mammals.* Translated by Simon Pleasaunce, edited by Sydney Anderson. New York: Simon and Schuster, 1983. Compact guide to individual species, using color symbols to show world location and habitat type.

Macdonald, David, ed. *The Encyclopedia of Mammals.* New York: Facts on File, 1984. Short factual paragraphs about each species, with general information about each family and order.

Nowak, Ronald. *Walker's Mammals of the World.* 2 vols. 6th ed. Baltimore: The Johns Hopkins University Press, 1999. A classic two-volume set, supplies details for each species, with notations about habitat, location, and special characteristics.

Voelker, William. *The Natural History of Living Mammals.* Medford, N.J.: Plexus, 1986. A good description of each family including natural history, ecology, and distinctive features.

MONOTREMES

Types of animal science: Reproduction, development
Fields of study: Biochemistry, cell biology, developmental biology, embryology, physiology, reproductive science

Monotremes represent an ancient developmental line of mammals. They retain many reptilian characteristics, including the laying of eggs. Monotremes are found primarily in Australia.

Principal Terms

BROOD POUCH: a temporary external pouch created by folding the skin of the abdomen together; used to carry young as they continue to develop

CLOACA: a bodily opening at the end of the gut into which both the waste disposal and reproductive systems open

ECHIDNA: a long-snouted, insect eating, egg-laying mammal; also known as the spiny anteater

MONOTREME: reptilelike mammals, distinguished from other mammals by the fact that they lay eggs and have a cloaca

PLATYPUS: web-footed, duck-billed, semi-aquatic, egg-laying mammal

---

The monotreme order of mammals is comprised of two families, Ornithorhinchidae and Tachyglossidae. The first family contains just one genus and species, that of the duck-billed platypus, while the second contains two genuses and two living species of echidnas (short-nosed and long-nosed). These animals have much of the conventional appearance of mammals, but they are also quite different in some respects from the rest of the mammals. Both the platypus and the echidnas lay eggs which are incubated and hatched outside the body of the mother.

Monotreme Anatomy

The monotremes are more closely related in evolution to the reptiles than are any other recent mammals. They are not the ancestors of the marsupials or the placental mammals, but rather they represent a distinct line of mammalian evolution. They possess certain mammalian features, such as hair and mammary glands, and they are warm-blooded. On the abdomens of the females, milk oozes from paired areas of tubular glands (they possess no teats for suckling) and is lapped up by the young. They have a four-chambered heart and skeletal features associated with mammals. Other features, such as the presence of a cloaca, along with features of the vertebrae and ribs, are very reptilian.

In male monotremes, the penis is attached to the ventral wall of the cloaca, and is divided at the tip into paired canals used only for the passage of sperm. In female monotremes, the oviducts, which carry the eggs from the ovary, open separately into the cloaca. In the oviducts, the eggs are fertilized and then covered with albumen (a protein like that found in chicken egg whites) and a flexible, sticky, leatherlike shell. This reptilian feature of a shell-covered egg is found in no other order of mammals.

Both living animals and fossil records indicate a range for the monotremes that includes only Australia, Tasmania, and New Guinea.

Duck-Billed Platypus

W. H. Caldwell, in 1884, first demonstrated that platypuses lay eggs, unlike other mammals. There are still many unknowns regarding platypus reproduction and development, especially in terms of biochemistry and endocrinology. Copulation generally takes place during the spring months of

August and September, and usually occurs in water. The female then withdraws to a rather complicated nesting burrow dug into the side of a streambed. Inside the burrow is a nesting chamber consisting of a bed of grass, reeds, and leaves. Generally, two sticky eggs are laid here. There is no pouch for the eggs to incubate in, and the exact mechanism of incubation remains unclear. It is likely that the female curls around the eggs and protects them with her body heat. Incubation is seven to ten days, after which the hatchlings use an egg tooth (much like reptiles) to emerge from the egg. At birth the hatchlings are about 17 millimeters (0.65 inches) long. They feed on ill-defined milk patches hidden under dense fur on the ventral side of the mother. Hatchlings have very large, strong forelimbs that are used for holding on to the fur over the milk patches. Because they are so well protected in the dense fur, hatchlings are nearly impossible to detect by looking at the mother.

As the young grow, the mother's mammary glands also grow. These mammary glands are very similar to those seen in other mammals, and the chemical composition of the milk is also similar. Until they emerge from the burrow, hatchlings feed on only the milk from their mother. Young platypuses emerge from the burrow for the first time in December or January, generally being thirty to thirty-five centimeters (twelve to fourteen inches) long. They will eventually grow to forty to forty-five centimeters as adult females, or fifty to fifty-five centimeters as adult males.

Echidnas

Previously known as spiny anteaters, echidnas have compact, rounded bodies covered with short, thick spines. They have elongated, slender snouts and strong limbs, and are powerful diggers. They have no teeth and, like other ant- and termite-eating mammals, have a long sticky tongue that reaches well beyond the snout. Termites, ants, and other small arthropods are swept into the mouth by its action.

Adult echidnas are solitary animals, except when it is time to mate. During the breeding season (late June to September), males will follow females for several weeks at a time before being allowed to mate. Often several males will be seen in single file following a female, this being referred to as an echidna train. It is not known exactly how males find breeding females, although it is likely that females secrete a scent that can be detected several miles away.

Females develop a brood pouch during breeding season. Following mating, a single fertilized egg is transferred from the cloaca to the pouch, although the actual mechanism for this transfer is not clear. The leathery-shelled eggs average 14-17 millimeters (0.55-0.65 inches) in length. The egg is incubated for ten days in the pouch before the hatchling appears. The young echidna (who is about 1.2-1.3 centimeters, or 0.5 inches, in length when it is born) remains in the brood pouch for an additional eight weeks until its spines develop. During this time it feeds on milk flowing from the mammary ducts onto tufts of hair within the pouch.

When the youngster emerges from the pouch, it is hidden by its mother in a protected spot. It appears that the young echidna then begins a short period of hibernation, during which continued development occurs. Echidnas are reproductively mature at one year of age.

—*Kerry L. Cheesman*

See also: Beaks and bills; Fauna: Australia; Lactation; Mammals; Reproduction; Reproductive strategies; Reproductive system of female mammals; Reproductive system of male mammals; Reptiles.

Bibliography

Augee, M. L., ed. *Monotreme Biology.* Mosman, Australia: Royal Zoological Society of New South Wales, 1978. The results of a research symposium on monotreme biology, this volume contains a series of short reports on various aspects of reproduction. Especially good is the review by Carrick and Hughes.

Grant, Tom. *The Platypus.* Rev. ed. Sydney, Australia: University of New South Wales Press, 1995. An easy-to-read guide to the biology, behavior, and life cycle of the platypus. Many good illustrations.

Rismiller, Peggy, and Roger Seymour. "The Echidna." *Scientific American*, February, 1991, 96-103. Well written for the lay reader, the article explores the biology and habits of the echidna in some detail.

Strahan, Ronald. *Mammals of Australia.* Washington, D.C.: Smithsonian Institution Press, 1995. An exhaustive, well-researched and written sourcebook.

Whitfield, Philip, ed. *The Simon and Schuster Encyclopedia of Animals.* New York: Simon & Schuster, 1998. Easy to read, but not a lot of information on any one species. Good illustrations, especially for children.

MOOSE

Types of animal science: Anatomy, classification, reproduction
Fields of study: Anatomy, systematics (taxonomy), zoology

Moose, the largest species of deer, are herbivorous ruminants. Once hunted almost to extinction, they are now a protected species.

Principal Terms

ARTIODACTYLS: hoofed mammals, including cattle, pigs, goats, giraffes, deer, and antelope
GESTATION: the term of pregnancy
HERBIVORE: an animal that eats only plants
KERATIN: a tough, fibrous protein which is a major component of hair, nails, hooves, and the outer covering of true horns
RUMINANT: a herbivore that chews and swallows plants, which enter its stomach for partial digestion, are regurgitated, chewed again, and reenter the stomach for more digestion

---

Moose are artiodactyls of the family Cervidae (deer). Almost all artiodactyls walk on two toes. Their ancestors had five, but evolution removed the first toe, and the second and fifth toes are vestigial. Each support toe—the third and fourth toes—ends in a hoof. Artiodactyls are herbivores and many, especially the Cervidae, are ruminants. This means that they chew and swallow vegetation, which enters the stomach for partial digestion, is regurgitated, chewed again, and reenters the stomach for additional digestion. This maximizes nutrient uptake from food.

Moose are the largest of deer. Like all cervids, male moose have solid, bony, branched antlers, which are shed and regrown each year. Moose live in northern Europe, eastern Siberia, Mongolia, Manchuria, Canada, Alaska, Wyoming, and the northeastern United States.

Physical Characteristics of Moose

Moose—called elk in Europe—have long, solid bodies. Males can be eight feet tall at the shoulder, ten feet long from muzzle tip to rump, and weigh up to 1,800 pounds. Females are 25 percent smaller overall. Moose heads, antlered in males, are angular, long, and have large muzzles. Their large eyes, as in other deer, are set back on their faces and on the sides of their heads. This optimizes their vision. Moose also have thick necks, humped shoulders, and long, strong legs that allow running speeds of up to forty miles per hour. Moose

---

Moose Facts

Classification:
Kingdom: Animalia
Subkingdom: Bilateria
Phylum: Chordata
Subphylum: Vertebrata
Class: Mammalia
Order: Artiodactyla
Family: Cervidae (deer)
Genus and species: Alces alces
Geographical location: Northern Europe, eastern Siberia, Mongolia, Manchuria, Canada, Alaska, Wyoming, and northeastern United States
Habitat: Mountains, forests, and near lakes or marshes
Gestational period: Seven to eight months
Life span: Fourteen years, but some live up to twenty years
Special anatomy: Antlers

Moose have palmate antlers, shaped like a fingered hand, rather than the spiky antlers of most other cervids. (Digital Stock)

fur is brownish-black, fading to grayish-brown on the belly.

Moose antlers are large and palmate, holding up to twenty branched points, which look like fingers on a hand. Antlers are important during mating, when a male lacking them would lose fights for mates and a chance for offspring. Antlers differ from true horns, which are pointed, permanent, bony structures often seen on heads of male ruminants. Horns have bone cores which are extensions of the skull's frontal bone. Atop the core is a skin layer, rich in a tough, fibrous protein, keratin. The high keratin content in this skin makes it tough and durable. Horns range from straight spikes to elaborately curved varieties. Animals having horns keep them for life Antlers are horns that are shed yearly and regrown. Like true horns, antlers grow out of skull bones. They arise from permanent frontal bone struc-

tures called pedicles. At first antlers have soft, velvetlike skin coverings. Instead of hardening, as in horns, the covering dies off and is rubbed away by the animal. Antlers function like horns, serving for self-protection and allowing males to develop ascendancy.

Moose are often found on mountains, in forests, and near lakes and marshes. They are herbivores lacking upper incisor and canine teeth, but having pads in their upper jaws to help lower teeth grind vegetable food. Adults eat up to fifty pounds of vegetation per day, including water plants, branches, twigs, and leaves from aspen, willow and birch trees, berries, and bark. A moose often stands up against young trees and bends them over to reach leaves at their tops. In winter, moose dig through snow to find grass, twigs, and other vegetation. Moose eat during the daylight hours and at night.

The Life Cycle of Moose

Male moose are solitary animals during spring and summer. In September and October, they fight for mates. A successful male may lead several females and babies all winter. In the spring, the male leaves his harem, returning to a solitary life. Moose gestation, seven to eight months, yields one or two offspring which nurse for four months and stay with the mother until she is ready to give birth again. A moose can mate when it is two years old. The life spans of moose can be up to twenty years.

Moose are so large and strong that they have few enemies. Bears and wolves may prey on them. However, they usually attack young, old, or sick moose and males weakened by hunger or mating battles. Human hunters are the greatest threat to moose. For example, during the moose mating season, male moose seek females. Moose hunters lure moose into shooting range by imitating the love call of a cow through a horn or cupped hands. At one time, hunters killed off almost all moose in the eastern United States. Currently, they are protected by law in the United States and Canada.

—Sanford S. Singer

See also: Antelope; Deer; Elk; Fauna: Asia; Fauna; Europe; Fauna; North America; Horns and antlers; Molting and shedding; Reindeer.

Bibliography

Franzmann, Albert, and Charles Schwartz, eds. *Ecology and Management of the North American Moose.* Washington, D.C.: Smithsonian Institution Press, 1998. Comprehensive coverage of the moose in North America from the Wildlife Management Institution, covering evolution, taxonomy, and all aspects of life and behavior. Photographs, maps, and graphs.

Geist, Valerius. *Moose: Behavior, Ecology, Conservation.* Stillwater, Minn.: Voyageur Press, 1999. The book contains interesting information on moose, illustrations, and bibliographical references.

Peek, James M., Craig Brandt, and Robin Brandt. *Moose.* Edison, N.J.: Chartwell Books, 1998. This book contains a lot of information on moose.

Silliker, Bill, Jr. *Moose: Giants of the Northern Forest.* Willowdale, Ontario: Firefly Books, 1998. This book describes the moose, its habits, range, and life cycle.

Van Wormer, Joe. *The World of the Moose.* Philadelphia: J. B. Lippincott, 1972. A solid book on moose, with illustrations and an extensive bibliography.

MORPHOGENESIS

Type of animal science: Development
Fields of study: Cell biology, embryology, genetics

Morphogenesis is the process, operating during the development of the organism and appearing through evolutionary history, by which patterns of form come into being in genetically controlled systems. By studying this multifaceted process, it is possible to understand underlying factors that regulate the appearance of patterns of structure in living organisms.

Principal Terms

BIFURCATION: the division of a Y-shaped and connected mesenchymal structure into a single proximal chondrogenic focus and two distal chondrogenic foci; this can lead to the formation of separate chondrification centers in a developing digit

CHONDRIFICATION: the process by which undifferentiated connective cells transform into chondrocytes (cells that make cartilage) and begin forming extracellular matrix

EPITHELIUM: the tissue that covers and lines all exposed surfaces of an organism, including internal body cavities such as the viscera and blood vessels

LIMB BUD: thickened epithelial cells along the lateral body fold that are underlain by mesoderm, creating a paddle-shaped extension from the trunk

SEGMENTATION: the division of a structure into linearly arranged segments; it can lead to the formation of somites, or it can lead to the formation of separate chondrification centers in a developing digit

ZONE OF POLARIZING ACTIVITY (ZPA): a region at the posterior base of the limb bud that seems to influence the distal development of pattern in a developing limb

Morphogenesis is the process that leads to the appearance of form in specific patterns, through the differential reproduction, growth, and movement of cells and tissues and interactions between tissues. Such movements are controlled by a variety of factors, including cell adhesion molecules (CAMs), the nature of the extracellular matrix (ECM), and the size of the cells themselves.

Most multicellular organisms begin their lives as fertilized eggs. Although this seemingly simple cell is really quite highly organized, it is still relatively uncomplicated when compared to the structural complexity of the parent. As each egg cell divides, the daughter cells become structurally and functionally distinct. This is the beginning of the process of cell differentiation, which can lead to complex multicellular organisms. As new cells appear, grow, mature, and divide, there is a point at which the eventual fate of these cells is determined. That is, at some point the eventual location, structure, and function of the descendants of such cells are fixed. This fate determination frequently takes place very early in the development of an organism. After a particular fate is determined, it is the process of morphogenesis that allows the potential fate to be realized. Also through the process of morphogenesis, patterns of form develop that increase the structural and functional complexity of an organism.

The Rules of Morphogenesis

Fate determination, morphogenesis, and pattern formation are not haphazard. In fact, these events are highly constrained, extremely stereotyped, and very predictable—so much so that models of morphogenesis have been formulated that rely on rules of development to establish the probabilities of particular forms developing. These rules take into account two types of conditions—initial and boundary conditions—that work together to produce pattern formation in a cascading process. Initial conditions are simply the cellular conditions that prevail in a certain part of an organism, such as cell size and number and the chemical makeup of cell membranes. Boundary conditions are the conditions that exist at boundaries of different types of tissue.

Only a few phenomena are responsible for setting the initial and boundary conditions within which a morphogenetic system operates. A variety of cellular phenomena are involved, including cell migration, cell division (producing new cell lineages), the rate of cell division, cell number, cell density, the orientation of daughter cells relative to parent cells, and even the places in which cell death occurs. Tissue-level phenomena also have an important impact on morphogenesis. In these cases, specific tissues possess certain properties and a limited ability to respond differently within the constraints allowed by those properties. For example, sheets of epithelial cells can become folded, but they seldom form a solid mass of cells. Finally, there are interactions that take place between different tissues. Among these interactions is induction, wherein one tissue will induce a specific response in an adjacent tissue. Examples of such induction are numerous and well known; they include the mesenchymal induction of dental epithelium during the formation of teeth.

Limb Morphogenesis

Although far from completely understood, one of the most thoroughly studied morphogenetic systems is the development of the tetrapod limb. (A tetrapod is a vertebrate with two pairs of limbs.) Because amphibians (such as frogs and salamanders) and chickens have large, easily obtainable eggs, they have been studied most. It will be useful to summarize the course of limb development, since it exemplifies many of the common features of morphogenesis and pattern formation.

Early in the formation of the limbs, a ridge forms along the flank of the embryo. Along this ridge, the apical ectodermal ridge (AER) develops as a thickened layer of epithelial cells. Undifferentiated connective tissue cells, called mesenchyme cells, accumulate underneath the AER and form the limb bud. These mesenchymal cells are derived from mesoderm that migrates into the limb bud following pathways of the ECM that are laced with CAMs, especially fibronectin. The AER is responsible for the formation of pattern in the proximodistal axis, which is a line from the base of a limb (the proximal end) to the outer end of the limb (the distal end).

Another focus of pattern formation also appears quite early in limb development; it is called the zone of polarizing activity (ZPA) and has influence over the anteroposterior axis of limb development. Together, the AER and the ZPA regulate a pattern of form that is shared by all tetrapods.

Under the influence of these pattern-generating centers, three types of mesenchymal chondrogenic foci will form: de novo condensations (which are unconnected), bifurcations, and segmentations. Segmentations and bifurcations both show recognizable connections, and they appear in a consistent and stereotyped fashion. Linear series of mesenchymal condensations showing connections with one another are called segmentations; a Y-shaped condensation with a single proximal focus connected to two distal foci is called a bifurcation.

In all tetrapods, a bifurcation leads to the formation of two skeletal elements in the region between the proximal bone in the arm (or leg) and the hand (or foot). In frogs and amniotes, the condensations at the base of the putative fourth digit, showing connection to the bone in the forearm (or lower leg), are the first to appear as the result of a bifurcation event. Each subsequent proximal digit condensation appears as a bifurcation. All distal condensations, after the proximal digital element forms, appear as the result of segmentation events. Thus, the development of limb elements is asymmetric and always emanates from the axis of the first digit to form chondrogenic foci.

The morphogenetic control of this pattern, which has been conserved throughout the long evolutionary history of tetrapods, is both simple and complex. It is simple in that only a few morphogenetic processes are responsible for generating the complex pattern of limb structure, so different in all the various tetrapods. It is complex in that so many phenomena can influence the ultimate structure of any particular limb. For example, the number of somites that contribute mesoderm to the limb bud mesenchyme can dictate the number of limb elements that will eventually form.

Studying Morphogenesis

There are two general methods for studying morphogenesis. In the older—but still widely used—approach, tissues are removed from an organism in order to be examined under a microscope. Tissues are often stained (a variety of staining materials may be used) to make them more readily visible, and they are usually cut into thin slices called sections. The purpose is to determine the position of individual tissues. If carefully staged materials have been used, it is possible to get some ideas about whether cells have moved and, if so, approximately how far they have moved. There are problems with studying morphogenesis in this way: Details can be missed, or, more worrisome, preconceived notions can bias observations. For example, a small set of sections may support a theory, but, because thin sections of tissue can vary greatly, many sections of the material under study may contravene a pet theory of development. One could simply reject the many conflicting sections, claiming them to be poorly prepared or somehow damaged material. Indeed, many sections are so rejected. With the advent of transmission electron microscopy (TEM) and scanning electron microscopy (SEM), more detailed observations could be made. New problems, however, accompany these methods. In TEM, the thinner sections which must be used increase the occur-

rence of variation, also increasing the chances of biased observations. SEM allows for the magnification of surfaces and a great increase in the depth of field, but it requires that materials be very carefully staged. In none of these methods is it possible actually to see individual cells move.

A second approach seeks to document cellular movements. Understanding such movements requires that individual cells be "marked." Marked cells are then grafted onto an unmarked host organism and followed. Both natural and artificial markers can be used. Pigment granules such as melanin and stained glycogen are examples of natural markers. These markers, however, can be affected by cellular activities, making it hard to follow the cells. Fortunately, there are other natural markers not so affected. For example, chick and quail cells stain differently, so the cells of one can be followed when they are transplanted into the other. One of the more frequently used artificial markers is tritiated thymidine. When it is introduced, this radioactive substance is permanently incorporated into the deoxyribonucleic acid (DNA) of the selected cells. Tritiated thymidine has several advantages: It works in any cell, it follows along with the cells, and it will be in all the offspring of the cell to which it was originally introduced. Unfortunately, as cells continue to divide, the concentrations of tritiated thymidine eventually approach undetectable levels.

Many systems have been studied using these methods, including the development of the tetrapod limb and vertebrate head, somite differentiation, and the appearance of the primary germ layers that eventually yield all other tissues.

Morphogenesis is the reflection of all the interactions that take place during the formation of a living organism and of the patterns of structure that characterize this organism. The study of morphogenesis is motivated by the desire to understand the appearance of patterns of structure. Such patterns are often conserved through evolutionary history. It is hoped that a clear picture of how these patterns are formed and regulated will show how the origin of novel morphologies is constrained. Thus, it will be possible to understand both the origin of novel forms and the maintenance of unchanging form.

—*Charles R. Crumly*

See also: Cell types; Determination and differentiation; Development: Evolutionary perspective; Growth; Histology.

Bibliography

Bard, Jonathan. *Morphogenesis: The Cellular and Molecular Processes of Developmental Anatomy.* New York: Cambridge University Press, 1990. A critical review of the physical mechanisms of morphogenesis. An appendix outlines three dozen areas where future research is needed. Clearly written, requiring only a minimal understanding of embryology. Extensive bibliography.

Bonner, John Tyler. *On Development: The Biology of Form.* Cambridge, Mass.: Harvard University Press, 1974. Updated from Bonner's earlier (1952) text on morphogenesis, this volume is also intended for the college-level student. Illustrations are rarely included. Endnotes and index.

De la Cruz, María Victoria, and Roger R. Markham. *Living Morphogenesis of the Heart.* Boston: Birkhäuser, 1998. An encyclopedic survey of the developmental anatomy of the heart, focusing on chicken embryos, with comparisons to other mammalian hearts. Written for medical professionals.

Edelman, Gerald M., and Jean-Paul Thiery, eds. *The Cell in Contact: Adhesions and Junctions as Morphogenetic Determinants.* New York: John Wiley & Sons, 1985. The subject of this book, intended for specialists, is the nature of the contacts that cells make and how they determine the fate of morphogenesis. Articles are contributed by researchers in this area and are grouped into five sections: membrane structure, determinants of animal form, cell adhesion molecules, extracellular matrix/cell-substrate adhesion, and specialized junctions. Most contributions are illustrated. Each article is followed by a literature cited section. The book ends with a subject index. Wessells (1977) is a more accessible college-level treatment of some of the same issues.

Hinchliffe, J. R., and D. R. Johnson. *The Development of the Vertebrate Limb: An Approach Through Experiment, Genetics, and Evolution.* Oxford, England: Clarendon Press, 1980. This volume is intended for specialists, but it can be understood by college-level students. It is a detailed and extensive treatment of limb development, divided into seven sections: the origin of the vertebrate limb, adaptation/diversity, descriptive and experimental embryology, regeneration, pattern, and mutants and the process of evolution. The well-illustrated text is followed by a long bibliography, an index, and plates.

Thomson, Keith S. *Morphogenesis and Evolution.* New York: Oxford University Press, 1988. This well-written, very short volume is intended for evolutionary biologists with a budding interest in development. Most of the examples chosen for discussion come from vertebrates. Nine chapters cover very broad topics including theories, the difference between pattern and process, pattern formation (both early and late), and the general properties of morphogenetic systems. The infrequent illustrations are all line drawings; the text is followed by references and a short index.

Trinkaus, John Philip. *Cells into Organs: The Forces That Shape the Embryo.* Englewood Cliffs, N.J.: Prentice-Hall, 1984. This exceptional book is intended as an upper-division college text that specialists might also find indispensable. It is an expanded version of an earlier edition of the same title. Fifteen chapters, illustrated by photographs and line drawings, extensively cover topics including cell movements, the structure of cell surfaces, cell adhesion, the environmental control of cell motility, mechanisms of morphogenetic movement, and the spread of cancer cells. Selected references follow each chapter. The author provides criticisms, suggests many avenues for future research, and notes many unsolved developmental problems.

Webster, Gerry, and Brian Goodwin. *Form and Transformation: Generative and Relational Principles in Biology.* New York: Cambridge University Press, 1996. Argues that the theory of Darwinian selection must be supplemented by a theory of biological form, based on a generative theory of organisms as developing and transforming entities of a distinctive type.

Wessells, Norman K. *Tissue Interactions and Development.* Menlo Park, Calif.: Benjamin/ Cummings, 1977. This brief but detailed text is intended for undergraduate college students. Eighteen chapters cover the main processes of development, including morphogenesis. Emphasis is on how tissues interact to bring about morphogenesis. The text is well illustrated with line drawings and numerous photographs. Each chapter closes with a brief "Concepts" section that outlines important points, followed by a references-cited section. The last chapter is a handy summation of the major points discussed. Glossary and index.

MOSQUITOES

Type of animal science: Classification
Fields of study: Anatomy, ecology, entomology, invertebrate biology, physiology, zoology

There are about 3,500 known species of mosquitoes, all of which are placed in the family Cuicidae of the insect order Diptera. The importance of mosquitoes to humans lies with their ability to feed on vertebrate blood and transmit organisms that cause disease.

Principal Terms

LABIUM: the sheath that contains the slender stylet-like mouthparts of the mosquito, including the mandibles, maxillae, and hypopharynx

SALIVA: the liquid containing enzymes secreted by the salivary glands that is injected into the host when the adult mosquito feeds

SIPHON TUBE: tube that extends from the rear of the larval abdomen at the air-water interface and allows the larva to breathe air

Although mosquitoes are found all over the world, as far north as the Arctic Circle, the vast majority of them live in the tropics and subtropics. The adults have wings and are able to fly, but the immature stages, or larvae, are wormlike and are confined to small bodies of water. Adult male mosquitoes do not have mouthparts that are capable of penetrating skin, so they feed primarily on nectar from flowers. However, adult females have piercing mouthparts that easily pass through the skin of larger animals so they can feed on their blood. About 200 million years ago they probably evolved from other insects that used their mouthparts to feed on plants.

The Life Cycle of Mosquitoes

Female mosquitoes lay their eggs in water or in areas that will be flooded with water after rainfall accumulates or the snow melts. The eggs hatch after a few hours of being submerged, and this is the reason for the usually large numbers of mosquitoes in the wet spring months. The larvae that hatch from the egg must remain in the water to develop, but they are only able to breathe when they come to the water's surface. At the end of the larval abdomen is a strawlike siphon tube that allows air to enter. By hanging with its head down and its siphon tube at the surface, the larva can continue to feed underwater and breathe at the same time. The larvae feed by filtering out small particles of food from the water and grow by periodically producing a new skin and molting the old skin. After three larval molts over the course of about one week, the larva changes into a comma-shaped pupa that does not feed but is transformed into the adult. The pupa comes to the water surface when its development is complete and the adult mosquito escapes from the pupal skin and flies from the water surface.

Adult mosquitoes often mate shortly after they emerge. The males of many mosquito species form swarms over trees and bushes into which the female enters. When she enters the swarm of males, her wingbeats produce a tone that attracts the male. Most females only mate once during their lives and store the sperm in a special sac that is used to fertilize the eggs just before they are laid. Males live only about two weeks and feed on sugars from plants. Female mosquitoes live for about a month. Unlike the males, they have needlelike mouthparts consisting of two elongated maxillae, two long mandibles, a labium, and a hypopharynx

1070

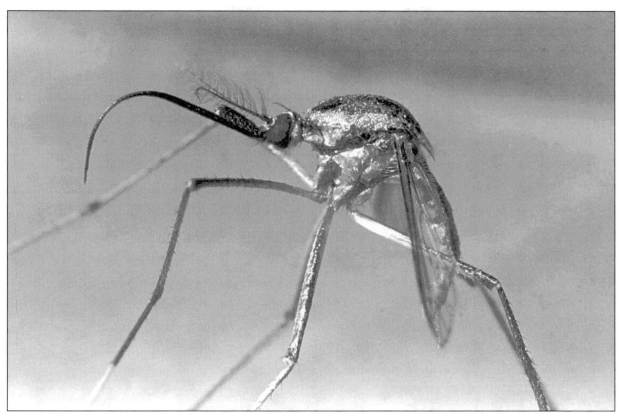

Only adult female mosquitoes have the piercing mouthparts that enable them to drink blood from mammals. (AP/ Wide World Photos)

that has a hollow tube running its length. When the mosquito feeds, the sheathlike labium that encloses the other mouthparts folds away and the six needlelike mouthparts then penetrate the skin. The tube within the hypopharynx is connected to the salivary glands and allows saliva to be injected into the animal being fed upon. The saliva prevents the blood from clotting in the mouthparts and may also lessen the pain so the mosquito can feed without being detected. The saliva that is injected also produces the common symptoms of itching and redness that usually follow the mosquito bite. Feeding on blood is necessary if the female is to develop her eggs. By using the protein contained in the blood, she is able to mature about one hundred eggs within two days of feeding. She can take another blood meal and develop another egg batch every two days. Feeding behavior usu-

ally occurs at dusk or during the evening hours, when its hosts are asleep and less likely to defend themselves. A few mosquito species do not feed on blood at all and acquire the protein they need to develop eggs from the larval stage.

Adult female mosquitoes have special sensory receptors on their antennae that can detect the odors of potential hosts. The carbon dioxide that animals give off as they breathe is one of the major signals the female uses to locate a host for a blood meal, but other chemicals produced by the skin are also important. The female mosquito may also use body heat and the visual image of the host to locate it. Darker colors are generally more attractive than are lighter colors. Common mosquito repellents are believed to act as chemical masks that disrupt the pattern of host stimuli so the female no longer recognizes the host.

Mosquitoes and Disease

Mosquitoes are not just nuisances when they feed on humans. Female mosquitoes are able to acquire parasites, such as viruses and protozoa, when they feed on an animal that already has an infection in its blood. Some of these parasites are able to infect the tissues of the mosquito and then invade its salivary glands. Once the salivary glands contain the parasites, they can be transmitted to an uninfected host the next time the mosquito injects its saliva and feeds. Many diseases of humans and other animals are transmitted by mosquitoes when they feed on blood. Malaria is the most important of these, causing over one million human deaths each year. Other important mosquito-transmitted diseases include dengue, yellow fever, and viral encephalitis, all caused by viruses. Heartworm is an important disease of dogs that is caused by a nematode worm transmitted by mosquitoes. Filariasis is a similar disease in humans that produces grossly enlarged limbs.

The best way to control mosquitoes is to limit the places larvae can breed. This involves preventing water from collecting in small containers and channeling water runoff so it does not accumulate in puddles that might last long enough for the larvae to develop into adults.

—Marc J. Klowden

See also: Antennae; Diseases; Ingestion; Insects; Metamorphosis; Molting and shedding.

Bibliography

Busvine, James R. *Disease Transmission by Insects.* New York: Springer-Verlag, 1993. A comprehensive review of the history of insect-borne disease.

Eldridge, Bruce F., and John D. Edman, eds. *Medical Entomology.* Boston: Kluwer, 2000. An excellent textbook covering all aspects of insects and the parasites they transmit to humans.

Harrison, Gordon. *Mosquitoes, Malaria and Man: A History of the Hostilities Since 1880.* New York: E. P. Dutton, 1978. This book tells the story of the attempts to control malaria and the people involved in the critical discoveries.

Klowden, M. J. "Blood, Sex, and the Mosquito: The Mechanisms That Control Mosquito Blood-Feeding Behavior." *BioScience* 45 (1995): 326-331. A description of the mechanisms that control blood-feeding by mosquitoes.

MOUNTAIN LIONS

Type of animal science: Classification
Fields of study: Anatomy, wildlife ecology

Large animals, usually classified in the small cat Felis *genus, mountain lions once ranged across the New World. However, human predation and habitat destruction have greatly reduced or eliminated many regional subspecies.*

Principal Terms

CARNASSIALS: pairs of large, cross-shearing teeth on each side of jaw

CREPUSCULAR: active at twilight or before sunset

EPIHYAL: presence or absence of this hyoid bone determines if a cat generally purrs or roars

HYOID BONES: series of connected bones at base of tongue

PAPILLAE: sharp, curved projections on tongue

Mountain lions, also known as American lions, catamounts, cougars, deer tigers, Florida panthers, and pumas, are classified in the genus *Felis* with small cats because they share a solid epihyal bone that restricts their ability to roar. Mountain lions purr and during mating emit harsh, frightening screams.

Among American felines they are second in size only to the jaguar. Adult male mountain lions weigh from eighty to over two hundred pounds, and are often nine feet long from nose to tip of tail, and up to thirty inches high at the shoulder. Females are about a third less in size. Cats in the equatorial regions are smaller and have thinner coats than those in the extreme north and south of the mountain lion range.

Mountain lion coats are uniform in color, varying from reddish bronze to brownish yellow or gray, with black markings around the mouth and eyes and on the tip of the tail. Kittens are born spotted, but these marks soon fade. Thirty subspecies of mountain lions, varying slightly in head shape, coat color, and size, occupy separate geographic regions.

Behavior

Kittens remain with their mothers for eighteen months to two years. Otherwise, mountain lions live solitary lives. Each female has a distinct hunting range; where such areas intersect, cats avoid each other. Males hunt over much larger tracts, sometimes covering hundreds of square miles. Each territory overlaps several female ranges, which the males check frequently for breeding opportunities. Older males face challenges from younger cats seeking to establish their own territories.

Mountain lions are nocturnal or crepuscular hunters; they are almost invisible when silently stalking their victims in dim light. Except for mothers and kittens, there are few reports of cats aiding each other during hunting. Deer are the major prey in North America, capybara and peccary the preferred quarry in equatorial America. When larger animals are unavailable, mountain lions eat rodents, rabbits, and beaver. Blending into their surroundings, cats creep slowly toward intended prey, closing with a furious forty-five mile per hour rush, sometimes leaping as much as forty feet to surprise their target. Sharp claws hold victims, as canine teeth sever their spinal column or windpipe. Powerful jaws, containing scissorslike carnassials and tongues with rasplike papillae, permit cats to harvest every speck of meat from their prize.

Relations with Humans

Native Americans, carrying carved cat images as fetishes, revered the mountain lion as the greatest of all hunters. Native rituals used lion skins and paws to ensure hunting success. Europeans were less respectful, treating big cats as dangerous vermin to be destroyed. From the seventeenth century through much of the twentieth century, governments offered bounties for lion skins. Humans were the only species consistently preying upon adult mountain lions. Hunters set iron traps, dug pits, and used dogs to chase and tree the cats.

As expanding human settlement made wild prey scarce, mountain lions found cattle, sheep, and horses irresistible. The cats tended to avoid humans, but rare attacks and killings frightened people and led to calls for the cats' extermination. John James

---

Mountain Lion Facts

Classification:
Kingdom: Animalia
Subkingdom: Bilateria
Phylum: Chordata
Subphylum: Vertebrata
Class: Mammalia
Subclass: Eutheria
Order: Carnivora
Family: Felidae (cats)
Genus and species: Felis concolor
Subspecies: F. c. cougar (eastern cougar), *F. c. coryi* (Florida cougar)
Geographical location: Original range was southern British Columbia to the Straits of Magellan and from the Atlantic Ocean to the Pacific Ocean
Habitat: Mountains, forests, deserts, and jungles
Gestational period: About three months
Life span: Up to twelve years in the wild, twenty or more years in captivity
Special anatomy: Large eyes with excellent night vision; jaws adapted to seizing and gripping prey, teeth designed for tearing and slicing flesh

Mountain lions prefer to stalk their prey during the twilight or early morning hours. (Corbis)

Audubon noted that by the 1840's, man had nearly eliminated mountain lions east of the Mississippi. By 1900, few mountain lions existed in North America east of the Rocky Mountain states.

Attitudes toward mountain lions began changing in the last decades of the twentieth century. Laws in western states banned or strictly limited hunting. However, as human intrusion into the mountain lion's habitat increased, cats occasionally attacked solitary hikers, joggers, and skiers, stimulating calls for removal of the predators. With great effort and expense, conservationists maintain a relict population of some seventy Florida panthers (*Felis concolor coryi*) in southwest Florida—a subspecies originally ranging from Louisiana to Florida. In Central and South America, where significant populations of mountain lions remain, destruction of habitat by expanding settlement has greatly reduced surviving numbers. Outside Florida, mountain lions are not technically an endangered species, but their long-term future remains precarious.

—Milton Berman

See also: Carnivores; Cats; Cheetahs; Fauna: Africa; Fauna: Asia; Groups; Jaguars; Leopards; Lions; Predation; Tigers.

Bibliography

Bolgiano, Chris. *Mountain Lion: An Unnatural History of Pumas and People.* Mechanicsburg, Pa.: Stackpole Books, 1995. Records native, pioneer, and modern reactions to mountain lions.

Kobalenko, Jerry. *Forest Cats of North America: Cougars, Bobcats, Lynx.* Willowdale, Ontario: Firefly Books, 1997. Extensive coverage of facts and legends about mountain lions; includes an annotated bibliography.

Maehr, David S. *Florida Panther: Life and Death of a Vanishing Carnivore.* Washington, D.C.: Island Press, 1997. Provides a detailed portrait of the biology, natural history, and current status of the Florida subspecies, with an assessment of its prospects.

Saunders, Nicholas J., ed. *Icons of Power: Feline Symbolism in the Americas.* New York: Routledge, 1998. Two chapters explore ritual and symbolic uses of mountain lions by Indians.

Young, Stanley P., and Edward A. Goldman. *The Puma: Mysterious American Cat.* 1946. Reprint. New York: Dover, 1964. Classic study of mountain lions with detailed information on their lifestyle and distribution.

MOUNTAINS

Types of animal science: Ecology, geography
Fields of study: Conservation biology, ecology, environmental science, wildlife ecology, zoology

Mountains cover one fifth of the earth's terrestrial surface, and are one of the most extreme environments in the global ecosystem. They are globally significant landforms that function as storehouses for irreplaceable resources such as clean air and water, and biological and cultural diversity, as well as timber and mineral resources.

Principal Terms

ASTHENOSPHERE: the region below the lithosphere where rock is less rigid than that above and below it

BIOCLIMATIC ZONE: a zone of transition between differing yet adjacent ecological systems

ENDEMISM: the occurrence of species only within narrow environmental ranges

LOCAL RELIEF: the elevation difference between the lowest and highest points in an area

SUCCESSION: directional change in communities of vegetation or animals

TECTONIC PLATE: tectonic plate theory suggests that the earth's surface is composed of a number of oceanic and continental plates which have the ability to move slowly across the earth's asthenosphere

---

Mountains are the most conspicuous landforms on earth. They are found on every continent, and have been defined simply as elevated landforms of high local relief, with much of the surface in steep slopes, displaying distinct variations in climate and vegetation zones from the base to the summit. The earth's mountain ranges have been created by the collision of tectonic plates. Associated with many of these mountain ranges are volcanoes. If the solidified magma of a volcano builds up, it can become a mountain; likewise, if the collision involves two oceanic plates, a string of volcanic mountains, called an island arc, can form on the ocean floor.

Mountain Habitat

Mountains are globally significant reservoirs of biodiversity. They contain rich assemblages of species and ecosystems. Because of the rapid changes in altitude and temperature along a mountain slope, multiple ecological zones are stacked upon one another, sometimes ranging from dense tropical jungles to glacial ice within a few kilometers. Many plant and animal species are found only on mountains, having evolved over centuries of isolation to inhabit these specialized environments. Mountains can also function as biological corridors, connecting isolated habitats or protected areas and allowing species to migrate between them. These extraordinary ecological conditions, coupled with many bioclimatic zones, have resulted in an extremely high number of ecological niches available for habituation in mountain ecosystems.

Mountain Fauna

Because of the great diversity in habitats within mountainous regions, with each region showing a different combination of environmental factors, total mountain fauna is relatively rich and the variety of small communities very great, in spite of the general severity of the mountain environment as a whole. Likewise, this diversity has resulted in a wide range of endemic species that have evolved over centuries of isolation from other genetic material. Rocky Mountain National Park typifies this diversity as a home to 900 species of plants, 250 species of birds, and 60 species of mammals. Some are easily seen and others are elusive, but all are part of the ecosystem in the park. On a global scale, mountain fauna diversity includes many species of ungulates, including elk, bighorn sheep, moose, and deer. Also included in mountain communities are many species of rodents. Rodent species may include beaver, marmots, squirrels, and chipmunks. Other mammalian animal life includes bear, canids, including coyote and wolf, and many species of felids, such as mountain lions and bob-

---

Major Mountain Ranges of the World

AFRICA: Atlas, Eastern African Highlands, Ethiopian Highlands
ASIA: Hindu Kush, Himalayas, Taurus, Elburz, Japanese Mountains
AUSTRALIA: MacDonnell Mountains
EUROPE: Pyrenees, Alps, Carpathians, Apennines, Urals, Balkan Mountains
NORTH AMERICA: Appalachians, Sierra Nevada, Rocky Mountains, Laurentides
SOUTH AMERICA: Andes, Brazilian Highlands

Mountain meadows provide habitats for many animal and bird species. (PhotoDisc)

cats. Mountain avian fauna includes many families of hummingbirds, bluebirds, hawks, falcons, eagles, and many more.

Threats to Mountains

Mountains are threatened in a variety of ways. There are constant threats from human activities, such as camping, hiking, and other recreational activities. Hikers create tracks in the soil that form erosion gullies. Likewise, hikers may trample on vegetation that has taken many years to grow. Commercial harvesting of trees in the lower forest zones of mountains is having an increasingly detrimental effect on biodiversity. Many countries have replanted indigenous trees with fast-growing coniferous trees, in an ill-fated effort to supply a growing human population with wood products. These hybrid forests are not nearly as beautiful as the native forests, but more to the point, they do not offer an environment conducive to the ecosystem that the native species supported. This problem creates a loss of wildlife, which becomes even more rare in these forests because of the decline of native vegetation. Global warming is another threat to mountain ecosystems. Snowlines are receding, and eventually, continued melting of glaciers and polar ice caps could lead to drying of major river systems which feed from them. Without question, human settlement and activities constitute the biggest threat to the mountain ecosystem.

—*Jason A. Hubbart*

See also: Chaparral; Ecosystems; Grasslands and prairies; Food chains and food webs; Forests, coniferous; Forests, deciduous; Habitats and biomes; Lakes and rivers; Marine biology; Rain forests; Savannas; Tidepools and beaches; Tundra.

Bibliography

Denniston, D. *High Priorities: Conserving Mountain Ecosystems and Cultures.* Worldwatch Paper 123. Washington, D.C.: Worldwatch Institute, 1995. This is a delightfully informative book. It is easy to read, with many relevant points regarding mountain ecosystem conservation.

Messerli, B., and J. D. Ives. *Mountains of the World: A Global Priority.* New York: Parthenon, 1997. This is a book for those readers who are more educated on the current environmental dilemmas concerning mountains and forestry. It is also a valuable source of general mountain ecosystem information.

Price, L. *Mountains and Man.* Berkeley: University of California Press, 1981. An easy-to-read, information-packed source concerning mountain-human interactions.

Sauvain, P. *Geography Detective: Mountains.* Minneapolis: Carolrhoda Books, 1996. This is an invaluable geographical perspective to mountains. It is loaded with easy-to-read information that is interesting and easily understood.

Stronach, N. *Mountains.* Minneapolis: Lerner, 1995. This book is filled with great information regarding general characteristics of mountains. The author also answers a wide array of questions regarding many mountain ecosystem specifics.

MULTICELLULARITY

Type of animal science: Evolution
Fields of study: Cell biology, evolutionary science, paleontology

By studying the fossil record, scientists have found that the first multicellular life appeared on the earth about 1 billion years ago. Before that time, only single-celled organisms existed. The appearance of multicellularity paved the way for the evolution of all higher organisms.

Principal Terms

EDIACARIAN (EDIACARAN) FAUNA: a diverse assemblage of fossils of soft-bodied animals that represents the oldest record of multicellular animal life on the earth

EUKARYOTIC CELL: a cell that has a nucleus with chromosomes and other complex internal structures; this is the type of cell which makes up all organisms except bacteria

FOSSILS: the remains of ancient life preserved in sediment or rock

MULTICELLULAR ORGANISMS: organisms consisting of more than one cell; there are diverse types of cells, specialized for different functions and generally organized into tissues and organs

PRECAMBRIAN EON: the earliest chapter of the earth's history, covering the time interval between the formation of the earth, about 4.6 billion years ago, and the beginning of the Cambrian period, about 570 million years ago

PROKARYOTIC CELL: a primitive cell that lacks a nucleus, chromosomes, and other well-defined internal cellular structures; only members of the kingdom Monera (such as bacteria) are prokaryotic cells—all higher organisms have eukaryotic cells

---

Multicellular organisms are those consisting of more than one cell. Three of the five kingdoms of living organisms are multicellular: the plants (Plantae), the animals (Animalia), and the fungi (Fungi). The other two kingdoms consist of single-celled organisms: the bacteria (Monera), which have primitive prokaryotic cells, and the protists (Protista), which have complex eukaryotic cells. Prokaryotic cells lack a nucleus and other internal cell structures and are found today only among the bacteria. Eukaryotic cells, on the other hand, contain a nucleus, other complex internal cell structures called organelles (such as mitochondria, which perform respiratory functions), and sometimes chloroplasts (which contain chlorophyll and perform photosynthesis). All multicellular organisms are composed of eukaryotic cells, so the eukaryotic cell must have evolved before multicellular organisms could develop. It is generally accepted that simpler types of organisms evolved first, followed by more complex organisms.

The Multicellular Kingdoms

Plants are multicellular organisms that have chlorophyll (a green pigment used for photosynthesis), plastids (internal structures on the cell that contain chlorophyll), and a cell wall that contains cellulose. Plants are sometimes called "primary producers" because they can manufacture their own food from carbon dioxide and water through a process called photosynthesis, using sunlight for energy and producing oxygen and organic matter (carbohydrates) as by-products.

Animals are multicellular organisms that cannot produce their own food and must feed on other organisms. They are "consumers." Metazoans have many types of cells, which are organized into tissues, and groups of tissues, which form organs. There are two primary embryonic tissue layers present in all metazoans (except the sponges). These are the ectoderm (outer layer) and the endoderm (inner layer). More advanced metazoans also have a third embryonic cell layer, the mesoderm, which lies between the other two layers.

Fungi (mushrooms and their relatives) possess cell walls like plants, but unlike plants, they lack chlorophyll. Although fungi appear plantlike, they cannot produce their own food because of the absence of chlorophyll, so they must feed by ingesting organic material and therefore are consumers. Because they are neither plants nor animals, the fungi are placed in a separate kingdom (Fungi).

Advantages and Origins of Multicellularity

Multicellularity probably evolved because it gave organisms some sort of advantage, assuring them of a greater chance of survival. Multicellularity allows organisms to become larger (which helps them to outcompete other organisms and provides a greater internal physiological stability), to have a longer life (because individual cells are replaceable), to produce more offspring (because many cells can be dedicated to reproduction), and to have a variety of body plans (which permits adaptation to various modes of life or environmental conditions). Specialization of cells for particular functions allows organisms to become more efficient.

Evidence for the origin of multicellularity comes from the fossil record, studies of the organization and biochemistry of living cells and organisms, and from studies of the embryonic and larval stages of animals. During the Archean eon (between 3.8 and 2.5 billion years ago), only single-celled, prokaryotic life (bacteria-like organisms) existed on the earth. Some of the prokaryotes were photosynthetic, including the cyanobacteria (or so-called blue-green algae). Some of these prokaryotes were colonial, with cells organized into structures such as chains, or filaments, or algal mats. These colonies differ from true multicellular organisms because they generally consist of only one type of cell rather than many types of cells. Colonies of blue-green algae formed moundlike structures called stromatolites, which were quite common during the Precambrian but are present only in a few areas today. Stromatolites appeared about 3 billion years ago but did not become abundant until about 2.3 billion years ago.

The Rise of Eukaryotic Organisms

Eukaryotic organisms appeared during the Proterozoic era. The eukaryotic cell probably evolved from prokaryotic ancestors some time before about 1.4 billion years ago. The oldest convincing fossils of eukaryotic cells are generally considered to be those from the 1.3 billion-year-old Beck Spring dolomite of California. Eukaryotic fossil cells have also been found in chert from the approximately 850 million-year-old Bitter Springs formation of Australia. The earliest eukaryotes were animal-like protozoans. This evolution occurred when photosynthetic prokaryotic cyanobacteria were ingested by protozoans and then developed a symbiotic (mutually beneficial) relationship with them. The evolution of the plantlike eukaryotes probably occurred by at least 1.4 billion years ago. (This date has been suggested because primitive multicellular algae fossils are present in rocks 1.3 billion years old.)

The eukaryotic cell was a prerequisite for the development of multicellular organisms. The plantlike eukaryotes are considered to be ancestral to the multicellular algae and higher plants. The protozoans are considered to be the ancestors of the metazoans (animals). The first multicellular organisms may have been algae. Fossils that appear to be primitive multicellular algae are known from the 1.3 billion-year-old sedimentary rocks of the Belt supergroup of Montana, and the 800 million- to 900 million-year-old Little Dal group of northwestern Canada. Multicellular algae can be found living today in both freshwater and marine environments.

Fossil fungi first appear in the fossil record in the 790 million- to 1,370 million-year-old Bitter Springs formation cherts of Australia. The fossil record of fungi is poor and not well known.

The Rise of Multicellular Animals

Multicellular animals evolved independently of the multicellular plants, probably arising from protozoan ancestors. The oldest evidence of metazoans (multicellular animals) in the geologic record is in the form of trace fossils. Trace fossils are imprints such as tracks, trails, or burrows made in sediment by moving animals. Over time, the sediment hardened into sedimentary rock as a result of compaction and cementation. The earliest trace fossils consist of simple trails and tubelike burrows. In some places, there is a succession of types of trace fossils from simple tubelike burrows in older rocks to more complex structures in younger rocks. This change suggests that the evolution and diversification of increasingly complex burrowing organisms occurred during the latter part of the Precambrian. The oldest trace fossils are less than one billion years old, and many scientists believe that it is unlikely that any trace fossils exist in rocks much older than about 700 million years old. The trace fossils appear in the geologic record just before the first appearance of soft-bodied metazoan fossils. Structures that resemble trace fossils, however, have been reported from much older rocks. Among these questionable traces are the one-billion-year-old Brooksella, which resembles a jellyfish. In addition, tubelike structures from the upper Medicine Peak quartzite in Wyoming have been dated at 2 billion to 2.5 billion years, at least 1 billion years older than the oldest known metazoans. The origin of these older traces is uncertain and may be a result of inorganic processes (such as dewatering of sediment), rather than of organisms.

One possible fossil metazoan that appears to be more than 850 million years old has been reported from the Tindir group of Alaska. This fossil is less than one millimeter long and appears to be a flatworm (phylum Platyhelminthes). Both the age and identification of this fossil have been disputed, but if valid, it is a very important find because some biologists theorize that the earliest metazoans would have been primitive flatworms.

The oldest unquestioned metazoan fossils are the imprints of a diverse assemblage of relatively well-developed, soft-bodied marine animals. More than half of the organisms appear to be some type of cnidarian or coelenterate (related to jellyfishes), about 25 percent appear to be segmented worms (related to annelids), and a small percentage appear to be arthropods (related to insects, crabs, and lobsters). Trace fossils are also present. This assemblage of soft-bodied fossils is called the Ediacaran fauna. It was discovered in 1946 in sandstones of the Pound subgroup in the Ediacara Hills of the Flinders Range in South Australia. The exact age of the Ediacaran fauna is uncertain because there are no nearby rocks of the proper type for radiometric dating. The Ediacaran fauna is clearly Precambrian, however, judging from its position in the geologic sequence. The soft-bodied Ediacaran fossils are separated from the younger fossil shells of the Cambrian (570 million years old) by a thick section of unfossiliferous rock (up to several hundred meters thick). Since the 1950's, fossils similar to the Ediacaran fauna have been found in rocks of approximately the same age on virtually every continent on the earth (with the possible exception of Antarctica). In some of these other areas, it is possible to date radiometrically the rocks associated with the fauna. These radiometric dates indicate that the early metazoan soft-bodied fossils range from about 620 million to 700 million years old. It is likely that the metazoans evolved some time prior to 700 million years ago because these fossils represent well-developed, complex animals.

Skeletonized faunas (animals with shells or other hard parts) did not appear until approximately 580 million years ago. The skeletonized faunas are represented by microscopic scraps, cones, tubes, and plates made of calcium phosphate or a hard organic material called chitin. It is not known exactly what types of organisms produced these skeletal remains, but the tiny fossils are so diverse and complex that it is assumed that the organisms must have had a long history of evolution during the Precambrian. The origin of skeletons was advantageous to marine organisms because hard parts provide protection against predators as well as the mechanical functions of support and muscle attachment.

Theories of the Origin of Multicellular Life

There are a number of theories to explain the origin of multicellular life. Most of the theories are derived from studies of various types of cells and living organisms, including advanced protozoans, early developmental stages (embryos), and larval stages. Four types of cells are central to these theories, and they are grouped into two categories: motile (capable of movement) and nonmotile (not capable of movement). The motile protists include flagellate cells (those with a whiplike "tail," or flagellum) and amoeboid cells (those such as Amoeba, which move by pseudopodia, or fingerlike extensions of the cell membrane). The nonmotile stages include coccine cells (those with many nuclei, sometimes called multinucleate cells) and sporine cells (those that divide and stick together to form multicellular aggregates).

There are many theories that have been proposed to explain the origin of plants, fungi, and metazoans (animals). Formerly, it was thought that plants evolved from prokaryotic algae (cyanobacteria, or blue-green algae), but it is more likely that plants arose from a eukaryotic ancestor, such as a flagellate cell. Flagellate algae are similar to flagellate protozoans, but it is not certain whether the algae evolved from the protozoan or vice versa. The presence of plastids (such as the chloroplasts that contain chlorophyll used for photosynthesis) may be the key feature separating plants from protozoans, fungi, and animals. According to a theory proposed by Lynn Margulis, plastids evolved from prokaryotic blue-green algae that were captured by eukaryotic cells. The sporine cell is another possible ancestor of the plants. Sporine cells appear to have had the capacity to evolve beyond the colony level and to produce complex tissue-level green algae and higher plants.

Fungi used to be considered as plants that had lost (or never evolved) chlorophyll. The discovery of a single-celled stage with flagellae among the more primitive fungi, however, suggests that fungi probably evolved from protozoans.

The ancestral multicellular organisms, which gave rise to all the more-complex living animals, are all extinct. The simplest multicellular animal living today is the sponge (phylum Porifera). The sponges are not considered to be ancestors of the more complex animals because their body organization and developmental history are very different. Sponges have no tissues, mouth, or internal organs. Instead, they consist of an aggregate of flagellate and amoeboid cells (and a few other types) roughly arranged in layers. The sponges may have evolved independently from the other metazoans. Sponges are classified as a distinct side branch of the animal kingdom (Parazoa), with a primitive multicellular grade of organization (no tissues). The remaining multicellular animals are grouped into the Eumetazoa.

Theories of Metazoan Origin

Several theories have been proposed to explain the origin of the metazoans. These theories can be placed into the following categories: evolution from single-celled protozoans; evolution from colonial protozoans; evolution from multinucleate coccine cells as a result of development of internal cell boundaries; and evolution from sporine cells. There are several versions of each of these theories, and there is no general agreement on which theory is best. Some researchers promote the colonial theory as the most widely accepted theory, whereas others claim no longer to take it seriously. Most experts agree that evolution of metazoans from colonial protozoans would seem to be easier than evolution directly from a single cell. Multicellularity may have arisen independently several times, in several different ways.

The colonial theory suggests that the metazoans evolved from flagellate or amoeboid protozoans that lived together in colonies, much like the modern green alga Volvox, which is shaped like a hollow sphere. From an original hollow spherical form, the shape of the ancestral metazoan changed as an indentation or invagination formed in the side. The indentation became larger, producing a double-walled "cup" (envision pushing one's thumb into the side of a deflated ball until that side becomes nested into the other side of the ball, forming a cuplike shape). The double-walled cup shape is referred to as a diploblastic body plan, meaning two layers of body tissue. These two layers are the ectoderm (outer layer) and the endoderm (inner layer). This process of indentation to produce a diploblastic (double-walled) form occurs in the embryos of many animals. The jellyfishes are a good example of animals with a diploblastic body plan. Nearly all groups of animals have ectoderm and endoderm (except the sponges), suggesting that nearly all groups of animals are related. Because the jellyfishes (phylum Cnidaria) have the simplest body plan, they are believed to be the most primitive. The diploblastic ancestral form has been called a gastrea. Ernst H. Haeckel, a prominent nineteenth century German biologist who studied animal embryos, believed that all bilaterally symmetrical animals evolved from a gastrea.

A second theory for the origin of metazoans suggests that the ancestral form was a bilaterally symmetrical animal resembling a flatworm. Some scientists believe that the complex organs and organ systems of metazoans are beyond the evolutionary potential of flagellate and amoeboid cells. The flatworm may have evolved from "cellularization" of a multinucleate coccine cell (formation of cell membranes around each of the nuclei) or from clumping of sporine cells. Most of the cells in the metazoans are sporine cells that stick together to form multicellular aggregates. Sporine protozoans do not exist, so it is hypothesized that sporine ancestors of the metazoans must have evolved from "preprotozoans." These hypothetical ancestors may have been solid balls of cells resembling the early stages of many embryos. At some point, the exterior cells may have developed (or redeveloped) flagellae and become specialized for locomotion, and the interior cells may have become specialized for digestion and reproduction. Such colonies of cells would have resembled the larval (immature) form of cnidarians, called a planula larva, and, hence, they are called planuloids. Planuloids are believed to have given rise to two groups of metazoans, the cnidarians (jellyfishes and their kin) and the flatworms. The primitive flatworms are believed to

have been ancestral to all other bilaterally symmetrical metazoans.

Evidence from Rocks

Theories to explain the origin of multicellular life have been developed by biologists as a result of studies of various types of cells and living organisms, including advanced protozoans, early developmental stages (embryos), and larval stages.

Geologists (scientists who study rocks) and paleontologists (scientists who study fossils) have a variety of techniques that they use to search for the evidence of life in Precambrian rocks (older than 570 million years). These include searching for fossil remains and chemical analysis of organic residues that are probably the breakdown products of once-living organisms.

The first step in the search for Precambrian life is to locate rocks of the proper age. Geologic maps exist for virtually all parts of the world. From an examination of these maps, it is possible to identify areas that contain rocks of the proper age. (The age of a rock is determined by radiometric dating.) Age, however, is not the only consideration. For fossil remains to be preserved, the rocks must also remain little altered from the way they were originally deposited. Metamorphism (geologic alteration caused by heat and/or pressure) has deformed many Precambrian rocks to the extent that any fossils that may have been present can no longer be recognized.

Assuming that undeformed rocks of the proper age can be located, the search begins for fossil remains. Unfortunately, most Precambrian rocks are not fossiliferous. Precambrian multicellular fossils are found in only a few places in the world. In Australia, soft-bodied Precambrian metazoan fossils are restricted to a few thin layers of sandstone in a sequence of Precambrian rock more than one thousand meters thick. In most places in the world, however, there is a thick section of unfossiliferous rock separating the Precambrian metazoan fossils from the shelly faunas in the Cambrian rocks. This unfossiliferous sequence of rock is an interval for which there is little or no information on the types of life that existed.

Before multicellular organisms appeared (prior to perhaps one billion years ago), only microscopic, single-celled organisms existed on the earth. Microscopic fossils of single-celled organisms are found by careful examination of fine-grained, dark-colored rocks such as black cherts. The black color of the rocks commonly indicates the presence of carbon, which is present in all living organisms and which may be preserved in some fossils. Very thin slices of rock are prepared and mounted on glass slides so that the organic matter can be studied. These slices of rock, called thin sections, are so thin that light can pass through them, and they are examined with a microscope. Much of the carbon in these rocks is present as amorphous (indistinct or shapeless) patches, but in some places, microscopic structures are present that appear to be the fossilized remains of single-celled organisms. Pieces of rock can also be prepared for examination using a scanning electron microscope. The search for microfossils is difficult and painstaking. Among the problems involved are the possibility of contamination by modern-day organic matter in the laboratory and the possibility that the microscopic structures may really be inorganic in origin.

Chemical tests are used to search for the products of biological activity, which may be preserved in rocks. In principle, rocks that have been influenced by biological activity should contain certain characteristic isotopic ratios. There are a number of problems inherent in searching for organic residues. Organic material may have been preserved in the rock, but it could easily have been altered subsequently by heat and pressure or by circulating fluids. In addition, circulating fluids can contaminate the rocks by introducing organic material from much younger rocks.

Multicellularity in the Evolutionary Process

Studying the origin of multicellularity helps one to understand the conditions that led to the evolution of plant and animal life on the earth. As one begins to understand how multicellular life evolved, one may begin to wonder about why it was such a slow process. It is known that the earth formed about 4.6 billion years ago and that the first cells appeared about 3.5 billion years ago, but that the first multicellular life did not appear until approximately 1 billion years ago. In other words, it took more than 3.5 billion years for multicellular life to develop. More than three quarters of the earth's history had passed before multicellular life ever appeared.

One may also begin to wonder about the conditions that promoted the origin of multicellular life. Of all the planets in the solar system, the earth seems uniquely suited to life. Two of the most important factors involved are the presence of liquid water (which requires a specific temperature range) and the presence of an oxygen-rich atmosphere. None of the other planets in this solar system has either of these two characteristics. Interestingly enough, the earth originally did not have liquid water or an oxygenated atmosphere. Geologic evidence suggests that the earth's early atmosphere was the result of volcanic outgassing and that it consisted of gases such as carbon dioxide, carbon monoxide, ammonia, methane, hydrogen sulfide, nitrogen, and water vapor. As the planet cooled from its original molten state, the water vapor in the atmosphere condensed to form liquid water, which fell to the earth as rain and accumulated to form the oceans, rivers, and lakes. There is abundant geologic evidence that the earth's early atmosphere lacked the free oxygen that is breathed today. In the absence of free oxygen, chemical evolution in the oceans or lakes led to the formation of organic compounds, or what has been called the "primordial soup." The first living cells, the prokaryotes, evolved in this organics-rich water. As time passed, some of the early prokaryotic cells became photosynthetic, which allowed them not only to produce their own food from water and carbon dioxide but also to produce oxygen as a waste product. Oxygen was toxic to these early organisms. In order to survive, the cells had to develop a mechanism to adapt to the presence of increasing levels of oxygen. The buildup of oxygen led to the development of the ozone layer in the atmosphere and to the appearance of the eukaryotic cell. As the percentage of oxygen in the atmosphere increased, it is believed that some threshold level was reached, and it became possible for the environment to support multicellular organisms. That allowed a rapid diversification of life on the earth.

Hence, it appears that multicellular life on the earth appeared as a result of some prehistoric accident that resulted in global atmospheric change— the buildup of a toxic waste product (oxygen) as a result of photosynthesis by early life-forms. One might also speculate on the possible global effects of the increasing waste products that humans are now producing. The thinning of the atmospheric ozone layer is but one manifestation of the way that life is presently changing the earth's fragile environment. Knowing that the formation of the ozone layer was probably essential to the appearance of multicellular life on the earth, it is alarming to speculate on the consequences of its destruction. Life as humans know it depends on an earth with environmental conditions in a precarious balance.

—*Pamela J. W. Gore*

See also: Adaptive radiation; Cell types; Cleavage, gastrulation, and neurulation; Development: Evolutionary perspective; Evolution: Animal life; Evolution: Historical perspective; Protozoa.

Bibliography

Ayala, Francisco J., and James W. Valentine. *Evolving: The Theory and Processes of Organic Evolution.* Menlo Park, Calif.: Benjamin/Cummings, 1979. This book covers the basic principles of evolution and contains a section on the geologic record, which is extremely important in studies of the origin of early life-forms. It is designed for a college-level audience but is clearly written and easy to read.

Bittar, E. Edward, and Neville Bittar, eds. *Evolutionary Biology.* Greenwich, Conn.: JAI Press, 1994. A collection of essays on the structure and dynamics of DNA, RNA, and protein, and more general topics on the origin and cellular basis of life.

Boardman, Richard S., Alan H. Cheetham, and Albert J. Rowell, eds. *Fossil Invertebrates.* London: Blackwell Scientific Publications, 1987. A textbook that is designed for use in college courses in invertebrate paleontology. It begins with a number of chapters on basic principles (such as ecology, paleoecology, evolution, classification, and fossil preservation), followed by a run-through of kingdom Protista and twelve invertebrate phyla of kingdom Animalia that are commonly preserved as fossils. Well illustrated with photographs and line drawings. Will be most useful for those who have had courses in high school biology and historical geology.

Cloud, Preston. *Oasis in Space: Earth History from the Beginning.* New York: W. W. Norton, 1988. This book is part of a series designed to make science more accessible to the general public. It is clearly written, well illustrated, easily understood, and up-to-date. Unfamiliar scientific terms are explained on first mention in the text, and the reader can easily locate definitions by using the index at the back of the book. A two-page time chart indicating major events in geologic history is in the front of the book. Earth history is clearly covered, in addition to many of the basic principles of geology that contribute to the interpretations.

Futuyma, Douglas J. *Evolutionary Biology.* 3d ed. Sunderland, Mass.: Sinauer Associates, 1998. A textbook aimed at advanced undergraduates and graduate students, emphasizing the history of life and microevolutionary processes. Comprehensive glossary, index, bibliography, abundant illustrations.

Minkoff, Eli C. *Evolutionary Biology.* Reading, Mass.: Addison-Wesley, 1983. A college textbook that deals with evolution in an interdisciplinary way. It is a good reference book that will be most useful to those who have had an introductory course in college-level biology, but it can also be understood by beginners because it clearly explains new terms and has a good index and a glossary.

Selander, Robert K., Andrew G. Clark, and Thomas S. Whittam, eds. *Evolution at the Molecular Level.* Sunderland, Mass.: Sinauer Associates, 1991. A collection of essays focused on four themes: bacteria and viruses, organelles, "selfish" genes, and nuclear multigene families.

Stanley, Steven M. *Earth and Life Through Time.* 2d ed. New York: W. H. Freeman, 1989. A textbook for introductory college historical geology classes that can be readily understood by those without previous training in geology. It clearly covers the origin of life and rise of multicellular organisms during the Precambrian. It also presents a clear picture of the interrelationship between the physical and biological history of the earth. It is amply illustrated and has a glossary, comprehensive index, and summary of the classification of major fossil groups in the back.

MUSCLES IN INVERTEBRATES

Type of animal science: Anatomy
Fields of study: Histology, invertebrate biology

Invertebrates, animals whose name means "without a backbone," make up more than 95 percent of all animal species. The internal organization of each of the invertebrate phyla differs drastically from one to another and from the vertebrate body plan. Variables such as skeletal elements and lifestyle affect the activity and survival of the group, but all invertebrates use muscles built on the same plan as those of vertebrates.

Principal Terms

ACTIN: one of the two major types of contractile proteins; it forms the thin myofibrils of the sarcomere

FAST MUSCLE: muscle cells that respond quickly to nervous impulses; in invertebrates, these muscle fibers have short sarcomeres and a low ratio of thin to thick myofibrils

MYOSIN: one of the two major contractile proteins making up the thick myofibrils

PARAMYOSIN: a structural protein associated with myosin myofibrils and thought to support them

SLOW MUSCLE: muscle cells that respond slowly to nervous impulses; in invertebrates, these muscle fibers have long sarcomeres and a high ratio of thin to thick myofibrils

TROPOMYOSIN: a double-stranded protein that lies in the grooves of actin myofibrils, blocking actin from attachment to myosin

TROPONIN: a globular protein composed of three subunits; one subunit binds calcium ions, and another draws tropomyosin away from actin, which allows myosin to form crossbridges constituting the third subunit

The most obvious characteristic of animals may be movement. Movement requires a mechanism. The mechanism allowing multicellular animals of all kinds to change position or posture is a muscle cell.

The Mechanism of Movement

A muscle cell, also called a muscle fiber, is specialized for contraction. Muscle is composed of many muscle fibers, and each muscle cell contains contractile molecules permitting the cell to shorten and thereby change its shape. Each muscle fiber moves by drawing its ends together or by contracting.

The contractile molecules are common to muscles of invertebrates and vertebrates. There are several kinds. Actin and myosin are the most prominent contractile molecules. Each actin molecule is rounded and forms myofibrils by assembling repetitive monomers to form a helix, a springlike shape. Two of these helices are coiled around each other like two intertwined strings of pearls.

Myosin is a more complex myofibril. Each myosin molecule is composed of a head and a tail. The tails of two myosin molecules are wound around each other helically, while the heads project from the same end but in different directions. In the presence of calcium ions, each head can swivel. If an actin molecule is exposed, the myosin head can attach to it, forming a crossbridge. Since they lie parallel to each other in the cell, when the myosin

1087

Synchronous flight muscles connected directly to the wings are found in many flying invertebrates, such as butter-flies and moths. (Rob and Ann Simpson/Photo Agora)

returns to its original position, it pulls the actin myofibril. The ends of the contractile unit move closer to each other, shortening the muscle cell.

Myosin tails lie parallel to each other, but at the opposite ends of the myosin myofibrils, the heads project in opposite directions. The heads lie close to the two end walls that attach the actin myofibrils. There is a small central area where only the oppositely facing tails of the myosin myofibrils are found.

A contractile unit is called a sarcomere. A sarcomere is a repeated unit within a muscle cell consisting of overlapping thick myofibrils (myosin) and thin myofibrils (actin). Since there are areas in which there is no actin and areas in which there is no myosin, some light stripes are seen crossing the myofibrils in a microscopic preparation of certain muscle fibers. Actin myofibrils con-nect to thick walls that mark the ends of a sarcomere. The myosin myofibrils are connected across the center of the sarcomere. During relaxation, actin is absent from the center, producing a light band. The regular pattern of light and dark areas along the length of the muscle fiber is characteristic of striated or skeletal vertebrate muscle. This vertebrate muscle type is the most studied in physiology and, therefore, is the best known.

The Specifics of Invertebrate Muscles
Invertebrate muscles have similar components, but the arrangement is less regular, the alignment of sarcomeres often being oblique. Actin and myosin myofibril arrangement is not readily delineated, but similar structures exist. A sarcoplasmic reticulum, similar to the endoplasmic reticulum of ordinary cells, is characteristic of vertebrate and

invertebrate muscle fibers. It is arranged in a series of flattened sacs over a sarcomere's myofibrils. The sarcoplasmic reticulum can quickly absorb calcium ions, keeping their concentration around the sarcomere low. Without calcium ions, the myosin heads cannot be cocked.

Transverse, or T, tubules extend along the thick end wall of each sarcomere. They conduct the nerve impulse throughout the sarcomere to initiate contraction rapidly and simultaneously. When the impulse reaches the sarcoplasmic reticulum, stored calcium ions are released. This release begins contraction. Contraction ceases only if adenosine triphosphate (ATP) is present. Because they no longer make ATP, dead animals' contracted muscles do not relax; the muscles become stiff (exhibit rigor mortis).

Other molecules are also typically found in muscle cells. Tropomyosin is a strand of molecules that lies in a groove of the helically wound actin myofibrils. It blocks the attachment of myosin heads to actin molecules. Troponin is a three-part molecule that rests along the actin myofibril. One subunit attaches calcium ions. This changes troponin's conformation so that another subunit attracts the tropomyosin molecule, unblocking the actin myofibril so that the cocked head of myosin molecules can attach to the actin.

Paramyosin is a molecule similar to myosin found in invertebrates. It seems to act as a "filler" for myosin, keeping the myosin myofibrils aligned so that they can interact with the actin. It may be the contractile protein of some invertebrate muscle types. It is prominent in the "catch" muscles of mollusks.

Muscle Response Rates

Muscles may vary in their response rates even though they look the same. "Fast" muscle fibers contract rapidly in response to a nerve impulse. "Slow" muscle fibers contract several times more slowly. Although both types use the same contractile machinery, the slow muscle cells have less sarcoplasmic reticulum. This means that the calcium ions necessary to initiate contraction are not released as quickly everywhere along the contrac-

tile fibrils and are not pumped away from them as quickly as they are in the fast muscle cells.

Most invertebrate muscles are synchronous. This means that each contraction is initiated by a nerve impulse. The rate of contraction of synchronous muscles is determined by the rate of passage of nerve impulses. Even in flight, these muscles usually contract only about thirty-five times per second. Synchronous flight muscles are connected directly to the wings and are found in insects such as grasshoppers, moths, butterflies, and dragonflies.

True flies and bees, as well as beetles and true bugs, have a specialized kind of flight muscle called an asynchronous muscle. In asynchronous muscles, every contraction is not initiated by a nerve impulse. The contraction rates, up to one thousand times per second, are so rapid that nerve impulses could not be received and acted on quickly enough to produce that contraction rate. In these muscles, contraction is initiated by a nerve impulse received and transmitted by the T tubules, but there is no sarcoplasmic reticulum. Even fibrils removed from the cell and placed in a solution containing calcium ions and ATP contract and relax in an oscillatory fashion. The rates seem to be regulated by the myofibrils rather than by the calcium concentration. The T tubules seem to signal the contraction to begin and later turn off the cycling behavior of the myofibrils.

Asynchronous muscles are not directly attached to the wings of the insect. Elevator muscles are attached to the roof of the thorax. Their contraction pulls the roof of the thorax down and elevates the wings. Contraction of the wing depressor muscles pulls down the wings, but this stretches the wing elevators, stimulating them to contract and allowing the thorax roof to "pop up." Raising the thorax shortens the wing depressor muscles and terminates the active state of the depressors. They relax until stretched again by the elevation of the wings. The elevator and depressor muscle contractions follow the same sequence of stretch → contraction → shortening → relaxation → stretch, but are out of phase with each other. The frequency of wing beats depends upon the mechanical properties of the thorax and wings and not upon the frequency of nerve impulses.

Unlike vertebrate animals, there is a correlation between sarcomere length and speed of contraction. Rapidly contracting arthropod muscle fibers have short sarcomeres and relatively low ratios of thin to thick fibrils. Slowly contracting muscle fibers have long sarcomeres and high ratios of thin to thick fibrils. Intermediate types of fibers can exist within the same muscle.

Variations of Invertebrate Muscle Types

Because invertebrates are so varied, there is no generalization to which an exception cannot be found. For example, lobsters and crayfish have two separate muscle fiber types in their tail musculature. The tails are important in swimming and, particularly, in escape maneuvers. They must flex and extend rapidly to evade predators or aggressors. The bulk of the tail muscles consist of short sarcomere, rapidly contracting flexors and extensors. Thin sheets of long sarcomere—slowly contracting flexors and extensors—lie near the carapace. They are used for postural adjustments and for slow movements.

Flexors and extensors are good examples of another principle of muscle action. Muscles contract and they shorten. This shortening moves a body part. Relaxation allows the muscle fibers to lengthen. The force needed to lengthen the relaxed muscle comes from the contraction of its antagonist, a muscle that produces movement in the opposite direction from that of the first muscle. For example, flexing the tail of a lobster or crayfish is performed by the tail flexors. Relaxation does not return the tail to its extended position. Contraction of the extensor muscles moves the tail away from the body and extends the tail again. Relaxation of the extensors may allow the tail to be less rigid in its extension, but the tail will not flex until the flexor muscles are contracted. Muscles work in antagonistic groups to produce opposite movements (flexion and extension) and to produce postural changes that allow a lobster or crayfish to maintain its position when the current changes direction or speed.

Studying Invertebrate Muscles

Numerous methods are used to study the muscles of the invertebrates. Initially, the contraction patterns of whole muscles were studied by electrically stimulating muscles and their nerves. As in all biology, however, work is being done on the molecular level.

Glycerinated muscle is a preparation used to study the molecular elements necessary for contraction. Soaking muscle cells in glycerin removes the cell membrane and sarcoplasmic reticulum, but leaves the contractile fibrils intact. These can be used to determine the effect of presence or absence of ATP, calcium, magnesium, or other factors that might influence the working of the fibrils. They are used much the same as a whole muscle preparation. The ends of the fibrils are attached to one point that does not move and to another that does move when the muscle fibrils contract and relax. The movement can be recorded on paper or on an oscilloscope screen. The tension generated, or the length of shortening, can be calculated.

Microscopic analysis of the structural organization of invertebrate fibers also adds to the store of knowledge. Most invertebrate muscles do not have the striated appearance of vertebrate skeletal muscles. The myofibrils vary in their arrangements and are often difficult to discern. Microscopic analysis shows that rapidly contracting fibers have low ratios of thin to thick fibrils and slowly contracting fibers have larger ratios of thin to thick fibrils.

Muscles do not function without the coordination of the nervous system. Its contribution is different from that in vertebrates. Invertebrate muscle fibers do not exhibit the all-or-none response characteristic of vertebrates. Invertebrate muscle fibers receive innervation from several nerve fibers. A single nerve may serve many muscle fibers. Any muscle fiber is served by more than one nerve. The contraction strength of invertebrate muscle fibers depends upon the number and types of nerves sending impulses to that muscle cell at any time.

The study of the biochemical composition of myofibrils is also important. Actin and myosin are similar in all species. Vertebrate myosin occurs in two forms. One form has a higher intrinsic ATPase activity and responds quickly; the other has a slower intrinsic ATPase activity and twitches slowly. Invertebrate muscle fibrils have no such differences between myosin molecules. Paramyosin is a molecule peculiar to invertebrates. It is thought to be part of a supporting structure for the myosin tails. Some invertebrate muscles have been found not to have troponin and to have different forms of tropomyosin.

Electrophysiological studies have shown that invertebrate muscle fibers receive excitatory and inhibitory nerve fibers. Excitatory nerve fibers secrete acetylcholine and cause contraction. Inhibitory nerve fibers secrete serotonin and dopamine. The relaxation produced by these fibers is thought to be mediated by cyclic adenosine monophosphate (cAMP), a derivate of ATP. The relaxation may be a result of the breaking of crossbridges stabilized by paramyosin.

Simple and Complex Arrangements

Invertebrate neuromuscular systems are organized along the same general principles as in vertebrates: They coordinate body movements. All muscle cells shorten after receiving a stimulus, but the shortening will produce motion only if there is a skeleton, a structure to which the muscle fiber is anchored and another part to which force can be applied.

Sedentary sea anemones are relatively simple coelenterates that escape from threats by withdrawing their soft tentacles and contracting toward the substrate protected by thick body walls. Their body wall is a cylinder composed of two layers of cells separated by mesoglea, a jellylike material that allows diffusion of nutrients, wastes, and gases. Two groups of muscles are embedded in the body wall. In the outer body wall, longitudinal muscles parallel the long axis of the body. Their contraction shortens the body wall and draws it downward toward its base; its attachment is to the substrate. The circular muscles ring the inside layer of the body wall. Their contraction lengthens the body by squeezing the contents in-

ward and narrowing the cylinder like a Chinese finger puzzle.

More complex invertebrates have more complex arrangements of muscles and skeletal parts. Their movements become more complex also. The best-known examples come from the larger and more numerous invertebrate phyla, the annelids, mollusks, and arthropods.

Annelids and Mollusks

Annelids have a hydrostatic skeleton like that of the sea anemone. Their bodies are divided into discrete segments, each with a fluid-filled cavity. Circular muscles of a segment contract, pressing against the fluid in the cavity extending it. This contraction of the circular muscles stretches the longitudinal muscles and sets the tiny bristles along the side of each segment into the substrate. The contraction of the longitudinal muscles squeezes the fluid into a shorter segment which stretches the circular muscles and widens the segment. The body is pulled forward by the bristles that had been set in the substrate.

One of the few examples of the use of hydrostatic pressure in place of muscles occurs in some spiders. The hind legs of jumping spiders have flexor muscles that bring the legs toward the body. They have no extensor muscles. Rapidly increased body fluid pressure straightens the legs, causing the spider to jump forward.

Mollusks include bivalves—sedentary clams, mussels, and oysters; gastropods—the slowly moving snails and slugs; and cephalopods—some of the quickest, most intelligent, and largest invertebrates, the octopods and squid. The muscular systems of these animals are as varied as their lifestyles.

The bivalves settle as adults in one place. They are unable to move about. Whenever a predator threatens them or environmental conditions change (for example, if the tide goes out), they must close their shells to shield their delicate body tissues. "Catch" muscles protect them by closing tightly without using much energy. These obliquely striated adductor muscles can maintain their contracted state for a long period. A short train of nerve impulses initiates a contraction that may last hours or days with no further nerve impulses. The catch muscle does not stretch readily, so prolonged pressure by a predator, such as a starfish, does not lengthen the muscle. The catch muscle remains contracted until a relaxation mechanism is activated by neural impulses in separate neurons. These catch muscles contain paramyosin surrounded by myosin molecules.

Arthropods

The arthropods are characterized by an exoskeleton, the joints of which allow movement. The skeleton is outside the body. It is composed of the cuticle, a hardened covering of chitin, with a thin, pliable hinge in the joints. Muscles cross the gap covered by the hinge. In arthropods, muscle tension is controlled largely or entirely by gradation of contraction within a motor unit. The degree of depolarization of these muscle fibers depends upon the frequency of impulses transmitted in motor neurons. Each fiber is a part of several motor units. In addition, inhibitory factors to arthropod muscles can prevent depolarization and, hence, contraction of the muscle (unlike vertebrate muscles). The neurons fire in short bursts to produce rapid movements.

Most rapidly contracting, short-sarcomere fibers are innervated by a phasic axon capable of producing rapid movements. The slowest, long-sarcomere muscle fibers are supplied only by the tonic axon that is active most of the time. These fibers provide for slow movements and postural adjustment. Muscle fibers with intermediate contraction times are innervated by both phasic and tonic axons. A muscle can, therefore, contract over a range of speeds and durations. An extreme example of this is the crustacean claw-opener muscle and the stretcher (extensor) of the proximal leg segment, which are both innervated by a single excitatory neuron. Separate inhibitory neurons supply each muscle so that they can be controlled independently.

In lobsters, the differentiation of the claws into a large crusher and a small cutter is correlated with muscular and neural activity. In their younger states, juvenile lobster claws have the same shape. Either claw can develop into the crusher. In the cutter claw, all the muscle fibers transform to fast fibers. In the crusher claw, all the muscle fibers become slow fibers. The change in shape and muscle type depends upon the presence of a manipulable environment and on the animal having unequal neuromuscular feedback to the central nervous system. The presence of a crusher on one side prevents the development of a crusher on the opposite side. Few generalizations describe invertebrates. Their muscles exhibit the same variety as the organisms that make up this diverse assemblage.

—*Judith O. Rebach*

See also: Anatomy; Arthropods; Crustaceans; Echinoderms; Exoskeletons; Histology; Insects; Invertebrates; Locomotion; Mollusks; Muscles in vertebrates; Physiology; Worms, segmented.

Bibliography

Anderson, D. T. *Atlas of Invertebrate Anatomy.* Sydney, Australia: University of New South Wales Press, 1996. A large collection of anatomical drawings of invertebrates, focusing on Australian examples.

Hickman, C. P., Jr., L. S. Roberts, and F. M. Hickman. *Integrated Principles of Zoology.* 11th ed. Boston: McGraw-Hill, 2001. A widely respected general textbook of zoology that offers some general background on the functioning of invertebrate muscle systems (pages 681-683).

Mader, Sylvia. *Biology: Evolution, Diversity, and the Environment.* Dubuque, Iowa: Wm. C. Brown, 1987. A clearly written text that treats the principles of the fundamental interactions of the musculoskeletal system.

Meglitsch, Paul. *Invertebrate Zoology.* 3d ed. New York: Oxford University Press, 1991. This well-known work describes all of the invertebrate groups and the attributes of each to give a fuller understanding of the interaction of physiology and environment on the evolution of organisms.

Purves, W. K., and G. H. Orians. *Life: The Science of Biology.* Sunderland, Mass.: Sinauer Associates, 1983. An excellent text that integrates discussion of vertebrate and invertebrate muscles so that they can be compared and contrasted effectively.

Schmidt-Nielsen, Knut. *Animal Physiology: Adaptation and Environment.* 5th ed. New York: Cambridge University Press, 1998. An excellent book that raises questions and provides answers. Chapter 11 relates specifically to the muscular system in vertebrates and invertebrates.

Starr, Cecie, and Ralph Taggart. *Biology: The Unity and Diversity of Life.* 9th ed. Pacific Grove, Calif.: Brooks/Cole, 2001. This clearly written text provides a summary of the functioning of invertebrate muscles.

Wells, Martin. *Lower Animals.* New York: McGraw-Hill, 1968. A slim, classic volume that contains much information on invertebrates and their physiology.

MUSCLES IN VERTEBRATES

Type of animal science: Anatomy
Fields of study: Histology, zoology

Vertebrate muscle tissue provides for many vital bodily functions. For these purposes, there are three major types of vertebrate muscle tissue: skeletal, cardiac, and smooth.

Principal Terms

INVOLUNTARY: functioning automatically; not under conscious control

MOTOR NEURON: a nerve cell that transmits impulses from the central nervous system to an effector such as a muscle cell

MOTOR UNIT: a motor neuron together with the muscle cells it stimulates

TISSUE: a group of similar cells that executes a specialized function

TWITCH: a rapid muscular contraction followed by relaxation that occurs in response to a single stimulus

VOLUNTARY: capable of being consciously controlled

The ability of vertebrates to move their bodies and many of the contents of their bodies is a feature of major importance for their survival. The movements result from the contractions and relaxations of tissues specialized for the active generation of force: the muscles. Although initially it may seem sufficient to have only one type of muscle tissue, reflection on the functional requirements makes it clear that more than one type of muscular tissue is probably necessary. For example, the movement of the limbs should be under the conscious, voluntary activation and control of the animal. Otherwise, unwanted and uncoordinated random limb movements would result, or desired and possibly vital movements would not be forthcoming. In addition, the control of the limb movements should be precise, with as wide a range as possible for the forces which can be generated.

On the other hand, consider the movement of the blood through the circulatory system. This job must be continuously performed every minute of the animal's life. It would obviously be better to have this function run automatically, without the need for conscious, voluntary activation and control. The need for precise control of the forces generated here is also not as great as for the limb movements, nor is the need for a wide range of forces as great. The flow of blood to the various organs can be much better regulated by varying the diameter of the blood vessels at or near their entrance to the organs, thereby varying the flow into the organs, while the pumping forces generated to propel the blood remain relatively constant.

For the functions of controlling blood flow into organs and the mixing and passage of food through the digestive tract, it is again most reasonable to have automatic operation of the muscles involved, without the need for conscious, voluntary activation and control. To control the huge number of such muscles consciously is an impossible task in any case. Also, these muscles need to be able to change their lengths greatly, and sometimes, to maintain a maximal contraction for very extended periods; however, very rapid actions are not as important.

The preceding considerations make it appear necessary to have three fundamentally different types of muscle tissue. In fact, vertebrates do have three types of muscle tissue, whose different characteristics match these three sets of operational requirements: skeletal muscle tissue, cardiac muscle tissue, and smooth muscle tissue.

Skeletal Muscle

The skeletal muscle tissue occurs in the form of muscles that are usually attached to bones and cause movements of the skeleton. Some skeletal muscles are attached to the skin or to other skeletal muscles. They are under the animal's conscious, voluntary control. Skeletal muscle cells are long, cylindrically shaped cells with rounded ends and, when viewed under a microscope, are seen to have thousands of alternating light and dark bands oriented perpendicular to their long axis. Because of this appearance, skeletal muscle cells are said to be striated. The cells of a muscle are arranged in a parallel fashion, with their long forms mostly following the long axis of the parent muscle.

When a nerve signal stimulates a skeletal muscle cell, the muscle cell contracts relatively quickly and then, just about as quickly, relaxes (a muscle twitch). The contraction results in a shortening and thickening of the cell, which pulls the two ends of the cell, and whatever is attached to them, toward each other. The duration of a skeletal muscle twitch varies from muscle to muscle. For example, a muscle cell from one of the muscles controlling eye movements may complete a twitch in about ten milliseconds, while a cell from the soleus muscle (found in the calf of the leg) will complete a twitch about ten times more slowly.

The strength of the overall contraction of a skeletal muscle depends on two factors: the number of individual muscle cells in the muscle which become stimulated and therefore contract (called multiple motor unit summation, or recruitment); and the frequency at which the stimulations, and therefore the contractions, occur (called temporal summation). Since most skeletal muscles are composed of hundreds of individual muscle cells, and since each of the nerve cells (motor neurons) which can stimulate the muscle makes contact with only a relatively limited number of the muscle cells (each motor neuron and the muscle cells it contacts forming a motor unit), it is possible to regulate the strength and precision of the muscle's contractions, voluntarily, over a very wide range. This is accomplished by varying the summa-tions—multiple motor unit (number of motor units activated) and temporal (the frequency of motor unit activation)—to meet the demands of a particular situation.

When two muscle twitches occur in rapid succession, it is possible for the muscle to begin the second twitch before it has completely relaxed from the first twitch. In this case, the contraction of the second twitch adds its strength to the force developed as a result of the first contraction; therefore, the second contraction develops more force than the first contraction. The result of continuing the frequency of such twitch-evoking stimuli is a somewhat jerky, oscillating contraction called incomplete tetanus. At high frequencies of stimulation (for example, about fifty twitches per second), there is no evidence of the muscle relaxing from any one twitch before the following twitch takes place, and the muscle makes a very smooth and sustained contraction called complete tetanus. If, following an initial single twitch-evoking stimulus of a skeletal muscle cell, a second stimulus is applied too quickly, there will be no further contraction of the muscle cell, regardless of how strong the second stimulus may be. The cell's ability to respond to this second stimulus is absent until enough time has passed to allow the cell to recover its excitability. The period during which the cell's excitability is absent is referred to as its refractory period. The refractory period of skeletal muscle cells is quite short (about five milliseconds). Skeletal muscles are responsible for the movements of the skeleton (skull, limbs, fingers, toes, and trunk) and for the variety of facial expressions that humans can produce. They also permit speaking, eye movements, breathing, chewing, and swallowing.

Cardiac Muscle

The second type of muscle tissue found in vertebrates is called cardiac muscle. It is the muscle tissue found in the heart. Microscopic examination of cardiac muscle cells reveals them also to possess striations similar to skeletal muscle; however, cardiac muscle is not under voluntary control. Hence, cardiac muscle is referred to as involuntary striated muscle. Cardiac muscle cells also differ from skeletal muscle in that the cardiac fibers branch and form interconnections with one another, forming a network of cells. The individual members of a network are joined by special types of cellular junctions called intercalated disks. The intercalated disks permit the excitation of a single cell, which must occur for the muscle cell to contract, to spread throughout the entire network of cardiac cells; therefore, the contraction of any one cardiac cell will result in the contraction of the entire network of cardiac cells. Thus, the heart's muscle tissue network operates as a functional unit, which is very important for the efficient development of the pressures necessary to push blood through the body's circulatory system. Another very important characteristic of cardiac muscle is its ability to contract in a spontaneous, rhythmical fashion without the need for neural or hormonal stimuli (although both nerves and hormones are present and function as modulators of cardiac activity, also at a subconscious level). This property is termed autorhythmicity, and it accounts for the involuntary nature of cardiac muscle. Cardiac muscle twitches are about ten to fifteen times longer than those of skeletal muscle, and the refractory period (about three hundred milliseconds) is about sixty times longer than that of skeletal muscle. The consequences of these traits are important. The long twitch maintains muscular pressure on the blood contents of the heart until most of it has been pumped out of the heart's chambers. The long refractory period prevents the development of tetanus, and allows the heart to relax between beats so that it can be refilled with blood.

Smooth Muscle

The third type of vertebrate muscle is called smooth muscle because of its lack of striations when viewed microscopically. Its basic and most important functions relate to the maintenance of stable internal conditions within vertebrate bodies. Examples of this are the regulation of blood flow to the various organs to supply them with the proper amounts of oxygen and nutrients, the maintenance of blood pressure during postural changes (such as from reclining to standing), the mixing and movement of ingested food within the digestive tract, and the directing of blood flow through or away from the skin to aid in body temperature regulation. Smooth muscle is also usually involuntary and often displays automaticity. Automaticity refers to the fact that smooth muscle often is stimulated to contract simply by being stretched, as, for example, occurs in the stomach following a meal, without the necessity of neural or hormonal stimulation. Nevertheless, involuntary nerves and hormones are both involved in regulating the actions of smooth muscle cells. The functions of smooth muscle are almost always such that contraction is the appropriate response to stretch of the organ containing the smooth muscle (as in the stomach example just mentioned).

Smooth muscle cells are very small: only about eight micrometers in diameter and between thirty and two hundred micrometers long. They are spindle-shaped, tapering toward each end. Although very small, these cells are able to contract to a length which is a much smaller fraction of their resting length than can either cardiac or skeletal muscle cells. Smooth muscle can also remain fully contracted for long time periods and consume very little energy. The speed of contraction of smooth muscle, however, is the slowest of all three muscle types.

Many smooth muscle cells are arranged as functional units which have the individual cells connected to each other by gap junctions. Gap junctions permit the passage of contraction signals from one cell to the next in the network in much the same way as it occurs in cardiac muscle. The result is that when one cell within the network begins to contract, a wave of contraction spreads throughout the entire network of smooth muscle cells.

Studying Vertebrate Muscle

Light and electron microscopes are frequently used to study muscle tissue. In particular, the electron microscope has made it possible to visualize the intricate and highly specialized internal structures of muscle cells, most of which are impossible to view with a light microscope, given their very small dimensions.

Many studies of muscle cells are actually studies of isolated parts of muscle cells. It is possible to break apart muscle cells and to separate many of their components from one another by the use of centrifugation. Centrifugation is the use of centrifugal force to separate objects by size and/or density. A centrifuge is a device in which samples of objects (in this case, muscle-cell structural components) that are to be separated are subjected to high centrifugal force generated by spinning the samples at great speed. The centrifugal force is the force that tends to impel an object outward from the center of rotation.

Once the various subcellular components are isolated, they can be subjected to a wide variety of biochemical tests to determine what type of molecules they contain (such as proteins, lipids, or carbohydrates), how much of each type they contain, and the exact chemical composition of these molecules (such as the amino acid sequence of a protein). In the case of muscle cells, much of the interest has centered on the large quantities of the proteins actin and myosin which they have been found to contain. Using isolated actin and myosin, scientists have learned much about their functions in muscle cells. In particular, it is now known that these two proteins are the molecules which generate the contractile forces which muscle cells are capable of producing. The precise molecular mechanisms are still unknown, but great progress has been made in discovering how actin and myosin accomplish the generation of contractile forces.

Studies of the electrical properties of muscle cells are performed using very sensitive amplifiers connected to fine glass pipette microelectrodes. These electrodes are formed from thin glass tubes which have been heated and pulled apart to produce a tapering of the wall of the tube. The final tip of the tapered tube is much smaller than a muscle cell. This makes it possible to insert the tip into the cell without killing it. When the electrodes are filled with a solution capable of conducting electrical signals, they can record the electrical activity that takes place in living, contracting, and relaxing muscle cells: their electrophysiological properties. These studies have revealed that all muscle cells possess an electrical voltage difference between their interior region and the surrounding environment. Just before a muscle cell begins to contract, this voltage difference decreases toward a zero value extremely rapidly. Combining such electrophysiological studies of muscle cells with simultaneous biochemical studies can be very rewarding.

Other techniques used in the study of muscle tissue include the observation of the relationships that various muscles have with other muscles, with nerves, and with bones, and the use of strain gauges to measure the strength of muscles when different experimental conditions occur (such as temperature changes, blood-flow variations, and pharmacological treatments). Even high speed X-ray motion pictures have been applied to the study of muscle tissue.

The Importance of Muscle
Vertebrate muscle tissue is involved in almost every bodily function. In particular, any function requiring the movement of some body part or parts needs to have a source of motive power. For this purpose, evolution has resulted in the development of the specialized muscle tissues found in all vertebrates, which account for up to 50 percent of their total body weight.

Because of the muscle-tissue characteristics of excitability (the ability to respond to stimuli), extensibility (the ability to be stretched), contractility (the ability to shorten actively and thereby generate force for the production of work), and elasticity (the ability to return to original length following extension or contraction), vertebrates have a wide range of important capabilities available which assist them in their survival.

For the muscular system, disease states or disorders resulting from improper nutrition, injury, or toxic substances are very often life-threatening conditions. This is most obvious for disturbances of the respiratory muscles or of the muscles involved in eating and swallowing; however, muscular disorders involving the heart, the circulatory system, the digestive tract, or any major group of skeletal muscles may also prove to be life-threatening.

The most common heart problems are caused by reduction or blockage of the blood supply to the heart muscle. Reduced blood supply usually is the cause of a reduced oxygen supply. The insufficient oxygen supply weakens the heart-muscle cells, causing the condition of ischemia. Complete interruption of the blood supply to an area of cardiac muscle tissue usually results in necrosis (death) of the affected muscle cells; the condition is referred to as myocardial infarction. The dead muscle cells do not regenerate but are replaced by scar tissue, which is not contractile. This results in decreased pumping efficiency by the heart. De-

pending on the size and location of the dead area, the result may range from barely noticeable to sudden death.

As a consequence of some bacterial infections, it is possible for the lining of the heart muscle to become inflamed. This can then result in abnormal irritation of some of the heart-muscle cells. These cells can then disrupt the normal autorhythmicity of the heart. Irregular heartbeats and/ or uncoordinated contractions among the heart-muscle cells ensues. These conditions are very serious and can lead to death.

—*John V. Urbas*

See also: Amphibians; Anatomy; Bone and cartilage; Endoskeletons; Fish; Flight; Heart; Histology; Locomotion; Mammals; Muscles in invertebrates; Physiology; Reptiles; Vertebrates.

Bibliography

Hildebrand, Milton. *Analysis of Vertebrate Structure.* 5th ed. New York: John Wiley & Sons, 2001. This college-level text describes muscle tissue, concentrating on skeletal muscles and presenting their anatomy, physiology, and evolution, as well as cardiac muscle. Smooth muscles are discussed in a variety of places throughout the text. The index is very complete; the glossary is adequate. The references for each chapter are good, but some are of an advanced nature.

Kardong, Kenneth V. *Vertebrates: Comparative Anatomy, Function, Evolution.* 2d ed. Dubuque, Iowa: Wm. C. Brown, 1998. A textbook written for upper-level undergraduates, with solid coverage of the muscular systems in different vertebrates.

Keynes, R. D. *Nerve and Muscle.* 3d ed. New York: Cambridge University Press, 2001. Covers all aspects of the nerve and muscle systems and their interrelations.

Netter, Frank H. *Musculoskeletal System: Anatomy, Physiology, and Metabolic Disorders.* Summit, N.J.: CIBA-GEIGY, 1987. Probably one of the best scientific illustrators, Netter has created stunning and extremely informative diagrams to help both the general reader and specialist in their study of the musculoskeletal system. High school readers may find the text somewhat difficult, but the illustrations alone make this book worth consulting. Highly recommended.

Tortora, Gerard J., and Nicholas P. Anagnostakos. *Principles of Anatomy and Physiology.* 9th ed. New York: John Wiley & Sons, 2000. Chapters 10 and 11 cover muscles and muscle tissue with good illustrations, excellent photographs, and well-written discussion. This college-level text has a very good glossary, an index, and a short list of titles for further reading.

MUTATIONS

Types of animal science: Development, evolution
Fields of study: Biochemistry, cell biology, developmental biology, embryology, evolutionary science, genetics

Mutations have been used to work out metabolic pathways, define genes and their controlling sites, understand how multicellular organisms develop, and study how organisms evolve.

Principal Terms

ALLELE: one of many possible sequences of a gene

CONTROLLING SITE: a sequence of nucleotides generally fifteen to sixty nucleotides long, to which a transcriptional activator or repressor binds

GENE: a sequence of one thousand to ten thousand nucleotides, which usually specifies a protein

MUTATION: a change in the nucleotide sequence of a gene or of a controlling site; changes in genes alter the protein, whereas changes in controlling sites determine where and how much of a protein is produced

---

In all living organisms, the hereditary information consists of two complementary strands of deoxyribonucleic acid, known as DNA. DNA strands are constructed of subunits called nucleotides that consist of a nitrogenous base, a deoxyribose sugar, and a phosphate. Generally, DNA strands consist of millions of nucleotides attached to each other like the rings of a chain. There are four different nucleotides. The two complementary strands are held together by hydrogen bonds between the bases. If there is an adenine (A) in one strand, it hydrogen bonds to a thymine (T) in the complementary strand. Similarly, if there is a guanine (G) in one strand, it hydrogen bonds to a cytosine (C) in the other strand. Thus, the amount of A is equal to T and the amount of G is equal to C.

The order of nucleotides in a strand specifies the order of amino acids in proteins.

Genetics is the study of how the information in DNA molecules is expressed and how DNA molecules account for the heredity of an organism. Changes in the sequence of nucleotides in DNA may alter an organism's proteins, which in turn may change one or more of an organism's traits. DNA changes are called mutations, and the organisms that harbor mutations are known as mutants. The commonly encountered trait or organism is referred to as the wild-type. The characterization of mutations and mutants has been and still is one of the best ways of discovering the function of genes and determining how organisms maintain themselves, evolve, and develop. A study of mutations and mutants also has shed light on numerous genetic diseases.

Heat: The Cause of Most Spontaneous Mutations

Most mutations are caused by the instability of the nucleotide bases. Sometimes bases hit by rapidly moving water molecules briefly alter their chemistry. These chemical changes are known as tautomeric shifts. Tautomeric shifts alter the distribution of electrons and protons in the bases so that bases in the complementary strands no longer pair normally. The redistribution of electrons and protons in the bases causes abnormal pairings to occur. For example, an abnormal adenine (A*) pairs with C and an abnormal guanine (G*) pairs with T.

When a DNA molecule is being replicated, spontaneous tautomer shifts can result in permanent mutations. Spontaneous mutations occur, for example, when an A in the template strand undergoes a tautomer shift (A→A*) just as the DNA polymerase reaches it. A cytosine pairs with the A* and becomes part of the new strand being synthesized by the DNA polymerase. When this new strand, with a C in it instead of a T, functions as a template, the complementary strand will have a G in it rather than an A. This type of tautomeric shift during DNA replication converts what normally would have been an A = T base pair in "granddaughter" DNA to a G = C base pair.

Chemicals That Cause Mutations

Mutations are induced by many chemical and physical agents that are called mutagens. Many chemicals act as mutagens. Nitrous acid, for example, diffuses into cells and removes amino groups from DNA bases. These chemically altered bases no longer base pair normally. When DNA is replicated or repaired, incorrect nucleotides are inserted opposite the chemically altered bases. Nitrous acid changes adenine to hypoxanthine, which pairs with cytosine. It also changes guanine to xanthine, which pairs with T.

Base analogues are molecules that closely resemble normal nucleotides and consequently are incorporated into DNA that is being repaired or replicated. A base analogue to thymine, such as 5-bromouracil (5BU), is efficiently incorporated into DNA. 5BU spontaneously undergoes tautomeric shifts at a high rate. The abnormal form of 5BU pairs with G rather than A. Thus, 5BU introduces many base pair transitions in newly synthesized DNA molecules.

The most potent mutagens are alkylating agents, such as nitrosamines, methyl bromide, and ethylene oxide. These mutagens attach methyl or ethyl groups (alkyl groups) to A and G. This causes A and G to undergo tautomeric shifts at a higher than normal rate.

High-Energy Electromagnetic Radiation and Particles

Ultraviolet (UV) light is a powerful mutagen. It generally penetrates cells but is readily absorbed by thymine and cytosine bases in DNA. When two thymines or two cytosines next to each other in a strand absorb UV light, they often react chemically with each other to form thymine dimers or cytosine dimers that distort the DNA. These distorted regions stimulate a repair system that cuts out the

---

The Use of a Mutation to Determine a Catabolic Pathway

Sir Archibald Garrod (1857-1936) was an English physician who studied the effects of a mutation in humans and discovered a catabolic pathway. Garrod's research into the cause of alkaptonuria, a condition where the urine turns black upon exposure to the air, led to the idea that alkaptonuria occurred in persons with two defective genes for the enzyme that eliminates homogentisic acid in the urine. Homogentisic acid turns black when oxidized by oxygen.

In a paper published in 1902, Garrod analyzed a number of families where alkaptonuria occurred, establishing that the trait was recessive and followed simple Mendelian genetics. This was the first account of recessive inheritance in humans. When Garrod fed alkaptonuric patients homogentisic acid, their urine blackened upon exposure to air and contained nearly the same amount of homogentisic acid

that they consumed. Normal individuals given homogentisic acid contained no detectable homogentisic acid in their urine. By feeding alkaptonuric patients various compounds that might give rise to homogentisic acid, Garrod discovered the metabolic pathway that led from the amino acids phenylalanine and tyrosine to homogentisic acid.

In his book published in 1908, *Inborn Errors of Metabolism*, Garrod concluded that alkaptonuria is caused by a block in the catabolic pathway that eliminates homogentisic acid. He reasoned that the catabolic block was caused by the lack of a specific enzyme. The fact that the trait followed Mendelian genetics suggested that a functional gene normally specified the enzyme involved in the breakdown of homogentisic acid. This was the beginning of the idea that genes specify enzymes.

Mutations Used to Determine an Anabolic Pathway

Between 1937 and 1941, George W. Beadle and Edward L. Tatum isolated fungal mutants that were unable to synthesize various vitamins and amino acids. They isolated these mutants to try to understand how genes control specific reactions in various metabolic (synthetic) pathways. The mutants would not grow on a simple medium unless supplied with the nutrient they were unable to synthesize. Beadle and Tatum induced large numbers of mutations in the fungus they were studying and isolated hundreds of mutants. Seven mutants were isolated, each carrying a defective form of a gene involved in the synthesis of the amino acid arginine. Using these mutants, Beadle and Tatum were able to order the chemical reactions by feeding the fungal mutants different metabolic intermediates. If the fungal mutants grew when a particular metabolic intermediate was provided, then the gene mutation affected a step leading to the synthesis of the intermediate. If the mutant did not grow, however, then the gene mutation affected a step that converted the intermediate into arginine. They concluded from their experiments that each gene controlled a different chemical reaction. They confirmed that a specific chemical reaction fails to take place in diploid organisms if both representatives of a given gene were defective. Beadle and Tatum's research strengthened the idea that genes specify enzymes. Their idea became known as the "one gene-one enzyme" hypothesis.

dimers and some DNA on either side and replaces the DNA with normal nucleotides. Excessive repair leads to an increased occurrence of spontaneous mutations. Sometimes a distortion in the template allows the DNA polymerase to add or to leave out nucleotides as it moves along the template during strand synthesis. This may explain how some additions and some deletions occur.

Very energetic electromagnetic radiation, such as X rays and gamma rays, as well as high-energy particles released from radioactive atoms, also induce mutations. These energetic mutagens easily penetrate cells and chemically alter many molecules in their path by stripping away electrons. Ions and radicals formed by these mutagens react with the DNA, causing bases to be released and DNA to break. DNA deletions, DNA transpositions, and DNA inversions may be promoted by DNA breakage.

When a gene is mutated, the protein the gene specifies generally becomes nonfunctional. In bacteria that have only one copy of each gene, traits are immediately altered by a mutation. On the other hand, in animals and plants that may have more than one copy of a gene, a mutation in only one gene may not produce a new trait because the wild-type (normal) gene often provides enough of the essential protein. When developing animals and plants are missing both genes, however, they may fail to develop or they may develop, but in a different way.

A few mutations are beneficial to the organism that acquires them and may make the organism better adapted to its environment. These beneficial mutations may make a protein work a little better or in a different way. Some mutations are also beneficial because they create diversity in a population. Diversity promotes the survival of a population by ensuring that some organisms survive if the environment drastically changes. A population that is too well adapted to a particular environment will not survive if there are significant changes in the environment. There have been at least five major mass extinctions during the history of life on earth, in some cases eliminating more than 85 percent of all species. The organisms that survived these mass extinctions were much less specialized than the organisms that did not.

Usefulness of Mutations

Mutations have been extremely useful in the study of organisms. Mutations allow scientists to understand what a particular gene and its product do. If the mutation eliminates the gene (and prod-

Mutations Used to Determine How Genes Are Regulated

In 1963, François Jacob and Jacques Monod published a classic paper in which they presented a model for how genes might be regulated. They used wild-type and mutant bacteria to demonstrate that certain enzymes are not produced when the genes that specified the enzymes are blocked by a repressor protein. Wild-type bacteria contain a repressor gene (R). A repressor specified by the repressor gene blocks DNA transcription (the synthesis of mitochondrial ribonucleic acid, mRNA) by binding to a controlling site (operator site, O) on the DNA that partially overlaps the ribonucleic acid (RNA) polymerase binding site (promoter site, P). If RNA polymerases are prevented from binding DNA, transcription is blocked. When transcription of a gene is inhibited, no enzyme can be synthesized. Jacob and Monod demonstrated that certain mutations (O) in the operator site could block repressor binding to the DNA. These controlling site mutations result in the continuous synthesis of mRNA and the enzymes the mRNA specified. These operator constitutive mutations (O) were shown to prevent repressor binding. Certain mutations in the repressor gene (R-) would eliminate the repressor and result in continuous synthesis of the enzymes regulated by the repressor. Other mutations in the repressor gene would create a super repressor (R) that failed to disengage from the operator site. This blocked the synthesis of the enzymes regulated by wild-type operators (O). Super repressors do not respond to the organic ligand that may bind to them and causes them to disengage from operator sites. Nevertheless, super repressors are unable to bind to constitutive operator sites. Researchers have discovered that many genes are also regulated by protein activators. Some promoter site mutations (P) increase the level of transcription, whereas other promoter site mutations (P-) abolish transcription.

Controlling sites, such as O and P, can be differentiated from genes, such as R, since controlling sites only affect genes on the DNA molecule where they are located (*cis* effect). Genes specifying repressors and activators affect genes both on the DNA molecule where the regulatory genes are located (*cis* effect) and also on other DNA molecules (*trans* effect). This is how genes became known as cistrons.

uct), scientists can guess what the gene does by looking at the affected organism. For example, if a mutation changes eye color (red to white), the affected gene most likely has something to do with pigment synthesis or deposition of the pigment in the eye.

The study of mutations and mutant organisms has helped scientists unravel anabolic (synthetic) and catabolic (degrading) pathways, determine how parental genes combine to produce new characteristics in progeny, clarify what genes are and what they do, establish how genes are regulated, and even decipher how multicellular organisms develop and evolve.

Mutations in Development and Evolution

The study of mutations and mutant organisms at the end of the twentieth century led to an understanding of how multicellular organisms develop and evolve. One of the most useful organisms in unraveling the development problem has been the small fruit fly *Drosophila*. Thousands of mutations that affect development of this organism have been characterized. Scientist found that a hierarchy of genes are involved in development. First, maternal genes are expressed. These genes activate gap genes and these, in turn, activate pair-rule genes. All of these gene catagories are known to be involved in regulating the expression of homeotic genes. Maternal, gap, pair-rule, and homeotic gene products all function as transcriptional activators and repressors. For example, the maternal gene product called bicoid stimulates its own synthesis, whereas it inhibits the synthesis of another maternal gene product called nanos.

Maternal Genes→Gap Genes→Pair Rule Genes→Homeotic Genes

This gene hierarchy is responsible for the anterior-posterior segmentation seen in *Drosophila*. Edward B. Lewis, Christiane Nüsslein-Volhard, and Eric Wieschaus shared the 1995 Nobel Prize in Physiology or Medicine for their studies of the genes that control *Drosophila* development.

Thomas Hunt Morgan

Born: September 25, 1866; Lexington, Kentucky
Died: December 4, 1945; Pasadena, California
Fields of study: Genetics
Contribution: Morgan's studies popularized the use of the fruit fly *Drosophila* for the study of animal genetics. He is credited with discovering the first sex-linked trait in *Drosophila* and with demonstrating how new characteristics could be passed on to successive generations. Morgan and his students showed that chromosomes exchanged genes, a process known as crossing over. In 1933, Morgan was awarded the Nobel Prize for Physiology or Medicine for his work on *Drosophila* genetics.

In his classic paper of 1910, Thomas Hunt Morgan described very rare white-eyed flies that appeared spontaneously in a red-eyed population. Because these mutants were always males, Morgan suspected that the gene controlling eye color was linked to the X chromosome that influenced the development of sex. The observation that white-eyed females could be produced from certain matings, however, indicated that the white-eyed trait was not limited to males.

Morgan's experiments demonstrated for the first time that a gene controlling eye color was linked (or limited) to the X chromosome. Red and white eyes are caused by different alleles of the same gene. These alleles on the X chromosome are represented in the following manner: X and X. Any gene linked to the X chromosome is called a sex-linked gene. In *Drosophila*, two X chromosomes generally result in a female fly, whereas one X chromosome results in a male. Morgan found that the R allele inducing red eyes is dominant over the r allele that allows white eyes to develop when it is the only allele a fly has.

The Y chromosome that pairs with the X chromosome in males (XY) lacks sex-linked genes. Thus, mating red-eyed females (XX) to white-eyed males (XY) results in all first filial (F1) generation flies, XX females and XY males, having red eyes. Matings between the F1 flies demonstrated that the red-eye-inducing allele and the white-eye-promoting allele always remained associated with the X chromosome. This suggested that the alleles were linked to the X chromosome.

In 1913, A. H. Sturtevant, working in Morgan's laboratory, reported on mutations linked together on a fruit fly's X chromosome. Sturtevant demonstrated that recombination between two X chromosomes could separate genes controlling different traits. In addition to using the alleles that determined eye color, Sturtevant used alleles that influenced wing formation. A normal wing forms under the influence of the L gene, but a miniature wing is associated with the l allele of the L gene. Genes linked on the X chromosome may be shown as follows: XL. Sturtevant observed recombination when he characterized the offspring from certain crosses. A female fly with red eyes and normal wings, XlXL, usually produces two types of eggs. One type of egg has the Xl chromosome, whereas the other type of egg has the XL chromosome. Very infrequently, when there is a crossover between the X chromosomes, rare eggs are produced with recombinant X chromosomes, one type of egg has the Xl chromosome, whereas the other type of egg has the XL chromosome. When these recombinant eggs fuse with a sperm carrying only a Y chromosome, recombinant male flies result, those that have normal eyes and normal wings (XLY) and those that have white eyes and miniature wings (XlY). By using various mutant flies, Morgan and Sturtevant discovered that they could both order a number of different genes on the X chromosome and determine how far they were from each other. The farther a gene is from another gene, the greater the number of recombinant offspring. The pattern of offspring was used to determine the sequence of genes on the X chromosome. Finding flies with mutations in different genes was essential for determining the sequence of genes and the distances between them.

—Jaime Stanley Colomé

Thomas Hunt Morgan popularized the use of the fruit fly Drosophila melanogaster *for the study of genetic mutations.* (Library of Congress)

Homeotic genes are found in all multicellular organisms. Homeotic genes similar to those found in *Drosophila* control the development of segments most visibly exemplified by the vertebrae and the bones in animals' appendages. Mutations in homeotic genes or their controlling sites affect the development of segments. Segments can be eliminated or modified by homeotic gene controlling site mutations.

One well-studied homeotic gene in *Drosophila* is the gene antennapedia, *antp*. Certain mutations in the controlling sites for the antennapedia gene result in legs developing rather than head antennae. Another homeotic gene is ultrabithorax, *ubx*.

Some mutations in the controlling sites for ultrabithorax gene result in a second pair of wings developing where the pair of halteres normally develop. Halteres are tiny, winglike appendages that all flies have, which promote stable flight. Other mutations in the controlling sites for *ubx* produce a second pair of winglike structures that are half haltere (anterior portion), half wing (posterior portion). By studying mutations and the altered traits, scientists have discovered that controlling site mutations change when and where proteins are synthesized. For example, if a protein is to be produced in seven segments along the anterior-posterior axis of an animal, there must be at least seven different controlling sites that can respond to the different activators and repressors produced in each segment.

Numerous studies suggest that antennapedia and ultrabithorax are transcriptional repressor-activators that not only repress the development of legs and wings, but also stimulate the development of antennae and halteres, respectively. The study of *Drosophila* mutants is beginning to clarify how antennae and mouth parts evolved from leglike appendages and how halteres evolved from wings. The study of genes and controlling sites has led to the understanding of their role in the maintenance, development, and evolution of every organism.

—Jaime Stanley Colomé

See also: Asexual reproduction; Breeding programs; Cleavage, gastrulation, and neurulation; Cloning of extinct or endangered species; Copulation; Courtship; Determination and differentiation; Development: Evolutionary perspective; Estrus; Fertilization; Gametogenesis; Hermaphrodites; Hydrostatic skeletons; Mating; Parthenogenesis; Pregnancy and prenatal development; Reproduction; Reproductive strategies; Reproductive system of female mammals; Reproductive system of male mammals; Sexual development.

Bibliography

Brennessel, Barbara. "Inborn Error of Metabolism." In *Encyclopedia of Genetics*, edited by Jeffrey A. Knight and Robert McClenaghan. Pasadena, Calif.: Salem Press, 1999. A short history of Sir Archibald Garrod's discoveries and a short discussion of other inborn errors of metabolism in humans.

Colomé, Jaime S. "Gene Regulation: Bacteria." In *Encyclopedia of Genetics*, edited by Jeffrey A. Knight and Robert McClenaghan. Pasadena, Calif.: Salem Press, 1999. Regulation of the lactose operon, arabinose operon, tryptophan operon and a flagellin operon is discussed.

Foran, John M. "Thomas Hunt Morgan: 1933." In *The Nobel Prize Winners: Physiology or Medicine*, edited by Frank N. Magill. Pasadena, Calif.: Salem Press, 1991. A history of Morgan's discoveries, in particular his work showing that one of the genes that controls eye color is linked to the X chromosome.

Fornari, Chet S. "Homeotic Genes." In *Encyclopedia of Genetics*, edited by Jeffrey A. Knight and Robert McClenaghan. Pasadena, Calif.: Salem Press, 1999. Discusses the function of homeotic genes.

Gliboff, Sander. "Gregor Mendel and Mendelism." In *Encyclopedia of Genetics*, edited by Jeffrey A. Knight and Robert McClenaghan. Pasadena, Calif.: Salem Press, 1999. A well-done history and summary of Mendel's research with wild-type and mutant pea plants.

Kalumuck, Karen E. "*Drosophila melanogaster*." In *Encyclopedia of Genetics*, edited by Jeffrey A. Knight and Robert McClenaghan. Pasadena, Calif.: Salem Press, 1999. A brief review of T. H. Morgan and Sturtevant's work with *Drosophila* followed by a discussion of the use of *Drosophila* to study development.

Kang, Manjit S. "One Gene-One Enzyme Hypothesis." In *Encyclopedia of Genetics*, edited by Jeffrey A. Knight and Robert McClenaghan. Pasadena, Calif.: Salem Press, 1999. A short history of the concept that each gene specifies a polypeptide.

Morgan, T. H. "Sex Limited Inheritance in *Drosophila*." *Science* 32 (1910): 120-122. Morgan demonstrates how a spontaneous mutation (sport) affecting eye color in *Drosophila* is passed to succeeding generations. He showed that a mutated form of the gene is recessive to the normal form of the gene and that the mutated and normal gene are linked to the X chromosome. Genes that are linked to the X chromosome are now known as sex-linked genes.

Sturtevant, A. H. "The Linear Arrangement of Six Sex-Linked Factors in *Drosophila*, as Shown by Their Mode of Association." *Journal of Experimental Zoology* 14 (1913): 43-59. Sturtevant provides evidence that genes are arranged in a linear sequence along the X chromosome. He was able to map the genes relative to each other and determine a distance between them.

Thompson, James N., and R. C. Woodruff. "Mutation and Mutagenesis." In *Encyclopedia of Genetics*, edited by Jeffrey A. Knight and Robert McClenaghan. Pasadena, Calif.: Salem Press, 1999. Discusses the processes of mutation and mutagenesis.

NATURAL SELECTION

Type of animal science: Evolution
Fields of study: Ecology, evolutionary science, genetics

Natural selection is the process of differential survival and reproduction of individuals resulting in long-term changes in the characteristics of species. This process is central to evolution.

Principal Terms

ADAPTATION: the process of becoming better able to live and reproduce in a given set of environments

EVOLUTION: any cumulative change in the characteristics of organisms or populations over many generations

FITNESS: the relative ability of individuals to pass on genes to subsequent generations

HERITABILITY: the extent to which variation in some trait among individuals in a population is a result of genetic differences

POPULATION: a group of individuals that occupy a common area and share a common gene pool

SPECIES: the group of all individuals or populations that interbreed or potentially interbreed with one another under natural conditions

Natural selection is a three part process. First, there must exist differences among individuals in some trait. Second, the trait differences must lead to differences in survival and reproduction. Third, the trait differences must have a genetic basis. Natural selection results in long-term changes in the characteristics of the population.

As one of the central processes responsible for evolution, natural selection results in both fine-tuning adaptations of populations and species to their environments and creating differences among species. The importance of natural selec-tion was first recognized by Charles Darwin, who provided the first widely accepted mechanism for evolutionary change. Natural selection is one of several processes responsible for changing the characteristics of populations and leading to an increase in adaptiveness. Other processes include genetic drift and migration. These processes inter-act with the processes responsible for producing variation (mutation and development) and those responsible for determining the rate and direction of evolution (mating system, population size, and long-term ecological changes) to establish the path of evolution of a species.

The Process of Natural Selection

Natural selection occurs through the interaction of three factors: variation among individuals in a population in some trait, fitness differences among individuals as a result of that trait, and her-itable variation in that trait. If those three condi-tions are met, then the characteristics of the popu-lation with respect to that trait will change from one generation to the next until equilibrium with other processes is reached. An example that dem-onstrates this process involves the peppered moth. It has two forms in the United Kingdom, a light-colored form and a dark-colored form; there is variation in color among individuals. Genetic analysis has shown that this difference in color is caused by a single gene; the variation has a herita-ble basis. The moth is eaten by birds that find their food by sight. The light-colored form cannot be seen when sitting on lichen-covered trees, while the dark-colored form can be seen easily. Air pol-lution kills the lichen, however, and turns the trees dark in color. Then, the dark-colored form is hidden and the light-colored form visible. Thus, differences in color lead to fitness differences. In the early nineteenth century, the dark-colored form was very rare. In the last half of the nineteenth century, however, air pollution increased, and the dark-colored form became much more frequent as a result of natural selection.

The characteristics of a population can be changed by natural selection in several ways. If individuals in a population with an extreme value for a trait have the greatest fitness on average, then the mean value of the trait will change in a consistent direction, which is called directional selection. For example, the soil in the vicinity of mines contains heavy metals that are toxic to plants. Individuals with the greatest resistance to heavy metals have the highest survivorship. Evolution leads to an increase in resistance. If individuals in a population with intermediate values for a trait have the greatest fitness on average, then the variation in the trait will be reduced, which is called stabilizing selection. For example, in many species of birds, individuals with intermediate numbers of offspring have the greatest fitness. If an individual has a small number of offspring, that parent has reduced reproduction and a low fitness. If the number of offspring is large, the parent will not be able to provide enough food for all the young, and most, or all, will starve, again resulting in reduced reproduction and a low fitness. Evolution leads to all birds producing the same, intermediate number of offspring. If individuals in a population with different values for a trait have the greatest fitnesses on average and intermediates have low fitness, then the variation in the trait will be increased. This is called disruptive selection. For example, for Darwin's finches, individuals with long, thin bills are able to probe into rotting cactus to find insects. Individuals with short, thick bills are able to crack hard seeds. Individuals with intermediate-shaped bills are not able to do either well and have reduced fitness relative to the more extreme types. Evolution leads to two different species of finch with different bills.

A Slow and Holistic Process

Natural selection is a slow process. The rate of evolution—that is, response to selection—is determined by the magnitude of fitness differences among individuals and the heritability of traits. Fitness differences tend to be small so that more fit individuals on average may have only a few more offspring than less fit individuals. Heritabilities of most traits are low to intermediate, meaning that most differences among individuals are not a result of genetic differences. So, even if one individual has many offspring and another has few offspring, they may not differ genetically and no change will occur. For example, if all the beetles in a population were between one and two centimeters in length and there was selection for larger beetles, it could take five hundred generations before all beetles were larger than two centimeters. Also, the direction of selection may change from one generation to the next, so that no net change occurs.

Natural selection does not act on traits in isolation. How a trait affects fitness in combination with other traits, called correlational selection, is important. For example, fruit flies lay their eggs in rotting fruit. Considered in isolation, a female should always lay as many eggs as possible. One fruit is not big enough for all the eggs she might lay, however, so she must fly from fruit to fruit. Flying requires energy, and the more energy that is used in flight the less that can be used to make eggs. So natural selection results in the division of energy between eggs and flight that yields the greatest overall number of offspring. This example demonstrates that the result of natural selection is often a trade-off among different traits.

By acting differently on males and females, natural selection results in sexual selection. This form of selection can explain differences in the forms of males and females of a species. In general, because male gametes, sperm, are much smaller and "cheaper" to produce than female gametes, eggs, more sperm than eggs are produced. As a result, it is possible for one male to fertilize many eggs, while other males fertilize few or no eggs. For example, a lion pride usually consists of

Charles Darwin

Born: February 12, 1809; Shrewsbury, England
Died: April 19, 1882; Downe, England
Fields of study: Entomology, evolutionary science, human origins, invertebrate biology
Contribution: Darwin was not the first philosopher or scientist to posit a theory of evolution, but his theories of natural and sexual selection provided much of the foundation for later scientific evolutionary theory.

Charles Robert Darwin had briefly studied medicine at the University of Edinburgh and attended Cambridge University, intending to prepare for the ministry, when he was offered a chance to sail on the HMS *Beagle* as a naturalist and companion to Captain Robert FitzRoy. The fifty-seven month voyage, from December 7, 1831, to October 2, 1836, allowed Darwin unique opportunities to explore fossils, fish and sea mammals, and coral reefs. Lengthy land excursions allowed him to examine land animals and fossils, primarily in South America.

Returning to England, he first published his findings as *Journal and Remarks*, volume 3 in the series *Narrative of the Surveying Voyages of H.M.S. "Adventure" and "Beagle" Between 1826 and 1836* (1839). This work was revised and published the same year as *Journal of Researches into the Geology and Natural History of the Various Counties Visited by H.M.S. "Beagle"* (1839). His findings caused him to question generally accepted assumptions about animal creation and to posit evolutionary change as occurring mainly through natural selection. In later works, he increasingly stressed the importance of sexual selection. Challenged by peers to examine individual species before generalizing about life as a whole, he began lengthy examinations of such life-forms as beetles and barnacles, which he published.

By the 1850's, despite his aversion to public controversy, he accepted the need to publish his general theories. They appeared in 1859 as *On the Origin of Species by Means of Natural Selection: Or, The Preservation of Favoured Races in the Struggle for Life*. Although he avoided discussion of human origin in this work, the controversy he dreaded was forthcoming. His ideas, however, were adopted by young scientists, most notably Thomas Henry Huxley, who sought to establish the natural sciences as disciplines separate

Charles Darwin's theory of evolution through natural selection revolutionized not only the scientific view of the world but popular understanding as well. (National Archives)

from the natural theology that then prevailed in universities. These scientists became his spokespersons, as he continued his experiments at his estate; his theories gained widespread acceptance. In 1871, he dealt directly with the origin of human life in *The Descent of Man*; he followed this, in 1872, with *The Expression of Emotions in Man and Animals*. These works clearly placed man within the animal kingdom, not the product of a separate creation.

Darwin also published on narrower topics involving animal life and fossils, and extensively revised *On the Origin of Species*, ultimately producing six revised editions in the quarter-century after its initial publication. He wrote a brief autobiography, published posthumously in 1887. He was awarded numerous honors in England and on the Continent. At his death, his work was so widely respected that, despite his religious skepticism, England honored him with burial in Westminster Abbey.

—Betty Richardson

one or a few males and many females. Other males are excluded, and they live separately; larger males are able to chase away smaller males. The thick mane on male lions helps to protect their throats when they fight other males. Thus, larger males with thicker manes father more cubs than other males, leading to selection on these traits. Only males are under natural selection since all females, regardless of size, will mate. The result is that males are larger than females and have manes.

Group Selection

Natural selection can occur not only among individuals but also among groups. This process is generally known as group selection; when the groups are composed of related individuals, it is called kin selection. Group selection operates the same way as individual selection. The same three conditions are necessary: variation among groups in some trait, fitness differences among groups because of that trait, and a heritable basis for that trait. For example, in Australia, rabbits introduced from Europe in 1859 spread rapidly during the next sixty years. In order to control the rabbits, a virus was introduced in 1950. At first, the virus was very virulent, killing almost all infected animals within a few days. After ten years, however, the virus had evolved to become more benign, with infected rabbits living longer or not becoming sick at all. Virulent strains of the virus grow and reproduce faster than benign strains. So, within a single rabbit, virulent strains have a higher fitness than benign strains. The longer a rabbit lives, however, the more opportunity there is for the virus to be passed to other rabbits. Thus, a group of benign viruses infecting a rabbit are more likely to be passed on than a group of virulent viruses. In this example, group selection among rabbits resulted in evolution opposite to individual selection within rabbits; however, group selection and individual selection can result in evolution in the same direction. In general, natural selection can act at many levels: the gene, the chromosome, the individual, a group of individuals, the population, or the species.

Natural selection is the primary process leading to adaptation of individuals. It involves many traits acting together, differences among males and females, and differences among levels. The interaction of all these processes of natural selection determines the path of evolution.

Measuring Natural Selection

Natural selection is investigated in two ways: by use of indirect measurements and direct measurements. The indirect methods involve observing the outcome of natural selection and inferring its presence. The direct methods involve measuring the three parts of the process and following the course of evolution. Although the direct methods are preferred, as they provide direct proof of natural selection, in most instances, only indirect methods can be used.

Indirect methods involve three kinds of observations. First, comparisons are made of trait similarities or differences among populations or species living in the same or different areas. For example, many species of animals living in colder climates have larger bodies than those living in warmer climates. It is inferred, therefore, that colder climates result in natural selection for larger bodies. Second, long-term studies are done of traits, in particular changes in a group in the fossil record. For example, during the evolution of horses, their food, grasses, became tougher and horses' teeth became thicker. It is inferred, therefore, that tough grass resulted in natural selection for thicker teeth. Third, comparisons are made of gene frequencies of natural populations, with predictions from mathematical models. Gene frequencies are measured using various techniques, including scoring differences in appearance, as with light-colored and dark-colored moths; using electrophoresis to observe differences in proteins; and determining the sequence of base pairs of deoxyribonucleic acid (DNA). The models make predictions about expected frequencies in the presence or absence of selection. Indirect methods are best at revealing long-term responses to evolution and general processes of natural selection that affect many species. The indirect methods suffer from the problem that often many processes will result in similar patterns. So, it must be assumed that other processes were not operating, or other predictions must be made to separate the processes.

Direct methods involve two kinds of observation. First, there is observation of changes in a population following some change in the environment. There are many types of environmental changes, including man-made changes, natural disasters, seasonal changes, and introductions of species into new environments. For example, from the changes in the peppered moth following a change in pollution levels, one can measure the effects of natural selection. The second type of observation is the direct measurement of fitness differences among individuals with trait differences. For example, individual animals are tagged at an early age and survival and reproduction are monitored. Then, statistical techniques are used to find a relationship between fitness and variation among individuals in some trait. Alternatively, comparisons of traits are made between groups of individuals, such as breeding and nonbreeding, adults and juveniles, or live and dead individuals, again using statistical techniques. For example, lions that breed are larger than lions that do not breed. Direct methods are best at revealing the relative importance to natural selection of the three factors (variation, fitness differences, and heritability). The direct methods suffer from two limitations. It takes a long time for evolution to occur. So, although one can measure natural selection, it is often not known if it results in evolution. Also, for many species, it is impossible or impractical to mark individuals and follow them through their lives.

Many methods can be used to study natural selection and evolution. Each method provides information about different parts of the process. Only through the integration of these methods can the entire process of evolution be revealed.

Adaptation and Evolution

Natural selection is the central process in adaptation and evolution. By understanding how the process operates and where its limits lie, scientists hope to determine why evolution has proceeded in the fashion that it has. Historically, it was only after Darwin presented his theory of natural selection that the idea of evolution became widely accepted in the nineteenth century. In the twentieth century, much of the work of evolutionary biologists during the 1930's, 1940's, and 1950's was to integrate the fields of genetics, ecology, paleontology, and systematics, using natural selection and evolution as the unifying concepts.

Knowledge of natural selection is still growing; many questions proposed by Darwin and others are yet to be answered. It is still not known to what extent organisms are well adapted to their environments or whether the evolution of the parts of the chromosome that are not translated into proteins are a result of processes that do not involve natural selection. Of the many theories of how natural selection works, it is still unknown which ones are the most important in nature and to what extent evolution is caused by natural selection at the level of the individual, the group, and the species.

Genetic engineering requires knowledge of natural selection. The addition of a new gene into an organism will result in natural selection on that gene and change selection on other genes. Efficiency will be gained if successful and unsuccessful outcomes can be predicted beforehand. If genetically engineered organisms are to be released into nature, scientists need to be able to forecast their fates, such as whether the organism will remain benign or will become a pest. Genes added to one organism could possibly spread to other, native species. The solutions to these dilemmas involve predictions of the outcome of natural selection.

An understanding of natural selection is critical for conservation biology. During the twentieth century, the rate at which natural areas are being destroyed and species are becoming extinct has accelerated tremendously. Conservation biology attempts to stop that destruction and preserve species diversity. For extinction of endangered species to be halted, it must be understood how

Alfred Russel Wallace

Born: January 8, 1823; Usk, Monmouthshire, Wales

Died: November 7, 1913; Broadstone, Dorset, England

Fields of study: Entomology, evolutionary science, population biology, zoology

Contribution: Wallace, a pioneer of the science of zoogeography, proposed a theory of evolution by natural selection in 1855 that predated and stimulated the publication of Charles Darwin's *On the Origin of Species* (1859).

Alfred Russel Wallace grew up in rural Wales and then in Hertford, England. His formal education was limited to six years at the Hertford Grammar School. From 1837 to 1844 Wallace worked in his brother William's surveying business. In 1844, Wallace taught at the Collegiate School in Leicester, England.

In 1848, Wallace and the entomologist Henry Walter Bates embarked on an expedition to Brazil. Wallace and Bates planned to collect and identify biological specimens and then pay for their trip by selling their collections. Wallace spent a total of four years exploring the Amazon River basin, collecting birds, butterflies and other insects.

Unfortunately, on the return voyage Wallace lost his precious collections when his ship caught fire and sank. Nevertheless, the expedition led to the publication of several articles and two books (*Palm Trees of the Amazon and Their Uses* and *Narrative of Travels on the Amazon and Rio Negro,* 1853). These reports attracted the attention of the Royal Geographical Society, which helped to fund his next expedition. For eight years (1854-1862) Wallace continued his research in the Malay Archipelago (Indonesia). Wallace's research on the geographic distribution of animals among the islands of the Malay Archipelago provided crucial evidence for his evolutionary theories and led him to devise what became known as Wallace's Line, the boundary that separates the fauna of Australia from that of Asia. By the time Wallace returned to England in 1862 he had collected over 125,000 animal specimens.

During an attack of a tropical fever, Wallace experienced a flash of insight in which he realized that natural selection could serve as the mechanism of evolution. Within a few days he completed his essay "On the Tendency of Varieties to Depart Indefinitely from the Original Type" and sent it to Charles Darwin for review and possible publication. Darwin was shocked to find that Wallace had developed a theory of evolution identical to that outlined in his own unpublished 1842 essay. Darwin's friends Charles Lyell and Joseph Hooker arranged for a joint presentation of the papers written by Wallace and Darwin and simultaneous publication in the August, 1858, *Proceedings of the Linnean Society.*

Graciously allowing Darwin to claim priority for the discovery of evolution by means of natural selection, Wallace continued to publish works on natural history and travel, including *The Malay Archipelago* (1869), *Contributions to the Theory of Natural Selection* (1870), *Geographical Distribution of Animals* (1876), and *Island Life* (1880). It was Wallace who called evolution by means of natural selection "Darwinism" in order to distinguish this theory from its predecessors. Unlike Darwin, however, Wallace continued to believe that natural selection could not account for the higher faculties of human beings.

—Lois N. Magner

Alfred Russel Wallace independently proposed a theory of evolution very similar to that of Darwin. (National Library of Medicine)

natural selection will affect these species given massive environmental changes. By discovering how evolution is occurring under natural conditions, researchers will learn how to design nature preserves to maintain species.

—*Samuel M. Scheiner*

See also: Adaptations and their mechanisms; Clines, hybrid zones, and introgression; Development: Evolutionary perspective; Ecological niches; Evolution: Animal life; Evolution: Historical perspective; Extinctions and evolutionary explosions; Gene flow; Genetics; Human evolution analysis; Nonrandom mating, genetic drift, and mutation; Population analysis; Population fluctuations; Population genetics; Population growth; Punctuated equilibrium and continuous evolution; Sex differences: Evolutionary origin.

Bibliography

Avers, Charlotte J. *Process and Pattern in Evolution*. New York: Oxford University Press, 1989. An intermediate-level college text that lays out the evidence for evolution and the processes that cause it. Chapters 6 and 7 present natural selection and its mechanisms and are a good introduction to the mathematical theories.

Bell, Graham. *Selection: The Mechanism of Evolution*. New York: Chapman and Hall, 1997. Clearly explains the processes of natural selection and explores its possible consequences. Offers many examples and gives an extensive review of the literature. Written for nonspecialists with some science background.

Brandon, Robert N. *Adaptation and Environment*. Princeton, N.J.: Princeton University Press, 1990. Explores the varying roles of environment at different levels of natural selection, the external, ecological, and selective environments.

Darwin, Charles. *On the Origin of Species by Means of Natural Selection*. London: J. Murray, 1859. The most important book written on natural selection. The basic premises are laid out, and data from natural and domestic species are presented. The ideas presented here are still being explored and tested.

Endler, John A. *Natural Selection in the Wild*. Princeton, N.J.: Princeton University Press, 1986. This book, although a somewhat more technical presentation, presents the best summary of the process of natural selection. Chapters 1 and 2 are a nontechnical description of the process and a discussion of the relationship between natural selection and evolution.

Futuyma, Douglas J. *Evolutionary Biology*. 3d ed. Sunderland, Mass.: Sinauer Associates, 1998. A textbook for an intermediate-level college course in evolution. Chapters 6 and 7 provide an overview of natural selection, its mechanisms, and its consequences. It is well illustrated and contains many examples.

Gould, Stephen J. *The Panda's Thumb*. New York: W. W. Norton, 1980. A compilation of columns from *Natural History* magazine, this book provides a very entertaining and accessible view of evolution and natural selection. This and Gould's other books are the nonscientist's best introduction to the subject.

Provine, William B. *Sewall Wright and Evolutionary Biology*. Chicago: University of Chicago Press, 1986. Sewall Wright was one of the most important figures in the development of evolutionary biology in the twentieth century. Besides describing his life, this book places his work within the context of the development of evolutionary theory. Chapters 7, 8, and 9 present an excellent overview of theories of evolution and natural selection and are understandable by a general audience.

Suzuki, David T., Antony Griffiths, Jeffrey Miller, and Richard Lewontin. *An Introduction to Genetic Analysis*. 4th ed. New York: W. H. Freeman, 1989. An intermediate-level college text that presents the experimental basis of the understanding of genetic processes. Chapter 23 focuses on natural selection and evolution and is useful for gaining a more mathematical understanding of the phenomenon.

NEANDERTHALS

Types of animal science: Anatomy, classification, evolution
Fields of study: Anthropology, archaeology, evolutionary science, human origins, physiology, systematics (taxonomy)

Neanderthals are the best-known extinct members of the human lineage. It is generally agreed that they were close relatives of modern humans, but the nature of the relationship is vigorously debated. They have been assumed to be a direct ancestor, a diseased member of the species, or an extinct side branch of the family tree.

Principal Terms

DEOXYRIBONUCLEIC ACID (DNA): the chemical that carries the instructions for all living things; closely related organisms have very similar DNA

GENUS: the first part of the scientific name of an organism; members of the same genus but different species are closely related, but cannot mate and produce fertile offspring

MITOCHONDRIA: subcellular structures containing DNA used to estimate the relationships between groups of organisms; the more similar the DNA, the more closely related the groups

SPECIES: the second part of the scientific name of an organism; members of the same species can mate and produce fertile offspring

SUBSPECIES: the third part of a scientific trinomial, assigned to one of two groups that can mate and produce fertile offspring, but that have some strikingly different characteristics

TAXONOMY: the science of classifying and naming living and fossil organisms, or the classification and scientific name of a living or fossil group

The fossil that gave the Neanderthals their name was found in a cave being quarried for limestone in Germany's Neander Valley in 1856. At least two Neanderthal fossils were discovered before the Neander Valley individual; however, neither was recognized as a member of an extinct human group until after the name "Neanderthal" was assigned. Many similar fossils have been found in scattered locations all over Europe and the Middle East since the Neander Valley discovery. Dates assigned to the various fossils indicate that the Neanderthals originated late in the Ice Age and became extinct a few thousand years before the last glacial retreat (from about 200,000 years ago to about 30,000 years ago). Thus the Neander Valley specimen lent its name to a fossil relative of modern humans that occupied Europe and the Middle East late in the Ice Age.

Though the Neanderthals were very similar to modern humans, they had several distinctive characteristics. Neanderthals were short and exceptionally stout-bodied with broad supportive bones and joints. This body form suggests a life filled with intense physical effort. Perhaps the compact body also helped them cope with cold stress under Ice Age conditions. Their brains were somewhat larger than modern human brains. That size may have compensated for the more massive total body size of the Neanderthals, since large-bodied organisms generally have larger brains. Their foreheads sloped up from their exceptionally heavy eyebrow ridges, their jaws extended forward beyond the plane of the face, and their chins were weakly developed. These and several other characteristics are used to define a fossil find as a Neanderthal.

Structure and Behavior

Rudolph Virchow's initial interpretation of the Neander Valley fossil as a diseased human was popular for a time. Virchow held that the fossil was a modern human whose unique features were the result of disease. However, as more fossils with the same characteristics were discovered all around Europe and the Middle East, this explanation became untenable. Later, misinterpretation of the characteristics of Neanderthal fossils led Marcellin Boule and others to interpret Neanderthals as stooped, bent-kneed, apelike subhumans with an animal nature to match.

Additional fossil discoveries, including evidence for toolmaking and burials, sometimes with flowers placed in the grave, caused anthropologists to rethink the presumed animal nature of the Neanderthals. Although the evidence for flowers has been challenged, the evidence for burials, presumably accompanied by mourning, is accepted by many anthropologists. In addition, fossils showed that some Neanderthals lived much of their lives with deformed limbs and other disabilities, which would have made it difficult or impossible for them to fend for themselves. Yet they apparently lived many years in that condition, suggesting the support of other members of a social group. Such behavior was not in keeping with Boule's picture of the Neanderthals as nonhuman animals.

Reinterpretation of the anatomic evidence also suggested that, instead of a bent-kneed, stooped posture, the Neanderthals walked on two rather straight legs and had hands capable of manipulating materials and making tools, much as modern humans do. All this indicated that the Neanderthals were more like modern humans than Boule's interpretation, and they came to be thought of in that light.

Taxonomic Relationship to Modern Humans

Neanderthals have always been recognized as close relatives of modern humans, but the specific taxonomy of the relationship is still a point of contention. They are placed in the same genus (*Homo*) as modern humans by almost all anthropologists, but researchers debate whether they were members of our species, *Homo sapiens* or belonged in their own species, *Homo neanderthalensis*. The discussion of the structure and behavior of Neanderthals bears directly on this question. If the Neanderthal characteristics were the result of disfigurement caused by disease, Neanderthals were simply aberrant humans and not especially interesting from the perspective of human evolution. However, if they were stooped, bent-kneed, and animal-like in behavior, they were probably a separate species, perhaps ancestral to modern humans, and therefore more interesting from the evolutionary perspective. On the other hand, if they were upright in stature, were skilled toolmakers, were supportive of their handicapped and elderly, and buried their dead with mementos such as flowers, they might earn the designation *Homo sapiens* and take on an even greater interest to the more modern members of that species.

Such arguments are part of the practical taxonomy of the Neanderthals, but the real key to species identification and species separation is (at least theoretically) interbreeding. If the members of two groups can mate with each other and produce fertile offspring, and if these offspring can produce fertile offspring, the two groups are generally considered to be members of the same species. Therefore, the real taxonomic question becomes: Could Neanderthals and early modern humans interbreed?

Because it is difficult to determine whether fossil groups interbred with one another, Neanderthal taxonomy has been primarily determined by anatomic and presumed behavioral characteristics, such as those already discussed. That taxonomy has vacillated with changing interpretations of those characteristics. Neanderthals have been placed in their own species (*Homo neanderthalensis*) for much of their history, but they have been identified as a human subspecies (*Homo sapiens neanderthalensis*) at other times. The latter designation implies that the Neanderthals and modern humans (*Homo sapiens sapiens*) were members of the same species and therefore could interbreed.

Determination of the Neanderthals' taxonomic position is an integral part of arguments over the mechanism of the origin of modern humans. There are two main hypotheses for that origin: the replacement hypothesis of Christopher Stringer and the multiregional hypothesis vigorously supported by Milford Wolpoff. The replacement hypothesis is also designated "out-of-Africa" because it assumes that a population of African origin expanded throughout Africa, Europe, and Asia and rapidly replaced the more primitive humanlike species living there, including the Neanderthals. Whether this replacement was by competition or by more direct and violent means is undetermined. The multiregional hypothesis suggests that the widespread, more primitive humanlike populations evolved into modern humans rather than being replaced by new immigrants. Both hypotheses hold that the more primitive populations also originated in Africa and spread to Europe and Asia at a much earlier date.

Because the Neanderthals are the best known and best understood early human group, an understanding of the Neanderthal relationship is critical to an understanding of the evolutionary history of humanity. A Neanderthal contribution to modern human ancestry would support the multiregional hypothesis, and the lack of such a contribution would be consistent with the replacement hypothesis.

Advances of the Late Twentieth Century
By the 1990's, the Neanderthals were well established as a group related to modern humans, but the questions remained: How close was the relationship? Did the two groups interbreed? Were Neanderthals a part of the evolutionary heritage of modern humans? During the 1990's, improved techniques and additional fossil discoveries led to greater understanding of the Neanderthals but little consensus on these questions. A few examples will illustrate the situation.

In a 1996 study, Jean-Jacques Hublin and several coworkers determined that Neanderthals found at an archeological site in France made bone tools and wore decorative emblems on their bodies, behaviors not uncovered with older Neanderthal fossils. They concluded that the Neanderthals were influenced by early modern humans who lived in the same area at the same time and that a reasonably elaborate cultural exchange must have occurred between the two groups. However, based on the strikingly different anatomy of the two groups' inner ears, they also concluded that the Neanderthals and modern humans did not interbreed. The investigators reasoned that if interbreeding had occurred, the two groups would have shared a common ear structure.

In 1997, Matthias Krings, Svante Paabo, and their colleagues isolated and engineered deoxyribonucleic acid (DNA) from the mitochondria of Neanderthal bones and compared it to DNA from modern human mitochondria. They found the Neanderthal DNA to be quite different from that of modern humans and concluded not only that the two groups were different species but also that Neanderthals were not ancestral to modern humans.

In 1998, Daniel Lieberman proposed that a reduction in the length of the sphenoid bone during embryology can explain most differences between the two groups' skulls. The sphenoid is a bone in the skull of both Neanderthals and modern humans, and Lieberman showed it to be shortened in modern humans but not in Neanderthals. He hypothesized that the impact of shortening the sphenoid resulted in the modern human skull characteristics, while the longer sphenoid resulted in the Neanderthal skull. Based on the fundamental nature of the change, he concluded that Neanderthals do not belong to the same species as modern humans and were probably not ancestral to modern humans.

In 1999, Cidàlia Duarte, Erik Trinkaus, and several colleagues discovered the buried remains of a four-year-old child in southern Spain. The skeleton was estimated to be about 24,500 years old, and they interpreted its anatomy to be a mixture of modern human and Neanderthal characteristics. Most anthropologists agree that southern Spain supported Neanderthal populations longer than other parts of the world, perhaps as late as 27,000 years ago, and that modern humans and Neanderthals coexisted in the region. Duarte, Trinkaus, and their group suggested that the skeleton they found demonstrated that the two groups did interbreed and that Neanderthals were part of the ancestry of *Homo sapiens*.

Significance

Consideration of this short list of studies in the 1990's demonstrates the state of knowledge about the Neanderthals' place in human evolution. Viewed alone, each study seems to clinch the position of its authors. In fact, the first three reinforce one another so well that Neanderthals would seem to be eliminated from direct participation in the evolution of modern humans. However, Duarte and Trinkaus's study would seem to clinch the opposite position, that Neanderthals were direct participants in the evolution of modern humans. This situation symbolizes the absence of consensus in the field. There are also established scientists with alternative viewpoints for each of these studies. Lieberman himself is a coauthor of a letter that criticizes his own conclusions about the sphenoid and points to the need for a better understanding of the development of primate skulls to help clarify the situation.

A number of anthropologists have pointed out that Krings and Paabo's conclusions are extrapolated from a single, short segment of the mitochondrial DNA and that more extensive studies, including studies of DNA from the nucleus, are

necessary before definitive conclusions can be drawn. In fact, nuclear DNA studies of modern humans have suggested that modern human DNA comes from a number of sources rather than a single African source as in the out-of-Africa hypothesis. Clearly, extensive DNA comparisons would be helpful; however, DNA from fossils is difficult to find and difficult to work with, so an extensive collection of such studies is not likely to accumulate.

Ian Tattersall, who rejects the Neanderthals as direct contributors to modern human evolution, has criticized Duarte and Trinkaus's data and their interpretation of the data. The verbal exchange has been bitter, not an unusual circumstance for disagreements in this field.

Although anthropologists have learned an enormous amount about the Neanderthals, their relationship to modern humans continues to escape consensus. This is, without question, a result of the difficulty of the problem and the tentative nature of the evidence. Most agree that the Neanderthals were a successful group closely related to modern humans. Everyone's hope is that more fossils, improved technology, and fresh insight will clarify the question because understanding the Neanderthals is likely to contribute to an understanding of humanity.

—*Carl W. Hoagstrom*

See also: Apes to hominids; Baboons; Cannibalism; Chimpanzees; Communication; Communities; Evolution: Animal life; Evolution: Historical perspective; Gorillas; Groups; Hominids; *Homo sapiens* and human diversification; Human evolution analysis; Infanticide; Learning; Lemurs; Mammalian social systems; Monkeys; Orangutans; Primates.

Bibliography

Akazawa, Takeru, Kenichi Aoki, and Ofer Bar-Yosef, eds. *Neandertals and Modern Humans in Western Asia*. New York: Plenum, 1998. A scholarly but understandable group of papers with contributions from many of the major workers in Neanderthal evolution and biology.

Ciochon, Russell L., and John G. Fleagle, eds. *The Human Evolution Source Book*. Englewood Cliffs, N.J.: Prentice Hall, 1993. The broad spectrum of human evolution is covered in this book, but a large section titled "The Neanderthal Question and the Emergence of Modern Humans" covers Neanderthals. Several of the most fundamental questions are dealt with in a clear and interesting fashion.

Fox, Richard G. "Agonistic Science and the Neanderthal Problem." *Current Anthropology*, supp. 39 (June, 1998). All articles in the supplement concern "The Neanderthal Problem and the Evolution of Human Behavior," the supplement's title. The interaction between Neanderthals and modern humans in Europe and in the Middle East and an archaeological consideration of the out-of-Africa model are the subjects of three other articles.

Shreve, James. *The Neanderthal Enigma*. New York: William Morrow, 1995. Written by a science writer rather than a scientist, this is an interesting account of the history of Neanderthals and Neanderthal studies.

Stringer, Christopher, and Robin McKie. *African Exodus*. New York: Henry Holt, 1996. A small, interesting book on the out-of-Africa hypothesis written by an anthropologist intimately involved with the development of the hypothesis (Stringer) and a science writer.

Tattersall, Ian. *The Last Neanderthal*. New York: Macmillan, 1995. An extensively illustrated account of what is known about Neanderthals and how it was learned. The prologue outlines two interesting, if imaginative, characterizations of the "last Neanderthal."

Tattersall, Ian, and Jeffrey H. Schwartz. "Hominids and Hybrids: The Place of Neanderthals in Human Evolution." *Proceedings of the National Academy of Science* 96 (June, 1999): 7. This commentary is a reply to the paper by Duarte, Trinkaus, and their colleagues (pages 7604-7609 of this same issue) in which they report the description of a 24,500-year-old fossil with characteristics of both Neanderthals and early modern humans. The two positions are clearly laid out in the article and the commentary.

Trinkaus, Erik, and Pat Shipman. *The Neanderthals: Changing the Image of Mankind*. New York: Alfred A. Knopf, 1992. An interesting and well-written account of the Neanderthals and the scientists who study them.

NERVOUS SYSTEMS OF VERTEBRATES

Type of animal science: Anatomy
Fields of study: Developmental biology, evolutionary science, neurobiology, physiology

The anatomy of vertebrate nervous systems determines many of the behaviors and adaptative capabilities of animals. Its study is a prerequisite to understanding how it functions.

Principal Terms

AXON: an extension of a neuron's cell membrane that conducts nerve impulses from the neuron to the point or points of axon termination

GRAY MATTER: the part of the central nervous system primarily containing neuron cell bodies and unmyelinated axons

INTERNEURON: a central nervous system neuron that does not extend into the peripheral nervous system and is interposed between other neurons

MYELINATED AXON: an axon surrounded by a glistening sheath formed when a supporting cell has grown around the axon

NEURAL INTEGRATION: continuous summation of the incoming signals acting on a neuron

NEURONS: complete nerve cells that respond to specific internal or external environmental stimuli, integrate incoming signals, and sometimes send signals to other cells

NUCLEUS (pl. NUCLEI): cluster of neuron cell bodies within the central nervous system

TRACT: a cordlike bundle of parallel axons within the central nervous system

WHITE MATTER: the part of the central nervous system primarily containing myelinated axon tracts

---

Animals must be able to coordinate their behaviors and to maintain a relatively constant internal environment, despite fluctuations in the external environment, in order to survive and reproduce. To do so, animals must monitor their external and internal environments, integrate this sensory information, and then generate appropriate responses. The evolution of the vertebrate nervous system has provided for the efficient performance of these tasks.

The Pattern of the Vertebrate Nervous System
Although the various vertebrates show differences in the organization of their respective nervous systems, they all follow a similar anatomical pattern. The nervous system can be partitioned conveniently into two major divisions: the peripheral nervous system (PNS) and the central nervous system (CNS). These divisions are determined by their location and function. The CNS consists of the spinal cord and the brain. The PNS, that part of the nervous system outside the CNS, connects the CNS with the various sense organs, glands, and muscles of the body.

The PNS joins the CNS in the form of nerves, which are cordlike bundles of hundreds to thousands of individual, parallel nerve-cell (neuron) axons (long tubular extensions of the neurons) extending from the brain and spinal cord. The nerves extending from the spine are called spinal nerves, while those from the brain are called cranial nerves. The elements of the PNS include sensory neurons (for example, those in the eyes and in the tongue) and motor neurons (which activate muscles and glands, thereby causing some sort of action or change to occur). Most nerves contain both sensory and motor axons.

Thus, the PNS can be divided into two major subdivisions: sensory (or afferent) neurons and motor (or efferent) neurons. There is very little information-processing accomplished in the PNS. Instead, it relays both environmental information to the CNS (sensory function) and the CNS responses to the body's muscles and glands (motor function). Sensory neurons of the PNS are classified as somatic afferents if they carry signals from the skin, skeletal muscles, or joints of the body. Sensory neurons from the visceral organs (internal organs of the body) are called visceral afferents.

The PNS motor subdivision also has two parts. One is the somatic efferent nervous system, which carries neuron impulses from the CNS to skeletal muscles. The other is the autonomic nervous system (ANS), which carries signals from the CNS to regulate the body's internal environment by controlling the smooth muscles, the glands, and the heart. The ANS itself is subdivided into the sympathetic and parasympathetic nervous systems. These are generally both connected to any given target and cause approximately opposite effects to each other on that target (for example, slowing or increasing the heart rate).

The CNS, where essentially all information-processing occurs, has two major subdivisions: the spinal cord and the brain. Virtually all vertebrates have similarly organized spinal cords, with two distinct regions of nervous tissue: gray and white matter. Gray matter is centrally located and consists of neuron cell bodies and unmyelinated axons (bare axons without the glistening sheaths called myelin, created by supporting cells wrapping around the axons). White matter contains mostly bundles of myelinated axons (white because they have glistening myelin sheaths around them). Bundles of axons in the CNS are called nerve tracts. Within the spinal cord, these are either sensory tracts carrying impulses toward the brain, or they are motor tracts transmitting information in the opposite direction.

Interneurons are neurons positioned between two or more other neurons. They accept and integrate signals from some of the cells and then influence the others in turn. Interneurons are particularly numerous within the gray matter. In the spinal cord, they permit communication up, down, and laterally. Most axons in the cord's tracts belong to interneurons.

The Vertebrate Brain

The brain of vertebrates is actually a continuation of the spinal cord, which undergoes regional expansions during embryonic development. The subdivisions of the brain show more variety among vertebrate species than does the spinal cord. The brain has three regions: the hindbrain, the midbrain, and the forebrain. Their structures are complex, and various systems of subdividing them exist. The major components forming the brain regions are the hindbrain's medulla oblongata, pons, and cerebellum; the midbrain's inferior and superior colliculi, tegmentum, and substantia nigra; and the forebrain's hypothalamus, thalamus, limbic system, basal ganglia, and cerebral cortex.

The hindbrain begins as a continuation of the spinal cord called the medulla oblongata. Most sensory fiber tracts of the spinal cord continue into the medulla, but it also contains clusters of neurons called nuclei. The posterior cranial nerves extend from the medulla, with most of their nuclei located there.

Also in the medulla, and extending beyond it through the pons and midbrain, is the complexly organized reticular formation. This mixture of gray and white matter is found in the central part of the brain stem but has indistinct boundaries. Essentially all sensory systems and parts of the body send impulses into the reticular formation. There are also various nuclei within its structure. Impulses from the reticular formation go to widely distributed areas of the CNS. This activity is important for maintaining a conscious state and for regulating muscle tone.

Prominent on the anterior (front) surface of the mammalian medulla oblongata are the pyramids: tracts of motor fibers originating in the forebrain and passing without interruption into the spinal cord to control muscle contraction. These tracts cross to the opposite side of the medulla before entering the spinal cord, which results in each side of the forebrain controlling muscle contraction in the opposite side of the body.

Many sensory fibers from the spinal cord terminate in two paired nuclei at the lower end of the medulla, the gracile and cuneate nuclei. Axons leaving these nuclei cross to the opposite side of the medulla and then continue as large tracts (the medial lemnisci) into the forebrain. Thus, each side of the brain gets sensory stimuli mostly from the opposite side of the body.

Immediately above the medulla is the pons. It contains major fiber pathways carrying signals through the brain stem, and a number of nuclei, including several for cranial nerves. Some pontine nuclei get impulses from the forebrain and send axons into the cerebellum, again with a majority crossing to the opposite side of the brain stem before entering the cerebellum.

On the dorsal (back) side of the medulla and pons is the cerebellum, an ancient part of the brain that varies in size among vertebrate species. The cerebellum forms a very important part of the control system for body movements, but it is not the source of motor signals. Its gray matter forms a thin layer near its surface called the cerebellar cortex and surrounds central white matter.

Vertebrates with well-developed muscular systems (for example, birds and mammals) have a large cerebellum, with several lobes and convex folding of its cortex. It is attached to the brain stem by three pairs of fiber tracts called cerebellar penduncles, which transmit signals between the left and right sides of the cerebellum and between the cerebellum and motor areas of the spinal cord, brain stem, midbrain, and forebrain. The cerebellum times the order of muscle contractions to coordinate rapid body movements.

The Midbrain and the Forebrain
The midbrain is the second major region of the brain. The midbrain's dorsal aspect, called the tectum, is a target for some of the auditory and visual information that an animal receives. The paired inferior colliculi form the lower half of the tectum.

They help to coordinate auditory reflexes to relay acoustic signals to the cerebrum. The two superior colliculi, the other half of the tectum, assist the localization in space of visual stimuli by causing appropriate eye and trunk movements. In lower vertebrates, the superior colliculi actually form the major brain target for visual signals. Connecting fiber pathways (commissures) link the individual lobes of each pair of colliculi.

The midbrain's tegmentum contains several fiber tracts carrying sensory information to the forebrain and carrying impulses among various brain-stem nuclei and the forebrain. Two cranial nerve nuclei concerned with the control of eye movements are also in the tegmentum. The reticular formation extends through the tegmentum and regulates the level of arousal. It also helps to control various stereotyped body movements, especially those involving the trunk and neck muscles. Finally, the tegmentum contains the red nucleus, which, in conjunction with the cerebellum and basal ganglia, serves to coordinate body movements. The substantia nigra functions as part of the basal ganglia to permit subconscious muscle control.

The forebrain, the final major area of the brain, differs from the lower areas in the more highly evolved functions it controls. It has a small but extremely important collection of about a dozen pairs of nuclei called the hypothalamus. These control many of the body's internal functions (such as temperature, blood pressure, water balance, and appetite) and drives (such as sexual behavior and emotions). Immediately above the hypothalamus lies the thalamus, another collection of more than thirty paired nuclei. The two thalami are the largest anterior brain-stem structures. Their ventral (front) parts relay motor signals to lower parts of the brain. The dorsal (back) parts transmit impulses from every sensory system (except olfaction, the sense of smell) to the cerebrum.

The limbic system is organized from a number of forebrain structures mostly surrounding the hypothalamus and thalamus. It determines arousal levels, emotional and sexual behavior, feeding behavior, memory formation, learning, and motivation. In general, the limbic system exchanges information with the hypothalamus and thalamus, and receives impulses from auditory, visual, and olfactory areas of the brain.

The basal ganglia function with the midbrain's tegmentum and substantia nigra, the cerebral cortex, the thalamus, and the cerebellum. These paired structures' functions are unclear, but it is known that they are important for adjusting the body's background motor activities, such as gross positioning of the trunk and limbs, before the cerebral cortex superimposes the precise final movements.

The cerebral cortex, like the cerebellum, is an ancient brain structure; however, it shows even more variation among vertebrate species than the cerebellum. It is formed into two hemispheres, which have olfactory bulbs projecting from their anterior (front) ends. The olfactory bulbs receive impulses from olfactory nerves for the sense of smell. The gray matter of the cerebral cortex is at the surface, enclosing the white matter (fiber tracts) beneath. The white matter connects various parts of the gray matter of one hemisphere with others within the same hemisphere and with corresponding parts in the opposite hemisphere. It also connects the cortical gray matter with lower brain structures. The ultimate control of voluntary motor activity resides in the motor areas of the cortex, although this control is heavily influenced by all the previously mentioned motor-control areas of the CNS.

Corresponding to each of the major senses (touch, vision, audition), there are primary sensory areas. These areas get the most direct input from their sensory organs by way of the corresponding sensory thalamic nuclei. Surrounding each primary area are association areas that receive a less direct sensory input but also more inputs from other sensory cortical areas. In general, the more intelligent an animal is, the larger are its association areas.

Studying the Nervous System

Many methods are used in studying vertebrate nervous systems. The level of description desired often determines the methods employed. For example, the gross structure visible to the unaided eye is usually investigated using the entire brain or spinal cord of the animal under study. It will then be photographed or drawn, sliced at various points either parallel to its long axis or across its long axis, and again photographed or drawn, until a complete series of such "sections" has been assembled.

To see finer structural details requires microscopes and very thin slices of nervous tissue (less than a millimeter in thickness). Preparation of such thin slices of this soft tissue requires that it first be either frozen or embedded in a block of paraffin wax. A special slicing device called a microtome is then used to produce the thin sections. For easy observation of different structural details (such as nuclei and fiber tracts), various chemical stains can be applied to the slices of tissue. These stains specifically color particular structural features green, blue, or some other color, thereby making them more visible.

Nervous tissue can be selectively and painlessly destroyed in an anesthetized animal by cutting a nerve trunk, inserting a fine wire electrode into the CNS and destroying tissue with electricity, or inserting a fine needle and then either injecting a chemical agent that kills nervous tissue or using a suction device to remove areas of tissue. The precision and reproducibility of wire or needle placement within the CNS is possible with a device called a stereotaxic frame. This instrument positions the animal's head and brain in an exact standard position. Then, wires or needles are inserted a certain distance away from (for example, behind, below, or to one side of) common landmarks on the skull. Stereotaxic atlases are books published by investigators for specific animals, with the exact coordinates in three-dimensional space for most CNS structures.

Following such procedures, the animal may be immediately and painlessly sacrificed, its brain or spinal cord removed, and the previously described thin slices prepared and stained. It may be necessary to allow the animal to recover from its surgical treatment, since several days or weeks must sometimes pass for the severed fiber tracts to degenerate. Then, following a painless lethal injection, the nervous tissue is prepared as above using special staining techniques, which reveal the pathways of degenerating nerve fibers in the tissue sections studied later under the microscope.

Although new techniques are constantly being developed, the preceding methods have revealed that the hundred billion neurons of the vertebrate nervous system form the most complexly organized structure known. Through this knowledge of the structure of the nervous system, it has become possible to study intelligently its functions, to diagnose its diseases, and to devise methods of treatment when it becomes damaged or diseased.

A Complex Structure

The vertebrate nervous system is the most complex structure known to humankind. The human nervous system, for example, has more than a hundred billion cellular elements, and perhaps a hundred trillion points of information exchange between these elements. It is impossible to understand the details of such a structure in the same way that one can understand the structure of a radio; however, the general organizational plan can be discovered through the application of modern neuroanatomical techniques.

It is a widely accepted tenet of physiology that in order to comprehend the functioning of an organ or an organ system, it is necessary to have a critical understanding of its structure. In fact, physiology and anatomy are inseparable because function (physiology) always reflects structure (anatomy): It is impossible for an organ to perform in any other way than its architecture permits.

The nervous system is the ultimate control and communication system in the vertebrate body. Its complexity allows the vast range of vertebrate behaviors, as well as the rapid and precise regulation of the body's internal environment. The most complex vertebrate nervous systems display self-consciousness, reasoning, and language capabilities.

There are many reasons for studying the organization of vertebrate nervous systems, ranging from purely theoretical (such as determining the mechanisms of memory recall or clarifying the evolutionary relationships among vertebrate species) to very practical (such as treatments for mental illnesses, precisely defining brain death, or designing better computers).

For many, the ultimate goal is to obtain a better understanding of the relationship between the brain and the mind. It has been proposed that the mind and mental processes are emergent properties that appear when a certain degree of organizational complexity within the nervous system has been reached. The individual elements of the nervous system (the neurons) are not the constituents that think or possess consciousness. These unique capabilities are achieved as a result of the specific connections between neurons and sensory organs, and among neurons themselves.

It does not matter whether one performs an analysis of a machine or of a nervous system; it can never be expected to reveal its soul or its consciousness—if they exist. All that can be done is to admire the intelligence of its designer or the wisdom of nature.

—John V. Urbas

See also: Anatomy; Brain; Development: Evolutionary perspective; Habituation and sensitization; Physiology; Reflexes.

Bibliography

Butler, Ann B., and William Hodos. *Comparative Vertebrate Neuroanatomy: Evolution and Adaptation.* New York: Wiley-Liss, 1996. Focuses on the nervous systems of nonmammalian vertebrates.

Hildebrand, Milton. *Analysis of Vertebrate Structure.* 5th ed. New York: John Wiley & Sons, 2001. This classic textbook is for college-level readers. Several chapters present material relevant to the vertebrate nervous system. The index is very complete, while the glossary is only adequate. The list of references for each chapter is good, but some are of an advanced nature.

Keynes, R. D. *Nerve and Muscle*. 3d ed. New York: Cambridge University Press, 2001. Covers all aspects of the nervous and muscular systems, and their interactions.

Simmons, Peter J., and David Young. *Nerve Cells and Animal Behavior*. 2d ed. New York: Cambridge University Press, 1999. A college-level text explaining the link between the nervous system and behavior. Extensive use of case studies in the functions of nerve cells, sense organs, sensory filters, motor systems, startle behavior, intra-specific signals, and hunting, hearing and echolocation.

NESTING

Types of animal science: Behavior, ecology
Fields of study: Conservation biology, ecology, environmental science, ornithology, wildlife ecology

Many animals build structures to house their eggs and their young. Egg-laying animals, such as birds, amphibians, reptiles, and fish, build temporary nests for each year's young, while social insects, such as bees, build nests, or hives, as permanent homes. The nests of mammals may be permanent or temporary, depending on species.

Principal Terms

INSECTIVORE: any animal that feeds on insects

MALLARD: a species of wild duck

ORNITHOLOGY: the scientific study of birds

PHEROMONES: chemicals excreted by animals into their immediate environments to identify their territorial influence to other organisms

TERRITORIAL BEHAVIOR: the combination of methods and actions through which an animal or group of animals protects its territory from invasion by other species

TERRITORY: any area defended by an organism or a group of organisms for purposes such as mating, nesting, roosting or feeding

Many animals build structures to house and protect themselves, their eggs, and their young. Nest sites include grasses, shrubs, and trees, but also cracks in trees, holes in the ground and banks, crevices in rocks, under the surface of the ground, within the nests of other larger animals, or even near wasp nests. The nests of birds are the most obvious and well known, but many amphibians, reptiles, fishes, social insects, and mammals also build nests of varying degrees of complexity and permanence.

Bird and Insect Nests

Larger bird nests are constructed of various kinds of material, such as mud, bark, roots, twigs, hair, feathers, grass, plant fibers, shed snake skin, spider webs, lichens, or even prey remains and human-made material such as shiny, light, metallic ornaments. Many nests are cup-shaped and open, while fewer are oval, round or ball-shaped, closed, with an entrance at the side or the roof. An example of the latter is the nest of the South American ovenbird (*Furnarius*), which is often built with mud on top of a fence or another exposed surface. Others build a domed, oven-shaped nest out of plant material (the North American ovenbird) or huge nests suspended from the ends of tree branches (thorn birds). Some oceanic gulls (*Rissa tridactyla*) nest on narrow cliff ledges. Grassland and tundra owls nest on the ground or on an elevated hummock.

Some birds utilize amazing skill to construct their nests. The weavers are capable of tying knots with strips of grass or palm leaves and can prepare an exceptionally tight and compact nest. Birds of the genus *Orthotomus* are known as tailorbirds due to their ability to manufacture nests that are built in a pocket made by sewing together the edges of one or more leaves, using plant fibers. Some others (the family Contingidae) prepare very primitive and weak nests in an apparent effort to avoid the attention of predators, since they are attended minimally by the parents. Finally, woodcreepers nest in holes, while vireos weave a cup between the arms of a forked branch.

Most bird species in an area construct a unique nest in a unique location but use specific construction materials transferred from a distance. Generally, female geese and robins build their own nests, but among several other species the nest is built only by the male, who may use it as a sexual attractant in courtship. In these species, the female chooses a nest and indicates her choice by adding the nest lining.

Woodpeckers create cavities in trees, thus supplying safe nesting sites for a large number of birds. These include owls, parrots, parids, and flycatchers. In some areas, forest managers protect pileated woodpecker cavity trees and employ strategies to encourage continued production of new cavity trees where even smaller woodpecker species may find a haven for nesting.

Some insects, such as termites and carpenter bees, excavate a tunnel through solid wood. When termite colonies accidentally separate from an original nest they may migrate or march to a new nesting site and develop supplementary reproductives.

Bird nests are most often made of plant material such as grasses, providing temporary homes for eggs and hatchlings. (NOAA Central Library)

Mammal Nests

Most rodents build underground residences with a central chamber, where they sleep, raise their young, and hibernate, and other chambers that serve as food storage quarters. Moles build the most complex shelters among all insectivores. The shelters have an underground nest chamber that is surrounded by concentric rings of tunnels that are interconnected by radiating ones. The presence of shallow surface tunnels allows the marking of their course and the accumulation of a large amount of earth indicates the location of the deep tunnel system. The mole uses its forefeet like a shovel to dig in a type of body movement that resembles that of a breaststroke swimmer. Shrews dig surface nests, but they also use the runs of other animals, such as rodents, and line them with plant material on which they place their offspring. Hedgehogs and solenodons construct nest chambers, usually during their breeding season, which warm them during low-temperature periods. Tree shrews build their nests of leaves and other vegetation among tree roots or in cavities of fallen timber, but they immediately desert them when they feel that their residence has been stalked or even detected by a predator.

Ground squirrels and kangaroo rats transport enough seeds and other food in their cheek pouches to last them for a whole season. Excavation of one rat den where a single five-ounce kangaroo rat resided produced nine underground storage chambers with a total of thirty-five quarts of seeds. Kangaroo rats also dig one-cubic-inch storage pits that they stuff with seeds. In one case, an area of fifty-five square feet adjacent to a den had close to nine hundred such pits. This can have a devastating effect on local crops, so-called rodent plagues. On the other hand, such underground activity tends to germinate seeds for wild grass in the arid steppe regions of Central Asia. Squirrels carry acorns and nuts to hollow trees and barns, as well as to holes in the ground. They also have an amazing memory of where the food is stored, which keeps them alive during the harsh winter. Finally, the North American pack rat or trade rat is attracted by bright, shining objects (like the magpie) and carries them home to its nest, where it stores them together with tree sticks and grass.

Cooperative breeding appears to exist in the form of communal nesting among several mammal species. Tree squirrels (*Sciuridae sciurini*), fox squirrels (*Sciurus niger*) and gray squirrels (*Sciurus carolinensis*) form kin clusters among unrelated members of their own species during all seasons, especially in winter. Gray squirrels have shown an intense female-female bond in the formation of groups.

Artificial Nesting

Humans living in suburban and rural areas can provide the appropriate housing for cavity nesting birds, and people do have the ability to become backyard bird specialists. Artificial structures have been used extensively in the United States and Canada to increase waterfowl production, although the users of the artificial materials are not always the animals for which they were intended. During a study conducted in 1995, in which artificial nesting structures were intended for mallard use, redheads were found to be using them for normal nesting. It is believed that a combination of elevated water surfaces and subsequent limited nesting in the emergent vegetation may have made these nesting sites attractive to the redheads, which has thus provided a new alternative for nesting redheads.

Another study on the excavation and use of artificial polystyrene snags by woodpeckers was performed in Eastern Texas over a period of five years in the early 1990's. Only half of the monitored downy woodpeckers (*Picoides pubescens*) appeared to use the artificial snags for cavity excava-

tion and later for nocturnal roosting, but not for nesting. None of the other six woodpecker species in the area excavated cavities in the artificial snags. Other animals, however, used these excavated cavities in the artificial snags. These include Carolina chickadees (*Parus carolinensis*), prothonotary warblers (*Protonotaria citrea*), southern flying squirrels (*Glaucomys volans*) and red wasps (*Polistes* sp.).

The human factor may affect, directly or indirectly, the breeding success of animals. Possible biological effects of electromagnetic fields attributed to high-voltage transmission lines are suspected to reduce the reproductive success of birds whose nests are nearby, as in the case of tree swal-

Territorial Behavior and Nesting

Animals appear to have a distinct notion of nesting and territory. The dog's persistent trend to leave its own scent during its evening walk is an example of territorial behavior, whereby an animal or group of animals, including insects, vertebrates, and other invertebrates, protects its territory from the invasion of other organisms. The boundaries may be clearly defined by the animal through sounds, such as a bird song, or scents, created by pheromones, which are secreted by specialized glands. The animal is often fierce in claiming its territory and may even fight and chase intruders to discourage them. Generally, the closer the requirements different species have of their territory, the less likely it is that they can coexist in the same area. Thus, species with similar requirements can sometimes live in the same area only if they differ in behavioral ways, such as in feeding and nesting patterns, as well as activity periods. However, should the food resources be limited in areas, such as the desert, direct competition may lead to violent confrontations. Invaders may attack eggs or young offspring to suppress their hunger.

Territorial behavior secures the ability for an animal to mate uninterrupted, raise its young in a less harmful area and keep other animals from devouring the food intended for the young. Often territories are temporary, as in the case of birds or mammals that protect the immediate feeding area until the young become old enough to be self-sufficient. That territory may even be identified as the nest itself, a phenomenon common with most birds. Territorial claims are often successful since the inhabitants of neighboring or overlapping territories may avoid each other, especially during the nonmating seasons.

lows. Similar effects have been postulated for the terns, gulls, and other birds that live in areas such as the Gulf Coast, where oil tankers, commercial fishing vessels, yachts, and other pleasure boats have degraded the environment. A debate and controversy occurred in the 1990's in the Pacific Northwest of the United States, where environ-mentalists lobbied for the preservation of the nesting habitat of the spotted owl to the detriment of the local logging industry.

—*Soraya Ghayourmanesh*

See also: Birds; Courtship; Displays; Habitats and biomes; Home building; Insect societies; Insects; Mammals; Reproductive strategies; Tool use.

Bibliography

Ananthaswamy, Anil. "That Nesting Instinct." *New Scientist* 167, no. 2250 (August 5, 2000): 14. This article discusses the research of United States and South African scientists on the ability of swarms of giant honeybees to retain a collective memory of home, even though old scouts that decided where to set up home may have died.

Doherty, Paul F., and Thomas C. Grubb. "Reproductive Success of Cavity-Nesting Birds Breeding Under High-Voltage Power Lines." *The American Midland Naturalist* 140, no. 1 (July, 1998): 122-128. This article discusses the effect of electromagnetic fields to the reproductive success of tree swallows and house wrens.

Dyes, John C. *Nesting Birds of the Coastal Islands: A Naturalist's Year on Galveston Bay.* Austin: University of Texas Press, 1993. Documents the year's cycle for birds on the rocky islands along the Texas Gulf coast, with special emphasis on the human pollution effect on the life and behavior of twenty species of terns, gulls, and other birds. Black-and-white photographs.

Heinrich, Bernd. "The Artistry of Birds' Nests." *Audubon* 102, no. 50 (September/October, 2000): 24-31. This article provides a good description of nest-building techniques of various bird species, including those that use tree holes created by woodpeckers as safe nesting sites.

Labauch, Rene, and Christyna Labauch. *The Backyard Birdhouse Book: Building Nest Boxes and Creating Natural Habitats.* Pownal, Vt.: Storey, 1999. A book that provides plans for building birdhouses for twenty-five cavity-nesting species with color photos, illustrations and explanations.

Skutch, Alexander F. *Helpers at Bird Nests: A Worldwide Survey of Cooperative Breeding and Related Behavior.* Iowa City: University of Iowa Press, 1999. This book discusses cooperative breeding that occurs among fifty different bird species and has an extensive bibliography and description of its evolution, kin selection, and demography.

NEUTRAL MUTATIONS AND EVOLUTIONARY CLOCKS

Type of animal science: Evolution
Fields of study: Ethology, genetics

A neutral mutation is a change in the nucleotide sequence of a gene that causes no change in the function of the protein coded by that gene. If most molecular mutations are neutral, as has been hypothesized, then they can be used as "clocks," because the number of changes in the gene (or gene product) is positively correlated with elapsed time.

Principal Terms

ALLELE: one of two or more alternate gene forms of a single gene locus

AMINO ACID: an organic molecule with an attached nitrogen group that is the building block of polypeptides

ELECTROPHORESIS: a technique for separating molecules when they are placed in an electrical field; the separation is usually based on their charge and weight

GENETIC CODE: the three-nucleotide base sequences (codons) that specify each of the twenty types of amino acids; there can be more than one codon for a particular amino acid

NUCLEIC ACID: an organic acid chain or sequence of nucleotides, such as DNA or RNA

PHYLOGENY: the evolutionary history of taxa, such as species or groups of species; order of descent and the relationships among the groups are depicted

POLYMORPHIC: a genotype or phenotype that occurs in more than one form in a population

The hypothesis has been advanced, primarily by Motoo Kimura of the National Institute of Genetics of Japan and Masatoshi Nei of the University of Texas, that the vast majority of poly-morphisms that occur at the molecular level are selectively neutral. Two major categories of poly-morphisms are involved. The first is attributable to changes in the nucleotide sequences of deoxyribonucleic acid (DNA). The second is isozymic variation, detectable by protein electrophoresis. Isozymic variation is usually caused by changes in the amino acid composition of the protein. Since amino acids are determined by the genetic code, this type of variation also ultimately depends on changes in DNA. The proponents of the neutral theory admit that most of the evolution that occurs on the nonmolecular level, such as changes in morphological and behavioral traits, is attributable to natural selection. On the molecular level, however, they believe that most of the changes are caused by chance.

Selection Versus Chance

Extensive variability has been found for both DNA sequences and isozymes within the majority of natural populations. Isozymic polymorphism, in which two or more variants of an enzyme occur within the same species, ranges from approximately 15 percent in mammals to approximately 40 percent in invertebrates. Isozymic variation, even within the same individual, ranges from about 4 percent in mammals to 14 percent in some insects. Variability in DNA sequences among individuals of the same species is even higher than those found for isozymes. Proponents of the neutral hypothesis hold that these levels of variability are too high to be attributable to selection, but instead, most variability at the molecular level is attributable to chance. The result, they say, has been a large amount of enzyme and DNA variability that is selectively neutral. They are neutral in the sense that their contributions to an organism's fitness are so small that their occurrence is attributable more to chance than to natural selection. Neutralists do not believe that most molecular mutations are neutral; they assume that most are harmful and are eliminated by natural selection. Rather, they believe that those that currently exist are adaptively equivalent. Proponents of the neutrality theory believe that changes in DNA and amino acid sequences are for the most part neutral, consisting primarily of the gradual random replacement of functionally equivalent alleles.

Although the neutral theory is able to explain much about molecular evolution, there are some issues that remain subjects of intense debate. It is known that some protein and gene variation is not neutral but, instead, under certain conditions, conveys selective advantages or disadvantages. In some organisms (for example, the Japanese macaque), there also appear to be more rare alleles than would be predicted by the neutrality hypothesis.

The Hypothesis of a Molecular "Clock"
If the neutrality theory of molecular evolution is correct, then changes in base sequences of DNA could act as evolutionary "clocks." This theory holds that because mutations change the DNA in all lineages of organisms at fairly steady rates over long periods of time, a clocklike relationship can be established between mutation and elapsed time. The number of base substitutions in the DNA is directly proportional to the length of time since evolutionary divergence between two or more species. The idea of molecular evolutionary clocks was forwarded in the early 1960's by Linus Pauling and Emile Zuckerkandl.

The molecular clock postulated by the neutrality hypothesis is not like an ordinary timepiece, which measures time in exact intervals. Rather, molecular changes occur as a stochastic clock,

such as occurs during radioactive decay. Although there is some variability for this type of clock (it is slower or faster during some periods than others), it would be expected to keep relatively accurate time over millions of years. A potential problem arises, however, because the rate of "ticking" for the molecular clock is not the same at every position along the DNA molecule. The rate has been shown to be slow for DNA sequences that directly affect the function of a protein (for example, those at an enzyme's active site), while the rate of change has been faster for positions on the DNA that are selectively neutral, that is, where they have little or no affect on the protein's function.

In a molecular clock, the number of changes in a DNA or protein molecule are the "ticks" of the clock. The number of "ticks," in turn, estimates the extent of genetic differences between two species. With this knowledge, scientists can reconstruct phylogenies. The phylogenies are usually depicted as branching patterns, which are based upon differences in DNA base-pair sequences or amino acid sequences. They depict not only the order of descent but also the degree of relatedness.

When choosing alternative phylogenetic hypotheses, biologists usually follow the principle of Occam's razor—the simplest theory is chosen over more complex ones. Thus, the phylogenetic tree that requires the fewest mutations is preferred over those that require more mutations. By "calibrating" the molecular clock with other, independent, events, such as those obtained from the fossil record, the actual chronological times of divergence can be estimated.

For example, humans and horses differ by eighteen amino acids in the alpha chain of the blood protein hemoglobin. It has been estimated from the fossil record that humans and horses diverged from a common ancestor approximately ninety million years ago. Other evidence suggests that half the substitutions took place since the time of divergence. Since nine amino acids have changed over a ninety-million-year period, the rate of amino acid substitution would equal approximately one every ninety million years.

Since mutation rates are known to be different for different genes, the ticking rate for different genes or proteins would not necessarily be the same. For example, the rate of substitution for the genes coding for the protein histone H4 is lower than that for the genes coding for the protein gamma interferon. Yet when nucleotide substitution rates are averaged for a number of different proteins, there does appear to be a marked uniformity in the rate of molecular change over time; the ticking of a number of clocks can be averaged, leading to more accurate estimates for divergence times.

Advantages and Disadvantages of the Theory
Much of the early work was done on sequence changes in proteins; however, there is a drawback to protein clocks. Their usefulness is limited because the genetic material itself is not being examined but, rather, a product coded by genes. This means that some of the changes in the genetic material may not be detected. For example, because of the redundancy of the genetic code, there could be a number of changes that occur in the nucleotide sequence of DNA that would not result in changes in the amino acid composition in a protein. Consequently, there has been great interest in directly examining the DNA itself.

Because of the advent of recombinant DNA techniques, molecular clocks can be based on changes in the genetic material. DNA-DNA hybridization also involves the comparison of DNA sequences, although on a broader scale. The DNA-DNA hybridization technique is attractive because it effectively compares very large numbers of nucleotide sites, each of which is effectively a single data point. One of the criticisms of molecular clocks is that most genes have not been found to "tick" with perfect regularity over long periods of time. During some periods, the rate of change (primarily because of mutation) may be fast, while at other periods it may be significantly slower. By comparing very large numbers of nucleotides, which represent many genes, DNA-DNA hybridization measures the average rates of change, which will produce more uniform estimates.

The concept of a molecular clock has been criticized on a number of grounds. First, it assumes that evolution in macromolecules proceeds at an approximately regular pace, whereas morphological evolution is usually recognized as occurring irregularly. It is also clear that the clock can tick at different rates among different macromolecules, whether they be proteins or DNA. Another problem is that the rate of the molecular clock varies among taxonomic groups. For example, the insulin gene has evolved much more rapidly in the evolutionary line leading to the guinea pig than in some other evolutionary lines. There are also notable differences among different parts of molecules. This variability was evident when sequences were compared among the first molecules examined in the light of the molecular clock hypothesis, notably hemoglobin molecules and cytochrome c. Another criticism is that a number of processes, known collectively as molecular drive, perturb the clock.

Recent data, however, suggest that nucleotide substitution rates in organisms as different as bacteria, flowering plants, and vertebrates are remarkably similar. For example, the average rate of substitution at "silent sites" (those at which mutation in the DNA produces no change in the amino acid encoded) is 0.7 percent per million years in bacteria, 0.9 percent for mammals, and 1 percent for plants. This relatively equal substitution rate across broad taxonomic categories would support the concept of the constancy of molecular clocks.

Testing the Hypothesis
A number of types of molecular clocks have been hypothesized. In the first group are techniques that directly estimate differences in the sequences of nucleic acids that make up DNA or in amino acid sequences (which are determined by DNA). Other methods are less direct, such as DNA-DNA hybridizations and immunological techniques. All the techniques ultimately assay genetic differences caused by base pair changes in DNA. Sequence comparisons and DNA-DNA hybridization techniques are now used more extensively.

In sequence comparisons, nucleic acid replacement in DNA or amino acid replacement in proteins are compared between species. Nucleic acid substitutions can be assayed by using restriction enzymes that only recognize specific base sequences or by direct sequencing. Amino acid substitutions can be assayed by traditional biochemical techniques, through automated sequencers, or by mass spectrometers. In both types of sequence analysis, the assumption is made that the greater the number of substitutions, the greater the evolutionary distance between the species. In DNA-DNA hybridization, DNA molecules from individuals from two species are separated into individual strands at high temperatures. The strands are then mixed at a lower temperature. This promotes joining of the strands from the different organisms. The extent of rejoining (how tightly they bond together) will be dependent on the degree of nucleotide pairing that occurs. If the nucleotide sequences between the two species are very similar, the DNA strands will bond very tightly; if there is little similarity, the bonds will be weak. The extent of bonding is measured by the temperature at which the new DNA duplex dissociates, or "melts." The higher the melting temperature, the greater the nucleotide similarity between the DNA strands from different species. The nucleotide similarity is presumed to be related to the evolutionary distance between the species.

Uses of the Theory
The use of molecules as clocks, in spite of their imperfect nature, has proven to be a valuable tool for inferring phylogenetic relationships among species and in estimating their times of divergence. Molecular data can be used independent of morphological and behavioral data for establishing evolutionary relationships. Similarly, divergence times estimated from the fossil record can be clarified through the use of molecular clock data. Molecular clocks have had a significant impact on evolutionary studies of organisms ranging from bacteria to humans, and molecular data have been instrumental in changing some long-held phylogenetic views. For example, the data obtained from DNA-DNA hybridizations in birds have

forced a major revision in bird taxonomy. Molecular clocks have been used to assign time scales to a large number of phylogenies. Some of the phylogenies are wide-ranging; approximate times of evolutionary divergence have been assigned to vertebrate species as diverse as sharks, newts, kangaroos, and humans. Others are more specific. Some of the best-known (and most controversial) work has been done on primates. In one set of experiments, the amino acid differences of serum albumin (a blood protein) were measured among different species of primates. By comparing the albumins of species whose divergence times were known from fossil evidence, researchers were able to "calibrate," or calculate, the mean rate of change for serum albumin. Previously, most anthropologists believed that humans and apes had diverged approximately twenty-five million years ago; the DNA-DNA hybridization data suggest a much more recent date of approximately five million years. Subsequent DNA studies have confirmed the latter estimate. This had led to a reevaluation of the primate fossil record and of the way in which primates have evolved, including humans.

Another group of researchers has used DNA-DNA hybridization data to calculate the divergence dates among different primate species. After calibrating the molecular clock with dates previously established from the fossil record, they estimated the following approximate divergence dates: Old World monkeys, 30 million years ago; gibbons, 20 million; orangutans, 15 million; gorillas, 7.7 to 11 million; and chimpanzees and humans, 5.5 to 7.7 million years ago. In contrast to earlier work, they concluded that humans and chimpanzees are genetically closer to each other than either are to gorillas. As with the serum albumin data, these new estimates of times and order of divergence led to a reexamination of primate evolution.

—*Robert A. Browne*

See also: Evolution: Historical perspective; Gene flow; Genetics; Hardy-Weinberg law of genetic equilibrium; Population analysis; Population genetics; Punctuated equilibrium and continuous evolution.

Bibliography

Avise, John C. *Molecular Markers, Natural History, and Evolution*. New York: Chapman and Hall, 1994. Focuses on the way in which molecular markers reveal evolutionary patterns and processes. Shows how studies of mitochondrial DNA in animals provide insights into historic processes of gene flow, speciation, and hybridization.

Ayala, Francisco J. "Molecular Genetics and Evolution." In *Molecular Evolution*, edited by F. J. Ayala. Sunderland, Mass.: Sinauer Associates, 1976. Ayala's well-written short introduction covers basic concepts such as the structure of DNA, the genetic code, estimating genetic variation in natural populations, and phylogenetics.

Dobzhansky, Theodosius, Francisco J. Ayala, G. Ledyard Stebbins, and James W. Valentine. *Evolution*. San Francisco: W. H. Freeman, 1977. Although dated, this book presents a broad perspective. The four authors are prominent in evolutionary studies. Aimed at beginning majors, this is a more interdisciplinary text than most. Chapter 2 is a good treatment of the genetic structure of populations, including a brief overview of molecular techniques for quantifying genetic variation. More detailed treatments of phylogenetic reconstructions and the use of molecular clock data are presented in chapter 9.

Easteal, Simon, Chris Collet, and David Betty. *The Mammalian Molecular Clock*. New York: Springer-Verlag, 1995. A useful introduction to the theory of the molecular clock in mammals.

Fitch, Walter M. "Molecular Evolutionary Clocks." In *Molecular Evolution*, edited by F. J. Ayala. Sunderland, Mass.: Sinauer Associates, 1976. An excellent introduction to the concept of evolutionary molecular clocks. The author was one of the first to advocate the use of molecular clocks.

Hartl, Daniel L., and Andrew G. Clark. *Principles of Population Genetics*. 3d ed. Sunderland, Mass.: Sinauer Associates, 1997. An advanced text that covers the topic in detail. The book is aimed at third and fourth year undergraduates as well as graduate students. It includes much of the data emerging from molecular methods. The authors are prominent researchers in the field. Detailed descriptions of the neutral theory and molecular clocks are presented in chapter 7.

Kimura, Motoo. *The Neutral Theory of Molecular Evolution*. Cambridge, England: Cambridge University Press, 1985. The originator and chief protagonist of the neutral theory has consolidated his thoughts and produced a book that, despite its difficult subject matter, is quite easy to understand.

_____. *Population Genetics, Molecular Evolution, and the Neutral Theory: Selected Papers*. Edited by Naoyuki Takahata. Foreword by James F. Crow. Chicago: University of Chicago Press, 1994. A collection of essays by one of the most prominent and influential theorists in the area of neutral mutations and population genetics. With introductory essays by Takahata.

Li, Wen-Hsiung. *Molecular Evolution*. Sunderland, Mass.: Sinauer Associates, 1998. Synthesizes developments in the field of molecular evolution since the 1980's. Written for advanced students and researchers in the field.

Nei, Masatoshi. *Molecular Evolutionary Genetics*. New York: Columbia University Press, 1987. An advanced text that summarizes and reviews developments in the field. The author is a prominent theorist. Very useful for those with a solid foundation in population genetics.

Strickberger, Monroe W. *Evolution*. 3d ed. Boston: Jones and Bartlett, 2000. An intermediate-level text that presents a detailed overview of evolution and evolutionary processes. Describes the numerous techniques that are used to reconstruct molecular phylogenies, including molecular clocks. The numerous diagrams are especially useful.

NOCTURNAL ANIMALS

Type of animal science: Behavior
Fields of study: Anatomy, behavior, physiology, zoology

Nocturnal animals are those active during the night. They are specially adapted to living where there is little or no light. Often, such creatures find their way, locate food, and detect danger in nearly total darkness.

At nightfall, many animals are just beginning to waken and to function. All of these night creatures are called nocturnal animals. Folklore often brands them as evil, inimical creatures. This is not so; they are merely different from diurnal animals, which are active during the day. Their activity during the hours of darkness is usually due to a combination of factors. First, the nocturnal animals may make use at night, without having undue competition, of food sources and habitats also used by diurnal animals. Second, they may be safe from many predators who would hunt them during the daylight hours. Finally, nocturnal animals may be predators which, at night, have much less competition for prey, or have a better chance to capture it, than in daylight.

Most nocturnal creatures are highly adapted to living under conditions in which there is little or no light. These adaptations enable nocturnal animals to find their way through the world, locate food, and detect danger in almost total darkness. Although the diurnal species outnumber the nocturnal creatures, almost every type of creature living on earth has a nocturnal version.

The owl's nocturnal vision is one hundred times more acute than that of humans. (Rob and Ann Simpson/ Photo Agora)

Regular Senses and Special Senses

Diurnal animals survive as long as they have a combination of senses that enables them to fit, successfully, into a daylight ecological niche. This usually includes ability to see adequately, to hear well, and to have a useful sense of smell. The ability to see, hear, and smell falls within a fairly broad range among earth's many successful diurnal animals. Furthermore, weaknesses in one sense, such as the nearsightedness of the rhinoceros, may be compensated for by enhancement of one or more of the other senses. In rhinos, both the ability to hear and to smell are very well developed.

Nocturnal animals live under conditions that are skewed far away from the normal visual range of diurnal organisms. They must operate under conditions where there is either very little light or even no light at all. For this reason, all animals that are nocturnal have at least one sense, sight, hearing, smell, taste, or touch, that is very highly developed, compared to those who are diurnal. These powerful sense adaptations enable them to survive well in the dark hours. The extent of such special sensory development varies, depending upon the organism, its needs, whether it also functions during the day, and the extent to which it is subject to predation by other organisms. In some cases, senses not seen among diurnal animals operate in nocturnal animals.

The large, carnivorous felines, such as lions, tigers, or leopards, are not threatened by many other organisms. They survive well by using a combination of keen eyesight and hearing, as well as an excellent sense of smell. To further aid these big cats, their eyes face forward, allowing very accurate judgment of distances when they hunt. They also rely quite heavily upon very acute reflexes, great strength, and the ability to run down most prey hunted.

An extreme in visual development is seen in owls, which are the ultimate nocturnal avian raptors and which function and hunt almost exclusively at night. These birds are gifted with superb vision, fine hearing, and a very wide visual and aural range. For example, the night vision of many owl species is one hundred times more sensitive than that seen in humans. In addition, owl hearing is very acute, aided in some cases by the possession of asymmetric skulls with the two ears at different places, further enhancing their hearing. Another adaptation that optimizes owl vision and hearing is the ability to turn the neck through 270 degrees. This gives owls the widest aural and visual range of all birds. It is therefore unsurprising that owls hear even the tiniest squeak or rustle made by their prey on the ground below them, and then very efficiently locate the prey by vision.

More on Bat Echolocation

An echolocating bat, needing to find prey or to navigate, emits ultrasonic waves of frequencies between twenty-five and eighty kilocycles per second. These sound waves bounce back from nearby objects. The patterns of the returning waves give a bat an accurate estimation of the size, shape, texture, and location of all objects scanned in this way. The process is very sensitive. Bats using echolocation can detect items one to two millimeters thick, located six to seven feet away.

Several adaptations of the auditory systems of bats optimize the effectiveness of its echolocation. One of them is the huge size—sometimes nearly as large as the bat's skull—and very complex folding of each of the bat's external ears (pinnae). This size, folding, and great pinna mobility collect returning sound echoes quite well and help the bat to pinpoint their sources.

Another feature of bat echolocation has to do with the muscles of the bat's inner ear, which contract briefly as the ultrasound emissions occur. The contraction is essential to counter the great intensity of the emitted ultrasound. If these muscles did not contract upon sound emission, the bat would deafen itself with its echolocation calls. Another aspect of the bat ear that aids echolocation is the light weight of the tympanic membranes and bones of the bat's middle ear. These auditory modifications, away from the norm in other rodents, allow faint echoes returning to the ear to cause vibrations that can be sensed and used.

Bats function only after the sun sets, most often feeding on insects captured while in flight. They have good hearing. However, their eyes are small and they possess relatively poor vision. Perhaps this is why they have developed a special sense, sonarlike echolocation, to pinpoint insect prey in the dark night skies and to navigate. In bat echolocation, sounds (often clicks) are emitted from the larynx or nose of the bat, depending upon species. These sounds then strike insects (and rocks or trees), and echoes bounce back to the bat's ears. The bat then uses the echoes to find its prey. Many bats have developed means to direct the sounds they make for echolocation. For example, bats with skin flaps on the nose often use them to direct sounds in the nasal passages. The excellent hearing found in bats is aided by their large, mobile, external ears. Bats are not "blind as bats." For example, they can use vision to navigate their way home, and often do. Some nocturnal birds, such as the nightjar, also use echolocation to capture their prey.

Rattlesnakes and many other snakes locate their prey in total darkness in another way that involves special sense organs. This process is most sensitive in poisonous snakes called pit vipers. These snakes have two heat-sensitive pits located on the sides of their heads. The pits help them to detect small animals which are their prey, such as rodents or birds. The detection is possible because the prey sought are warmer than their surroundings. The pit viper heat sensors are large groups of nerve endings so sensitive that they can detect the body heat of a rodent over a foot away. The heat discrimination of the pit nerve endings is huge and they can identify a temperature difference of less than one hundredth of a degree.

Nocturnal Animal Habitats

People generally think of nocturnal animals as living in the wild. However, many live cheek-by-

---

Nocturnal Animals

BATS sleep in caves or trees by day and seek food soon after sunset. Most eat insects, which they capture while in flight, via echolocation. Here, sound is emitted in flight, strikes insects, and then echoes bounce back. The bat uses the echoes to find its prey.

GERBILS, JERBOAS, and other rodents burrow in the ground for safety from predators and to escape the heat of their desert environment. They come out at night to eat insects and plants. Their large eyes give them good night vision and their huge ears pick up very faint sounds.

LEOPARDS and other big cats stalk herbivore prey in the dark. They depend on their ability to stalk prey silently or to ambush them, very acute hearing, good night vision, and their capacity for quick movement.

NIGHTJARS are nocturnal insect-hunting birds. They fly with mouths open and eat bugs in their paths. They, like bats, use echolocation to find food.

OWLS of woods, forests, and barns roost during the day. At night they hunt mice, voles, and small birds. First, an owl listens for tiny squeaks or rustling sounds from the ground. When these signals occur, the owl attacks silently. Its predation is aided by night vision one hundred times more light-sensitive than in humans, and a wide field of vision.

SCORPIONS and some SPIDERS are nocturnal, sleeping in burrows by day. At night they catch prey in similar ways. Scorpions await the close approach of insect prey. Then they move quickly, catch prey in their pincers, paralyze, and kill them. Trapdoor spiders speed out of partly open, sand-covered, silk burrow doors, seize, and kill insects that come close.

TARSIERS are small nocturnal primates. They sleep by day, and at night eat leaves, bugs, and frogs. Nocturnal existence protects them from predation. Large, sensitive, forward-facing eyes and very acute hearing via large, mobile ears help them to obtain food and to move effortlessly in dark trees.

jowl with humans in parks, gardens, and empty lots in urban and suburban areas. Just a few examples are badgers, raccoons, deer, and foxes. People often see them and view them as pests or problems because they raid gardens and garbage cans, and some kill pets. Another problem associated with nocturnal animals is that they can carry severe, contagious diseases (for example, foxes and raccoons may carry rabies) which can be fatal to people. Finally, many nocturnal species are on the verge of extinction. These range from insects to the big cats. A few nocturnal species, such as tigers, are finally being covered by international conservation agreements, perhaps just in time. The others should be helped to survive, too.

—*Sanford S. Singer*

See also: Adaptations and their mechanisms; Bats; Bioluminescence; Competition; Ears; Ecological niches; Ethology; Eyes; Habitats and biomes; Hearing; Owls; Predation; Rhythms and behavior; Sense organs; Sleep; Urban and suburban wildlife; Vision.

Bibliography

Boot, Kelvin. *The Nocturnal Naturalist*. North Pomfret, Vt.: David and Charles, 1985. This useful text gives useful information on nocturnal animals and wildlife-watching.

Brown, Vinson. *Knowing the Outdoors in the Dark*. Harrisburg, Pa.: Stackpole, 1972. This compendium shows many species of nocturnal animals, giving their habits, describing how to recognize them, and telling where to find them. It also has a large bibliography.

Kappel-Smith, Diana. *Night Life: Nature from Dusk to Dawn*. Boston: Little, Brown, 1990. This nice book covers nocturnal animals geographically, in an interesting way. Its chapters are Arizona, North Dakota, Hawaii, Connecticut, and Louisiana. They give a picture of nocturnal species of hot and dry, cold, temperate, and wet habitats.

Lawlor, Elizabeth. *Discover Nature at Sundown*. Harrisburg, Pa.: Stackpole, 1995. A guide to nocturnal animal watching that emphasizes enhancing awareness of one's own senses. Offers guidance on tools needed, as well as the animals to be seen.

Pettit, Theodore S. *Wildlife at Night*. New York: G. P. Putnam's Sons, 1976. This brief book is a valuable beginning to thinking about nocturnal animals, dealing most with mammals, birds, and invertebrates.

Popper, Arthur N., and Richard R. Fay, eds. *Hearing by Bats*. New York: Springer-Verlag, 1995. This specialized, edited text has a lot of information on bat hearing and echolocation. It also speaks to mammal hearing in general and has an excellent bibliography.

NONRANDOM MATING, GENETIC DRIFT, AND MUTATION

Type of animal science: Evolution
Fields of study: Ecology, ethology, genetics, population biology

Nonrandom mating, genetic drift, and mutation are three mechanisms, besides natural selection and migration, that can change the genetic structure of a population.

Principal Terms

ALLELE: alternative forms of a gene for a particular trait

ASSORTATIVE MATING: a type of nonrandom mating that occurs when individuals of certain phenotypes are more likely to mate with individuals of certain other phenotypes than would be expected by chance

GAMETE: a haploid sex cell that contains one allele for each gene; sperm and egg cells are gametes that fuse to form a diploid zygote

GENETIC VARIATION OR DIVERSITY: the total number and distribution of alleles and genotypes in a population

GENOTYPE: the complete genetic makeup of an organism, regardless of whether these genes are expressed

HETEROZYGOTE: a diploid organism that has two different alleles for a particular trait

HOMOZYGOTE: a diploid organism that has two identical alleles for a particular trait

INBREEDING: mating between relatives, an extreme form of positive assortative mating

PHENOTYPE: the expressed genetic traits of an organism

Evolution is a process in which the gene frequencies of a population change over time, and nonrandom mating, genetic drift, and mutation are all mechanisms of genetic change in populations. These mechanisms violate the assumptions of the Hardy-Weinberg model of genetic equilibrium by increasing or decreasing the frequency of heterozygote genotypes in the population.

Nonrandom mating occurs in a population whenever every individual does not have an equal chance of mating with any other member of the population. While many organisms do tend to mate randomly, there are some common patterns of nonrandom mating. Often, individuals tend to mate with others nearby, or they may choose mates that are most like themselves. When individuals choose mates that are phenotypically similar, positive assortative mating has occurred. If mates look physically different, then it is negative assortative mating. Population geneticists use the term "assortative" because it means "to separate into groups," usually in a pattern that is not random. The terms "positive" and "negative" refer to the probability that mated pairs have the same phenotype more or less often than expected by chance. Two color varieties of snow geese (*Chen hyperborea*), blue and white, are commonly found breeding in Canada, and they show positive assortative mating patterns based on color. The geese tend to mate only with birds of the same color; blue mate with blue and white with white. Since a bird's color (phenotype) is determined by the presence of a dominant blue color allele, matings between similar phenotypes are also matings between similar genotypes. Matings between similar genotypes cause the frequency of individuals that are homozygous for the blue or the white allele to be greater, and the frequency of heterozygotes to be less than if mating were random and in Hardy-Weinberg equilibrium. Negative assortative mating increases the frequency of heterozygote genotypes in the population and decreases homozygote frequency. Assortative mating does not change the frequency of the blue or white alleles in the goose population; it simply reorganizes the genetic variation and shifts the frequency of heterozygotes away from Hardy-Weinberg equilibrium frequencies.

Inbreeding is the mating of relatives and is similar to positive assortative mating because like genotypes mate and result in a high frequency of homozygotes in the population. In assortative mating, only those genes that influence mate choice become homozygous, but inbreeding increases the homozygosity of all the genes. High homozygosity means that many of the recessive alleles that were masked by the dominant allele in heterozygotes will be expressed in the phenotype. Deleterious or harmful alleles can remain hidden from selection in the heterozygote, but after one generation of inbreeding, these deleterious alleles are expressed in a homozygous condition and can substantially reduce viability below normal levels. Low viability resulting from mating of like genotypes is called inbreeding depression.

Genetic Drift and Mutation

Genetic drift, like positive assortative mating, reduces the frequency of heterozygotes in a population, but with genetic drift, the frequency of alleles in a population changes. Nonrandom mating does not change allele frequency. Genetic drift is sometimes called random genetic drift because the mechanism of genetic change is random and attributable to chance events in small populations, such that allele frequencies tend to wander or drift. Statisticians use the term "sampling effect" to describe observed fluctuations from expected values when only a few samples are chosen, and it is easy to observe by tossing a coin. A

fair coin flipped a hundred times would be expected to produce approximately fifty heads and fifty tails, plus or minus a few heads or tails. Yet, if the coin is flipped only four times, it is not too surprising to get four heads or four tails. The probability of getting either all heads or all tails on four consecutive flips is one out of eight, but the probability of getting all heads or all tails decreases to much less than one in a billion as the sample size increases from four to a hundred tosses. Similarly, it is much easier for nonrandom events to occur in small populations than in large populations. If a population has two alleles with equal frequency for a particular trait, then the result of random mating can be simulated by tossing a coin. The frequency of each allele in the next generation would be determined by flipping the coin twice for each individual, since sexually reproducing organisms have two alleles for each trait, and counting the number of heads and tails. In a small population, only a few gametes, each containing one allele for the trait, will fuse to form zygotes. Chance events can cause the frequencies of alleles in a small population to drift randomly from generation to generation; often one allele is lost from the population.

In small populations with fewer than fifty mating pairs, alleles may be eliminated in fewer than twenty generations by random genetic drift, leaving only one allele for a particular trait in the population. Thus, all individuals would be homozygous for the remaining allele and genetically identical. Theoretically, in any finite population random genetic drift will occur, but it is usually negligible if the population size is greater than a hundred. Sometimes, disasters or disease may drastically reduce the population size, causing a bottleneck effect. The bottleneck in population size reduces genetic variability in a population because there are only a few alleles and results in random genetic drift. Many islands and new populations are established by a small group of founders that constitute a nonrandom genetic sample because they have only a fraction of the alleles from the original large population. Founder effects and bottleneck effects are phenomena that result in a loss of heterozygosity and decreased genetic variability because of the chance drift in allele frequency away from Hardy-Weinberg equilibrium values in small populations.

Mutations are any changes in the genetic material that can be passed on to offspring. Some mutations are changes at a single point in the chromosome, while at other times, pieces of the chromosome are removed, extra pieces are added, or pieces are exchanged with other chromosomes. All these changes could result in the formation of new alleles or could change one allele into a different allele. The random mistakes in the chromosomes occur at the molecular level, and only later are the changes in information or alleles translated into phenotypic differences. Thus, mutation is the ultimate source of genetic variability and is random with respect to the needs of the organism. Most mutations are lethal and are never expressed, but nonlethal mutations provide the necessary variation for natural selection. Even though mutations are very important for evolution, they have only a small effect on allele and genotype frequencies in populations because mutation rates are relatively low. If an allele makes up 50 percent of the gene pool and mutates to another allele once for every hundred thousand gametes, it would take two thousand generations to reduce the frequency of the allele by 2 percent. The net effect of mutations is to increase genetic variability, but at a very slow rate.

Studying Genetic Variability

Population geneticists use a wide variety of laboratory, field, and natural experiments to investigate genetic variability. Natural experiments are situations that have developed without a scientist intentionally designing an experiment, but conditions are such that scientists can test a theory. Researchers have used known pedigrees or ancestral histories of zoo animals and have found that mortality rates of inbred young are often two to three times higher than for noninbred young. Population geneticists use pedigrees to calculate the probability that two alleles are identical by descent; this research provides an index of the amount of inbreeding in a population.

The study of random genetic drift is usually carried out in the laboratory. Scientists often use small organisms that reproduce quickly, such as fruit flies (*Drosophila melanogaster*), to conserve space and save time. In a 1956 study of eye color conducted by Peter Buri, after only eighteen generations and sixteen fruit flies per population, more than half of the 107 populations started had only one of the two alleles for eye color.

Mutations are so rare that even fruit flies reproduce too slowly for scientists to study the effects of mutations on populations, even though much is known about the mechanism of mutation by studying *Drosophila*. Small bacterial growth chambers can hold many millions of bacterial cells, and, since they reproduce quickly, even mutations that occur in only one in a million cells can be detected. In 1955, it was found that mutation rates were very low in bacteria until caffeine was added to the growth chamber, whereupon mutation rates increased tenfold. Any chemical or type of radiation that can cause mutations is called a mutagen. Electrophoresis has also been a useful tool for the study of nonrandom mating, genetic drift, and mutations, because allele and genotype frequencies can be determined from samples of the population and unique alleles can be identified.

The Dangers of Inbreeding

Most governments and religions forbid marriages between close relatives because matings between first cousins result in a 20 percent decrease in heterozygosity; for those between brothers and sisters, there is an 80 percent decrease in heterozygosity. The decrease in heterozygosity and genetic variation and increase in homozygote frequency often result in inbreeding depression because deleterious recessive alleles are expressed. All inbreeding is not undesirable; many of the prizewinning bulls and pigs at state fairs have some inbreeding in their pedigrees. Most breeds of dogs were produced by breeding close relatives so that the offspring would have particular traits.

Zookeepers and others that breed and protect rare and endangered species must continually be concerned about the negative effects of both inbreeding and genetic drift. Most zoos are lucky if they have two or three pairs of breeding adults, and total population sizes are usually very small compared to those of natural populations. These conditions mean that inbreeding may reduce the vigor of the population and genetic drift will reduce the diversity of alleles in the population, thus reducing the chances of survival for the captive species. There is hope for rare and endangered species if independent inbred lines are crossed, thus reducing the effects of inbreeding depression, and if breeding adults from other zoos or populations are traded occasionally, thus increasing the effective population size.

Mutations are the ultimate source of genetic variation and so are very important in the study of evolution, but the population-level effects of one mutation are difficult to study because of the low frequency of natural mutations. Certain nonlethal mutations may have little evolutionary impact but may be important medically because spontaneous mutations result in hemophilia or dwarfism (achondroplasia) in more than 3 out of 100,000 cases. As exposure to background radiation and chemical levels increases, mutation rates are likely to increase, as well as the incidence of mutation-related diseases.

—*William R. Bromer*

See also: Convergent and divergent evolution; Evolution: Historical perspective; Hardy-Weinberg law of genetic equilibrium; Mutations; Population genetics; Punctuated equilibrium and continuous evolution; Reproduction.

Bibliography

Ayala, Francisco J. *Population and Evolutionary Genetics: A Primer*. Menlo Park, Calif.: Benjamin/Cummings, 1982. This book introduces required basic genetics and mathematics in the process of explaining the concepts and theories of evolution at the genetic level. Ayala's writing style is very clear and he uses only high school algebra throughout the text. An excellent section on probability and statistics.

Crow, J. F. *Basic Concepts in Population, Quantitative, and Evolutionary Genetics*. New York: W. H. Freeman, 1986. A college-level text that emphasizes the theory of genetics, with plenty of examples from the empirical literature that enhance the text. The reader should have an understanding of basic genetics and calculus. Classical and molecular genetics are also discussed.

Fisher, R. A. *The Genetical Theory of Natural Selection*. New York: Dover, 1958. Fisher's terse writing style makes this important book difficult to read, but it provides one of the best descriptions of the effects of mutation and natural selection on genetic variability.

Hartl, Daniel L. *A Primer of Population Genetics*. 3d ed. Sunderland, Mass.: Sinauer Associates, 2000. This introductory-level text for college students is written so that only basic algebra skills are required. Most of the problems are illustrated with numerical examples that are often based on actual data. Chapters on genetic variation and causes of evolution provide a good introduction to basic population genetics. There is also a section on molecular genetics.

Mettler, L. E., T. G. Gregg, and H. E. Schaffer. *Population Genetics and Evolution*. 2d ed. Englewood Cliffs, N.J.: Prentice-Hall, 1988. A well-written college text for courses in population genetics. Parts require some basic math, genetics, and chemistry. Most of the modern theories and hypotheses are given a historical perspective with a brief chronology of important papers on the topic.

Real, Leslie A., ed. *Ecological Genetics*. Princeton, N.J.: Princeton University Press, 1994. A collection of papers by five different writers based on a symposium series, offering a balance of comprehensive review of an aspect of the field of population genetics and ecology, and specific reviews of the author's own case studies.

Wilson, E. O., and W. H. Bossert. *A Primer of Population Biology*. Sunderland, Mass.: Sinauer Associates, 1971. Uses simple mathematical models to present advanced topics in a self-teaching format. Chapter 2 provides the basics of population genetics with a number of problems dealing with mutation and genetic drift.

NOSES

Types of animal science: Anatomy, behavior, development, evolution
Fields of study: Anatomy, biochemistry, biophysics, cell biology, developmental biology, genetics, herpetology, histology, immunology, neurobiology, ornithology, pathology, physiology, zoology

The nose is the major access channel for the sense of smell. The nose signals danger by detecting the smell of spoiled food, smoke, or natural gas. The nose also warms and humidifies the air as it moves onto lungs. It filters out particles and bacteria found in the inspired air and so prevents these particles from entering the lungs.

Principal Terms

MUCUS: the watery material covering the internal nasal structures that aids in humidification, warming, and particle filtration
OLFACTION: the sense of smell
SEPTUM: the bony structure that divides the nose into two sections
TURBINATES: bony structures that define the internal nasal anatomy

---

In vertebrates, the back of the throat is connected to the outside air through a passageway called the nose. The outside opening of the nose is referred to as the external naris, whereas the opening from the nose to the back of the throat is called the internal naris. In reptiles and amphibians, air drawn into the lungs passes through the external naris and into a tubelike structure which is the simplest form of a nose. In animals like the salamander, the nasal air passageway is a straight tube. Air exits this tube via the internal naris where it proceeds into the back of the throat. In some amphibians, such as the bullfrog, the channel between the external and internal naris has a bony bump on the floor of the nasal passageway. This bump, called the ementia, apparently functions like a baffle plate, so that the incoming air stream is deflected. As a result of this deflection, air moving through the nose is turbulent. This turbulence likely increases the ability to detect smells as well as improves the efficiency of the other nasal functions.

In mammals, the internal anatomy of the nose is defined by bony structures called turbinates. These turbinates produce multiple and convoluted air flow paths for the inspired air. As in the frog, these convoluted flow paths facilitate the various nasal functions.

In mammals the nose is divided into two halves along the midline by a bony structure called the nasal septum. In some mammals, such as the rat, the septum is incomplete, and so there is some mixing in the nose of air that comes into the left and right nostrils.

Air entering the nose travels in the airspace between the turbinates and the nasal septum. The surface of these nasal structures has a very rich blood supply, and so blood flow to these areas can be quickly changed. By changing the amount of blood flow to the nasal structure, the diameter of the nasal airspace itself can be quickly changed. Because of the speed and amount of diameter change that can occur in these areas they are called nasal "swell spaces."

Heat, Humidity, and Purification

When cold air enters the nose, it causes the blood flow to the swell spaces to increase dramatically. This causes a swelling of the nasal tissue and so reduces the size of the airspace through which the incoming air must travel. Because the air passageways are now more narrow, more heat can be transferred from the blood stream to the incoming air. Thus, the cold air is effectively warmed before

All canids, including wolves, have exceptionally keen senses of smell. (PhotoDisc)

it enters the lungs. When air that is warmer than the body temperature enters the nose the reverse happens, and the nasal air passageways are made wider.

The material covering the turbinates and nasal septum is called the mucus. This mucus layer is mostly composed of water and serves to humidify the incoming air. When the air is dry, water is evaporated from the mucus into the inspired air. When the air is dry, a considerable quantity of water can be lost through the nose. For animals living in dry, desert conditions, the nasal humidification process is critical because it is necessary to save every drop of water while still humidifying the incoming air. When these animals take air into the lungs, the inspired air passes through the extensive turbinate structures of the nose where it is humidified by the evaporation of water from the nasal mucus. As evaporation takes place, the surface of the mucus is cooled. When it is time to discharge the air from the lungs, the expired air passes over the cooled surface of the mucus. As a result, water in the inspired air is now picked up by the cooled surface and so it is not lost in the expired air. This type of water conservation method has been seen in kangaroo rats living in the deserts of Australia and in certain birds, such as the cactus wren, which inhabit warm, dry areas.

In addition to supplying water for humidification purposes, the nasal mucus serves as a trap for particulate matter. Smoke and dust particles and even airborne bacteria are trapped in the mucus of the nose. Beneath the nasal mucus is a layer of cells that contain hairlike protrusions called cilia. As the cilia of these cells beat, they create a wavelike action in the mucus. The mucus is thus moved through the nose and to the back of the throat. Once in the throat the mucus is swallowed and the particles and bacteria are dealt with in the stomach. The movement of the nasal mucus is an ongoing process and so trapped particles are continually being removed from the nose. There are also white blood cells and enzymes found in the mucus that destroy bacteria.

When particles (and sometimes even air) touch parts of the nose a sneeze occurs. In this case, breathing is stopped and air is forcibly expelled from the lungs at a high flow rate. The sneeze is an attempt to expel the particles from inside the nose.

However, when the odorant is very foul or stings the nose, breathing may be temporarily stopped. This protects the lungs from potentially damaging chemicals. The detection of potentially harmful chemicals occurs both through the smell receptors as well as through pain receptors that are found in the nose.

—*David E. Hornung*

See also: Anatomy; Beaks and bills; Bone and cartilage; Brain; Circulatory systems of invertebrates; Circulatory systems of vertebrates; Claws, nails, and hooves; Digestive tract; Ears; Endoskeletons; Exoskeletons; Eyes; Fins and flippers; Immune system; Kidneys and other excretory structures; Lungs, gills, and tracheas; Muscles in invertebrates; Muscles in vertebrates; Nervous systems of vertebrates; Physiology; Reproductive system of female mammals; Reproductive system of male mammals; Respiratory system; Sense organs; Skin; Smell; Tails; Teeth, fangs, and tusks; Tentacles; Wings.

Bibliography

Association for Chemoreception Sciences. http://www.achems.org. This Web site contains a description of current work in the field as well as a discussion of smell disorders.

Getchell, T. V., R. L. Doty, L. M. Bartoshuck, and J. B. Snow, eds. *Smell and Taste in Health and Disease*. New York: Raven Press, 1991. Discussion of the clinical aspects of smell problems.

Gibbons, Byron. "The Intimate Sense of Smell." *National Geographic* 170, no. 3 (1986): 321-361. An excellent overview of the anatomy, physiology, and psychology of the sense of smell. The making of perfumes, use of dogs for tracking, and the history of smell are all well covered.

Vroon, Piet. *Smell: The Secret Seducer*. New York: Farrar, Straus and Giroux, 1997. A cultural history and compendium of odd facts about noses.

NUTRIENT REQUIREMENTS

Type of animal science: Physiology
Fields of study: Biochemistry, physiology

All animals depend on external sources of nutritional raw materials for energy, growth, maintenance, and functioning.

Principal Terms

CARBOHYDRATE: an organic molecule containing only carbon, hydrogen, and oxygen in a 1:2:1 ratio; often defined as a simple sugar or any substance yielding a simple sugar upon hydrolysis

LIPID: an organic molecule, such as a fat or oil, composed of carbon, hydrogen, oxygen, and sometimes phosphorus, that is nonpolar and insoluble in water

MINERAL: one of the many inorganic elements other than carbon, hydrogen, oxygen, and nitrogen that an organism requires for proper body function

PROTEIN: an organic molecule containing carbon, hydrogen, oxygen, nitrogen, and sulfur and composed of large polypeptides in which over a hundred amino acids are linked together

VITAMIN: an organic nutrient that an organism requires in very small amounts and which generally functions as a coenzyme

---

Food, used to provide material for production of new tissue and the repair of old tissue as well as used as an energy source, is obtained from a variety of plant, animal, and inorganic sources. Regardless of the source, food must provide its consumer with a sufficient amount of the essential nutrients. A nutrient is any substance that serves as a source of metabolic energy, raw material for growth and repair of tissues, or general maintenance of body functions.

General Nutritional Requirements

Animals differ widely in their specific nutritional needs, depending on the species. Within any given species, those needs may vary according to variations in body size and composition, age, sex, activity, genetic makeup, and reproductive functions. A small animal requires more food for energy per gram of body weight than does a larger animal, because the metabolic rate per unit of body weight is higher in the smaller animal. Likewise, an animal with a cool body temperature will have less energy needs and require less food than an animal with a high body temperature. An egg-producing or pregnant female will require more nutrients than a male. In order for an animal to be in a balanced nutritional state, it must consume food that will supply enough energy to supply power to all body processes, sufficient protein and amino acids to maintain a positive nitrogen balance and avoid a net loss of body protein, enough water and minerals to compensate for losses or incorporation, and those essential vitamins that are not synthesized within the body.

Activities such as walking, swimming, digesting food, or any other activity performed by an animal require fuel in the form of chemical energy. Adenosine triphosphate (ATP), the body's energy currency, is produced by the cellular oxidation of small molecules, such as sugars obtained from food. Cells usually metabolize carbohydrates or fats as fuel sources; however, when these carbon sources are in short supply, cells will utilize proteins. The energy content of food is usually measured in kilocalories, and it should be noted that the term "calories" listed on food labels is actually kilocalories (1 kilocalorie = 1000 calories). Cellular metabolism must continually produce energy to maintain the processes required for an animal to remain alive. Processes such as the circulation of blood, breathing, removing waste products from the blood, and in birds and mammals, the maintenance of body temperature, all require energy. The calories required to fuel these essential processes for a given amount of time in an animal at rest is called the basal metabolic rate (BMR). For a resting human adult, the BMR averages from thirteen hundred to eighteen hundred kilocalories per day. As physical activity increases, the BMR increases.

Energy balance requires that the number of calories consumed for body maintenance and repair and for work (metabolic and otherwise) plus the production of body heat in birds and mammals be equal to the caloric intake over a period of time. An insufficient intake of calories can be temporarily balanced by the utilization of storage fats, carbohydrates, or even protein, and will result in a loss of body weight. On the other hand, an excessive intake of calories can lead to the storage of energy sources. Animals normally store glycogen, but when the glycogen stores are full, food molecules, such as carbohydrates and protein, will be converted to fats.

Nutrient Molecules

Proteins are composed of long chains of amino acids and serve a number of important functions in all living organisms, but they are primarily used as structural components of soft tissues and as enzymes. Proteins can also be utilized as energy sources if they are broken down into amino acids. Animal tissues are composed of about twenty different amino acids. The ability to synthesize amino acids from other carbon sources, such as carbohydrates, varies among species, but few, if any, animal species can synthesize all twenty required amino acids. Those amino acids that cannot be synthesized by an animal, but are required for the synthesis of essential amino acids, are the so-called essential amino acids, and must be included in the diet. Humans, for example, require nine essential amino acids. Both plant and animal tissues can serve as protein sources, but animal protein generally contains larger quantities of the essential amino acids.

Carbohydrates are primarily used as immediate sources of chemical energy, but they can also be converted to metabolic intermediates or fats. Some carbohydrates are also structural components of larger molecules. For example, the nucleic acids deoxyribonucleic acid (DNA) and ribonucleic acid (RNA) contain the sugars deoxyribose and ribose, respectively, as an integral component of their structure. Most animals can also convert proteins and fats into carbohy-

---

Feeding Mechanisms

Since most animals cannot absorb nutrients directly from their environments, they must exert energy in order to obtain food. Although animals show tremendous diversity in their methods of obtaining food, most feeding habits can be classified in one of three different types. Many animals, particularly those living in aquatic environments, feed on particulate matter, the free-floating material made up of plankton, microscopic aquatic life-forms, and organic remains of dead and decaying plants and animals. Some of the animals utilize a technique referred to as suspension feeding, in which the food material is drawn into the digestive tract by currents created by external structures, such as cilia or setae. Other particulate feeders feed on deposits of detritus (decaying organic matter) that accumulate at the bottom of lakes or oceans.

A second feeding method involves the consumption of food masses. Most of these animals utilize specialized adaptions that allow them to capture and manipulate solid food. Herbivores have special adaptions for cutting and crushing or grinding plant tissues, while carnivores such as predators must have the ability to capture, hold, and either swallow the prey whole or shred it into smaller pieces. Feeding on liquids is a third type of feeding habit. This method, which involves the sucking of fluids from a host plant or animal, is utilized primarily by parasites, but it can also be observed in certain insects.

drates. The principle sources of carbohydrates are the sugars, starches, and cellulose in plants and the glycogen stored in animal tissue.

Lipids are an important and essential component of all biological membranes. In addition, several animal hormones, such as the sex hormones, are lipoidal in nature. Fats and lipids are also especially suitable as concentrated energy reserves, because each gram of fat supplies twice as much energy as a gram of carbohydrate or protein and does not have to be dissolved in water. Hence, animals commonly store fat for times of caloric deficit when energy expenditure exceeds energy uptake. Some animals, such as migratory birds and hibernating mammals, store large quantities of fat to offset the times that they are not actively feeding. Lipid molecules include fatty acids, monoglycerides, triglycerides, sterols, and phospholipids.

All animals require an adequate supply of essential inorganic minerals. Carbonate salts of the metals calcium, potassium, sodium, and magnesium as well as some chloride, sulfate, and phosphate are important constituents of intra- and extracellular fluids. Calcium phosphate is present as hydroxyapatite, a crystalline material that gives hardness and rigidity to the bones of vertebrates and the shells of mollusks. Certain metals, such as copper and iron, are required for oxidation-reduction reactions and for oxygen binding and transport. The catalytic function of many enzymes requires the presence of certain metal atoms. Animals require moderate amounts of some minerals and only trace quantities of others.

Animals require a variety of vitamins, diverse and chemically unrelated organic substances. Vitamins primarily function as coenzymes for the proper catalytic activity of essential enzymes. As with amino acids, the ability to synthesize differ-

---

Functions of Essential Minerals and Vitamins

MINERALS

CALCIUM: nerve and muscle function, blood clotting, bone and tooth formation

CHLORINE: acid-base balance, gastric juice

CHROMIUM: associated with glucose and energy metabolism

COBALT: component of Vitamin B_{12}

FLUORINE: maintenance of tooth structure

IODINE: component of thyroxine, a thyroid hormone

IRON: component of essential enzymes and electron carriers in energy metabolism

PHOSPHORUS: transfer of chemical energy, bone and tooth formation, nucleic acid synthesis

POTASSIUM: proper nerve function, acid-base balance, water balance

SELENIUM: component of essential enzymes and functions in association with vitamin E

SODIUM: proper nerve function, water balance, acid-base balance

SULFUR: component of certain amino acids

COPPER, MAGNESIUM, MANGANESE, MOLYBDENUM, ZINC: components of essential enzymes

VITAMINS

VITAMIN A (CAROTENE): formation of visual pigments, maintenance of certain membranes

VITAMIN B_1 (THIAMINE), Vitamin B_2 (RIBOFLAVIN), NIACIN: coenzymes for certain enzymes in energy metabolism

VITAMIN B_6 (PYRIDOXINE): coenzyme for amino acid synthesis and fatty acid metabolism

VITAMIN B_{12} (CYANOCOBALAMIN): required for nucleoprotein synthesis

VITAMIN C (ASCORBIC ACID): vital to collagen formation, serves as an important antioxidant

VITAMIN D (CALCIFEROL): increases calcium absorption from gut, bone and tooth formation

VITAMIN E (TOCOPHEROL): maintains pregnancy in mammals; serves as an important antioxidant

VITAMIN K (NAPHTHOQUINONE): required for synthesis of a protein necessary for blood clotting

BIOTIN: coenzyme for enzymes associated with protein synthesis

FOLIC ACID: required for nucleoprotein synthesis and formation of red blood cells

PANTOTHENIC ACID: forms part of coenzyme A associated with energy metabolism

ent vitamins from other carbon sources varies among species. Those essential vitamins that cannot be synthesized by the animal itself must be obtained from other sources, primarily from plants but also from dietary animal flesh or from intestinal microoganisms. Vitamin C (ascorbic acid) can be synthesized by many animals, but not by humans. Vitamins K and B are produced by intestinal bacteria in humans. Vitamins such as A, D, E, and K are fat soluble and can be stored in fat deposits within the body; however, water soluble vitamins such as vitamin C are not stored and are excreted through the urine. Hence, the water soluble vitamins must be consumed or produced continually in order to maintain adequate levels.

Although not commonly thought of as a nutrient, water is tremendously important and comprises up to 95 percent or more of the weight of some animal tissue. Water is replaced in most animals by drinking, ingestion with food, and to some extent, by the metabolism of carbohydrates and lipids.

—D. R. Gossett

See also: Biology; Cannibalism; Carnivores; Cold-blooded animals; Digestion; Digestive tract; Food chains and food webs; Herbivores; Hibernation; Ingestion; Metabolic rates; Migration; Osmoregulation; pH maintenance; Predation; Ruminants; Scavengers; Thermoregulation; Warm-blooded animals; Water balance in vertebrates.

Bibliography

Campbell, N. A., L. G. Mitchell, and J. B. Reece. *Biology: Concepts and Connections.* 3d ed. San Francisco: Benjamin/Cummings, 2000. An outstanding introductory biology text that gives a clear concise description of nutrient requirements.

Carr, D. E. *The Deadly Feast of Life.* Garden City, N.Y.: Doubleday, 1971. A keen insight on what and how animals eat.

Fox, I. S. *Human Physiology.* 6th ed. Boston: WCB/McGraw-Hill, 1999. An excellent treatment of the physiology of nutrition in humans, but applicable to other mammals.

Jennings, J. B. *Feeding, Digestion, and Assimilation in Animals.* 2d ed. New York: St. Martin's Press, 1972. An excellent comparative approach to nutrition among animals.

Randall, David, Warren Burggren, Kathleen French, and Russell Fernald. *Animal Physiology: Mechanisms and Adaptations.* 4th ed. New York: W. H. Freeman, 1997. An advanced text that gives an excellent description of the nutrient requirement in animals.

Weindrach, R. "Caloric Restriction and Aging." *Scientific American* 274 (1996): 46-52. This article gives a very good discussion of how organisms from protists to mammals live longer on well-balanced but low-calorie diets.

OCTOPUSES AND SQUID

Type of animal science: Classification
Fields of study: Marine biology, zoology

Octopuses and squid belong to the order Octopoda, members of which have eight tentacles covered with sucker pads. They have two eyes and a well-developed nervous system and are considered among the most intelligent of all invertebrates.

Principal Terms

INVERTEBRATES: animals without backbones
TENTACLES: a long flexible arm or projection

Octopuses and squid belong to the order Octopoda, which is in the class Cephalopods, phylum Mollusca, which includes snails and clams. They are related to other mollusks but have no internal or external shell. There are more than one thousand species of octopuses and squid alive today. They have two eyes and eight (octopuses) to ten (some squid) tentacles that are attached directly to the head. The tentacles are covered with sucker pads that help the animal move along the ocean floor, its normal place of residence. They have the most complex brain of the invertebrates, and also have long- and short-term memories, much like vertebrates.

The genus *Octopus* contains several different species, which vary greatly in size. The smallest is *Octopus arborescens*, a species that averages two inches in length. The largest species, *Octopus dofleini*, commonly known as the giant Pacific octopus, can grow to sixteen feet in length and weigh up to three hundred pounds, with a tentacle span of thirty feet. The best known and most widely distributed species is *Octopus vulgaris*, a medium-sized octopus found in every ocean. This species is from two to three feet in length and lives in holes on the ocean floor. It feeds mainly on crabs, lobsters, and other crustaceans. It is very intelligent: One Japanese scientist taught an octopus how to open a sealed jar in order to get at its contents, a spiny lobster.

An octopus can change its skin color quickly when it is frightened or threatened. Its changed color provides camouflage to help it blend in with the background so that the octopus becomes almost invisible to predators. Scientists have discovered that octopuses change colors not only for defense, but also to reveal their moods and emotions. Angry members of the species turn a deep red in color, while during mating season, both males and females display stripes and colors that reflect their inner excitement. Some species squirt out clouds of a dark inky substance, which hangs in the water for several seconds. This inkblot has the same size and shape as the octopus, and while it draws the attention of the predator, the octopus escapes. When an enemy, such as a moray eel, attacks an octopus, it may lose one of its eight arms in the fight, and while the eel watches the twitching arm the injured octopus can swim away. The tentacle grows back very quickly.

Squid are cephalopods and belong to the order Teuthoidea (ten-armed), which has many species. They are found in every ocean and range in size from less than one inch to more than sixty-five feet in length. Adult squid of some species can race through the water at speeds of up to twenty-five miles per hour. They are aggressive hunters, equipped with two more tentacles than octopus. The squid use these extra arms to catch their food. Some squid hunt in packs and use their ability to change color to lure their prey. They are also bioluminescent, which means they can light up.

Cephalopod Facts

Classification:
Kingdom: Animalia
Subkingdom: Eumetazoa
Phylum: Mollusca
Class: Cephalopoda
Subclass: Nautiloidea (chambered nautilus), Coleoidea (octopuses and squid)
Orders: Sepioidea (cuttlefishes), Teuthoidea (squid), Octopoda (octopuses)
Suborder: Three suborders of Octopoda—Palaeoctopoda (finned), Cirrata (webbed), Incirrata (round bodied)
Geographical location: Every ocean, but mainly the Indian Ocean and the South Pacific
Habitat: The ocean floor
Gestational period: Varies by species from forty days to one year
Life span: One to three years for octopuses; five years for squid; fifteen to twenty years for chambered nautilus
Special anatomy: Eight to ten tentacles, usually with suction cups at the end; round, sacklike bodies for octopuses

---

Actually, bacteria living under their skin produce the light in squid that live close to the surface. Deep-water squid, on the other hand, living thousands of feet below the surface, make their own light. These species have light-producing organs called photophores that make two chemicals: a protein called luciferin and an enzyme known as luciferase. When the two chemicals are mixed, the enzyme breaks down the protein, releasing a pale, blue-green light. This is the same process used by fireflies. Some squid species can squirt out clouds of glowing bacteria when they are endangered. This light show distracts the predator and helps the squid escape.

Squid, Cuttlefish, and Nautilus

The giant squid, *Architeuthi*, is the largest invertebrate on the earth, having eyes the size of automobile headlights and weighing up to one thousand pounds. Despite their size, no one has ever seen one of these animals in its natural environment. Although these monsters of the deep have been found in the nets of commercial fishermen, in the stomachs of sperm whales, and washed ashore in Australia, no scientific information has been gathered on the species by direct observation. Perhaps the major reason so little is known about giant squid is that they live so deep in the ocean. Another reason is their enormous speed and rapid acceleration. Because of this speed they can easily evade their major predators, tuna and sharks. Some apparently even have the ability to leave the water for short periods and glide through the air like flying fish to escape their enemies.

The cuttlefish, genus *Sepia*, is a cephalopod that can eject clouds of ink to confuse enemies and can swim almost as rapidly as squid. It most remarkable characteristic, however, is its ability to change color, which biologists believe is used by the species as a method of communication. Cuttlefish can display thirty different color patterns that range from very light to very dark. They can also exhibit zebralike stripes and patterns that look like a pair of dice. They can change colors quickly to escape predators by blending into the background. Color changes are also used to attract mates and outshine rivals during courtship. Cuttlefish have ten arms or tentacles, with two longer than the other eight. Most of the time the two longer arms are tucked under their bodies, but if supper swims by they shoot out to grab the shrimp or fish and drag it to their mouths.

Next to the octopus, the cuttlefish is probably the most intelligent invertebrate in the sea. In experiments, cuttlefish have been taught to recognize which colored disks signal food and which do not. Baby cuttlefish know from the moment of birth how and when to bury themselves in the sand, how to squirt an inky substance to confuse predators, and how to get away quickly when they encounter larger animals. The ink of the cuttlefish, which is called sepia (the genus name) was the original dark pigment used in India ink. This ink was used in quill pens in England and France in the 1600's and is still used today by artists for drawing and lettering.

The chambered nautilus is one of five species in the genus *Nautilus*. It is a living fossil. The surviving species are found only 60 to 1,500 feet below the surface of the Indian and Pacific oceans. Five hundred million years ago, nautiloids were the masters of the sea. At that time there were more than 2,500 species, some with straight shells and others with coiled shells like the modern nautilus. They are nocturnal feeders, coming to the surface only on moonless nights to feast on small fish and shrimp. They are the only members of the class Cephalopoda to still have an external shell. They have eighty to one hundred arms surrounding their heads. Each arm can touch and taste food. The nautilus gets its name from the thirty to thirty-eight walled chambers that are found inside its shell. The animal lives in the outermost chamber and uses the others to float in the water.

The nautilus is an endangered species, mainly because of the beauty of its shell. Philippine Island fishermen catch about five thousand living animals every year to sell to shell collectors. They are also hunted extensively off the coasts of India and Indonesia.

Growth and Reproduction

Most cephalopods, including the giant squid, grow quickly and die after a short life. The great Pacific octopus, for example, which is only one-twentieth the size of the giant squid, lives just two to three years. It is estimated that the giant squid lives for no more than five years. This means that its growth to adulthood is extremely fast. When it is a baby, it is less than an inch long, but within three years it has reached almost sixty feet in length. Growing at such a rapid rate requires huge amounts of food. Giant squid eat enormous quantities of fish and other squid, which they catch with the suckers on their tentacles and crush in their massive mouths.

Octopuses and squid have separate sexes and reproduce internally. In some species, modified

The tentacles of an octopus are attached directly to its head, leading to the name "cephalopod," meaning "head-foot." (Digital Stock)

sucker discs at the tip of one of its tentacles distinguish the male. This arm is used to remove a packet of sperm from within his body cavity and insert it into the cavity of the female. Within two months after mating, the female attaches strands of clustered eggs to the ceiling of her dwelling under the sand. The number of eggs laid varies greatly from species to species, with the common octopus laying anywhere from 200,000 to 400,000 very tiny eggs. Other species lay as few as 150 eggs. Once the eggs are laid, the mother gently caresses the eggs with her suckers to keep algae and bacteria from growing on them. She also squirts them with streams of water to keep them clean. As the baby inside the egg matures, the mother's gentle caresses become more violent to help the developing octopus break away from its egg house. Most mothers do not eat after laying their eggs and die shortly after their eggs have hatched. Baby octopuses in species like *Octopus vulgaris* are carried about in water currents for about a month before they settle to the bottom, where they start to feed. Babies in other species look just like a miniature adult and immediately sink to the bottom where they start living. On average, the survival rate for a baby octopus is extremely tiny. Only about one or two out of 200,000 eggs will survive to become an adult.

The chambered nautilus cements its eggs to rocks or coral on the seafloor. They take almost a year to hatch. A baby nautilus is only about an inch long when it hatches from its egg case but grows into an adult quickly. It is the longest-living cephalopod, with some adults living fifteen to twenty years. Unlike the octopus, the chambered nautilus does not die immediately after reproduction.

Cephalopods make up a large part of the diet of whales, seals, fishes, and seabirds.

Cephalopod Evolution

The earliest ancestors of octopuses and squid were mollusks with thick shells that protected them from their enemies. Of the estimated one thousand species of cephalopods still living, only the chambered nautilus and cuttlefish have remnants of an internal shell. Cephalopods began losing their shells in the Triassic period, 245 to 208 million years ago. The earliest squid appeared in the Jurassic period, 208 to 144 million years ago. The oldest known octopod, *Palaeoctopus newboldi*, now extinct, comes from the Cretaceous period and is at least 140 million years old.

—Leslie V. Tischauser

See also: Camouflage; Deep-sea animals; Defense mechanisms; Intelligence; Invertebrates; Learning; Marine animals; Marine biology; Mollusks; Shells; Tentacles.

Bibliography

Ellis, Richard. *Monsters of the Sea*. New York: Alfred A. Knopf, 1994. A complete natural history of cephalopods with a special emphasis on giants of the species. Written for nonspecialists.

_____. *The Search for the Giant Squid*. New York: Lyons Press, 1998. An exciting discussion of recent scientific efforts to search for and study the habits and habitat of this monster of the deep.

Lane, Frank W. *The Kingdom of the Octopus*. New York: Sheridan House, 1960. A brief history of the species, written for children. Well illustrated.

Ward, Peter. *The Natural History of Nautilus*. Boston: Allen & Unwin, 1987. A detailed illustrated history of this living fossil.

OFFSPRING CARE

Types of animal science: Behavior, reproduction
Fields of study: Ethology, ornithology, reproduction science

In most species, one parent alone or neither parent cares for the offspring. In some species, however, both parents jointly care for their offspring. In a few species, several adults share parenting duties or take turns "babysitting."

Principal Terms

ALLOPARENTING: performance of parenting duties by an individual not the parent of the offspring (though usually a relative)

ALTRICIAL: the condition of being weak and relatively undeveloped at birth (or hatching) and thus dependent upon parental care for a prolonged period

CUCKOLD: a partnered male who is helping his mate to raise offspring which are not genetically his own

NEST-PARASITE: also called brood-parasite; an individual (or species) that lays its eggs in the nest of another individual (or species) and does no parenting at all

PRECOCIAL: the condition of being strong and relatively well developed at birth (or hatching) and thus not particularly dependent upon parental care

SEX-ROLE REVERSAL: generally used to refer to species in which the male does most of the parenting

VIVIPAROUS: characterized by live birth (as opposed to egg-laying)

In the animal kingdom, there are species in which neither parent cares for offspring, species in which one parent cares for offspring, species in which both parents care for offspring, and species in which individuals other than parents help care for offspring. Patterns of caring for offspring relate to several key factors.

Precocial Species Versus Altricial Species

The most important factor determining the number of caretakers, as well as the quality and duration of parental care, is simply how much care the offspring need. In species in which young are born well developed and capable of surviving on their own, there is little reason for parents to provide help. Hatchling fish and baby turtles, for example, are born or hatched looking just like miniature adults, and are fully capable of moving about and feeding themselves. Young of such species are referred to as precocial, and usually get no parental care at all. Young of most invertebrate species as well are capable of surviving without help, and so are typically left to fend for themselves.

On the other hand, in species in which young are born or hatched quite helpless with no chance of surviving on their own, it should not be surprising that one or both parents are likely to stay around. Young of such species are referred to as altricial. Dogs and cats are excellent examples of species with altricial offspring; so are humans. Generally speaking, mammals are more altricial than most groups of animals, and all mammalian young initially depend upon their mother for food (milk), delivered by means of lactation.

Compared to other animals, birds, too, are relatively altricial—especially raptors (birds of prey) and songbirds. When altricial birds first hatch they are featherless and unable to regulate their own body temperature; one or both parents must regularly warm the hatchlings just as they earlier warmed the eggs. Furthermore, altricial birds hatch with their eyes closed and they are not strong enough or coordinated enough to leave the nest to feed or to flee from danger. Because baby birds need tremendous amounts of food and there are usually quite a few offspring, in most species of birds, both parents provide care.

Prey Species Versus Predatory Species

Although the young of birds and mammals are quite helpless compared to the young of animals in other taxonomic groups, there are variations. As a rule, prey species tend to be more precocial than predatory species and, therefore, to require less parental care.

Young of species which are herbivorous (vegetarian) but are in constant danger of being eaten by carnivorous (meat-eating) species cannot afford to be completely helpless even if they are dependent upon their mother for food. Grazing mammals, such as zebra and deer, are vulnerable to lions and wolves, so young of these species must be able to stand up and run just a few minutes after birth: they are born with open eyes, well-developed muscles, and a full coat of hair. Although mammalian young need to nurse from their mother, they are able to start grazing fairly quickly, so typically it is only the mother who pro-

vides support, and her care is not exceptionally prolonged.

Among birds the same pattern is found: Herbivorous species, such as ducks and chickens, that are preyed upon by mammalian, avian, or reptilian predators, hatch with open eyes, the ability to walk, run, or swim, and a coat of downy feathers so that even though they cannot yet fly, they can leave the nest without need of further incubation. Offspring of these species can usually eat on their own almost immediately after hatching, and therefore can often get by with only one parent to show them what to eat and to protect them until they can fly.

Compared to young of prey species, young of predatory species tend to be born in a much more altricial state, and therefore, to need more parenting. Lion cubs and wolf pups do not have open eyes or the strength and coordination to leave their nest, den, or lair until a few weeks after birth. Even after they can move about and are weaned from mothers' milk, they are still too uncoordinated to hunt successfully and must be fed by one or both parents. In fact, species such as lions and wolves, which not only have altricial young, but frequently have large litters, often re-

"Noncooperative" Breeding

In Australia, birds called choughs (pronounced "chuffs") live in extended family groups that cooperate to defend territories and raise young that may need provisioning for as long as two years. Some of the youngest helpers are not much older than their younger siblings, and when provisioning their charges, often cheat. If no older adult is looking, a young helper may pretend to feed its sibling, but at the last minute, swallow the food itself.

In the southwestern United States live acorn woodpeckers, a communal species in which a group of up to a dozen birds shares common nest-holes and food repositories for storage of acorns. Despite their seeming cooperation, when more than one female lays eggs in the same nest-hole, they initially take turns destroying or removing one another's eggs

from the hole and placing them in the food repository for later consumption. Over a third of eggs laid get destroyed in this fashion.

Some birds lay their eggs in another bird's nest and do not do any parenting at all. Facultative or opportunistic nest-parasites (also called brood-parasites) are species such as ducks, which usually make their own nest and raise their own young, but which, in a good season, may lay extra eggs in a neighbor's nest or, in a bad season, may forego nesting but lay a few eggs elsewhere on the chance that another parent will manage better. Obligate nest-parasites are species such as the Australian and European cuckoos, which never make a nest and never raise their own offspring, but which lay their eggs in the nest of birds of a different species.

cruit other adults to help raise their offspring. Amongst African (but not American) lions, related females form groups called prides that hunt together and help take care of one another's offspring. Wolves form hunting packs, and all members (both male and female) help to feed the young. This behavior is called aunting when it is done by a female relative, or more generally, alloparenting. Alloparenting is also seen among some rodents and primate species.

As with mammals, carnivorous and insectivorous birds come into the world in a more altricial state than their vegetarian brethren, and therefore need more care. In the majority of bird species, both parents cooperate to raise young and, as with some of the most altricial mammalian species, if there is a large brood, parents of some species recruit helpers. In birds, helpers are usually the parents' offspring from a previous brood or season, and are thus siblings or half-siblings of the young they are helping to raise. Species that use this extended family system are referred to as cooperative breeders—although some are less cooperative than others.

Quantity Versus Quality Parenting

A third factor relevant to parenting is the number of offspring, either sequentially or in litters, that an individual produces. All individuals have a limited life span and a limited amount of energy, and during that life span they can allocate that energy either to producing a large number of offspring or to providing intensive care for a smaller number of offspring. Thus, while animals with altricial young that require intensive care do not have an option of producing huge numbers of offspring, species with precocial young do.

Most invertebrates and many vertebrates (other than birds and mammals) take the "quantity" strategy: They produce large numbers of offspring and provide no parental care at all. A large majority of the offspring of these species die before reaching adulthood, but a small percentage survive and reproduce. Extreme examples of taking a "quantity" strategy are the semelparous species, species that can only reproduce once in their lifetime. Salmon are well-known for this form of reproduction. Many spiders and insects, too, die before their only batch of young are even hatched.

At the other extreme is the strategy of having a small number of well-cared-for offspring, each of which has a high probability of survival. All altricial species are constrained to the "quality" strategy, but some precocial species opt for "quality" as well. Alligators and crocodiles are excellent mothers, protecting their eggs before they hatch, then transporting and protecting the young afterward. Some species of amphibians, fish, and even insects are devoted parents.

Maternal Care Versus Paternal Care

In mammals, if only one parent is necessary, it is always the mother who is committed to caretaking because she is the one who must provide the offspring with their first food through nursing. No other animals, however, nurse their young, so in other species, if only one parent is necessary, it does not necessarily have to be the mother who becomes the caretaker.

In single-parent species, whether it is the mother or father who becomes the caretaking parent depends, to a great extent, on parental certainty. In viviparous species (species that give birth to live young rather than lay eggs) the parent that gives birth is, for certain, one of the two genetic parents, and is therefore the parent most likely to care for the young if they need it. With rare exception, that means that mothers become the caretaking parent in viviparous species. In egg-laying (oviparous) species, parental certainty depends on whether fertilization was internal (as in birds and many invertebrates) or external (as in most fishes and amphibians). If fertilization is internal, then again, it is only the mother who is certain to be a genetic parent of the eggs she lays, and who is therefore most likely to take on the role of caretaker when only a single caretaker is needed. If, on the other hand, fertilization is external, then both parents are equally likely to be the genetic parent of any young that later hatch from eggs at the breeding site, and if parenting is needed, male and female are equally likely

Fathers Extraordinaire

Because mothers are the single parent in most single-parent species, those species in which the father does most of the parenting are referred to as sex-role-reversed. Typical of role-reversed species is a cluster of sex differences that are reversed in many ways other than just parenting: Females tend to be larger, more aggressive, and more flashy than the males, and they may be the more "promiscuous" sex, mating with a series of stay-at-home fathers in a mating system called polygyny.

Polygyny and exclusive paternal care have been best studied in birds. Other types of animals, however, also provide examples of exclusive paternal care. In several species of invertebrates, fish, and amphibians, females lay their eggs and leave, while males remain to guard the eggs after they have been fertilized, protecting them against the vagaries of weather, water currents, and predation. Even after the eggs hatch the male may continue to guard the young or to transport them: Some tiny Amazonian frogs carry their tadpole offspring on their back if the small pool of water where they hatched starts to dry up. In other species, males carry fertilized eggs on their back, in their mouth, or in a specialized pouch. In the case of pipefish, seahorses, and some frogs, the male's tissue actually provides nutrients to the developing embryos. In these rare creatures, one could truly say that it is the male that becomes pregnant.

to become the caretaker. The result is that while most species that have external fertilization do not remain with their eggs or provide any parental care at all, in those that do, factors other than parental certainty determine which sex becomes the guardian.

In species with internal fertilization, the probability that a particular male is the father (or one of the fathers) of a particular set of offspring depends upon how many males the female mated with and when. In the few species that have been closely studied, if a mother needs help to raise her offspring, males seem to expend effort in proportion to the probability that they are actually the genetic father. This is because males that are parenting or otherwise providing resources for offspring that are not genetically their own are wasting their effort in terms of reproduction.

A male that is caring for another males' offspring is sometimes referred to as a cuckold. In some human cultures, to call a man a cuckold (or the equivalent) is a huge insult. However, in humans and other animals, when a male provides care for a female's offspring even though they are not genetically his, he is sending her a signal that he is a good provider, and perhaps he is making it more likely that she will mate with him in the future.

The Spectrum of Parental Behavior

Most parents are good to most of their offspring most of the time. Parents provide food and sometimes shelter; they protect their offspring from danger and chase away predators; they may place their offspring in safe refuges; they may even teach their offspring necessary information or skills such as how to hunt, the location of productive feeding sites, or traditional migration routes.

At some point, however, there may be conflict between parents and offspring; conflict is especially common over how much care the parents provide versus how independent the offspring have become. Avian parents, for example, may displace their offspring from the nest to make way for a new brood; mammalian mothers may resort to force to wean their maturing offspring and so be able to nurse a new infant or litter. Alternatively, offspring may be sexually mature and ready to leave, but are manipulated by their parents to remain in the family acting as helpers.

The most extreme forms of parental manipulation involve neglect and what is called tolerated siblicide. In particularly bad times when resources are scarce, parents may provide care for only one or a few offspring, allowing the others to die. Even in good times, parents with very large broods or litters may neglect the smallest and weakest young. Parents of some species allow older siblings to kill, and perhaps eat, their younger siblings; in fact, this behavior is the norm in some species.

Humans tend to equate parenting with moral goodness; among other animals it equates simply to survival and reproduction. Different species provide parental care—or not—as it is needed, in order to maximize the probability that at least a few offspring will survive to maturity. In some vertebrate species parenting may be associated with intense emotions and bonding, as it is in humans, but across the animal kingdom there is no one right way to parent; what works is what works.

—*Linda Mealey*

See also: Asexual reproduction; Birth; Breeding programs; Cleavage, gastrulation, and neurulation; Cloning of extinct or endangered species; Copulation; Courtship; Determination and differentiation; Development: Evolutionary perspective; Estrus; Fertilization; Gametogenesis; Hermaphrodites; Hydrostatic skeletons; Mating; Parthenogenesis; Pregnancy and prenatal development; Reproduction; Reproductive strategies; Reproductive system of female mammals; Reproductive system of male mammals; Sexual development.

Bibliography

Clutton-Brock, T. H. *The Evolution of Parental Care*. Princeton, N.J.: Princeton University Press, 1991. Extremely comprehensive but somewhat technical, this book covers prenatal parenting issues, such as egg-laying versus live birth, egg size, brood size, and gestation length, before addressing postnatal parental care.

Hrdy, Sarah. *Mother Nature*. New York: Pantheon, 1999. Written by a renowned feminist anthropologist, this popular book focuses on primate mothers.

_____, ed. "Natural-Born Mothers." *Natural History* 104, no. 12 (December, 1995): 30-43. A lovely five-article special section of the monthly glossy from the American Museum of Natural History.

Mealey, Linda. *Sex Differences: Developmental and Evolutionary Strategies*. San Diego, Calif.: Academic Press, 2000. A user-friendly, but information-packed textbook on sex differences in the behavior of humans and other animals.

Sluckin, W., and M. Herbert, eds. *Parental Behavior*. New York: Basil Blackwell, 1986. The opening chapter provides an overview of parenting across species. Subsequent chapters cover parenting in birds, rodents, ungulates (hoofed animals), carnivores, and primates.

Tallamy, Douglas W. "Child Care Among the Insects." *Scientific American*, January, 1999, 72-77. Short synopsis of why some insects care for offspring while others do not. Generous use of examples and color photos.

Wynne-Edwards, Katharine E., and Catherine J. Reburn. "Behavioral Endocrinology of Mammalian Fatherhood." *Trends in Ecology and Evolution* 15, no. 11 (2000): 464-468. Somewhat technical but short summary of the relationship between hormones and parenting behavior in males (including humans).

OMNIVORES

Types of animal science: Anatomy, classification
Fields of study: Anatomy, entomology, ornithology, marine biology, zoology

Omnivores are animals that eat both plants and animals. They are found in all types of animals, including arthropods, fish, birds, and mammals. Omnivore diets may vary seasonally.

Principal Terms

CARNIVORE: a flesh-eating animal

CARRION: dead animals

DIURNAL: active during the day

HERBIVORE: an animal that eats only plants

HERMAPHRODITE: an organism having male and female reproductive systems

RADULA: a tonguelike, toothed organ for grinding food

Many animals are either herbivores, who eat only plant food, or carnivores, who eat only the flesh of other animals. The preference for one type of food or the other depends largely on the type of digestive system that the animal has, and the resources it can put into its "energy budget." Meat is generally easier to digest and requires a less complex digestive system and a relatively short intestinal tract. However, in order to get meat, carnivores have to invest a lot of time hunting their prey, and the outcome of a hunt is always uncertain. The food of herbivores is much easier to obtain, since plants do not move and all the herbivore has to do is graze on the grasses, leaves, or algae readily available around it. However, the cellulose that plants are made of is very tough to digest, and thus herbivores must have a much more complex and lengthy digestive tract than carnivores. Many herbivores are ruminants, with multipart stomachs, who have to chew and digest their food more than once in order to get adequate nutrition from it.

Carnivores and herbivores are also vulnerable to a loss of their food source. Herbivores whose digestive systems are specialized to process only one type of food will starve if that food becomes scarce due to drought or some other climatic change. Carnivores often have specialized hunting patterns that cannot be changed if the prey (usually herbivores) become scarce due to loss of their own food source.

Omnivores maximize their ability to obtain food by having digestive tracts capable of processing both plant and animal food, although they are usually not capable of digesting the very tough plant material, such as grasses and leaves, that many large herbivores eat. Omnivores may also be scavengers, eating whatever carrion they may come across. Omnivores often lack the specialized food-gathering ability characteristic of pure carnivores and herbivores. Many animals often thought of as carnivores are actually omnivores, eating both plants and animals.

Types of Omnivores

Omnivores can be found among all types of animals, living on land and in water. They include fishes, mollusks, arthropods, birds, and mammals.

Most insects are either herbivores, such as grasshoppers, or carnivores such as mantises. However some, such as yellow jacket wasps, are omnivores, eating other insects, fruit, and nectar. Omnivorous snails and slugs eat algae, leaves, lichens, insects, and decaying plant and animal matter. Their main organ for eating is called a radula, a tonguelike, toothed organ that is drawn along rocks, leaves, or plants to scrape off food; it

Raccoons eat everything they can get their paws on, including fish, insects, birds, plants, crustaceans, and carrion. (PhotoDisc)

is also used to bore holes through shells of other mollusks, to get to their flesh.

Omnivorous fish include the common carp, goldfish, catfish, eels, and minnows. Since a fish's food is often suspended in the medium through which the fish swims—water—being able to gulp up whatever comes into its mouth is an efficient way for a fish to eat. Similarly, bottom-feeders (fish that suck up material from the floor of whatever body of water they inhabit) also benefit from not needing to sort through the material before they ingest it.

Many birds are omnivores, such as robins, ostriches, and flamingos. The pink or red color of flamingos occurs because they eat blue-green algae and higher plants which contain the same substances that make tomatoes red. They also eat shrimp and small mollusks.

Mammal omnivores include bears, members of the weasel family, such as skunks, the raccoon family (raccoons and coatimundis), monkeys, apes, and humans. Raccoons and coatis, found only in the Americas, eat insects, crayfish, crabs, fishes, amphibians, birds, small mammals, nuts, fruits, roots, and plants. Like other omnivores, they also eat carrion. Bears eat grass, roots, fruits, insects, fishes, small or large mammals, and carrion.

—*Sanford S. Singer*

See also: Baboons; Bears; Carnivores; Chimpanzees; Digestion; Digestive tract; Ecosystems; Food chains and food webs; Gorillas; Herbivores; Hominids; Ingestion; Lemurs; Metabolic rates; Monkeys; Nutrient requirements; Orangutans; Plant and animal interactions; Predation; Raccoons and related mammals; Shrews; Skunks; Teeth, fangs, and tusks; Weasels and related mammals.

Bibliography

Kay, Ian. *Introduction to Animal Physiology*. New York: Springer-Verlag, 1999. An introductory textbook on animal physiology that takes a comparative approach. Includes coverage of the gastrointestinal systems of carnivores, herbivores, and omnivores.

Lauber, Patricia. *Who Eats What?* New York: HarperTrophy, 1995. A juvenile book that very clearly illustrates food chains and food webs, showing the place of omnivores in the wider ecological picture.

Llamas, Andreu. *Crustaceans: Armored Omnivores*. Milwaukee: Gareth Stevens, 1996. Written for a juvenile audience, focuses on crustaceans and their lifestyle, including their food habits.

McGinty, Alice B. *Omnivores in the Food Chain*. Logan, Iowa: Powerkids Press, 2002. Another book written for younger readers, focuses on the place of omnivores within the food chain and the ecosystem.

OPOSSUMS

Type of animal science: Classification
Fields of study: Anatomy, zoology

Opossums are omnivorous mammals which are the only native marsupial in North America.

Principal Terms

ARBOREAL: living in trees
CARRION: rotting carcasses
MARSUPIUM: pouch
OPPOSABLE: positioned opposite of other objects
PREHENSILE: wrapping

Opossums represent the oldest surviving mammal family. Ancestors resembling modern opossums lived on earth at the same time as the dinosaurs. Scientists have located seventy-million-year-old opossum fossils. The Virginia opossums are found in the United States and Canada, while relatives representing seventy species of the Didelphidae family live in Central and South America.

Anatomy

The Virginia opossum is the largest species of the Didelphidae family. They are about the size of a domestic cat and measure from 0.3 to 0.8 meters (1 to 2 feet) in length, with a 22 to 50 centimeter (9 to 20 inch) long tail. Their weight varies from 2 to 5.5 kilograms (4 to 12 pounds). Other opossums are much smaller, averaging 7.5 centimeters (3 inches) in length with a 5 to 10 centimeter (2 to 4 inch) long tail.

Opossums have varying lengths and thicknesses of fur, in shades of white, gray, brown, and black.

Some opossums have stripes. Opossum eyes are black, and their ears are usually hairless. Their faces have a mask or are white. They have a pink nose at the end of a long, pointed, whiskered snout which has fifty sharp teeth. Their four feet and tail are also pink and hairless. Each forefoot has five toes with claws. Opossums have an opposable, thumblike toe on their hind feet that can grasp objects and cling to branches. They are arboreal animals and agile climbers. Their prehensile tails are used for balance.

Life Cycle

Male opossums attain maturity at eight months, and females are sexually mature between six and nine months. Female opossums can produce two litters annually. As many as fifty-six offspring

Opossums have prehensile tails that allow them to hang from branches and other objects. (Corbis)

1163

may be in a litter, but, because female opossums can only nurse twelve to fifteen newborns in their marsupium, most newborn opossums die.

Born blind, each newborn opossum, which is almost embryonic and as small as a bean, crawls from the birth canal near their mother's tail and across her stomach to her pouch. They attach to nipples inside the pouch, where they nurse for two to three months. This nursing period provides them with immunities to diseases. The babies stay inside the pouch when the mother leaves the den to forage. As they grow and the pouch becomes full, the young opossums sometimes briefly leave the pouch, then return for nourishment. From the age of three to five months, the babies ride on their mother's back. While traveling this way, the young opossums gain scavenging and survival skills.

Behavior

Opossums tend to be solitary, nomadic animals that can range over 30 to 96 acres (0.5 to 1.5 square miles) daily. They nest in hollow trees and other animals' burrows. Opossums are nocturnal, foraging at night. Because they are adaptable, opossums can live in a variety of habitats and are frequently found in urban areas which formerly were wooded. They adjust their scavenging and living habits to find food and shelter whether they are in a rural setting or in the middle of a city. Extremely cold weather is the primary environmental condition that deters opossums from otherwise suitable habitats. Opossums are hardy and immune to most diseases. They are the mammal most resistant to rabies.

Opossums are omnivorous and eat a variety of insects, especially crickets, beetles, and cockroaches. They also consume snails, slugs, snakes, worms, birds, and rodents in addition to carrion and eggs. Berries, fruit, and vegetables appeal to opossums, particularly when overripe. Opossums have keen senses of smell, vision, and hearing to locate sustenance and clean up organic wastes in their territories.

The life expectancy of opossums is one to three years in the wild and as many as ten years when

---

Opossum Facts

Classification:
Kingdom: Animalia
Subkingdom: Metazoa
Phylum: Chordata
Subphylum: Vertebrata
Class: Mammalia
Subclass: Eutheria
Order: Marsupialia
Suborder: Didelphoidea
Family: Didelphidae
Genera: Didelphis (opossums, with three species); *Marmosa* (mouse opossums, with forty-seven species); *Monodelphus* (short-tailed opossums, with fourteen species); six other genera
Geographical location: North, Central, and South America
Habitat: Forests, grasslands, mountains, and swamps
Gestational period: Twelve to fourteen days
Life span: From one to three years in the wild, up to ten years in captivity
Special anatomy: Opposable thumblike toe on the hind foot, prehensile tail, pouch

---

kept in captivity. In addition to being preyed upon by wildlife and domesticated animals, opossums are killed by humans for sport, fur, and meat, or by accident with vehicles. Baby opossums often survive automobile impacts which kill their mothers. Opossums show a variety of defense mechanisms: They move more slowly than their predators and often spray a foul-smelling secretion to thwart attacks. Virginia opossums feign death by becoming limp when frightened and unable to escape. They sometimes hiss or growl, exposing their teeth. Opossums occasionally fight and bite. Opossums hide in brush-covered areas that are difficult for predators to access. They can make sounds, including screeches, but are usually quiet. Scientists have gauged opossums' ability to learn and distinguish objects as greater than that of dogs and almost equivalent to pigs.

The National Opossum Society (http://www.opossum.org) offers information about how people can rehabilitate injured opossums and raise orphans. The group also addresses the controversial use of opossums as laboratory research specimens.

—*Elizabeth D. Schafer*

See also: Fauna: Australia; Fauna: North America; Kangaroos; Marsupials; Omnivores; Reproductive strategies; Scavengers; Urban and suburban wildlife.

Bibliography
Allen, Thomas B., ed. *Wild Animals of North America*. Washington, D.C.: National Geographic Society, 1995. Popular account of the history, folklore, and cultural activities associated with opossums.

Gardner, Alfred L. "Virginia Opossum." In *Wild Mammals of North America: Biology, Management, and Economics*, edited by Joseph A. Chapman and George A. Feldhammer. Baltimore: The Johns Hopkins University Press, 1982. Thorough, scientific description of opossum anatomy, behavior, and life cycle, with a lengthy bibliography.

Hartman, Carl G. *Possums*. Austin: University of Texas Press, 1952. A classic text regarding opossum development and opossum-human interaction.

Krause, W. J. *A Review of Histogenesis/Organogenesis in the Developing North American Opossum (Didelphis virginiana)*. 2 vols. Heidelberg, Germany: Springer-Verlag, 1998. Scholarly discussion of the anatomical development of opossums from fertilization to maturity.

Seidensticker, John, with Susan Lumpkin. "Playing Possum Is Serious Business for Our Only Marsupial." *Smithsonian* 20, no. 8 (November, 1989): 108-119. Describes the author's research with Virginia opossums.

ORANGUTANS

Type of animal science: Classification
Fields of study: Anatomy, conservation biology, wildlife ecology, zoology

Orangutans are large apes with orange to reddish-brown shaggy hair. They have very long arms and no tail. They survive solely in the jungles of Borneo and Sumatra, where their habitat has been destroyed to the point that they are listed as a threatened species.

Principal Terms

ARBOREAL: living completely or primarily in the trees

BIPEDAL: walking on only two feet, as humans do

QUADRUPEDAL: walking on all four feet

SEXUAL DIMORPHISM: the occurrence of anatomic and physiologic differences that distinguish males from females of a particular species

---

The name "orang-utan," commonly written in the hyphenated form, comes from two Malay words: *orang* meaning "person," and *hutan* meaning "forest or jungle." Thus, Malaysian *orang-utan* means "persons of the jungle." Since these animals are very humanlike and live secretive lives in the dense jungle, the origin of the name makes sense. Orangutans are considered to be a threatened species. Less than twenty thousand are believed left in the wild.

Orangutans are the second largest of the apes, and show marked sexual dimorphism. Males may grow to be 220 pounds (100 kilograms) and females about half that. The arms of a full-grown male may reach a span of 7 to 8 feet (2.1 to 2.4 meters), and their hands are longer than any other primate. These arms and hands are ideally adapted for the arboreal life. Comparatively, the legs are short and weak; there is no external tail. While adult males with arms extended (swinging through the jungle) may appear to be enormous, when standing erect on the ground they rarely exceed 4.5 feet (1.3 meters) in height. Females, by contrast, reach only 3.5 feet (1.1 meters).

Habitat

Two to three million years ago, orangutans lived as far north as China and as far south as Java. As land bridges formed during the Ice Age, orangutans moved south in search of a warmer climate. Today they can only be found on the Indonesian islands of Borneo and Sumatra.

Orangutans are the only truly arboreal apes, spending most of their life in forest trees. Their anatomy is well suited for this lifestyle. They walk

---

Orangutan Facts

Classification:
Kingdom: Animalia
Phylum: Chordata
Subphylum: Vertebrata
Class: Mammalia
Order: Primate
Family: Hominidae
Genus and species: Pongo pygmaeus
Geographical location: Sumatra and Borneo
Habitat: Dense rain forests, particularly lowland forests
Gestational period: Eight to nine months
Life span: Thirty-five years in the wild, fifty years in captivity
Special anatomy: Orange to red-brown hair; arm length exceeds torso length

---

up trunks using irregularities in the bark to give a grip to fingers and toes, and proceed silently through the middle stories of the forest. These middle stories are especially well suited for horizontal travel, where densely growing trees poke up into the canopy. Often vines are used for quickly moving up and down to get to the next horizontal branches. Orangutans do not jump; they climb and walk the branches on all four legs. They may also sit, recline, or hang in a variety of positions, including suspended from both feet or from one foot and one hand.

On the ground, orangutans are normally quadrupedal, although they occasionally walk in the bipedal position. Their weight is borne by clenched fists with the palm touching the ground (unlike gorillas and chimpanzees). Their walk is similar to that of a dog, with diagonally opposed limbs moving forward together.

Orangutans live alone, in pairs, or in small family groupings. They build nests in the trees from groups of small branches, bent or broken and laid across one another, then lined with smaller branches that are patted down into a circle of approximately three feet in diameter. Nests are placed ten to one hundred feet above the ground and are difficult to spot. Nests may be built new each night when animals are moving about, but may remain intact for several months after being built.

While moving and at rest in trees, the orangutan grasps vegetable and animal matter within its reach, testing each one as food. It prefers a variety of jungle fruits as its principal diet, but also eats or chews an infinite variety of buds and leaves, flowers, bark, epiphytes, canes and roots, honey, and even fungi. It forages and eats at leisure, picking fruits with cupped hands and spitting seeds and shells back out of its mouth.

Orangutans satisfy most of their need for water by taking it in with their moist food. When on the ground they drink from a stream or lake by bending over from a standing position. They have also been seen to squat down and use their hands to spoon the water into their mouths.

Reproduction and Development

Orangutans reach maturity at about ten years of age, and can begin breeding at any time thereafter. Mating begins with a male singing a song, a low hum that increases to a deep roar before decreasing again. Often playful wrestling, touching, and other acts precede mating. Mating occurs in trees, face to face, usually in a hanging position. Mating may occur repeatedly over a period of several weeks. The pair then separates and each goes its own way. Males play no role in parenting.

Female orangutans have a menstrual cycle similar to that of a human female. It lasts twenty-nine days, with a slight flow of blood for three to four days. Pregnancy lasts nearly nine months, and newborns generally weigh 2.25 to 3.25

Orangutans are very shy primates that spend virtually all of their lives in trees. (PhotoDisc)

Biruté Galdikas

Born: May 10, 1946; Weisbaden, West Germany

Fields of study: Conservation biology, systematics (taxonomy), wildlife ecology, zoology

Contribution: Galdikas has devoted her life to studying the endangered orangutan and preserving its rain-forest habitat. She is the world's leading authority on these mysterious apes, whom she has studied for nearly thirty years in the jungles of Borneo.

Biruté Galdikas was born in West Germany not long after the end of World War II. Her parents met at a refugee camp after fleeing their Lithuanian homeland during the war. Two years after her birth, her father moved the family to Canada so he could work in the copper mines. Galdikas grew up in Toronto and enjoyed visiting the zoo. She became particularly fascinated by the primates, and noted that the behavior and expressions of the orangutans reminded her of humans. She hoped that one day she could learn more about them.

Her family moved to Vancouver, and then to Los Angeles. Galdikas entered UCLA to study psychology and anthropology. She received her B.A. degree in 1966 and entered graduate school to study anthropology. It was while she was a student there in 1969 that she met the renowned paleontologist Louis Leakey. It did not take long for her to convince him to support her efforts to study the wild orangutans (Dr. Leakey already supported the work of Jane Goodall with chimpanzees and Dian Fossey with gorillas). With Dr. Leakey's support for her project, Galdikas set out for Borneo in 1971, and set up a jungle camp that she called Camp Leakey.

From Camp Leakey and various jungle outposts, Galdikas has studied the orangutan in its natural environment for thirty years. She has rescued many babies that were taken illegally from the jungle, restored them to health, and has released more than two hundred back to the wild. She has marveled at the mysteries surrounding orangutan social life, and has studied their behavior, speech, reproduction, and life cycle in the wild. She has collected enough data to radically update our understanding of the forces that shape the lives of orangutans in the wild. Galdikas and her colleagues have even taught sign language to orangutans in the hope of being able to learn what was important to them. By hiring many local people to help with her work at Camp Leakey, she has also helped the government and the peoples of Borneo to gain a greater respect for the orangutan and the need to preserve its habitat.

Amidst her studies in the jungles of Borneo, Galdikas wrote about her observations and prepared her doctoral thesis. In 1978, she received a Ph.D. in anthropology from UCLA. When not engaged in field work in Indonesia, she teaches at Simon Fraser University in British Columbia and at Indonesia's Universitas Nasional.

Biruté Galdikas formed and continues to run the nonprofit Orangutan Foundation International. Its purpose is to prevent the extinction of orangutans and their rain-forest habitat by raising money to fund research, conservation, and educational programs. She continues her scientific work to learn more about these solitary primates.

—Kerry L. Cheesman

pounds (1.1 to 1.6 kilograms). The infant clings to the mother's fur and the mother holds the infant with one arm, usually over the hip. This leaves three limbs for traveling and feeding. The youngster nurses for two to three years, with solid food (chewed up fruit) being added periodically. By the end of the first year the youngster begins to explore away from its mother, but remains within eyesight at all times.

By age four, a youngster is pretty much on its own, feeding itself and roaming freely. At that point, the mother is able to mate again. Offspring have only about a 50 to 60 percent survival rate in the jungle, with accidents and disease striking many young orangutans.

Orangutans are shy animals that have only one natural enemy: humans. Currently, orangutans are protected by law in all of their territory, but poaching and illegal logging continue to threaten the survival of the species. Unless large areas of undisturbed jungle are set aside as sanctuaries, the orangutan may have a hard time surviving in the future.

—*Kerry L. Cheesman*

See also: Apes to hominids; Baboons; Cannibalism; Chimpanzees; Communication; Communities; Evolution: Animal life; Evolution: Historical perspective; Fauna: Africa; Gorillas; Groups; Hominids; *Homo sapiens* and human diversification; Human evolution analysis; Infanticide; Learning; Lemurs; Mammalian social systems; Monkeys; Primates.

Bibliography

Galdikas, Biruté. *Reflections of Eden: My Years with the Orangutans of Borneo*. Boston: Little, Brown, 1995. An enjoyable, easy-to-read biography of the world's leading orangutan authority. Filled with pictures and stories of individual animals.

Galdikas, Biruté, and Nancy Briggs. *Orangutan Odyssey*. New York: Harry N. Abrams, 1999. Exquisite photographs of the work at Camp Leakey and beyond. Clear, easy-to-read text.

Kaplan, Gisela, and Lesley Rogers. *The Orangutans: Their Evolution, Behavior, and Future*. Cambridge, Mass.: Perseus, 2000. An inclusive, yet readable summary of what we know about the biology, behavior, and intellectual capacity of orangutans. An excellent source of information.

Macdonald, David, ed. *The Encyclopedia of Mammals*. New York: Facts on File, 1984. Good description of orangutans, with comparison to other primates.

Russon, Anne. *Orangutans: Wizards of the Rain Forest*. Willowdale, Ontario: Firefly Books, 2000. A report of the author's ten years working with orangutans in Borneo. Good history of orangutans; excellent reference list.

OSMOREGULATION

Type of animal science: Physiology
Fields of study: Biophysics, cell biology

Osmoregulation is the ability an organism must have to adjust its internal concentrations of solutes and of water so that it can maintain an osmotic pressure appropriate to its functioning.

Principal Terms

EURYHALINE: the ability of an organism to tolerate wide ranges of salinity

HYPEROSMOTIC: describes a solution with a higher osmotic pressure, one containing more osmotically active particles relative to the same volume, than the solution to which it is being compared

HYPOOSMOTIC: a solution with a lower osmotic pressure, fewer osmotically active particles relative to the same volume, than the solution to which it is being compared

ISOSMOTIC: a solution having the same osmotic pressure, the same number of osmotically active particles relative to the same volume, as the solution to which it is being compared

OSMOCONFORMER: an organism whose internal osmotic pressure approximates the osmotic pressure of its environment; such an organism is also referred to as "poikilosmotic"

OSMOREGULATOR: an organism that maintains its internal osmotic pressure despite changes in environmental osmotic pressure; such an organism is also referred to as "euryosmotic"

STENOHALINE: the inability of an organism to tolerate wide ranges of salinity

Osmotic pressure of a solution is the measure of the tendency of water to enter a solution from pure water. Osmoregulation, the regulation of osmotic pressure, is vital to every organism. The phenomenon collectively is called "osmosis." It is the difference of hydrostatic pressure that must be created between that solution and pure water to prevent any net osmotic movement of particles in the water when the solution and pure water are separated by a semipermeable membrane. Hydrostatic pressure is a measuring device: It is a means of assessing the tendency of a solution to take on water osmotically.

Only dissolved solutes contribute to osmotic pressure. The number of individual particles determines the strength of osmotic pressure. Each particle makes a roughly equal contribution to osmotic pressure. The same number of molecules of a substance such as sodium chloride (table salt), which ionizes in water to release two ions (one sodium ion and one chlorine ion) display twice as much osmotic pressure as the same number of molecules of glucose, which retains its molecular form in water. Cells and other suspended materials do not contribute to osmotic pressure.

Several characteristics of solutions depend upon the number of particles in the solution. These are called "colligative properties." Increasing the number of particles of solute impairs the ability of the solvent to change state. The colligative properties are the freezing point, the boiling point, the osmotic pressure, and the vapor pressure. Only freezing point depression and vapor pressure are used to determine osmotic pressure.

Osmoticity refers to the osmotic pressure of solutions. Isosmotic solutions have equal osmotic pressures. A hypoosmotic solution has an osmotic pressure lower than the solution to which it is being compared; a hyperosmotic solution is one with a greater osmotic pressure than the solution to which it is being compared. These solutions can be body fluids, environmental liquids, or laboratory solutions.

The terms used to describe the changes in volume of cells exposed to solutions of differing concentrations are often confused with those comparing osmotic pressure. Changes of cell volume are described by the term "tonicity." Solutions isotonic to a cell cause no change in cell volume. Hypotonic solutions will cause the cell to swell as water diffuses into the cell; the cell may even burst. Hypertonic solutions will cause a cell to shrink as water diffuses across the cell membrane into the solution.

The Challenges of Osmoregulation

This discussion reveals some of the problems that an organism encounters in the environment as concentrations of water and salts vary. Most organisms attempt to regulate both their volume and their ion content. If volume is not regulated, the chemicals within the cell will become too dilute to react or too concentrated to interact. If ions are not regulated, chemical reactions will be affected by inappropriate levels of ions, which may change the electrochemical properties of the cellular solution. Thus, there are independent challenges to volume regulation, to ion regulation, and to osmotic regulation. The homeostatic physiological responses to all three types of challenges are interconnected but distinct.

Osmoregulation is the regulation of the ratio between all dissolved particles, regardless of their chemical nature as ions or molecules, and water. All organisms are exposed to osmotic stress. Any organism incurs obligatory water losses. These occur during respiration, urination, and defecation. The organs most often thought of as participating in osmoregulation are the kidneys. They are intimately concerned with the elimination or conservation of water. Some salts are also found in urine. The proportions of the salts excreted in urine may be different from those in the body fluids because the kidney can retain required ions while eliminating less desirable ones.

Freshwater organisms are in danger of dilution, and they excrete great quantities of dilute urine. Saltwater organisms usually produce small quantities of isosmotic urine, which preferentially excretes divalent ions such as magnesium and sulfate. Other surfaces lose water, causing desiccation. The composition of the diet also influences the need for excretion of urine. Nitrogenous wastes from protein metabolism must be eliminated in urine, often as urea, which requires water for its excretion.

Carbon dioxide released during metabolism of carbohydrates and fats is eliminated by the respiratory organs. In terrestrial organisms, air leaving the lungs is usually saturated and some water is lost on expiration. The respiratory organs of aquatic organisms are gills. Their surfaces must be permeable to water. Freshwater organisms gain water through them and hypoosmotic marine organisms lose water through them.

The metabolism of carbohydrates produces what is known as metabolic water, which can be used to prevent desiccation. Metabolic water produced from the metabolism of fats is lost because of the higher rate of respiration required to supply the oxygen needed in fat oxidation.

Salt Loss

Preformed water is present in any food. Even the driest seeds contain a small amount of water. The nutrients also always include salts. The presence of great quantities of salts may require urinary loss of water in excess of the preformed water found in the food.

Although feces may appear to be solid, they contain some water that was not absorbed in the gut. The presence of salts and other solutes in the digesta may also draw water from the hypoosmotic body fluids into the gut. One of the reasons that humans cannot drink seawater, in fact, is that the magnesium ions in ocean water increase the permeability of the gut and increase water loss, because the seawater is hyperosmotic to body fluids. More water is lost than can be gained.

Salts can be lost from the body by means other than urine formation and defecation. Marine reptiles and birds have salt glands located on the head. Since neither can produce hyperosmotic urine, these glands allow the elimination of salt with a minimum loss of water as the secretion may be four to five times as concentrated as body fluids. The cloaca of birds and the rectal glands of sharks also have the capacity to excrete salts.

One of the most fascinating mechanisms of osmoregulation is found in elasmobranch fish—the sharks, skates, and rays. Their body fluids are hyperosmotic but hypoionic to seawater. Blood salt concentrations are below those of seawater. Excess osmotic pressure is supplied by two molecules: urea and trimethylaminoxide (TMAO). Urea is toxic to most organs, but some organs resist its deleterious effects. Others would be harmed but are apparently protected by the TMAO. Retention of both urea and TMAO minimizes enzymatic disturbances by urea and allows the elasmobranchs to avoid the salt gain associated with hypoosmotic body fluids.

Organisms exposed to environmental variations have two choices: they can maintain internal constancy or homeostasis at the expense of metabolic energy, or they can allow their internal conditions to follow that of the environment. Organisms that maintain their internal osmotic pressure despite changes in external osmotic pressure are called osmoregulators. These euryosmotic organisms are protected from environmental changes. Their metabolism can continue to function, but much of the energy will be used to maintain their body fluids at the appropriate osmotic pressure.

Organisms that allow their osmotic pressure to follow that of the environment are called osmoconformers. These poikilosmotic organisms often have a limited tolerance for such changes. They are stenohaline. They may be less vigorous at salinities other than their optimal levels. The adults of such groups (for example, mollusks such as oysters and mussels) may be found in salinity extremes not tolerated by their young. These populations must be maintained by immigration of young spawned in more favorable salinity conditions.

Hormonal Regulation of Osmotic Pressure

The internal osmotic pressure is affected by the hormones present in the body fluids. In invertebrates such as annelids, mollusks, and arthropods, neuroendocrine changes are seen upon changing the osmotic pressure of the environment. These changes indicate that nervous and endocrine systems are at work regulating the osmotic pressure of the organism. In most invertebrates, the biochemical nature of these hormones is unknown. Some freshwater pulmonate snails, however, produce an antidiuretic hormone and a neurosecretory factor associated with electrolyte balance. Depending upon the demands placed on them, insects such as grasshoppers and cockroaches can synthesize diuretic or antidiuretic hormones.

The best-known hormonal factors in ion regulation are studied in vertebrates. The pituitary gland produces antidiuretic hormone (ADH), which promotes water retention in terrestrial vertebrates. In fish and amphibians, ADH may induce urine formation and increase water loss through diuresis.

The adrenal gland also produces hormones that influence ion retention. In mammals, aldosterone increases reabsorption of sodium in the kidney and promotes the excretion of potassium. In nonmammalian vertebrates, extrarenal glands maintain salt and water balance by affecting the gills and intestines of fishes, the urinary bladder and skin of amphibians, and the salt glands of elasmobranchs, reptiles, and birds.

Measuring Osmoregulation

Osmoregulation involves the balancing of water and solutes in the body so that the animal can continue to function. Because the presence of particles influences certain physical characteristics of the solution, these colligative properties can be used to determine osmotic pressure of solutions. Colligative properties change with increasing numbers of particles in solution: The osmotic pressure increases; the boiling point increases; the freezing point decreases; the vapor pressure decreases.

Freezing point depression and vapor pressure can be used to measure a solution's osmotic pressure. Freezing point is used most often. It works for the same reason that salt is spread on ice on sidewalks in winter. The salt lowers the freezing temperature of the water. Body fluids are much more dilute than the salt and water mixtures that melt ice, but the salinity of the ocean (approximately thirty-five parts salt for each thousand parts of solution) causes it to freeze as much as 1.6 degrees Celsius lower than pure water. Only marine organisms have body fluid osmotic pressures in that range. Terrestrial organisms have much less salt and therefore much lower osmotic pressures in their body fluids. Because most body fluids are so dilute, a large sample may be required to determine freezing point depression. When only small volumes of body fluid exist, the determination becomes more difficult.

Vapor pressure determinations are also used in osmometry. Usually, a small amount of the fluid being studied is tested in a capillary tube. In one ingenious method, the capillary tube is placed in a solution more concentrated than the experimental fluid. The higher osmotic pressure of the reference solution pushes a meniscus up the tube. The rate of movement of the meniscus depends upon the difference in concentrations between the experimental and the reference solutions.

Another ingenious method of using vapor pressure to determine the osmotic pressure of an experimental solution requires enough fluid to fill a depression. A glass plate with capillary tubes filled with reference solutions of known osmotic concentrations is mounted over the experimental fluid. The reference tube that exhibits no movement is at equilibrium with the experimental fluid. One of the rigors of this method is that all movement must stop.

Another method involves capturing a precise volume of experimental fluid in a capillary tube. The shape of drops of the same volume of reference solutions is compared to that of the experimental solution. Those of the same concentration will have the same shape because their vapor pressures are exerting equal force on the drop. Thermocouples are also used in vapor pressure determinations. This procedure is delicate and costly and is used infrequently. All these procedures are difficult and require patience. Now, electronic instruments analyze the constituents of solutions and allow easier calculation of the osmotic pressures of solutions than ever before.

Freshwater Versus Saltwater Environments
All organisms experience osmotic stress. There is no environment in which the osmotic pressure and the ion composition exactly match the requirements of the cells. Every organism must expend metabolic energy to maintain appropriate water and ion concentrations.

Freshwater organisms are hyperosmotic to their environment. They risk losing scarce ions through their permeable gills and in their urine and also tend to take up water through their gills or other surfaces and in their food. They face the problem of dilution of their body fluids by the environment.

Marine organisms are often isosmotic to the salt water they inhabit. They must change the concentrations of some ions, however, in order to attain this state. Magnesium is present in greater amounts in seawater than is desirable in their body fluids and must be eliminated. These organisms ion regulate even though they are not in danger of volume changes.

Marine organisms that evolved from freshwater or terrestrial ancestors are often hypoosmotic to seawater. They are in danger of desiccation as water from body fluids diffuses into the hyperosmotic ocean water. They also must regulate the types of ions that are retained and eliminated from their bodies.

Marine organisms may also be exposed to freshwater when they enter rivers, which dilute the salt content of the incoming tidal water. Under these conditions, the water is brackish—not as salty as the sea, but not as pure as freshwater. The criticality of this situation depends upon whether the organism is tolerant or intolerant of salinity changes. Organisms that can live in only a narrow range of salinities are called stenohaline. Organisms that are tolerant of wide ranges of salinities are called euryhaline.

Terrestrial organisms are always hyperosmotic to their environment, so they continually face desiccation in the air. They also must adjust the ion composition of their body fluids, because the foods that they eat may not have inorganic ions in the desired ratios and because some ions are always lost in urine.

One example of the influence of these effects concerns the interaction of oysters and the protistan parasite known as MSX. (MSX stands for "multinucleate sphere unknown," which refers to the protista *Haplosporidium nelsoni*.) The MSX organism survives in osmotic pressures greater than 0.4 osmolar. Oysters are osmoconformers which grow in saline, brackish, and nearly fresh water. At osmotic concentrations less than 0.4 osmole, oysters can survive and are unaffected by MSX. When rainfall is abnormally low, however, the salinity of brackish water increases, and oysters which were protected in low-salinity water are ex-

posed to higher-salinity water, which allows the MSX organism to infect them.

Organisms exposed to tides may protect themselves from exposure to variations in osmotic pressure by sealing themselves off, the way snails and bivalve mollusks do. Others may move offshore to more saline waters or onshore, away from the increasing salinity. Worms that burrow in the sediments of salt water are protected from transient changes in salinity because there is little exchange of solutes with the overlying salt water.

The vertebrates adapted to their various environments by using hormones to regulate salt and water balance. Because of the differing demands of aquatic and terrestrial environments, in different groups, the same hormone may have opposite effects, but that effect is always to maintain the optimal osmotic pressure to ensure survival.

—*Judith O. Rebach*

See also: Kidneys and other excretory mechanisms; Lakes and rivers; Marine biology; Nutrient requirements; pH Maintenance; Thermoregulation; Tidepools and beaches; Water balance in vertebrates.

Bibliography

Brown, A. D. *Microbial Water Stress Physiology: Principles and Perspectives*. New York: John Wiley & Sons, 1990. Explores the means by which organisms adapt to survive acute and chronic water stress.

Gilles, R., E. K. Hoffmann, and L. Bolis, eds. *Volume and Osmolality Control in Animal Cells*. New York: Springer-Verlag, 1991. A comprehensive review and summary of current literature in the field of osmolality and volume control in animals in terms of both organic and inorganic ions.

Hadley, Neil F. *Water Relations of Terrestrial Arthropods*. San Diego, Calif.: Academic Press, 1994. Brings together data on the physiology and anatomy of arthropods in the context of their adaptations to terrestrial environments. Written for advanced students and researchers in the field.

Krogh, August. *Osmotic Regulation in Aquatic Animals*. Reprint. New York: Dover, 1965. Reprint of the 1939 edition, a classic work in the field. It is clearly written and covers all aspects including techniques used to determine osmotic pressure.

Prosser, C. Ladd. *Adaptational Biology: Molecules to Organisms*. New York: John Wiley & Sons, 1986. This text, by the leader in the field of adaptational physiology, is one of the most comprehensive in the field.

Smith, Homer W. "The Kidney." *Scientific American* 188 (January, 1973): 40-48. A classic exploration of the structure and function of the kidney and its role in osmoregulation.

Strange, Kevin, ed. *Cellular and Molecular Physiology of Cell Volume Regulation.* Boca Raton, Fla.: CRC Press, 1994. A textbook on cellular osmoregulation and membrane transport physiology.

OSTRICHES AND RELATED BIRDS

Type of animal science: Classification
Fields of study: Anatomy, ornithology, physiology, population biology

The ostrich and its relatives are famous for their exotic appearance and for being flightless birds. They are the oldest type of bird alive today, dating back eighty million years to the age of dinosaurs.

Principal Terms

MOLT: to shed old feathers and grow new ones

MONOGAMOUS: having only one mate

NOMADIC: travels distances in search of food

OMNIVOROUS: eating both animal and plant foods

POLYGAMOUS: having more than one mate

PRECOCIAL: in birds, those that are down-covered, fully developed, and active at birth

---

Ostriches belong to a striking group of flightless birds known as ratites, that also includes emus, cassowaries, rheas, and kiwis. Ratites have flat, smooth breastbones that lack a keel to which flight muscles could attach. They are thus unable to fly, and have weak wing muscles. They do use their wings, spreading them out to help them cool off, and also to splash water when bathing. Ratite feathers are different from those of flying birds. The individual strands are not interlocked, and thus they are soft and billowy and air passes right through them. Their plumes have long been admired by humans and used for decoration and adornment. As with other birds, their feathers function as protection from the elements, and ratites preen, spreading waterproofing oil to their feathers with their beaks. They also molt once a year. Ratites have heavy, strong bones and powerful leg muscles, and are able to run swiftly. They are omnivores, feeding on a variety of grasses, plants, seeds, fruit, insects, and small animals. There are many farms in the United States that raise ostriches, emus, and rheas for their feathers, hide, meat, and oil. In their native countries, ratites are hunted or raised for their feathers and as food. Ostriches have also been tamed for riding and for pulling carts.

Ostrich Characteristics

Ostriches have long necks and legs and are the largest living birds. Males stand eight feet tall, and weigh three hundred pounds. The ostrich can take strides of twenty-five feet and outrun pursuers at speeds of forty miles per hour. If cornered, the ostrich has a powerful kick that can maim an enemy. It has two toes on each foot, and a razor-sharp toenail that both grips the ground while running and can slash the flesh of its enemy.

Male ostriches are black with white plumes on their tail and wings. Females are grayish-brown. The head and legs are featherless. The neck is covered with down and is red or grayish. The ostrich has huge eyes with long protective lashes and has keen eyesight for spotting danger a long way off. It can make loud hissing and roaring noises.

Ostriches are native to Africa; they are nomadic and graze on open savanna. They often follow herds of zebras or antelope, catching insects and small animals stirred up by their hooves. They swallow sand and stones to help grind up their food. Contrary to popular belief, they require a regular water supply.

When mating, male ostriches make a booming call and perform a courtship dance for the females. They are polygamous, taking three or more

Ratite Facts

Classification:
Kingdom: Animalia
Phylum: Chordata
Subphylum: Vertebrata
Class: Aves
Orders: Struthioniformes (ostriches); Casuariiformes (emus, cassowaries); Rheiformes (rheas); Apterygiformes (kiwis)
Families: Struthionidae (ostriches, one genus, six species); Dromaiidae (emus, one genus, three species); Casuariidae (cassowaries, one genus, four species); Rheidae (rheas, two genera, nine species); Apterygidae (kiwis, one genus, seven species)
Geographical location: Southern Hemisphere—Africa (ostrich), Australia (emu and cassowary), New Guinea (cassowary), South America (rhea), New Zealand (kiwi)
Habitat: Semidesert and open plains (ostrich, emu, rhea); forests (cassowary, kiwi)
Gestational period: One breeding cycle per year; incubation varies from forty days (ostrich) to eighty days (kiwi)
Life span: Ostriches average forty years in the wild, up to eighty years in captivity; others average ten to twenty years
Special anatomy: Unkeeled breastbone, tiny wings, unbarbed feathers, solid bones, strong muscular legs

hens as mates. The male scratches a shallow pit into which each female lays up to a dozen eggs, for a total clutch size of up to thirty eggs. This communal nesting behavior is unusual among birds. The male shares incubation with one dominant female. The male sits at night and the female during the day. Ostriches lay the largest eggs of all living birds, seven inches long and three pounds. The eggshell is very tough and hard for predators to crack open. The parent will sometimes lay with its neck outstretched on the ground when danger threatens, giving rise to the legend that they bury their heads in the sand. They may also feign injury to lure predators away from the nest. Newborn chicks are precocial and instinctively know how to search for food. They are full adults by three years of age.

Ostrich Relatives

Rheas live in flocks on grasslands in South America. They are similar to ostriches in behavior and appearance, although they have three toes on their feet, as do most of the other ratites. They are brownish in color and can be five feet tall. They are polygamous, but only the male incubates the eggs.

Emus live on plains in the Australian outback, and flock nomadically according to rainfall patterns and the resulting food supply. The emu is the second largest flightless bird, nearly six feet tall and eighty-five pounds. It has brown feathers and a loose, moplike tail. Emus are monogamous.

Cassowaries live in the rain forests of New Zealand and northeastern Australia. They are solitary and territorial, pairing only to mate. They feed primarily on fruit fallen from trees. Cassowaries have black, loosely hanging feathers, and the wings are composed only of quills. They have bright blue heads and colorful wattles. A distinctive bony crown on the head called a casque helps them push through the dense forest, and is also used to turn over litter in search of food.

Kiwis are elusive, nocturnal birds that live in the forests of New Zealand. They are the smallest ratites, about the size of a chicken. They have round, brown-feathered bodies, short legs, four toes, and run by placing one foot directly in front of the other. Their long, slender beaks have nostrils at the very tip, and are used to probe the ground to locate worms, beetles, spiders and larvae by smell. Males have a shrill, whistling mating call. Females lay only one or two eggs that are enormous in proportion to their body size.

—*Barbara C. Beattie*

See also: Birds; Fauna: Africa; Fauna: Australia; Fauna: South America.

Bibliography

Arnold, Caroline. *Ostriches and Other Flightless Birds*. Minneapolis: Carolrhoda Books, 1990. A good introduction for the intermediate reader that focuses mainly on ostriches.

Baskin-Salzberg, Anita, and Allen Salzberg. *Flightless Birds*. New York: Franklin Watts, 1993. Covers all flightless birds for the juvenile reader, clearly explaining ratite characteristics with a section on each type.

Deeming, Denis C., and D. Charles Deeming, eds. *The Ostrich: Biology, Production and Health*. New York: Oxford University Press, 1999. A state-of-the-art presentation on the raising of ostriches, written by specialists from academia and industry.

Drenowatz, Claire, ed. *The Ratite Encyclopedia: Ostrich, Emu, Rhea*. San Antonio: Ratite Records, 1995. Provides in-depth information on ratite anatomy and physiology as well as the various aspects of ratite farming, such as rearing, farm design, and record management.

Gould, Stephen Jay. "Tales of a Feathered Tail." *Natural History* 109, no. 3 (November, 2000): 32-42. An insightful discussion of the evolutionary theory regarding the descent of flightless birds from dinosaur ancestors.

Ostriches are one of the few species of flightless birds, but they can run very fast on their long legs. (Corbis)

OTTERS

Types of animal science: Behavior, classification, ecology
Fields of study: Ecology, wildlife ecology, zoology

Otters are found in many parts of the world where there is clean, flowing water. Otters are large members of the weasel family, the Mustelidae, which includes mink, skunks, and badgers.

Principal Terms

CARNIVORE: an animal that feeds on other animals for its diet

DELAYED IMPLANTATION: a process of delaying the implantation in the uterine wall of the fertilized egg

DORSAL: the back portion of an animal

NOTOCHORD: a dorsal, flexible, rodlike structure extending the length of a vertebrate's body; serves as an axis for muscle attachment

PELAGE: a mammal's fur coat

PREDATOR: an animal that preys on other animals for its food

Playful and proud, river otters are born to have fun. At least, it appears that river otters are having fun, because of their playful antics with each other, as well as their facial and body expressions. By any measure, otters are extremely curious and intelligent mammals.

River Otters

Adult river otters reach a length of about 127 centimeters, including their tails, which are nearly one third of that length. Adults weigh between five and fourteen kilograms. Females are slightly smaller than males.

River otters have a dense, short fur with great water resistance. Air is trapped beneath their dense furry coats and acts as insulation against the cold water, where these mostly aquatic mammals feed. Otters roll and rub themselves on sand, rocks, old logs, and even snow in their coat grooming activities. These carnivorous mammals mostly feed on fish. Crayfish also serve as diet items when abundant. Researchers have found that river otters feed directly on fish proportion-

Otters are playful, intelligent, and very curious. They spend much of their lives in and near water. (Corbis)

ally to their availability and inversely to the fish's swimming ability.

Shelters that have been abandoned by other animals are frequently used by river otters, such as old beaver dens or riverbank excavations. On occasion, river otters use rock piles and log jams as dens. Some investigators have discovered nests along river and stream banks which river otters had constructed of aquatic vegetation.

Mating activity usually occurs in the water, although there are reports of mating on land. The breeding season is usually late winter to spring. Litter size is between one and six cubs. The newborn otters have full pelage, but their eyes are not open and they have no erupted teeth. Females wean their young at about three months of age.

Several predators, including bobcats, foxes, and alligators, have been observed dining on river otters. Humans also have made their impact on otter populations in at least three ways: habitat destruction, water pollution, and overtrapping.

Sea Otters

Another interesting otter is the sea otter. These seafaring mammals are confined to the northern Pacific Ocean. Being the largest in the family Mustelidae, sea otters weigh between twenty-seven and thirty-eight kilograms and attain a length of about 148 centimeters. Unlike seals and whales, which have a fat layer (blubber) for insulation, sea otters rely on air trapped beneath their densely packed pelage. Some researchers have estimated the number of hairs in a sea otter's coat to be 800 million.

Otter Facts

Classification:
Kingdom: Animalia
Phylum: Chordata
Class: Mammalia
Order: Carnivora
Family: Mustelidae
Subfamily: Lutrinae
Genus and species: Lutra canadensis (North American river otter), *L. lutra* (European river otter), *L. felina* (marine otter), *L. provocax* (southern river otter), *L. longicaudis* (neotropical river otter), *L. umatrana* (hairy nosed otter); *Hydrictis maculicollis* (spot-necked otter); *Lutrogale perspicillata* (Indian smoothed-coated otter); *Aonyx cinera* (oriental short-clawed otter), *A. capensis* (Cape clawless otter); *Pteronura brasiliensis* (giant otter); *Enhyda lutris* (sea otter)
Geographical location: Until the eighteenth century, North American river otters were found in all major water courses in the United States and Canada; presently, river otters are scattered in several river systems across the United States and Canada, and many river otters have been reintroduced into river systems where they once flourished
Habitat: Exclusively aquatic, adaptable to many nonpolluted aquatic environments
Gestational period: 288 to 375 days, with delayed implantation
Life span: Ten years, with a maximum of twenty-three years
Special anatomy: Webbed feet

Sea otters, just as their freshwater cousins, have delayed embryo implantation. Females usually produce one pup in a litter. The pup is unable to swim or dive until it is two to three weeks old.

Food preferences of sea otters include abalones, sea urchins, clams, and crabs. It is a common practice of sea otters to eat while floating on their backs. Furthermore, sea otters use rocks to open the shells of their diet items, making them a member of a small group of animal tool users.

Unlike other members of the Mustelidae, sea otters lack functioning anal scent glands. Also, unlike most carnivores, sea otter teeth are adapted to crushing their prey, rather than tearing.

—*Sylvester Allred*

See also: Fur and hair; Lakes and rivers; Marine animals; Skunks; Tool use; Weasels and related mammals.

Bibliography

Buchsbaum, Ralph, et al. *The Audubon Society Encyclopedia of Animal Life*. New York: Crown, 1982. An old but useful reference with many colorful photo of mammals and other animal life from around the world.

Chanin, P. *The Natural History of Otters*. New York: Facts on File, 1985. A fact-filled book with nothing but otters. One of a series of books on specific natural histories of mammals.

Chapman, J. A., and G. A. Feldhamer, eds. *Wild Mammals of North America*. Baltimore: The John Hopkins University Press, 1982. An excellent and extensive reference covering all aspects of mammals living in North America.

Feldhamer, G. A., L. C. Drickamer, S. H. Vessey, and J. F. Merritt. *Mammalogy: Adaptation, Diversity, and Ecology*. Boston: WCB/McGraw-Hill, 1999. This is an excellent mammalogy textbook with many examples, illustrations, taxonomy, and references.

Hall, E. Raymond. *The Mammals of North America*. 2 vols. 2d ed. New York: John Wiley & Sons, 1981. An extremely wonderful and fully illustrated text, well worth pursuing.

OWLS

Type of animal science: Classification
Fields of study: Anatomy, ornithology, physiology, zoology

Any of the Strigiformes, a group of birds with highly specialized characteristics for nocturnal activity, including soft feathers and enhanced hearing and eyesight.

Principal Terms

ASYMMETRICAL EARS: in some species of owls, the ears are of unequal size and located unequally on the sides of the head

ASYNCHRONOUS: an uneven event; in hawks and owls, the staggered hatching of eggs that results in a nest of different-aged young

FACIAL DISK: the distinctive concentric circles of feathers that encircle the eyes of owls, helping direct sound toward the ears

OWL PELLETS: compacted packets of undigested prey that is regurgitated. Owl pellets may be used to determine food habits

OWLS: any of the strigiformes, a group of birds with highly specialized characteristics for nocturnal activity, including soft feathers and enhanced hearing and eyesight

NEST BOX PROGRAMS: construction and placement of nest boxes in suitable habitat to provide nesting platforms for specific birds of prey

PAIR BOND: close relationship between a male and female for breeding purposes

With their rounded facial disk encircling their large, forward-looking eyes, owls are the most recognizable of birds. Their unique traits also include superb auditory abilities and soft feathers for silent flight. Sharp talons for catching and killing prey and powerful bills for tearing flesh complete their basic characteristics. Most owls are colored in drab shades of brown, buff, and gray, either spotted or streaked, which helps conceal them during daylight hours. Woodland owls tend to be darker, while those of open country are lighter and paler. Thus, the eastern North American race of great horned owl (*Bubo virginianus virginianus*) is much darker than the pale northern race (*B. v. lagophonus*) of interior Alaska and the Yukon. A few smaller owls have rounded, eyelike disks on the back of their head to deter predators. Although once thought to be the nocturnal kin of hawks and eagles, owls are actually most closely related to the frogmouths and nightjars. The similarities between hawks and owls result from the evolutionary convergence of morphological features that facilitate their roles as avian hunters of live animals.

The 205 species of owls are a widespread and successful group that occupies virtually all habitats on all continents, from tundra to tropics, and are even found on most oceanic islands. They range in size from the forty-gram sparrow-sized elf owl (*Micrathene whitneyi*) of the Southwest desert to the formidable eagle owl (*Bubo bubo*) of Eurasia, which may reach 0.6 meter in length and weigh forty kilograms.

All the owls are placed in a single order, the Strigiformes, in which two owl families are recognized. The family Tytonidae includes sixteen species of barn (*Tyto* sp.), grass, and bay owls (*Phodilus* sp.), defined by small, dark eyes set in a narrow skull with a heart-shaped facial disk. The other 189 species are loosely grouped in the family

Strigidae, all of which have rounded skulls and large, wide-set eyes in a concentric facial disk. Owls of both families are named for their plumage colors or patterns (the tawny owl, *Strix aluco*, the black-and-white owl, *Strix nigrolineata*, the spectacled owl, *Pulsatrix perspicillata*, and the spotted owl, *Strix occidentalis*), habitats (the barn owl, *Tyto alba*, and jungle owllet, *Glaucidium radiatum*), size (great gray owl, *Strix nebulosa*, and little owl, *Athene noctua*), power and strength (eagle owls, *Bubo* sp.), presence of ear tufts (great horned, long-eared, *Asio otus*, and short-eared owls, *Asio flammeus*), or for their distinctive songs (screech owls, *Otus* spp., saw-whet owls, *Aegolius* sp., and barking owl, *Ninox connivens*).

Throughout history, owls have been alternately revered and feared. To the ancient Greeks, the solemn owl was the bird of wisdom and a companion of their warrior goddess, Athena. The Romans attached more ominous signs and portents to the ghostly cries of owls in the night. During the Middle Ages, owls were thought to be the companions of witches and the harbingers of evil and death. Many Native American tribes placed owls on a higher footing. The Arikara Plains Indians had secret owl societies, in which initiates were adorned with facial masks of owl feathers, while the Pimi Indians believed spirits of departed warriors assumed the shape of owls. Thanks to enlightened conservation efforts, owls are at long last recognized as important, interesting, and beneficial birds and all are protected by law.

Hunting and Food

All owls are predators, hunting a variety of animals commensurate with their size and strength. Woodland owls mostly hunt by the perch-and-pounce method, but hawk owls (*Surnia ulula*), short-eared owls, and other open country species may forage, harrier-like, over fields and meadows in search of prey. Burrowing owls (*Speotyto cunicularia*) are more terrestrial than most, spending a good deal of time running across the ground pursuing insects and small mammals. Most owls have broad wings—shorter in woodland species that maneuver in vertically complex habitats, longer and more hawklike in species that hunt open

The owl's soft, fluffy feathers allow it to fly in complete silence through the night. (Corbis)

country or are migratory. Bird-chasing owls, such as pygmy owls (*Glaucidium* sp.) and brown owls (*Ninox* sp.) have longer wings and tails for agile flight.

Owls use their combination of large eyes, superb hearing, and silent flight to hunt and catch prey. Their relatively large wings and small bodies give owls a low wing loading which, combined with soft, fluffy feathers, enables the quiet flight that makes owls such efficient nocturnal hunters. The large eyes of owls are densely packed with light-gathering rods for seeing in very low light, while the overlapping fields of their binocular vision enable precise parallax judgment of distance to prey. Head bobbing movements seen in many owls also help estimate distance and angle to prey.

If light is absent or nearly so, owls can continue to hunt, substituting ears for eyes. Studies by ornithologist Roger Payne have shown that barn owls, for example, can locate prey in total darkness entirely by sound. When an owl hears rustlings of prey, it turns its head toward the sound, using the facial disk of flattened feathers to direct and amplify faint sounds toward the ears set on either side of the wide, flat face to pinpoint the location of the source. The asymmetrical ears of some species—the right ear is larger and higher on the skull than the left—permits determination of vertical and horizontal direction to the sound. If the sound reaches the higher right ear first, then the source is from above; if it reaches both ears at the same time, the source lies straight ahead. By turning its head, the owl can determine the precise distance and angle of the flight path to the prey to within 1.5 degrees. When flying toward prey, the head is forward of the body to detect prey movements and make minor adjustments. Once within striking range, the owl extends its legs and spreads the talons in a wide oval to snare prey. Most prey are killed by the powerful, slashing talons, but larger animals may be dispatched by a bite to the back of the neck.

Owls tend to be opportunistic in their food habits, hunting a wide variety of mammals, birds, and other vertebrates. Rabbits, rats, and mice are staples of many of the large and medium-sized

Owl Facts

Classification:
Kingdom: Animalia
Phylum: Chordata
Subphylum: Vertebrata
Class: Aves
Order: Strigiformes
Families: Tytonidae (barn owls); Strigidae (owls), with subfamilies Buboninae (true owls, eighteen genera, eighty-three species), Striginae (long-eared and dish-eyed owls, six genera, twenty-six species)
Geographical location: All continents except Antarctica
Habitat: Virtually all habitats
Gestational period: Around thirty days, depending on species
Life span: Depending on species, three to ten years
Special anatomy: Rounded facial disk that directs sounds toward the ears; soft feathers for soundless flight; sharp talons and beak; large eyes with densely packed rods to enhance night vision; ability to turn the head up to 270 degrees; some species have asymmetrically spaced ears to assist in locating prey by sound

owls, but birds, lizards, snakes, insects, fish, and occasionally carrion are also consumed when available. The smallest owls, such as the elf owl, tropical screech owls, and boobook owls (*Ninox boobook*) are mostly insectivorous, while the larger and more powerful owls take hares, rabbits, and other medium-sized mammals and birds. The Eurasian eagle owl is a champion hunter, fully capable of killing chamois and foxes. With their long, bare shanks, the fishing owls (*Ketupa* sp.) of Africa and Southeast Asia spear fish from woodland streams or wade in the shallows to search for frogs, crabs, and crayfish.

Although some owls have a reputation for taking chickens and other poultry, most owls help keep injurious rodent populations in check. This awareness has led to the establishment of barn owl nest box programs in some inner cities to provide nest sites for an urban owl population to control rodents.

Small prey is swallowed whole but larger prey is broken into pieces. Prey is digested to a semiliquid consistency and then passed on to the intestinal tract. The undigested remains, mostly fur, feathers, bones, teeth, and other indigestible substances, are compacted into a small ball which is regurgitated in a reflexive, choking motion that casts the pellet out. The process takes several hours, so that today's pellet contents represent yesterday's meal. Owl biologists collect and analyze the pellets to determine food habits and impact of owl hunting on different species.

Owl Seasons

Tropical owls may breed at any time of year, but temperate owls commence breeding activity in late winter or early spring. Males claim a territory with territorial and courtship songs and postures. Songs of almost all owls consist of a series of hoots or wailing cries that echo ghostly through the night skies. Many owl species pair for life. Courtship may involve alternate duetting, billing and cooing, and mutual preening. Males of many species present food offerings, both as a courtship gift and to display their hunting ability.

Following courtship the female selects a suitable nest site. Although snowy owls (*Nyctea scandiaca*) of the Arctic select a spot beneath a clump of tundra sedge, most owls choose secluded tree hollows, cavities in cliffs, rock outcrops, or caves. Some species may also appropriate hawk or squirrel nests. Nest improvement usually consists of scraping a shallow hole or lining the nest with a few breast feathers. The oval or nearly oval eggs are usually laid at two-day intervals, but incubation begins with the first egg, resulting in a nest of different-aged young. The female incubates the eggs and the male brings food to the female or to a delivery site near the nest. The male may replace the female on the nest for short periods while she hunts, but his role in incubation is unclear.

Recently hatched young are typically fed pieces of food by the female. Later the adults simply deposit food at the nest and allow the young to pick and tear at it. Nest defense is weak or nonexistent during the early stages of nesting, but intensifies when young are in the nest. Defense varies from alarm calls and vigorous bill clacking in smaller species to aggressive and determined attacks by great horned owls and other large species. The female is usually most active in nest defense.

The fledged young typically remain in company of the adults for a few weeks before dispersing in search of new territories, usually in late summer or early fall. This is the most dangerous period of their lives, as they must perfect their hunting skills while avoiding enemies.

Longevity varies greatly; great horned owls and other large species may live ten or more years but the life span of smaller owls is usually only a few years. Other than humans, owls have few enemies. Larger owls prey on smaller owls while ravens and crows steal an occasional egg. Humans continue to be the main threat to owl populations. About thirty-five species are currently listed as threatened or endangered. Pollutants, disturbance, and collisions with vehicles and structures all take their toll of owls, but habitat loss factors most heavily. Several island races have disappeared following habitat alteration or introduction of exotics, and island populations continue to be at risk.

—Dwight G. Smith

See also: Beaks and bills; Birds; Claws, nails, and hooves; Domestication; Feathers; Flight; Hearing; Molting and shedding; Nesting; Predation; Respiration in birds; Vision; Wildlife management; Wings.

Bibliography

Bent, Arthur C. *Owls*. Part 2 in *Life Histories of North American Birds of Prey*. 1938. Reprint. New York: Dover, 1961. Bent exhaustively summarized all of the available reports of ecology, behavior, and food of North American owls. Still an important reference.

Burton, James A., ed. *Owls of the World: Their Evolution, Structure, and Ecology*. Rev. ed. London: Peter Lowe, 1992. An entertaining but still very informative book written for a more general audience. Contains excellent summaries of the natural history of the world's owls.

Clark, Richard J., Dwight G. Smith, and Leon Kelso. *Working Bibliography of Owls of the World*. Washington, D.C.: National Wildlife Federation, 1978. All the books, papers, articles, and news about owls ever published (to 1978) is presented in this bibliography, which also includes summaries of owl taxonomy and distributional status across the globe.

Del Hoyo, Josep, Andrew Elliott, and Jordi Sargatal, eds. *Barn-Owls to Hummingbirds*. Vol. 5 in *Handbook of the Birds of the World*. Barcelona, Spain: Lynx Edicions, 1999. A stunningly illustrated tour de force of all of the world's owls. Photos and plates make this massive, oversized volume a must-see for anyone even remotely interested in owls. Lengthy and learned introductory chapters precede the section on barn and bay owls of the world and another equally impressive chapter introduces the typical owls. What makes this volume exceptional is that each species account is by noted field researchers, most of whom have actually conducted research on the particular species.

Johnsgard, Paul A. *North American Owls: Biology and Natural History*. Washington, D.C.: Smithsonian Institution Press, 1988. Detailed accounts of all species of North American owls written in an authoritative and entertaining manner. Natural history and ecology is emphasized throughout.

Konig, Claus, Friedhelm Weick, and Jan-Hendrik Becking. *Owls: A Guide to the Owls of the World*. New Haven, Conn.: Yale University Press, 1999. An authoritative treatment of all of the world's owls. The book includes an overview describing why owls are really owls and detailed accounts of every recognized owl species. Each account succinctly describes identification, vocalizations, distribution, habitat, measurements and weight, habits, food, and breeding, status and conservation, and lists recent references. There are also chapters on evolution of owls, how to study owls, and owl conservation. Excellent color plates of owls are a highlight of the book.

PACKS

Types of animal science: Behavior, ecology
Fields of study: Ecology, ethology, population biology, wildlife ecology, zoology

Packs are social groups of carnivores, consisting of one or more families. They are generally highly cooperative for the purposes of rearing young and capturing food, particularly prey that is larger than any one member of the pack. These social packs are often called prides, clans, troops, schools, or packs.

Principal Terms

ANTIPREDATOR BENEFITS: benefits that come from actions that protect individuals from being killed

CARNIVORES: animals that eat the flesh of other animals

DOMINANCE HIERARCHIES: ranks of individuals within a group

Packs occur whenever animals group together in highly organized social systems for traveling, hunting, feeding, and sleeping, usually with bonds of attachment between all members. These social units tend to be fairly stable in composition, in comparison to most herds. This stability results in a social hierarchy, commonly called a dominance hierarchy, in which individuals are ranked in order of the number of other animals below them. Dominance hierarchies are maintained by dominant animals threatening subordinates, but fights are rare. Generally, subordinates engage in appeasement behaviors or avoid the dominant individual(s) altogether.

Dominance characteristics can be linked to body size, age, or weaponry. However, in some packs, dominance is inherited. This is common in baboon and macaque monkey troops, and in hyena clans, and is dependent upon the mother's rank. Females can move up in status every time a younger sister is born; however, several subordinates can form a coalition to challenge an individual above them in status.

Advantages to dominant animals are many. Greater access to food is one common benefit, and in some packs only the dominant pair breeds while the subordinates help to rear their young. Costs to dominants are that they are frequently challenged and run a greater risk of being killed or injured.

Packs and Predation

Predatory methods differ among species that form packs. Lions, who do not display endurance running, rely on stalking their prey and then rushing. During the rushing phase, lions will target any individual that appears to be slower or weaker than others in the prey herd. This tactic, also used by wolves, requires the predator to approach the prey very closely prior to rushing. In contrast, African wild dogs are known as coursers, chasing their prey for many miles until they can either drive it into other pack members or exhaust it. Hunting success in these dogs is very high. Hyenas vary their hunting tactics based on prey species. For attacks on wildebeest, they rush the herd, run for a while, stop to choose a target, and then resume the chase. As the chase continues, more and more hyenas join, and they generally take down their prey when it turns or runs into a lake or stream. They rely upon sheer numbers to overwhelm their target.

Theoretically, the formation of packs allows individuals within the group to exploit more food resources than they would be able to do on their own. In some pack species, this is the case: Cooperative hunts of lions are more successful than those by solitary lions, and there is some evidence to suggest the same for African wild dogs and hyenas. However, the cooperation does not necessarily increase the amount of food available for each member of the pack; in some cases, the nutritional intake per individual decreases with an increase in the size of the pack. Nonetheless, packs may be more able to protect their kills from predators and thus may be more capable of consuming prey completely than could one individual. Packs may also be more successful at driving other packs off from a kill; lions are known for relying on the kills of hyenas for up to half of their food. Hyenas form clans of up to sixty individuals, but these clans break up when food is scarce. Similarly, lions may form larger prides when food is common, but may split into pairs or threesomes as food becomes scarce.

Packs and Rearing of Young

If food per individual is not increased with increasing pack size, then other reasons must be present for the establishment and continuation of a pack. Suggested benefits include better defense of cubs against infanticide by outside coalitions of males that take over the pack (common in lion prides), higher reproductive success because all pack members share in feeding the offspring, and providing food for young while they grow to maturity, thereby ensuring that offspring will survive.

Rearing young in a group can benefit the young because of opportunities to learn from more than

---

Wolf Packs

Wolves are one of the best-known examples of animals that form packs for cooperative hunting and rearing of cubs. Most packs consist of fewer than eight members, although two or more packs can temporarily come together. When this happens, a pack may accept one member of another pack, while rejecting and chasing other members away. The packs always include a breeding pair of adults, some pups, subadults, and other adults, some of which may be allowed to breed in addition to the dominant pair. Packs can produce up to six cubs per year, and more than one female can reproduce, but often the dominant female will interfere with the care of the subdominant female's cubs.

Wolves prey on deer, moose, buffalo, sheep, caribou, and elk. The stimulus for beginning a hunt is apparently the completion of consumption of a previous kill. Wolves typically pursue prey by direct tracking, although they can occasionally come upon prey by chance encounter. Once they have located their prey, which usually occur in herds (except moose), they begin stalking until they get within feet of their targets. They then rush the target and chase it. Chases can last for less than a mile or for long distances over several days. They focus on the weakest targets: young, old, and sick prey. The wolves can injure a prey animal and then leave it alone for several days, resuming harassment until the prey can no longer rise and run. The pack then attacks from all sides to kill.

Wolves have been described as hunting in an extremely cooperative manner. However, what is more generally observed is that wolves in a pack attack the rear quarters of their prey, and the weight of many wolves biting and slashing on this area helps to bring the prey down. Wolves apparently have a low hunting success rate, which means that they must hunt frequently and test many individuals within a herd before finding one that displays some characteristic likely to be disadvantageous. While many have postulated that wolves form packs to facilitate food capture and that an increase in the pack size will help to increase the amount of food captured per individual, there is, in fact, a negative relationship between pack size and food available per wolf. Packing behavior may instead be more dependent on cooperation between kin where adults can more efficiently share their food with offspring, and young can obtain more food by remaining with their parents than they could by hunting on their own.

Baboon Troops

Hamadryas baboons form groups consisting of a dominant male, one to four females and their offspring, and some accessory males. A number of groups can come together to form bands of forty to fifty individuals. At night, these bands can coalesce to form troops of one hundred to over seven hundred individuals. Groups seem to form for mating purposes and care of the young. Bands are distinct feeding units and troops seem to avoid protection against nocturnal predators. A troop may range over an area comprising three to six square miles, and baboons move in troops to cover this vast area, gathering food, and congregating at water holes. While on the move, the subordinate males take the lead, followed by older juveniles and nonreproductive females. Females with infants and the most dominant males take up the middle positions, and other subdominant males bring up the rear. If a predator approaches the troops from any direction, it must encounter males. Many observers have reported that leopards, cheetahs, lions, and dogs can be run off by the males. Antipredator benefits seem to be one of the major benefits of this species' sociality, but adoption of infants and protection of juveniles from harassment also are known. In contrast to hamadryas baboons, gelada baboons have a social organization that is more similar to that of herding ungulates in that it varies with food availability and consists of distinct types of herds: harem herds, bachelor herds, and juvenile herds. Antipredator benefits dominate.

In both hamadryas and gelada baboons, play behavior by infants and juveniles serves to teach them all the important skills that must be learned and practiced. Thus, in addition to providing protection against predators, grouping in baboons serves to facilitate learning in offspring. Finally, grouping provides advantages because these species' food sources are clumped, and safe sleeping areas are limited.

---

one adult. It can also provide the young with practice in certain tasks that later prove important when the offspring are on their own. Cooperative hunting can provide young with the opportunity to learn hunting skills from their elders. Generally, this benefit occurs in longer lived species that produce only a few young per year per female.

The formation of packs tends to occur in groups where kin form the nucleus of the association. Dominance hierarchies are common in such packs and serve to stabilize the relationships among group members. Benefits are then realized in terms of hunting larger prey, spreading food resources between the older and younger generation, help in rearing young from all or most members of the pack, and dissemination of skill learning from adults. In a few packs, associations can serve an antipredatory function or can allow the defense of resources that would otherwise be taken by some other species or group.

—*Kari L. Lavalli*

See also: Communities; Competition; Demographics; Groups; Herds; Hierarchies; Learning; Mammalian social systems; Offspring care; Predation; Reproductive strategies; Territoriality and aggression.

Bibliography

Drickamer, L. C., S. H. Vessey, and D. Meikle. *Animal Behavior: Mechanisms, Ecology, and Evolution.* 4th ed. Dubuque, Iowa: Wm. C. Brown, 1996. Describes the mechanisms, ecology, and evolution that support a comprehensive study of behavior.

McFarland, D. *Animal Behavior.* 3d ed. Harlow, England: Longman, 1999. This textbook provides an excellent summary of why animals should socialize and how social organizations are formed.

Mech, L. D. *The Wolf: The Ecology and Behavior of an Endangered Species*. Minneapolis: University of Minnesota Press, 1981. This book presents one of the most comprehensive reviews of wolf biology based on a series of research articles from a symposium on the wolf's ecology.

Packer, C., D. Sheel, and A. E. Pussey. "Why Lions Form Groups: Food Is Not Enough." *American Naturalist* 136, no. 1 (July, 1990): 1-19. This article examines the alternative explanations for why lions find it beneficial to live in prides.

Schmidt, P. A., and L. D. Mech. "Wolf Pack Size and Food Acquisition." *American Naturalist* 150, no. 4 (October, 1997): 513-517. This article examines the amount of food per wolf in packs of variable size to determine whether packs form to increase the amount of food acquired, or for other reasons.

Whittenberg, J. I. "Group Size and Polygyny in Social Mammals." *American Naturalist* 115, no. 2 (February, 1980): 197-222. This article examines the different benefits that may contribute to grouping in social mammals from a mathematical viewpoint and then correlates the results to real-world examples.

PAIR-BONDING

Type of animal science: Behavior
Fields of study: Anthropology, ethology, ornithology, zoology

Pair-bonding is the formation of attachments or relationships of various durations for purposes of reproduction and care of the young, relationships that may appear analogous to marriage and family in humans.

Principal Terms

BOND, BONDING: the tie or relationship between opposite-sex partners in a pair bond

BONDING BEHAVIORS: behavior patterns that establish, maintain, or strengthen the pair bond

CONSORT PAIR, CONSORTSHIP: a temporarily bonded pair within a polygamous group

LONG-TERM PAIR BOND: pair bonding that continues beyond a single reproductive period

MONOGAMY: exclusive pair-bonding between one male and one female

PAIR, PAIRING: may refer to mating, sexual coupling (or copulation) or to formation of a pair bond, depending upon the context

POLYGAMY: a mating system in which one male mates with several females (polygyny) or one female mates with several males (polyandry)

PROMISCUITY: a mating system in which sexual partners do not form lasting pair bonds, where their relationship does not persist beyond the time needed for copulation and its preliminaries

In the animal world, there is a great variety of ways in which males and females pair or associate for reproductive purposes (called mating systems). While some animals pair with more than one member of the opposite sex (polygamy), others pair exclusively with only one mate (monogamy). While some animals associate only long enough to copulate (promiscuity), other animals form pair bonds that last for varying lengths of time, from one reproductive period (until young leave the nest) to a lifetime. While some pair-bonding has been observed in all vertebrate classes, and even in some invertebrates (some crabs and insects), it is particularly common in birds, while infrequent in fish and mammals.

Pair-Bonding in Birds

Most bird species are monogamous, although some are polygamous, and a few may be promiscuous. In some birds, such as prairie chickens, male and female pair only for copulation, after which the female is on her own to nest and parent. However, most birds pair bond, remain with their mates, and cooperate in some way until their young can leave the nest and survive independently. Some birds, such as song birds, pair bond for only a single breeding season, while others, such as swans, geese, penguins, and albatrosses, exhibit pair-bonding for more than one season, sometimes for life.

The American robin, for example, is mostly monogamous while it pair-bonds and shares parenting for one breeding season. Typically, in the spring, male robins are the first to migrate north. When the females arrive about a week later, the males have already selected territories and call or carol as though advertising for a mate. Courtship, involving three types of songs (only the males sing) and feeding, leads to mating and pair-bonding. The female builds a nest and lays her clutch of usually three or four eggs. While the female incubates the eggs, the male stands guard nearby, though he may help with the incubation as well. After the eggs hatch, both parents forage and feed the young. When the young leave the nest, they still need approximately two weeks of feeding and parental care before they are mature enough to survive independently. While the father provides this care, the mother repairs the nest and lays another clutch of eggs. American robins typically have two or three broods per year. In this case, the pair bond usually endures for the breeding season, but not through migration south for the winter and the nonbreeding season.

Pair-Bonding in Fish

Most fish do not form lasting pair bonds. An interesting exception, however, is the seahorse. At the beginning of the seahorse breeding season, males and females court for several days. The female produces eggs, and when they reach maturity, she deposits them into an egg pouch on the male seahorse's trunk. Then she swims away. Male seahorses typically remain in a habitat of approximately one square meter, while females range over an area that may be a hundred times as large. During the male's pregnancy, the female returns every day for a short, five- to ten-minute morning visit. During this visit, the seahorse pair exhibit social interaction and bonding behaviors reminiscent of courtship. Afterward, the female swims away until the next day, when she returns for another visit. After two to three weeks of pregnancy, the male gives birth to a few dozen or more baby seahorses during the night. When his mate returns in the morning for their daily visit, the male is ready for courtship, and the mating and reproductive cycle starts over again. Females have been observed to refuse to mate with other males during their mate's pregnancy, and so appear to be monogamous for the breeding season. In laboratory experiments, where mates were separated during pregnancy and the female interacted with another male, the original pair bond was broken. Thus, the social interaction and bonding behaviors that occur during the seahorse couple's short

Birds are the most likely of all animals to pair-bond for life. (Adobe)

daily visits appear to be important factors in maintenance and longevity of pair-bonding.

Pair-Bonding in Mammals

Polygamy is the most common mating system among mammals. Within polygamous (or promiscuous) groups, such as baboons and chimpanzees, a male and female may pair temporarily and separate themselves somewhat from the group, while engaging in social grooming and sexual activity. Temporary pairings that do not endure to the end of a reproductive cycle are referred to as consort pairs or consortships in the primate literature. When pair-bonds in a polygamous system are more lasting, the social structure is sometimes called a harem.

Only a small minority (3 to 5 percent) of mammals exhibit monogamous pair-bonding. Gibbons, the smallest of the apes, are particularly interesting because they are the exceptional case, and exhibit not only long-term monogamous pair-bonding but a social organization somewhat analogous to the human nuclear family. Gibbons pair at eight to ten years of age, and have five to six offspring, spaced about three years apart, over their ten- to twenty-year reproductive lifetime. Gibbon offspring remain with their family groups until they approach or reach sexual maturity, when they may leave voluntarily or be evicted by the same-sex parent. Gibbons are territorial, and family members cooperate as needed to defend both territory and mate or family. Just prior to sunrise every morning, mated males sing solo songs that can be heard up to a kilometer away, seemingly identifying their territory as occupied. Later in the morning mated females sing their own songs, and join their mates in singing duets, which appear to publicize both territory and pair-bonding.

Pair-Bond Formation and Maintenance

Formation of a pair-bond usually involves the behavior patterns of courtship and mating. Usu-ally, the male initiates pair formation, while the female decides whether a bond is formed or not. How long a pair-bond endures depends upon various factors. Many animal pairs maintain and strengthen their relationship by continuing the bonding behaviors that were initially used in courtship. Some animals maintain close physical proximity, groom each other, communicate with movement (display) or with vocalizations (call or song), or share food, nests, and territory. Reproductive success and dependency of young also appear to maintain pair-bonding.

—*John W. Engel*

See also: Birth; Breeding programs; Communication; Copulation; Courtship; Estrus; Grooming; Mating; Pregnancy and prenatal development; Reproduction; Reproductive strategies; Reproductive system of female mammals; Reproductive system of male mammals; Sexual development; Vocalizations.

Bibliography

Birkhead, T. R. *Promiscuity: An Evolutionary History of Sperm Competition and Sexual Conflict*. London: Faber & Faber, 2000. This book reviews research on promiscuity and explores how competition between males and choice by females operate after insemination has taken place.

Black, J. M., ed. *Partnerships in Birds: The Study of Monogamy*. New York: Oxford University Press, 1996. Provides an overview and synthesis of research and theory on pair bonding in birds. Fourteen case studies are used to compare and contrast the variety of pair bonds exhibited by birds.

Elia, I. *The Female Animal*. New York: Henry Holt, 1988. Provides an overview and analysis of the female animal's evolution and behaviors, including bonding and reproduction.

Lott, D. F. *Intraspecific Variation in the Social Systems of Wild Vertebrates*. New York: Cambridge University Press, 1991. This book provides a very comprehensive review of the literature on within-species variation in mating and parental-care systems.

Sussman, R. W. *Lorises, Lemurs, and Tarsiers*. Vol. 1 in *Primate Ecology and Social Structure*. Needham Heights, Mass.: Pearson, 1999. This book provides a review of the literature and analysis of the social behaviors and organization of lorises, lemurs and tarsiers.

PALEOECOLOGY

Type of animal science: Ecology
Fields of study: Evolutionary science, paleontology, zoology

Paleoecology is the study of ancient organisms and their relationships to one another and to their environments. The characteristics of ancient environments may be determined by examining rock and fossil features.

Principal Terms

CATASTROPHISM: a belief that the earth and its features are the result of sudden, violent upheavals that occurred in the relatively recent past

DENDROCLIMATOLOGY: the study of tree-ring growth as an indicator of past climates

FOSSIL: any trace of the presence of an ancient organism, whether track, imprint, or preserved remains

PALYNOLOGY: the study of pollen and spores; also called paleopalynology

RADIOCHRONOMETRY: the determination of the age of an object using radioactive isotope decay rates

TAPHONOMY: the series of events between the death of an organism and its preservation as a potential fossil; also, the study of such events

THANATOCOENOSIS: an assemblage of fossil species from a particular environment; a fossilized community; compare to biocoenosis, an assemblage of living species

UNIFORMITARIANISM: the belief that the earth and its features are the result of gradual biological and geological processes similar to the processes that exist today

---

Paleoecology is the study of ancient environments. As a field of science, it is most closely related to paleontology, the study of fossils. It is also related to paleoclimatology, paleogeography, and a number of other areas of study dealing with the distant past. All these disciplines have a handicap in common: Because they deal with the past, scientists are unable to apply the usual scientific criteria of direct observation and measurement of phenomena. When ecologists study present environments, direct measurements of environmental variables such as temperature, wind speed, current, soil moisture, and biological interactions can be taken. The organisms may be observed in their habitats—nesting, competing, eating, and being eaten. In order to make any conclusions about the past, scientists must assume at least one statement to be true without direct observation: The processes that exist in the modern universe and on the modern earth existed in the past.

Although one cannot observe any ancient falling objects, one assumes that the law of gravity existed in the past and that objects fell down; since rainfall occurs today, it must have occurred in the past. Similarly, paleontologists must assume that ancient plants and animals had tolerances to temperature, moisture, and other environmental parameters similar to those of modern organisms. This belief that "the present is the key to the past" is called uniformitarianism. It has been a key concept of the biological and earth sciences for almost two hundred years. Uniformitarianism does not include the belief that the ancient earth was like the modern earth in its life-forms or geography. During the early part of the nineteenth century, another worldview dominated: catastrophism, the belief that the earth is relatively young and was formed by violent upheavals, floods, and other catastrophes at an intensity unlike those of modern earth. Many catastrophists explained the presence of fossils at high elevations by the biblical Flood of Noah. The uniformitarian viewpoint prevailed and, although admitting that local catastrophes may be important, their long-term, earthwide importance was denied. It is interesting to note that catastrophism was revived in the 1980's to explain certain important events. The rapid extinction of the large dinosaurs at the close of the Mesozoic era has been attributed to the climatic changes associated with an alleged encounter between the earth and a comet—certainly a catastrophic event.

Types of Evidence About the Past
One of the most intensively investigated paleoecological problems has been the changing environments associated with the ice ages of the past million years. Analysis of pollen from bogs in many parts of the world indicates that there have been at least four advances and retreats of glaciers during that period. Evidence for this is the changing proportions of pollen from tree species found at the various depths of bogs. In North America, for example, spruces (indicators of cool climate) formerly lived much farther south than they do now. They were largely replaced almost eight thousand years ago by other tree species, such as oaks, which are indicative of warmer climates. This warming trend was a result of the latest glacial retreat.

Tree ring analysis not only enables paleoecologists to date past events such as forest fires and droughts, but also allows them to study longer-term cycles of weather and climate, especially those of precipitation and temperature. In addition, trees serve as accumulators of past mineral levels in the atmosphere and soil. Lead levels in tree wood showed a sharp increase as the automobile became common in the first half of the twentieth century because of lead additives in gasoline. Tree rings formed since the 1970's have shown a decrease in lead because of the decline in use of leaded fuels. Tree-ring analysis has also been a valuable tool for archaeologists' study of climatic

changes responsible for shifting patterns of population and agriculture among native Americans of the southwestern United States.

Fossil evidence is the chief source of paleoecological information. A fossil bed of intact clam shells with both valves (halves) present in most individuals usually indicates that the clams were preserved in the site in which they lived (autochthonous deposition). Had they been transported by currents or tides to another site of deposition (allochthonous deposition), the valves would have been separated, broken, and worn. Similarly, many coal beds have yielded plant fossils that indicate that their ancient environments were low-lying swamp forests with sluggish drainage periodically flooded by water carrying a heavy load of sand. The resulting fossils may include buried tree stumps and trunks with roots still embedded in their original substrate and numerous fragments of twigs, leaves, and bark within the sediment.

Certain dome- or mushroom-shaped structures called stromatolites are found in some of the most ancient of earth's sedimentary rocks. These structures may be several meters in diameter and consist of layers of material trapped by blue-green algae (cyanobacteria). Such structures are currently being formed in shallow, warm waters. Uniformitarian interpretation of the three-billion-year-old stromatolites is that they were formed under similar conditions. Their frequent association with mud cracks and other shallow- and above-water features leads to the interpretation that they were formed in shallow inshore environments subject to frequent exposure to the air.

Relative oceanic temperature can be estimated by observing the direction in which the shells of certain planktonic organisms coil. The shell of *Globigerina pachyderma* coils to the left in cool water and to the right in warmer water. *Globigerina menardii* shells coil in an opposite fashion—to the right in cool water and to the left in warmer water. Uniformitarian theory leads one to believe that ancient *Globigerina* populations responded to water temperature in a similar manner. Sea-bottom core samples showing fossils with left- or right-coiling shells may be used to determine the relative water temperature at certain periods. Eighteen-thousand-year-old sediments taken from the Atlantic Ocean show a high frequency of left-handed *pachyderma* and right-handed *menardii* shells. Such observations indicate that colder water was much farther south about eighteen thousand years ago, a date that corresponds to the maximum development of the last Ice Age.

Fossil Shape and Content

Fossil arrangement and position can be a clue to the environments in which the organisms lived or in which they were preserved. Sea-floor currents can align objects such as small fish and shells. Not only can the existence of the current be inferred, but also its direction and velocity. Currents and tides can create other features in sediments which are sometimes indicators of environment. If a mixture of gravel, sand, silt, and clay is transported by a moving body of water such as a stream, tide, or current, the sediments are often sorted by the current and deposited as conglomerates—sandstones, siltstones, and shales. Such graded bedding can be used to determine the direction and velocity of currents. Larger particles, such as gravel, would tend to be deposited nearer the sediment source than smaller particles such as clay. Similarly, preserved ripple marks indicate current direction. Mud cracks in a rock layer indicate that the original muddy sediment was exposed to the atmosphere at least for a time after its deposition.

Certain minerals within fossil beds or within the fossil remains themselves can sometimes be used to interpret the paleoenvironment. The presence of pyrite in a sediment almost always indicates that the sedimentary environment was deficient in oxygen, and this, in turn, often indicates deep, still water. Such conditions exist today in the Black Sea and even in some deep lakes, with great accumulations of dead organic matter.

The method of preservation of the remains of the fossilized organism can be an indication of the environment in which the creature lived (or died). Amber, a fossilized resin, frequently contains the embedded bodies of ancient insects trapped in the resin like flies on flypaper. This ancient environ-ment probably contained resin-bearing plants (mostly conifers), and broken limbs and stumps that oozed resin to trap these insects. Mummified remains in desert areas and frozen carcasses in the northern tundra indicate the environments in which the remains were preserved thousands of years ago.

Marks made on fossil parts by other organisms offer indirect evidence of the presence and activity of other species that might not have left fossil remains. Predators and scavengers can leave such marks on bones and shells by boring, scratching, and gnawing. One of the most controversial taphonomic problems in paleoecology is distinguishing between tooth marks left by animal scavengers and predators on bones and those marks left by the stone and bone tools of early human ancestors.

Studying Paleoecology

Fossils, especially fossil assemblages (thanatocoenoses), are the most commonly used indicators of ancient environments. The use of any fossil in interpreting the past must be subject to several qualifications. The fossil record is sparse for most groups of organisms because fossilization itself is a relatively rare event. Rapid burial of the remains and the presence of hard body parts (wood, shells, bones, and teeth) are only two of several fossilization prerequisites that must usually be met. This means that terrestrial organisms and soft-bodied organisms are seldom fossilized. Events leading to fossilization after the death of an organism (taphonomy) usually destroy the soft tissues through decay and scavenging and often disrupt and distort the remaining hard parts through transportation and weathering. An additional taphonomic problem is encountered when clumps or clusters of fossil remains are located. Without careful study, it is difficult to determine whether these assemblages are truly representative of the groupings of the organisms in life or if they are simply coincidental aggregations of such items as shells and limbs that were swept together by currents or wind and thus not indicative of the living situation and environment. Because of limitations on the interpretation of ancient environments by the use of fossilized body parts, trace fossils are often more reliable indicators of environmental conditions. Trace fossils are preserved tracks, burrows, trails, and other indirect indications of the presence of an organism. The presence of marine worm burrows, for example, can indicate environmental factors such as salinity and depth. Such traces are not transported from one site to another because transportation results in their destruction. Whenever these imprints are found, therefore, paleoecologists are able to make some inferences about the environment in which they were formed.

One of the most important methods to be mastered by paleoecologists is stratigraphy—the science of correlating and determining the age of rock layers with those of the fossils contained within these layers or formations. Rock layers or strata are not usually connected over large regions. While they might have been deposited as sediments at the same time and under the same conditions, subsequent erosion has usually made the layers discontinuous. Stratigraphers attempt to correlate discontinuous rock strata by measuring and describing them and by noting the presence of unique fossils called index fossils. If two strata are correlated, then they were probably deposited during approximately the same period, although there may be a gradation of conditions. For example, there may be a layer of sediment deposited at the same time, but under nearshore conditions at one spot and under offshore conditions at another. Relative ages are determined by using the law of superposition: Older rocks lie beneath younger rocks. One can say that a certain stratum is older than, the same age as, or younger than another layer, depending upon their relative positions. Absolute ages (estimated age in years before the present) are determined by measuring the amounts of certain radioactive elements within igneous rocks. Such radiometric age determinations are of less value for sedimentary rocks since they give the age of the minerals of the rock, not the age of the rock itself.

Uses of Paleoecological Research

Paleoecological data are applicable to other, related paleo-fields of the earth and life sciences. The study of fossils, paleontology, is enhanced by the inclusion of information about the fossil organisms' environments and relationships with other organisms. Paleontologists should attempt to reconstruct ancient environments because organisms did not exist alone or in vacuums: They lived in dynamic biological communities. Paleogeography relies heavily on paleoecological information to discern the locations, directions, and time intervals of glaciation, deposition of sediments, temperature, and other environmental variables. This information has been used to determine the past positions of continents and has been a valuable contribution to scientists' knowledge of continental drift.

Paleoclimatologists, who study ancient regional and planetwide conditions, must make use of local bits of paleoecological information to see the big picture of climate. One of the major concerns of paleoclimatology is the recognition of planetary climatic cycles and associated environmental and biological cycles. If there is a repeated recurrence of global environmental change, then predictions about future climatic change become more accurate and probable.

—*P. E. Bostick*

See also: Apes to humans; Communities; Demographics; Dinosaurs; Ecology; Ecosystems; Evolution: Animal life; Evolution: Historical perspective; Extinction; Fossils; Habitats and biomes; Human evolution analysis; Paleontology; Plant and animal interactions; Predation.

Bibliography

Arduini, Paolo, and Giorgio Teruzzi. *Simon and Schuster's Guide to Fossils*. New York: Simon & Schuster, 1986. An excellent introduction to the subject of fossils and fossilization. Good discussion of paleoecology. Well illustrated with color photographs and symbols indicating the time periods and habitats of the organisms. One in a series of nature guides.

Bennett, K. D. *Evolution and Ecology: The Pace of Life*. New York: Cambridge University Press, 1997. Argues against Darwin's thesis that evolutionary processes visible in ecological time can predict macroevolutionary trends.

Board on Earth Sciences and Resources, Commission on Geosciences, Environment, and Resources, National Research Council. *Effects of Past Global Change On Life*. Washington, D.C.: National Academy Press, 1995. A geophysical perspective on paleoecology, intended to provide scientific information for policymakers.

Brett, Carlton E., and Gordon C. Baird, eds. *Paleontological Events: Stratigraphic, Ecological, and Evolutionary Implications*. New York: Columbia University Press, 1997. A compilation of twenty essays devoted to *Lagerstätten* and epiboles and their importance in understanding paleoecology.

Davis, Richard A. *Depositional Systems: A Genetic Approach to Sedimentary Geology*. Englewood Cliffs, N.J.: Prentice-Hall, 1983. An advanced but understandable textbook about sediments and their environmental aspects of deposition. Glossary and extensive bibliography.

Dodd, J. Robert, and Robert J. Stanton, Jr. *Paleoecology: Concepts and Applications*. 2d ed. New York: John Wiley & Sons, 1990. A paleoecology textbook, with numerous tables and references.

Newton, Cathryn, and Lèo Laporte. *Ancient Environments*. 3d ed. Englewood Cliffs, N.J.: Prentice-Hall, 1989. One in a series of earth science paperback books. Valuable reading list classified by subjects and chapters. College and advanced high school level. Illustrated with line drawings and a few photographs.

Shipman, Pat. *Life History of A Fossil: An Introduction to Taphonomy and Paleoecology*. Cambridge, Mass.: Harvard University Press, 1993. Perhaps the best reference available on the fossilization process (taphonomy). Written at the advanced college level, this is a well-illustrated book with graphs, maps, line drawings, and photographs. Glossary and further references.

PALEONTOLOGY

Types of science: Anatomy, classification, ecology, evolution
Fields of study: Anatomy, anthropology, evolutionary science, human origins

Paleontology is the branch of geology that deals with prehistoric forms of life through the study of fossil animals, plants, and microorganisms. Many bursts of evolution and many extinction events have been uncovered by this field.

Principal Terms

EVOLUTION: the science that studies the changes in astronomical bodies, the earth, and living organisms and creates explanations of how changes occur and how things are related.

FOSSILS: the hardened, petrified (rocklike) impressions of animal, plant, and microbial remains

GEOLOGICAL PERIODS: the twelve divisions in successive layers of sedimentary and volcanic rocks, which are differentiated by the distinctive fossils present within each division

MASS EXTINCTIONS: five events in the history of life that resulted in the disappearance of more than 75 percent of all species

PALEONTOLOGY: the branch of geology that deals with prehistoric forms of life through the study of fossil animals, plants, and microorganisms

PLATE TECTONICS: the branch of geology that studies the fragmentation of the earth's crust into plates (crust fragments), the movement of plates, and the subduction or uplifting of plates

RADIOISOTOPES: unstable elements that decay into stable forms at a constant rate, which are used to determine how long ago volcanic rocks solidified

When organisms die, their soft parts usually decompose into their constituent elements, leaving no trace. The harder parts, such as wood, exoskeletons, bones, teeth, and shells, may survive many years, but eventually even they decompose. Sometimes, however, organisms are buried in an avalanche, are covered by volcanic ash, or sink into mud. If the burial inhibits decomposing microorganisms and prevents the rapid replacement of the buried tissue with inorganic material, a fossil is likely to form.

There are a number of different ways in which fossils form. Most fossils form through a process of mineralization. In this type of fossilization, buried material is slowly replaced by minerals such as calcium carbonate ($CaCO_3$), silicate (SiO_2), pyrite (FeS_2), sulfate ($Ca [SO_4] - H_2O$), and iron-phosphate ($Fe_3[PO_4] - H_2O$). These mineralized remains can be separated from the surrounding sediment and may superficially appear to be the actual remains. Sometimes fossils can form when the remains completely decompose. Organisms encased in a highly compacted sediment or covered by volcanic ash that quickly hardens, imprint into the covering a replica of their surfaces. When the organisms degrade, the space fills with minerals that harden. This is how surface replicas of organisms form. Mineralization and replication are responsible for most fossils. In some cases, organisms crawling through or stepping in sediment leave imprints of their passing, called trace fossils. Imprinted sediment explains some fossils.

Microorganisms, plants, and animals often get caught in tree sap (resin) that ages over millions of years, first into copal and then into amber. Amber

1199

Continued on page 1206

The Leakey Family

Louis Leakey

Born: August 7, 1903; Kabete, Kenya

Died: October 1, 1972; London, England

Fields of study: Anthropology, evolutionary science, human origins, paleontology

Contribution: African paleontologist-anthropologist whose most notable works were the characterization of a nearly complete 1.8-million-year-old skull of *Australopithecus boisei* and the piecing together of a 1.8-million-year-old skull of *Homo habilis*.

Mary Leakey

Born: February 6, 1913; London, England

Died: December 9, 1996; Nairobi, Kenya

Fields of study: Anthropology, evolutionary science, human origins, paleontology

Contribution: African paleontologist-anthropologist whose most notable works include the discovery of 3.6-million-year-old humanlike footprints made by *Australopithecus afarensis*, the piecing together of the 1.8-million-year-old skull of *Australopithecus boisei*, and the characterization of the nearly complete 1.5-million-year-old *Homo erectus* fossil known as Turkana boy.

Richard Leakey

Born: December 19, 1944; Nairobi, Kenya

Fields of study: Anthropology, evolutionary science, human origins, paleontology

Contribution: African paleontologist-anthropologist, son of Louis and Mary, whose most notable works include the piecing together of a 1.9-million-year-old skull of *Homo habilis*, the reconstruction of a 1.8-million-year-old skull of *Homo rudolfensis*, and the unearthing and characterization of Turkana boy.

Maeve Leakey

Born: 1948

Fields of study: Anthropology, evolutionary science, human origins, paleontology, zoology

Contribution: African zoologist-paleontologist, wife of Richard, whose most notable works include her reconstruction of 1.8-million-year-old *Homo rudolfensis* and her characterization of 4.1-million-year-old *Australopithecus anamensis*. In early 2001, her announcement of the discovery of a new 3.5-million-year-old fossil hominid,

Kenyanthropus platyops, threw the previous understanding of human evolution into confusion.

Louis Seymour Bazett Leakey was born in Kenya to English missionary parents. He trained as an anthropologist at the University of Cambridge, where he developed the dream of looking for evidence of ancient humans in Africa. Mary Douglas Leakey was born in England and studied archaeology, specializing in stone tools. After their marriage in 1936, Mary and Louis left England for Africa, establishing their home in Kenya.

In 1959, Mary discovered the first traces of a very ancient apelike animal at a site in Olduvai Gorge in northern Tanzania. Louis and Mary pieced together the fossilized fragments, recreating the nearly complete skull of a 1.8-million-year-old bipedal apelike creature now known as *Australopithecus boisei*. This animal had a brain capacity of approximately 500 cubic centimeters. At the time, Louis thought this apelike creature was the "missing link" between apelike creatures and humans because of the nearby discovery of very primitive paleoliths (stone tools) dated between 1 and 2.5 million years ago. Then in 1971, the skull of 1.8-million-year-old *Homo habilis* was pieced together by Mary and Louis. It was discovered in the same strata as *A. boisei*. *Homo habilis* had a brain capacity of approximately 600 cubic centimeters and appeared to be more closely related to modern humans than *A. boisei*. Richard Leakey characterized two very important *Homo* skulls: one of a 1.9-million-year-old *Homo habilis*, with a brain capacity of only 510 cubic centimeters, and a second of 1.8-million-year-old *Homo rudolfensis*, with a brain capacity of 775 cubic centimeters. These and other discoveries suggest that the paleoliths were produced by one of the *Homo* populations and not by any of the *Australopithecus* populations. Turkana boy's fossil remains, characterized by Richard and Maeve, suggest that this 1.5-million-year-old *Homo erectus* was about 4.5 feet tall and would have grown to about 5.5 feet. He had a brain capacity of about 880 cubic centimeters.

In 1975, Mary Leakey discovered the traces of humanlike footprints in hardened volcanic ash at Laetoli, near Tanzania's Olduvai Gorge. These footprints have been dated at 3.6 million years by the potassium-argon method. Since no *H. habilis* fossils this old have been discovered, the footprints must be of an *Australopithecus*, possibly related to *Australopithecus afarensis*, whose fossil remains have been discovered in the area and in much of East Africa. Although these animals lived between 3.8 and 2.7 million years ago, they were fully erect, and their footprints appear very human. Maeve Leakey discovered the fossil remains of 4.1-million-year-old *Australopithecus anamensis*. Her discovery of this very apelike creature has increased the span of time in which bipedal apelike creatures evolved.

The discoveries made by the Leakey family helped to establish the fact that a number of different biped species of *Australopithecus* and different species of *Homo* were living together in Africa 1.8 million years before the present. Locomotion on two legs was the first human trait to develop, at least 3.6 million years ago, but it developed first in a number of nonhuman animals, various species of *Australopithecus*. The crude stone tool culture first characterized by Louis and Mary, as well as the proximity of *Homo habilis* fossils suggests that *H. habilis* or closely related species such as *H. rudolfensis* were the first animals to be both bipeds and toolmakers. The Leakey discoveries indicate that some early human population closely related to *H. habilis* or *H. rudolfensis* was the ancestor of *H. erectus*, although Maeve Leakey's discovery of a new genus, *Kenyanthropus platyops*, announced in early 2001, suggests that the lineage of human ancestry in the middle Pliocene is more complex than previously thought. The Leakeys' discoveries also established that Africa, over a period of four million years, was the birthplace of numerous species of both *Australopithecus* and *Homo*.

—*Jaime Stanley Colomé*

The Leakey family (left to right, Richard, Mary, and Louis) have been a major force in the discovery of hominid ancestors in Africa. (Win Parks, National Geographic Society)

Periods and Extinctions

Era	Period	Date (millions of years ago)	Newly dominant fossils	Extinctions	Most Likely Causes
Cenozoic	Quaternary	2-present	Recent mammals	One lesser extinction at the T-Q boundary.	
	Tertiary	65-2	Diversified small and large mammals	Two lesser extinctions: one each at around 30 million years and around 10 million years before the present. Mass extinction that eliminated around 75 percent of all species. All dinosaurs except for birds, many land reptiles, all marine reptiles (plesiosaurs and ichthyosaurs), and all flying reptiles (terosaurs) went extinct.	Separation of North and South American land masses from European and African land masses. This altered oceanic currents and atmospheric streams, which produced glaciers. Glacier formation caused the sea level to drop around 200 feet. Volcanic activity may also have contributed to glacier formation. Deep water may have been anoxic because of poor circulation. A huge meteor impact may have contributed to the mass extinction.
Mesozoic	Cretaceous	145-65	Diversified dinosaurs, flowering plants	Three lesser extinctions: one at the J-K boundary and one each at around 110 million years and around 90 million years before the present.	

| Jurassic | 205-145 | Diversified early dinosaurs, first birds, small early mammals | Lesser extinction at around 138 million years before the present. Seed-ferns all went extinct. Mass extinction that eliminated about 80 percent of all species. | Breakup of Pangaea initiated. Newly forming continents in the northern hemisphere, equatorial continent, and South Pole continent. This altered oceanic currents and atmospheric streams. The temperature was increasing during the T-J transition, yet sea levels fell slightly (around 50 feet), suggesting some glacier formation. Volcanic activity may have contributed to glacier formation. Deep water may have been anoxic because of poor circulation. |
| Triassic | 250-205 | Early dinosaurs, diversified reptiles: mammal-like reptiles, marine reptiles; cycads, diversified ginkgoes | Mass extinction that eliminated around 95 percent of all species. All trilobites, all armored fishes, and all reptilelike amphibians went extinct. | Formation of Pangaea completed. A massive continent from the Northern hemisphere to the South Pole. Water level dropped around 300 feet suggesting huge glacier formation at the poles. Extensive volcanism lowered the temperature, as did altered water currents and atmospheric streams. Volcanism produced huge amounts of toxic sulfur dioxide and acid rain as well as carbon dioxide. Deep water may have been anoxic because of poor circulation. |

Periods and Extinctions—continued

Era	Period	Date (millions of years ago)	Newly dominant fossils	Extinctions	Most Likely Causes
Paleozoic	Permian	290-250	Diversified reptiles, decline of reptilelike amphibians	One lesser extinction occurred at the C-P boundary.	
	Carboniferous	360-290	Diversified sharks, bony fishes; early reptiles, reptilelike amphibians; seedless ferns, seed ferns, ginkgoes, pine forests	At least one lesser extinction occurred. Mass extinction that eliminated around 80 percent of all species. Most armored fishes and most jawless fishes went extinct.	Movement of northern continents toward Gondwana at the South Pole. Glaciers formed, and sea level dropped around 100 feet due to altered oceanic currents and atmospheric streams. Volcanic activity may have contributed to glacier formation. Deep water may have been anoxic because of poor circulation.
	Devonian	400-360	Diversified armored jawless and bony fishes; amphibians; insects; seedless, spore-forming tree ferns	At least two lesser extinctions and six minor extinctions occurred.	

Period	Dates (millions of years ago)	Life Forms	Extinctions	Causes
Silurian	440-400	Armored jawless fishes, first bony fishes, first fishes with jaws (placoderms and sharklike fishes); simple land plants	At least one lesser extinction occurred. Mass extinction that eliminated around 85 percent of all species. Most Burgess shale organisms went extinct. Trilobites survived, but just barely.	Fusion of some northern continents with each other at the equator. Movement of Gondwana toward the South Pole. Massive glaciers formed and sea level dropped 150 feet. Glaciers were caused by altered oceanic currents and atmospheric streams. Volcanic activity may have contributed to glacier formation. Deep water may have been anoxic because of poor circulation.
Ordovician	500-440	Diversified arthropods, first jawless fishes, primitive chordates, filter feeders, mollusks	At least two lesser extinctions occurred.	
Cambrian	540-500	Burgess shale fossils: early arthropods, animals with mineralized shells like the brachiopods, animals with notochords	At least five lesser extinctions occurred in which different groups were affected to different degrees.	Unknown causes.
Precambrian Proterozoic	600-540	Ediacaran fossils: sponges, jellyfish, segmented wormlike animals, colonial filter feeders	Lesser extinction that eliminated up to 50 percent of all species near the end of the Precambrian.	Unknown causes.
Archaeozoic	1000-600 3800-1000	Microscopic animals, algae Bacteria		

and the organisms encased in it are often referred to as fossils. Ethanol or isopropyl alcohol dropped on copal makes its surface sticky but has no effect on amber. Copal is usually less than 2 million years old, whereas the oldest ambers are about 150 million years old.

The hard parts (skin, cellulose walls, wood, and exoskeletons) of organisms are generally preserved in copal and amber, although some tissues are replaced by minerals such as pyrite (FeS_2). The pyrite turns portions or all of the fossil black. Completely blackened fossils retain little of the original organism. The fossils in copal and amber have given paleontologists a picture of life during different ancient (Jurassic and Cretaceous) and recent (Tertiary and Quaternary) periods.

Radioisotopes Used to Date Fossils

The radioisotope often used to date very old fossils is potassium-40 (^{40}K), which decays into argon-40 (^{40}Ar) by the conversion of a proton into a neutron. It takes 1,300 million years for one half of a ^{40}K sample to decay into ^{40}Ar. By using the half-life for ^{40}K($T_{1/2} = 1300$ million years) and measurements of ^{40}K and ^{40}Ar in volcanic rock, the volcanic rock can be dated. If, for example, a sample of volcanic rock contains equal amounts of ^{40}K and ^{40}Ar, then the volcanic rock would be 1,300 million years old. Fossils are dated by determining the age of volcanic rock below and above the fossils. The age of the fossils would be between the ages determined for the volcanic rock.

Carbon-14 (^{14}C) and carbon-12 (^{12}C) measurements are used to date fossils less than thirty thousand years old. ^{14}C decays into ^{12}C by losing two neutrons. The half-life for this decay is 5,700 years. Generally, in living organisms the ratio of radioactive ^{14}C to nonradioactive ^{12}C remains constant because fresh ^{14}C replaces the lost ^{14}C. When an organism dies, however, ^{14}C no longer replaces the ^{14}C lost, and so the ratio of ^{14}C to ^{12}C decreases in the surviving tissue. The smaller the ratio of ^{14}C to ^{12}C, the older the sample is. A fossil may be dated by directly measuring the age of the fossil, by determining the age of organic material closely associated with the fossil, or by determining the age of organic material both below and above the fossil. The errors become quite large for dates greater than forty thousand years.

Mass Extinctions

The five mass extinctions during the last five hundred million years have been very important in the evolution of life because they eliminated nearly all of the most highly adapted species that existed before each extinction. This allowed many of the surviving, more generalized species to evolve into highly specialized species that filled vacated and new environments. The species that arose after each mass extinction were very different from those they replaced.

In most cases, mass extinctions were caused by the plate tectonics that fused and fragmented continents. Colliding continents created shallow seas and mountains. Fragmenting continents initially created shallow seas. In some cases, these shallow seas developed into deep oceans. The fusion and the fragmentation of continents drastically changed oceanic currents and weather patterns, which often lowered temperatures both in the ocean and on land. The movement of plates to and from the equator and poles also affected the world's environments. Extensive volcanic activity occurred both when small continents collided and fused with each other and when massive continents fragmented and the volcanic debris blown into the upper atmosphere absorbed and reflected light, resulting in a cooling of the earth. Large decreases in temperature, changes in rainfall, and environmental loss were deleterious for highly adapted species.

Huge comets, asteroids, or meteoroids colliding with the earth may have played a minor role in any of the mass extinctions. The mass extinction that marked the end of the Cretaceous period about sixty-five million years before the present, which eliminated more than 75 percent of all species and all of the dinosaurs except the birds, may have been helped along by a meteoroid about ten miles across. Although this impact spewed millions of tons of earth into the atmosphere, it is not known whether the debris was sufficiently fine to remain in the atmosphere long enough to drastically alter the temperature and weather. Many biologists favor the idea that plate tectonics, not a meteoroid, ended the Cretaceous period. During the late Cretaceous period, the supercontinent Pangaea was fragmenting into northern and southern continents. In addition, these continents themselves were fragmenting from north to south, and an ocean was forming between them, slowly becoming the Atlantic Ocean.

At least nineteen lesser extinctions and five minor extinctions (Devonian and Devonian-Carboniferous transition) occurred during the last 600 million years. In some cases, the lesser extinctions eliminated up to 50 percent of all species. This allowed new species to evolve into the environments vacated.

Geological Periods and Dominant Fossils

The preceding table, titled Periods and Extinctions, characterizes each of the geological periods by listing some of the newly dominant fossils (organisms) that appeared during each period. The timing of the five major and nineteen lesser extinctions is indicated, along with the most likely cause of the extinctions.

Dating the different layers in sedimentary rock, some many hundreds of feet thick, establishes the great age of the fossils in the layers. The fossils in the different layers of sediment clearly illustrate the gradual and in many cases the rapid disappearance of most early forms of life as well as the repeated appearance of totally new forms of life. Numerous extinctions induced by the evolving earth eliminated countless species and allowed countless other species to evolve. Geology (plate tectonics, dating of strata, and paleontology) and evolution have produced a complex picture of the evolving earth and life.

—*Jaime Stanley Colomé*

See also: Apes to humans; Communities; Demographics; Dinosaurs; Ecology; Ecosystems; Evolution: Animal life; Evolution: Historical perspective; Extinction; Fossils; Habitats and biomes; Human evolution analysis; Paleontology; Plant and animal interactions; Predation.

Bibliography

Dingus, Lowell, and Timothy Rowe. *The Mistaken Extinction: Dinosaur Evolution and the Origin of Birds*. New York: W. H. Freeman, 1998. For all its information, this book is easy to read. It contains a comprehensive discussion of dinosaur fossils from the Jurassic and Cretaceous periods and how they relate to fossil and modern birds. The many photographs and outstanding figures and drawings make this an outstanding reference book.

Gould, Stephen Jay. *Wonderful Life: The Burgess Shale and the Nature of History*. New York: W. W. Norton, 1989. A very readable book, containing extensive discussion of the Cambrian period fossils found in the Burgess Shale of Canada. The book contains numerous informative photos and interpretive drawings of the fossils.

Hallam, Anthony, and P. B. Wignall. *Mass Extinctions and Their Aftermath*. New York: Oxford University Press, 1997. An extensive discussion of mass extinctions and what caused them.

Johanson, Donald C., and Blake Edgar. *From Lucy to Language*. New York: Simon & Schuster, 1996. This is a picture book of ape and human fossils. Concise analyses and discussions place these fossils into a temporary evolutionary framework. Clearly explains the contributions made by the most famous paleontologists of the twentieth century. Beautiful and informative photographs on every page.

Leakey, Richard, and Roger Lewin. *Origins Reconsidered: In Search of What Makes Us Human*. New York: Doubleday, 1992. Richard Leakey reconsiders the evolutionary relationship between the fossilized *Australopithecus* species and *Homo* species that have been discovered during the twentieth century, with special attention to *A. boisei* (*Zinganthropus boisei*), *A. afarensis* (Lucy), *H. habilis*, *H. erectus* (Turkana Boy), *H. erectus neandertal*, and *H. sapiens sapiens*. The remains of *A. boisei*, *H. habilis*, and *H. erectus* (Turkana Boy) were discovered by members of the Leakey family.

Vickers-Rich, Patricia, and Thomas Hewitt Rich. *Wildlife of Gondwana*. Bloomington: Indiana University Press, 1999. Although Laurasia and Gondwana formed from the fragmentation of Pangaea during the Triassic period, Gondwana rocks range from the Carboniferous to the Jurassic. This paleontology book correlates plate tectonics and the accompanying environmental changes with extinctions and the appearance of new life forms. The book contains many useful diagrams, beautiful pictures of fossils, and photos of animals and plants now living in Australia.

PANDAS

Type of animal science: Classification
Fields of study: Anatomy, conservation biology, physiology, reproduction science, wildlife ecology

The giant panda, with its distinctive black and white coat, is one of the most easily recognized animals in the world. It is extremely specialized for a bamboo diet. Despite efforts by many conservationists, the panda is highly endangered.

Principal Terms

CARNIVORE: animal dependent on an animal diet

DELAYED IMPLANTATION: an extended period after fertilization when an embryo stops developing, and before it attaches to the uterine wall and resumes embryonic development

HERBIVORE: animal dependent on a plant diet

RADIAL SESAMOID: forefoot wrist bone

Giant pandas are members of the bear family, and resemble other bears in size and shape. In contrast, the red panda, which lives in the same habitat as the giant panda, is closely related to and resembles the raccoon. The black-and-white giant panda coat is recognizable to people all over the world. Its legs, ears, eye patches, and a band across the shoulders are black, while the rest of the coat is white. Pandas have large jaws and broad teeth, with an increased number of cusps that help them chew tough bamboo stalks. The wide jaw contributes to the large, round shape of the panda's head. The black eye patches create an illusion of very large eyes. These features, along with short legs, give the panda a cute, infantlike appearance to humans, which contributes to their enormous popularity. Another remarkable feature is the "thumb." Pandas have the same five digits of other bears, plus a sixth digit, an opposable thumb, which is actually a modified wrist bone (sesamoid). The thumb allows them to grasp bamboo with considerable dexterity.

Diet and Reproduction

Giant pandas are the most nearly herbivorous of the bears. Ninety-nine percent of their diet consists of bamboo. They also eat other plants and meat that they can scavenge, and will eat a variety of foods in captivity. However, pandas live in areas once covered by vast bamboo forests, and their jaws, teeth, paws, and behavior are all adapted to eating bamboo. Nonetheless, they have a short, simple digestive tract similar to those of other bears, a sign of their carnivorous ancestry, which is not well-adapted to digesting the leaves and fiber of bamboo. Accordingly, pandas can only digest 21 percent of the bamboo that they consume, whereas ruminants such as cows digest up to 60 percent of the plant material that they eat. Because of this inefficiency, pandas consume 12 to 15 percent of their body weight in bamboo each day, and must spend twelve to fourteen hours each day eating.

Pandas mate in the spring between March and May, with cubs born in late summer. The total gestational period varies from 87 to 165 days. Cubs are born very small, between three and five ounces. Combined with hormonal data, these characteristics suggest that pandas have a delayed implantation. That is, after fertilization, the embryo remains free-floating in the uterus for several months before attaching to the uterine wall. After attachment, pregnancy is only about forty days, resulting in small newborns. Delayed implantation also occurs in some other bear species. Pandas have between one and three cubs at a time. However, they usually raise only one cub, which the mother nurtures intensively for several months.

Status and Distribution

The giant panda is an endangered species, with fewer than 1,500 individuals remaining. Panda habitat once covered an area of roughly 450,000 square miles in southeastern China, ranging from central China, to Hong Kong, into Burma and Vietnam. Today, they are found within only a 5,400-square-mile area. Even within this area, they are separated into many subpopulations, which prevents interbreeding. Within their range, pandas live at elevations above human settlements (four to eight thousand feet) to the upper edges of bamboo for-

The panda's distinctive black-and-white coloration makes it one of the most recognized wild animals. (Corbis)

est (ten to eleven thousand feet). The continuing expansion of human farms and villages has forced them from the lower elevations, which further contributes to the fragmentation of their populations. Small subpopulations are at a high risk for inbreeding, which reduces the genetic variability and individual fitness, placing panda survival in doubt.

Because of the giant panda's endangered status and charismatic traits, extensive efforts are being made to prevent its extinction. It is considered a national treasure in China, and killing one is punishable by death. Western conservationists, including the World Wildlife Fund, which uses the panda as its symbol, are also aiding the preservation efforts. Captive panda breeding, mostly in China, has been a focus of these efforts. Unfortunately, breeding programs have never produced enough cubs to introduce into the wild, or even to sustain the captive populations. Still, many scientists and conservationists around the world are working to ensure a future for the giant panda.

—*Laura A. Clamon*

See also: Bears; Breeding programs; Endangered species; Fauna: Asia; Grizzly bears; Polar bears; Raccoons and related mammals; Zoos.

Panda Facts

Classification:
Kingdom: Animalia
Subkingdom: Bilateria
Phylum: Chordata
Subphylum: Vertebrata
Class: Mammalia
Order: Carnivora
Family: Ursidae (bears)
Genus and species: Ailuropoda melanoleuca (giant panda)
Geographical location: China
Habitat: Bamboo forests
Gestational period: Probably a three-month embryonic diapause, followed by a forty-day gestation
Life span: Twenty to twenty-five years; known record, thirty years
Special anatomy: Coloration is white with black splotches; opposable "thumb"

Bibliography

Brown, Gary. *The Great Bear Almanac*. New York: Lyons & Burford, 1993. A comprehensive, general source of bear facts.

Gould, Stephen Jay. *The Panda's Thumb: More Reflections in Natural History*. New York: W. W. Norton, 1980. This is a classic essay on evolutionary significance of the panda's thumb.

Kallen, Stuart A. *Giant Pandas*. Minneapolis: Abdo & Daughters, 1998. For young readers, this book describes physical characteristics, habitat, and behavior of pandas.

Schaller, George B. *The Last Panda*. Chicago: University of Chicago Press, 1993. This book details a leading conservationist's experiences in trying to preserve the panda.

Stirling, Ian ed. *Bears*. San Francisco: Sierra Club, 2000. Although written for a general audience, the contributors are among the world's leading bear researchers. It includes fantastic photographs.

PARROTS

Types of animal science: Anatomy, classification, reproduction
Fields of study: Anatomy, ornithology, zoology

Parrot species include macaws, cockatoos, true parrots, parakeets, and lories. They are kept as pets for their beautiful coloring and ability to learn to talk.

Principal Terms

PLUMAGE: the feathers of birds
ZYGODACTYL: having two toes pointing backward and two toes pointing forward

The hundreds of parrot species are vivid-colored members of the bird family Psittacidae. They belong to five classes: macaws, cockatoos, true parrots, parakeets, and lories. In parrot plumage, reds and greens often predominate, but blue, purple, yellow, and black also appear.

Parrots inhabit warm South and Central America, southern North America, Africa, Madagascar, Indonesia, and southern and Southeast Asia. They live in lowland tropical or subtropical and mountain forests. Parrot sizes range from three-inch New Guinea pigmy parrots to South American macaws, over three feet long.

Macaws, the largest parrots, have long, pointy tails. Cockatoos of Australia and Indonesia are white, with colored crests and other touches of yellow, red, or pink. True parrots are smaller, square-tailed, and have many green feathers. Parakeets, smaller than most true parrots, have long, pointy tails. Lories have red or orange bills, in-

Parrots and their relatives are distinguished by their curving beaks, which they use to help them climb. (Adobe)

stead of gray bills like true parrots. In most species, males and females look similar, but males are more brightly colored.

Physical Characteristics of Parrots

The most noticeable features of parrots, beyond color, are their down-curved, hooked bills, thick, muscular tongues, and short legs. The bills have strong grasping ability that helps parrots to climb well. Parrot feet are zygodactyl, meaning that the two outer toes of the foot point backward and grip in the opposite direction to the two forward-pointing inner toes. Because of this, parrots walk awkwardly. However, zygodactyly makes them excellent climbers.

Parrots eat seeds, fruits, and nuts. Australian lories also eat pollen and nectar. The thick, muscular tongues of most parrots manipulate nuts and seeds, breaking them open as needed. Longer lorie tongues have brushlike tips for eating pollen and nectar. Most parrots find their food in trees, using feet and bills to navigate search areas.

The Lives of Parrots

Parrots are social birds that often live in flocks. Their loud voices are harsh and used in constant communication. Parrot breeding seasons depend on the geographic location of their habitat and the food they eat. Species living outside the tropics, where food supply changes seasonally, have yearly mating seasons. Those in tropical regions breed at irregular intervals when food is available.

Most parrots pair for life. Males attract mates by hopping, bowing, wagging tails, and flapping wings. After mating, females lay two to eight small white eggs. A mated pair does not part after breeding. They eat together and groom each other year round. Most parrots nest in holes in trees, termite mounds, and rock or ground tunnels. Others lay eggs in large grass or twig nests. Females incubate eggs for eighteen to thirty-five days, while males supply mates with food. Parrots are born blind and dependent on their parents. Young leave the nest after 1 month in smaller species and after 3.5 months in larger species. Some parrots live for sixty to eighty years.

Parrot Facts

Classification:
Kingdom: Animalia
Subkingdom: Bilateria
Phylum: Chordata
Subphylum: Vertebrata (have backbone)
Class: Aves
Order: Psittaciformes
Family: Psittacidae, with subfamilies Nestorinae (keas, one genus, 5 species); Psittrichasinae (vulturine parrots); Kakatoeinae (cockatoos, five genera, 14 species); Micropsittinae (pygmy parrots, one genus, 4 species); Trichoglossinae (lories, two tribes, fourteen genera, 21 species); Strigopinae (owl parrots); Psittacinae (true parrots, five tribes, fifty-four genera, 126 species)
Geographical location: South and Central America, southern North America, Africa, Madagascar, Indonesia, and southern and Southeast Asia
Habitat: Lowland tropical or subtropical and mountain forests
Gestational period: Eggs hatch in seventeen to thirty-five days
Life span: Forty to eighty years, in captivity
Special anatomy: Down-curved bills, muscular tongues, zygodactyly

---

Some Representative Parrot Species

African gray parrots (*Psittacus erithacus*) of Central and Western Africa grow to one-foot lengths and one-pound weights. They have gray bodies, black wingtips, and red tail feathers. They eat fruit, seeds, nuts, and berries, nesting in holes in trees. Females lay about four eggs and incubate them for a month, while males feed them. Chicks are fed by both parents. They fly in 2.5 months and parents feed them for 5 more months. These birds form flocks of up to thirty-six individuals. In captivity they live for up to eighty years.

Princess parrots (*Polytelis alexandrae*) live in the scrub land of central and western Australia. They nest in eucalyptus tree holes and eat acacia buds, seeds, berries, and fruit. They are high-altitude fliers, who travel widely seeking food. Their flocks contain up to twenty-four birds. Full-grown, they are fifteen inches long including the tail, and weigh around four ounces. Back, belly, and wing plumage is olive green and yellow; tail feathers are violet; throats are pink; bills are red-orange; and heads are light blue. Breeding occurs between September and December. Females lay four to six eggs and incubate them for three weeks. Young can fly at three months old. An endangered species, they are protected by law.

Indonesian salmon-crested cockatoos (*Cacatua moluccensis*) have plentiful, pink-tinted white plumage. Atop their heads are crests of salmon-red feathers, raised to show desire to mate. They eat berries, seeds, nuts, fruits, and insects. Breeding season is in November, and after mating they pair for life. Nests are in tree hollows. Females lay four to seven white eggs. Both birds incubate them for a month. After hatching, young remain in the nest for three months, and then live on their own. Salmon-crested cockatoos live for sixty years in captivity.

Wild parrots are pests. For example, farmers see cockatoos as nuisances because they eat crops. An interesting side note is that parrots are very ingenious. This is due to their great intelligence, estimated to equal that of porpoises and primates.

Parrots are liked as pets, due to their attractive coloring and ability to learn to talk. The popularity of pet parrots has brought some species close to extinction. In most countries, laws regulate their capture, export, and import. However, the laws are difficult to enforce.

—Sanford S. Singer

See also: Beaks and bills; Birds; Domestication; Feathers; Flight; Molting and shedding; Nesting; Respiration in birds; Wildlife management; Wings.

Bibliography

De Grahl, Wolfgang. *The Parrot Family: Parakeets, Budgerigars, Cockatiels, Lovebirds, Lories, Macaws.* New York: Arco, 1981. Contains much data on Psittacidae, with illustrations.

Higgins, P. J., ed. *Handbook of New Zealand, Australian, and Antarctic Birds.* Vol. 4, *Parrots to Dollarbirds.* New York: Oxford University Press, 1999. An up-to-date, comprehensive, and accurate synthesis of the ornithology of the area, including distribution maps, sonograms of calls and songs, and diagrams of life cycles. Extensive references, black-and-white illustrations.

Juniper, Tony. *Parrots: A Guide to Parrots of the World.* New Haven, Conn.: Yale University Press, 1998. A comprehensive work on parrots, including illustrations, natural history, maps, and a bibliography.

Sparks, John. *Parrots: A Natural History.* New York: Facts on File, 1990. An interesting illustrated book which covers the natural history of many parrot species.

PARTHENOGENESIS

Type of animal science: Reproduction
Fields of study: Embryology, reproductive science

Parthenogenesis is a process of asexual reproduction found mostly in lower animals. The egg divides and develops normally in the absence of fertilization.

Principal Terms

ARTHROPOD: a highly diverse phylum of animals that contains the crustaceans, insects, and spiders

DIPLOID: a cell containing two sets of chromosomes, usually one derived from the father and one derived from the mother; the normal condition of all cells except reproductive ones

HAPLOID: a cell (egg or sperm normally) containing one set of chromosomes, half the number found in nonreproductive cells

MEIOSIS: a special process of cell division consisting of two nuclear divisions in rapid succession, resulting in four haploid cells

OVUM: the female sex cell, or egg; usually undergoes meiosis following fertilization by a sperm

---

The term parthenogenesis comes from the Greek, meaning "to create from a virgin." Hence, it is sometimes popularly referred to as virgin birth. Parthenogenesis is a form of reproduction in which an ovum develops into a new individual without having been fertilized by a sperm. It is akin to other forms of asexual reproduction in that the offspring inherits all genetic information and traits from a single parent. The offspring is, therefore, genetically identical to the mother, and may be considered to be (in the broader sense of the word) a clone. The offspring may or may not look identical to the mother, since environment plays a large role in outward appearance. In fact, offspring may be male as well as female in those species where gender is determined by factors other than genetics alone, or by genetic factors different from those that occur in humans.

Natural Parthenogenesis

Parthenogenesis as a mode of reproduction is common in the animal kingdom up through the arthropods. In some, such as the aphid, parthenogenesis is the primary method of reproduction. Aphid eggs laid in the fall survive the winter and hatch as wingless females. These aphids mature in two weeks and reproduce parthenogenetically, giving live birth to daughters only, who in turn give live birth to the next generation of daughters. This cycle of parthenogenesis repeats itself through the spring and summer months until finally, in the fall, the next to last generation gives live birth to both females and males. These mate with each other and, rather than giving live birth, produce eggs that will hatch the following spring to start the cycle of parthenogenesis once again.

Other insects, such as honeybees and most species of ants, use both parthenogenesis and sexual reproduction side by side. In these species, unfertilized eggs give rise to haploid male drones, whereas fertilized eggs lead to diploid female workers and queens. The only insect known to be completely female (no males have ever been reported) is the white-fringed weevil. Since no males exist for fertilization, all reproduction is by parthenogenesis alone.

Ernest Everett Just

Born: August 14, 1883; Charleston, South Carolina
Died: October 27, 1941; Washington, D.C.
Fields of study: Cell biology, developmental biology, embryology, invertebrate biology, marine biology, reproduction science
Contribution: A pioneer in the fields of marine biology, cell biology, and fertilization, Ernest E. Just challenged many of the leading biological theories of his day. He significantly increased our understanding of the process of artificial parthenogenesis and the physiology of cell development.

Following graduation from Dartmouth College (1907) with a degree in zoology and special honors in botany and history, Ernest Just was offered a teaching position at Howard University. An excellent teacher with clear leadership skills, he was appointed head of the new Zoology Department in 1912. At Howard he also served as a professor in the Department of Physiology at the Medical School. He was awarded the first National Association for the Advancement of Colored People (NAACP) Spingarn Medal for exceptional contributions, in 1915, and received his doctorate in experimental embryology from the University of Chicago in 1916.

In 1909, Just was invited by Dr. Frank Lillie to spend his summer doing research at the Marine Biological Laboratory at Woods Hole, Massachusetts. His research there over the next twenty years, and the fifty papers which he published, led to international acclaim in several scientific fields. His work included the subjects of fertilization, experimental parthenogenesis, hydration and dehydration in cells, cell division, and the effects of magnetic fields and ultraviolet rays on chromosome number in animals and in altering the organization of the egg with respect to polarity. He successfully challenged Jacques Loeb's theory of artificial parthenogenesis, pushing the understanding of the process well ahead. Much of his early work is reported in his book, *Basic Methods for Experiments on Eggs of Marine Organisms* (1939), and in *General Cytology* (1924), which he coauthored with several of the other prominent biologists of the day, including Frank R. Lillie, T. H. Morgan, M. H. Jacobs, and E. G. Conklin.

A gifted, versatile, and productive researcher, Dr. Just was often referred to as a scientist's scientist.

Ernest Everett Just was a leading researcher, prior to World War II, in the fields of cell biology and parthenogenesis. (Associated Publishers, Inc.)

Even so, he experienced much of the racism of this period in history, and found the walls of prejudice and discrimination in America too high to climb. It was for that reason that he chose to move his research to Europe, and spent most of his time after 1933 working in Germany, France, and Italy.

While he was living in France, his masterpiece, *The Biology of the Cell Surface* (1939), was published in Philadelphia. This important book summarized his life's work on small marine animals, and provided new knowledge for people interested in biological research. Unfortunately, Germany invaded France in 1940, and Dr. Just was imprisoned by the Germans. His health had begun to fail before his imprisonment, and it worsened during this time. After he was released, he returned to the United States, but was too ill to continue his research. He died of cancer a year after his return.

Dr. Ernest Everett Just was recognized in 1997 by a U.S. Postal Service commemorative stamp (Black Heritage Stamp Series) for his contributions to biology.

—*Kerry L. Cheesman*

Parthenogenesis has also been reported in aquatic organisms, such as rotifers, marine nematodes, and some mollusks. It rarely occurs in higher orders of animals, except in certain birds. A good example of avian parthenogenesis is the turkey, where it appears to be fairly common. The work of M. W. Olsen and others in the 1960's and 1970's established the fact that parthenogenesis is fairly common in most fowl, including chickens, and usually occurs in eggs that are incubated for a few days in the absence of fertilization. Thus, there may be economic implications associated with parthenogenesis in these birds.

Two processes collectively referred to as abortive parthenogenesis are known to occur in the animal kingdom. In gynogenesis, a sperm enters the egg in the normal way, but the egg begins to develop without including the nucleus (genes) of the sperm. In androgenesis, the nucleus of the egg fails to begin development following fertilization, and the genes of the sperm take over the role of directing early development. Neither of these is considered a normal reproductive process.

Most naturally occurring parthenogenesis is associated with a modification of the meiotic process that restores the diploid condition to the egg. Little evidence exists concerning the stimuli needed or the process required to produce haploid individuals from haploid eggs (as in bees and ants).

Artificial Parthenogenesis

The phenomenon of natural parthenogenesis was discovered in the eighteenth century by Charles Bonnet. In 1900, Jacques Loeb (and T. H. Morgan, in a separate study) accomplished the first clear case of artificial parthenogenesis when he pricked unfertilized sea urchin eggs with a needle and found that normal embryonic development ensued. Through further studies he claimed that both butyric acid and hypertonic seawater (water with an excess of salt in it) were necessary to stimulate parthenogenesis.

Several years later, E. E. Just demonstrated that Loeb's understanding of what he had done was in error, and showed that hypertonic seawater alone could stimulate parthenogenesis. He also demonstrated that the larvae produced looked no different than those produced from eggs that had been normally fertilized by sperm in the ocean. Through this and other experiments on sea urchins, Just advanced the knowledge of the cellular events of parthenogenesis tremendously.

Since then other investigators have been able to stimulate parthenogenetic development in silk moths, starfishes, marine worms, and various amphibians by pricking them, shaking them, changing the temperature, or washing them in acids, alkalis, salt solutions, or other fluids. In turkeys and chickens, factors such as temperature, viruses, additives in feed, and even the age of the hen have all been shown to increase the incidence of artificial parthenogenesis.

In mammals, there is no good evidence that complete development can occur by stimulated parthenogenesis, although a few reports have appeared over the years. For instance, in 1936, Gregory Pincus induced parthenogenesis in rabbit eggs through temperature change and chemical agents, but development was not complete. Abortive or incomplete parthenogenesis does appear to be a normal, or at least common, occurrence in some species of mammals, although the reason for such developmental quirks remains unclear. No reliable reports of human parthenogenesis experiments have been published.

—*Kerry L. Cheesman*

See also: Asexual reproduction; Birth; Breeding programs; Cleavage, gastrulation, and neurulation; Cloning of extinct or endangered species; Copulation; Courtship; Determination and differentiation; Development: Evolutionary perspective; Estrus; Fertilization; Gametogenesis; Hermaphrodites; Hydrostatic skeletons; Mating; Pregnancy and prenatal development; Reproduction; Reproductive strategies; Reproductive system of female mammals; Reproductive system of male mammals; Sexual development.

Bibliography

Encyclopedia of the Animal World. London: Elsevier, 1972. A clearly written children's and youth encyclopedia, with a good article on parthenogenesis.

Hayden, Robert. *Seven African American Scientists.* Rev. ed. Baltimore: Twenty-first Century Books, 1992. A clear, well-written chapter on E. E. Just describes his contributions to parthenogenesis and cell biology.

Olsen, M. W. *Avian Parthenogenesis.* Washington, D.C.: U.S. Department of Agriculture, 1975. A detailed description of research on parthenogenesis in birds; somewhat technical for many readers.

Savage, Thomas F., and Elzbieta I. Zakrzewska. "A Guide to the Recognition of Parthenogenesis in Incubated Turkey Eggs." Corvallis: Oregon State University, 2000. www.orst.edu/dept/animal-sciences/poultry/. A good summary of research on turkey parthenogenesis and its implications for farmers and hatcheries. Includes an extensive reference list.

Waldbauer, Gilbert. *The Handy Bug Answer Book.* Detroit: Visible Ink Press, 1998. Clearly written for children, this book answers questions about reproduction and development and many other topics.

PELICANS

Type of animal science: Classification
Fields of study: Anatomy, ornithology, physiology, population biology, wildlife ecology

Pelicans, unmistakable due to their large size, long bills, and enormous throat pouches, have existed for forty million years.

Principal Terms

BROOD: to cover the young with the wings
CREST: tuft of feathers on the head
MANDIBLE: the upper or lower portion of the bird's bill
THERMALS: rising currents of warm air
TOTIPALMATE: having all four toes fully webbed

Pelicans are large water birds that live on seacoasts or in warm inland water habitats. There are seven generally recognized species. The coastal brown pelican lives from North Carolina to the Gulf of Mexico and the Caribbean, and from British Columbia to Chile, as well as the Galápagos Islands. The American white pelican inhabits inland habitats in the western, central, and southeastern United States, Mexico, and Canada. The remaining species live in Australia, Europe, Asia, and Africa.

Physical Characteristics of Pelicans

Pelicans rank among the largest of living birds, ranging in size from four to six feet in length, and weighing from four to sixteen pounds. The brown pelican is smallest, and the large eastern great white pelican has a wingspan of up to ten feet. They have short legs and broad, fully webbed feet, which act as powerful paddles in the water, and cause them to walk with an awkward waddle on land.

The pelican is famous for its huge, featherless throat pouch that is attached to the lower mandible. The pouch stretches as it fills with water when the bird is fishing, and can hold nearly three gallons. The pouch also functions as an evaporative cooling mechanism. The pelican opens it mouth and flutters its pouch, which keeps air flowing over the moist surface.

Plumage color varies among species and according to age. Pelicans can be predominantly white, black, brown, or gray, with markings on the head, wingtips, underfeathers, or tail. The legs and feet are orange, brown, or black, and the bill and pouch are reddish, orange, or black. During the breeding season, these body parts change color, and many pelicans develop a yellow patch on the chest, a distinctive crest, and a bright ring around the eyes. The American white pelican also grows a noticeable horny knob on its beak.

Feeding and Other Behaviors

Pelicans feed on many species of saltwater and freshwater fish, from tiny anchovies to fish weighing over a pound. Small crayfish, salamanders, frogs, and snakes are also consumed. The pelican thrusts its head and neck underwater and uses its pouch as a dip net to scoop up its prey. It drains out the water, then tilts its head back and swallows the fish whole. Many pelicans are social fishers, swimming in a circle to close in on the school, then all thrusting and dipping at once. Brown pelicans are solitary feeders, utilizing a spectacular plunge dive from about twenty feet above the water, with neck outstretched and bill pointed down. They hit the water hard, stunning the fish and trapping them in the pouch. Air sacs beneath

Pelicans have large, featherless pouches under their beaks that are used for scooping fish out of the water and for evaporative cooling. (PhotoDisc)

ship, the male uses various behaviors such as bowing, stretching, and pouch displaying to attract a female. Both engage in nesting, the male often gathering sticks and bringing them to the female to incorporate into the nest. Nests are built on the ground or in trees. One to three eggs are laid, and take thirty days to hatch. The chicks are born featherless, and the parents brood them, protecting them from the elements until their feathers come in. Unless the food supply is abundant, only the strongest chick survives. Young chicks eat the parent's regurgitated food. Older chicks feed by sticking their

the pelican's skin cushion its dive and help it to surface quickly.

Pelicans are graceful fliers. They take off against the wind, beating their wings and pumping their feet simultaneously, hopping until they are airborne. They fly at a relatively slow speed, with their necks retracted and their heavy bills tucked in and resting on the breast, and often glide on thermals to conserve energy. Pelicans regularly fly in flocks in a V-formation, flapping their wings in unison. They often travel a great distance in search of food.

Pelicans sleep standing or sitting on their bellies, with the head twisted back and the beak tucked into its feathers. Self-care activities include muscle exercises such as body shaking, wing flapping, tail wagging, leg stretching, bill throwing, and yawning. Pelicans groom themselves by splash bathing, preening with their beaks, and by rubbing their heads over the body to distribute waterproofing oil to their feathers.

Reproduction

Pelicans are warm weather birds, migrating in large flocks to nest in huge colonies. During court-

---

Pelican Facts

Classification:
Kingdom: Animalia
Phylum: Chordata
Subphylum: Vertebrata
Class: Aves
Order: Pelicaniformes
Suborder: Pelecani
Family: Pelecanidae
Genus and species: Pelecanus onocrotalus (eastern great white pelican), *P. erythrorynchos* (American white pelican), *P. occidentalis* (brown pelican), *P. rufescens* (pink-backed pelican), *P. philippensis* (spot-billed pelican), *P. crispus* (Dalmatian pelican), *P. conspicillatus* (Australian pelican)
Geographical location: North and South America, eastern Europe, Asia, Africa, Australia
Habitat: Coastal areas or inland lakes and rivers
Gestational period: Incubation lasts thirty days
Life span: Eight to twenty years
Special anatomy: Webbed feet; long, pointed bill with retractable throat pouch; air sacs under the skin; large wingspan

heads in the parent's pouch and throat. By twelve weeks of age, the chicks can fly and begin to hunt for themselves. By one year they have their full plumage. They begin reproduction at three to four years of age.

Endangerment

Pesticides, oil spills, habitat destruction, entanglement in fishing lines, and human disturbance have affected pelican populations in various parts of the world. In 1970, the U.S. Fish and Wildlife Service listed the brown pelican as an endangered species due to the heavy usage of dichloro-diphenyl-trichloroethane (DDT) and endrin. When pelicans ate DDT-contaminated fish, they produced thin-shelled eggs that crushed during incubation. Endrin was toxic. Populations improved after the pesticides were banned, but the birds are still endangered. The spot-billed pelican of Asia and the Dalmatian pelican of Europe and China also face difficulties.

—*Barbara C. Beattie*

See also: Beaks and bills; Birds; Domestication; Feathers; Flight; Molting and shedding; Nesting; Pollution effects; Respiration in birds; Wildlife management; Wings.

Bibliography

Johnsgard, Paul A. *Cormorants, Darters, and Pelicans of the World*. Washington, D.C.: Smithsonian Institution Press, 1993. A comprehensive treatise for the serious student, reflecting the current state of knowledge of all pelican species.

Patent, Dorothy Hinshaw. *Pelicans*. New York: Clarion Books, 1992. Written for young adults and up, this book provides accurate coverage and many illustrations, with emphasis on the North American white and brown species.

Stone, Lynn. *The Pelican*. New York: Dillon Press, 1990. Clear treatment for juvenile and young adult audiences, focusing on the North American species.

Williams, Ted. "Lessons From Lake Apopka." *Audubon* 101, no. 4 (July/August, 1999): 64-72. An eye-opening account of the destructive chain of events caused by farming and pesticide use which led to a massive die-off of pelicans and other fish-eating birds.

PENGUINS

Type of animal science: Classification
Fields of study: Anatomy, ornithology, physiology, population biology

Penguins, highly specialized for their flightless aquatic existence, have origins dating back sixty million years. Their unique lifestyle and behavior have long fascinated observers.

Principal Terms

CRÈCHE: penguin chicks grouped together for warmth and safety
KRILL: small, shrimplike sea creatures
MOLT: to lose feathers and grow new ones
ROOKERY: a nesting penguin colony

Penguins are flightless marine birds that dwell only in the southern hemisphere. They do not inhabit the Arctic, where polar bears live. There are seventeen generally recognized species of penguins. Six species, the Adelie, gentoo, chinstrap, rockhopper, king, and emperor penguins, live in the cold environments of the Antarctic region. The rest live in subantarctic and temperate regions. The macaroni, fiordland, Snares, erect-crested, yellow-eyed, fairy, and royal penguins live off the coasts of New Zealand and Australia and nearby islands. The Magellanic and Humboldt penguins live off the coast of South America. The African penguin lives off the southern coast of Africa, and the Galápagos penguin is native to the Galápagos Islands. Penguins spend much of their lives in the ocean, coming to shore mainly to breed.

Physical Characteristics

All penguins are black with white undersides, and are commonly described as wearing tuxedos. This color pattern acts as camouflage when the penguin is swimming, protecting it from predators. From underneath, the white belly blends with the bright water surface, and from above, the black back is indistinguishable from the dark water. Penguin species can be grouped according to common characteristics. Banded penguins have black and white stripe patterns on their chests and heads. The crested penguins all have bright yellow or orange plumes on their heads. Brushtail penguins have long stiff tail feathers. The king and emperor penguins have bright yellow and orange chest and head patches, and the yellow-eyed penguin has a yellow crown. The fairy penguin's feathers are bluish.

The emperor penguin is the largest, at nearly four feet tall and seventy-five pounds. The small fairy penguin is sixteen inches tall and about three pounds. All have solid, heavy bones that help them dive deeply into the water. They have streamlined bodies that move smoothly through the water as they pump their strong, flipperlike wings and steer using their webbed feet and tails as rudders. Penguins can hold their breath for many minutes at a time, and they frequently leap out of the water, porpoiselike, to take in more air.

On land, penguins walk with an awkward sideways waddle. Because their short legs are set back on their bodies, they stand erect and must hold out their flippers for balance. Penguins often toboggan themselves by flopping on their bellies and pushing with their flippers and feet.

Penguin feathers are tiny and stiff, overlapping to form a waterproof coat. An underneath layer of down helps to trap warm air and protect the penguin from the cold water and wind. Penguins of the Antarctic region have an insulating layer of blubber. Those in temperate climates often have to cool themselves down by ruffling their feathers and holding out their flippers. They can control the flow of blood to their unfeathered areas, such as the feet and under their flippers, which helps regulate their body temperature. Penguins preen their feathers regularly, to spread waterproofing oil from a gland near the tail.

Feeding Behavior and Enemies

Penguins are carnivores. They eat many types of small sea creatures, such as fish, squid, and krill. After locating a school they snatch quickly with their sharp beaks. The tongue and upper palate are covered with stiff spines that grip the slippery food and assist in moving it toward the throat. Penguins make several catches per dive, swallowing the prey whole along with some seawater.

Their specialized salt glands above each eye help them drain the extra salt they ingest.

Penguins usually enter and exit the water in large groups, to protect themselves from predators who often lurk near the shore. Their main enemies are sea lions, leopard seals, and killer whales. On land, adult penguins are safe. Petrels, skuas, gulls, and sheathbills hunt babies and eggs.

Reproduction

Most penguins follow an annual breeding cycle that begins in the spring, but timing varies according to species and climatic conditions. The Galápagos penguin will breed any month that the water temperature is right, and sometimes twice a year. Emperors begin their cycle in autumn, so there is a good food supply when their chicks hatch in spring. King penguins only reproduce twice every three years, because they follow a fifteen-month cycle.

Some penguins nest on the shore and others travel many miles inland to reach their rookeries, and they return to the same ones each year. Penguins are social, and one rookery may have thousands of penguins in closely spaced nests. They often squabble over nesting materials, mates, and territory. Nests are built of grass or stones on the ground, in rock crevices, or in burrows. The male engages in an ecstatic display to attract a female, pointing his beak, flapping his flippers, and squawking. Penguins usually mate for life, and in subsequent years the pair will greet each other affectionately. Two eggs are laid, and the parents take turns incubating for a few weeks at a time while the other leaves to feed. The incubating parent does not eat, and often loses a great deal of weight. Each penguin has a brood patch, an area of bare skin on its

---

Penguin Facts

Classification:
Kingdom: Animalia
Phylum: Chordata
Subphylum: Vertebrata
Class: Aves
Order: Sphenisciformes (penguins)
Family: Spheniscidae
Genus and species: Six genera and twenty-five species, including *Eudyptes chrysocome* (rockhopper penguin), *E. pachyrhychus* (fiordland penguin), *E. robustus* (Snares penguin), *E. sclateri* (erect-crested penguin), *E. chrysolophus* (macaroni penguin), *E. schlegeli* (royal penguin); *Spheniscus magellanicus* (Magellanic penguin), *S. humboldti* (Humboldt penguin), *S. mendiculus* (Galápagos penguin), *S. demersus* (African penguin); *Pygoscelis adeliae* (Adelie penguin), *P. antarctica* (chinstrap penguin), *P. papua* (gentoo penguin); *Aptenodytes patagonicus* (king penguin), *A. forsteri* (emperor penguin); *Megadyptes antipodes* (yellow-eyed penguin); *Eudyptula minor* (fairy penguin)
Geographical location: Along the coasts of Antarctica, Australia, New Zealand, South Africa, Chile, Peru, Argentina, and the Galápagos Islands
Habitat: Oceans and coasts in both cold and temperate latitudes
Gestational period: Incubation varies by species from thirty-three to sixty-four days
Life span: Twenty to thirty years
Special anatomy: Aerodynamic body shape; flippers; webbed feet; short, stiff, overlapped feathers; spiked tongue

The penguin's black-and-white coloration serves as camouflage, with the white underbelly making the swimming penguin difficult to locate from underneath while its black back blends into the water when seen from the top. (Digital Stock)

lower belly, that allows for better heat transfer to the eggs. Incubation varies from five weeks for the small fairy penguin to nine weeks for the emperor.

The king and emperor penguins are exceptions to the nesting rule. They lay only one egg, which they cradle on their feet instead of building a nest. They cover it with a flap of skin to keep it warm. Kings take turns incubating, while with emperors, only the male incubates.

Chicks are born down-covered, except for emperor chicks, which are naked. The parents brood them while they are young and feed them regurgitated food. When the chicks get too large for brooding, they huddle in crèches while the parents leave to hunt for food. When the chicks are grown and go off on their own, the parents molt. They cannot go into the water without their full coats of feathers, so they fast during this two- to four-week period.

—*Barbara C. Beattie*

See also: Beaks and bills; Birds; Domestication; Fauna: Africa; Fauna: Antarctica; Fauna: South America; Feathers; Flight; Molting and shedding; Nesting; Respiration in birds; Wildlife management; Wings.

Bibliography

Chester, Jonathan. *The World of the Penguin*. San Francisco: Sierra Club Books, 1996. A photojournalistic portrait, with descriptions of each species.

Lynch, Wayne. *Penguins of the World*. Willowdale, Ontario: Firefly Books, 1997. An entertaining and well-illustrated reference covering all aspects of penguin life.

Nagle, Robin. *Penguins*. New York: Gallery Books, 1990. Informative and thorough, geared toward the young adult and popular audience.

Stone, Lynn M. *Penguins*. Minneapolis: Lerner, 1998. Accurate, complete coverage for the juvenile audience.

Williams, Tony D. *The Penguins: Spheniscidae*. New York: Oxford University Press, 1995. A scholarly and authoritative review of the species.

pH MAINTENANCE

Type of animal science: Physiology
Field of study: Biochemistry

The maintenance of pH, which means "potential of hydrogen," is the regulation of acidity in the body fluids of animals; pH is the measure of acidity. Conditions that are too acidic or are not acidic enough will alter proteins such as enzymes and render them less effective. It is, therefore, vital that living organisms maintain proper acidity in body fluids.

Principal Terms

ACIDOSIS: a body fluid pH of less than 7.4 at 37 degrees Celsius

ALKALOSIS: a body fluid pH greater than 7.4 at 37 degrees Celsius, the opposite of acidosis

ANAEROBIC METABOLISM: metabolism in the absence of oxygen that leads to the production of lactic acid, a strong acid

PARTIAL PRESSURE: the pressure exerted by a specific gas in a mixture of gases such as the atmosphere; it is analogous to concentration

pH: the negative logarithm of the hydrogen ion concentration, with higher hydrogen ion concentrations indicating lower pH; the pH scale goes from 0 to 14, with a pH of 7 being neutral, values below 7 indicating acidity, and values above 7 indicating alkalinity

STRONG ACID: an acid that dissociates almost completely into its component ions; hydrochloric acid, for example, dissociates almost completely into hydrogen ions and chloride ions

WEAK ACID: an acid that does not dissociate to a great extent; carbonic acid, for example, dissociates to produce some ions, but most of the molecules remain in their original forms

---

The maintenance of pH is a collection of processes that regulate the hydrogen ion concentration of body fluids. The total body fluids account for about 60 percent of body mass. The body fluids are divided into several compartments. The largest volume of body fluid is in the intracellular compartment, and this consists of about 40 percent of body mass. The extracellular fluid consists of interstitial fluid (about 15 percent of body mass) and several specialized compartments. The largest of the specialized compartments is the vascular space, or blood volume. About half of the blood volume is in blood cells that belong to the intracellular volume. The other half is plasma, which makes up about 5 percent of the body mass. There are several other small extracellular spaces, such as the cerebrospinal fluid and the aqueous and vitreous humors of the eyes, that comprise smaller percentages of the extracellular volume. Each of these compartments is regulated in order to maintain normal pH. The pH levels of the different compartments can be very different. In mammals, the extracellular pH as measured in the plasma is generally 7.4. The specialized fluids within the extracellular compartment can differ. For example, the cerebrospinal fluid has a pH of 7.32. Intracellular pH can vary from cell type to cell type but tends to be lower than extracellular pH. For example, skeletal muscle has an intracellular pH of 6.89, while red blood cells have a pH of 7.20.

pH Maintenance

The first line of defense in the maintenance of pH

is through the use of chemical buffers. Buffers are chemicals that resist pH change by absorbing hydrogen ions from acidic solutions or contributing hydrogen ions to alkaline solutions. They are made by producing a mixture of a weak acid and the salt of a weak acid. An example of a buffer is given by the bicarbonate buffer system. The components are H_2CO_3 (weak acid) and $NaHCO_3$ (salt of the weak acid). When a strong acid such as HCl is added to a solution containing this buffer pair, the H^+ released from the highly dissociated HCl combines with HCO_3^-; that is completely dissociated from the $NaHCO_3$ to form weakly dissociated H_2CO_3. This prevents the large pH drop that would otherwise occur because the H_2CO_3 is a weak acid. This buffer system is primarily extracellular. The body fluids also contain intracellular buffer systems which can resist pH changes when acids are introduced.

Acids can be introduced in several ways. Exercise and oxygen deprivation cause a buildup of lactic acid as a result of anaerobic metabolism. Metabolism of sulfur-containing amino acids causes the production of sulfuric acid, another strong acid. Both aerobic and anaerobic metabolism cause the buildup of carbon dioxide, which combines with water to form carbonic acid; even though this is a weak acid, the large accumulation of carbon dioxide can lower pH.

Acids produced by any of these means are buffered by one of the buffer systems in the body. In addition to the bicarbonate ($NaHCO_3$) buffer system described above, there are two intracellular buffers that participate in pH maintenance. In the phosphate buffer system, monosodium phosphate (H_2NaPO_4), a weak acid, dissociates to produce H^+ and $HNaPO_4^-$. The H^+ is exchanged for Na^+ by the kidneys or buffered on protein, leaving the salt disodium phosphate (HNa_2PO_4). This salt dissociates into Na^+ and $HNaPO_4^-$. The $HNaPO_4^-$ can then absorb H^+ ions which have dissociated from strong acids to form H_2NaPO_4, which is weakly dissociated. The phosphate buffer system is the least important of the buffer systems because there is so little phosphate in the body. The most important buffer system in the body is pro-

tein. Some of the individual amino acids in proteins can act as weak acids and accept H^+ ions; the amino acid histidine is the most significant of these under the temperature and pH conditions of the body. Protein buffering is primarily an intracellular phenomenon, as the protein concentration in extracellular fluid is relatively low. The intracellular protein buffer makes up about three quarters of the total body buffering capacity.

Acid-Base Disturbances

The most important extracellular buffer is the bicarbonate buffer system discussed above. This system is aided directly by a second line of defense against pH disturbances. This is called physiological buffering. Physiological buffering refers to the fact that the supply of the most important components of the bicarbonate buffer system, carbon dioxide and bicarbonate, can be controlled. The organs responsible for this control are the lungs (carbon dioxide) and kidneys (bicarbonate). For example, during exercise, the buildup of lactic acid consumes bicarbonate (HCO_3^-) and lowers pH. This is called a metabolic acidosis. The low pH stimulates the rate of breathing, which in turn results in increased elimination of carbon dioxide by the lungs. This lowers the partial pressure (concentration) of carbon dioxide in the blood, reducing the amount of carbonic acid (H_2CO_3) that can dissociate to form H^+. The reduction of carbon dioxide is called a respiratory alkalosis. This respiratory alkalosis compensates for the metabolic acidosis caused by the depletion of bicarbonate. The kidneys provide the final correction for the initial metabolic acidosis. These organs accomplish this in two ways: reabsorption of bicarbonate, and secretion of H^+ into the urine in exchange for Na^+. When bicarbonate concentration in the extracellular fluid is reduced by a metabolic acidosis, the hormone aldosterone, produced by the adrenal glands, stimulates the secretion of H^+ and reabsorption of bicarbonate. This raises both the pH and bicarbonate concentration back to normal. The increased pH relaxes stimulation of breathing, allowing breathing rate to slow and the partial pressure of carbon dioxide to return to normal.

The above is an example of one of the four basic types of acid-base disturbance: metabolic acidosis (bicarbonate concentration reduced), metabolic alkalosis (bicarbonate concentration increased), respiratory acidosis (an increase in the partial pressure of carbon dioxide), and respiratory alkalosis (a decrease in the partial pressure of carbon dioxide). As before, metabolic acidosis is compensated for by respiratory alkalosis. This is part of a general pattern in which one condition is compensated for by the opposite condition. Another example would be a respiratory acidosis, lowering the pH and stimulating bicarbonate retention and H+ excretion in the kidneys to produce a compensatory metabolic alkalosis.

Body Temperature and Ion Exchange
Temperature also affects pH. This is usually of no consequence to warm-blooded animals such as mammals and birds, which have a constant body temperature. It is, however, of great consequence to cold-blooded animals such as fish, amphibians, and reptiles, in which body temperatures vary with the environmental temperature. There is a decrease in pH by about 0.015 for each degree increase in temperature. This means that the pH at 5 degrees Celsius is 7.88, or 0.48 pH units higher than at mammalian body temperature (pH 7.40 at 37 degrees Celsius). This does not mean that cold-blooded animals do not regulate pH. While the regulation differs from that of warm-blooded animals in that pH varies, it is still regulated in the sense that it varies in a very predictable manner. By controlling partial pressures of carbon dioxide and bicarbonate ion concentrations in much the same way that mammals do, cold-blooded vertebrates maintain their body fluid pH at a constant 0.6 to 0.8 pH units higher than the pH of pure water at the same temperature. It is, thus, the relative alkalinity that is regulated, rather than the pH.

While the kidneys of mammals are intimately involved in pH maintenance, the situation in lower vertebrates can be much different. Most fish use ion exchange transport systems in their gills to regulate pH. Specialized cells in the exterior lining of the gill surfaces transport sodium ions (Na^+)

from the water bathing the animals into the blood in exchange for H^+. Similarly, chloride ions (Cl^-) are transported into the animal in exchange for HCO_3^-. When fish become acidotic, they increase Na^+/H^+ exchange and decrease Cl^-/HCO_3^- exchange. This results in a net elimination of H^+ and a net conservation of HCO_3^- to elevate the pH of the body fluids. While the kidneys of fish do participate in pH maintenance, they seem to play only a minor role, usually about 5 percent of the total regulatory effort. The kidneys of amphibians also appear to play a minor role in pH maintenance. The major regulatory organs in these animals are the skin and urinary bladder. Aquatic salamander larvae increase Na^+/H^+ exchange and decrease Cl^-/HCO_3^- exchange across their skin to regulate internal pH. Toads, which are primarily terrestrial and thus seldom in contact with water, appear to use their urinary bladders for pH regulatory ion exchanges. Relatively little is known about the role of the reptilian kidney in pH maintenance. The alligator kidney plays a major role in acid elimination. The urinary bladder of turtles also participates in pH regulation in much the same manner as the urinary bladder of the toad. The bird kidney plays a major role in pH maintenance, and its urine pH values are similar to the pH values of mammalian urine.

Studying pH Maintenance
The regulation of pH is usually studied by measuring the pH and the partial pressure of carbon dioxide of the blood. This measurement of the regulation of extracellular pH is a reflection of the overall acid-base status of the individual being studied. The first requirement is to be able to sample blood from an undisturbed, resting subject. This usually means an indwelling arterial cannula, through which blood from an artery can be sampled. The blood is then injected into a chamber in a blood-gas analyzer containing a glass pH electrode that measures the pH of the plasma. Additional blood is injected into another chamber that contains oxygen and carbon dioxide electrodes that measure the partial pressures of these two gases. Bicarbonate concentration can be cal-

culated from the mathematical relationship among pH, the partial pressure of carbon dioxide, and the bicarbonate concentration when the former two are known. When making these kinds of measurements, it is important to measure the pH and partial pressure of the blood at the same temperature as the animal. Thermostated electrodes are used for this. The calculation used to estimate bicarbonate concentration requires the use of two constants which are also temperature-dependent. It is important to use the constants that are appropriate to the temperature of the animal and the measurement conditions.

Alternatively, bicarbonate concentration can be approximated by measuring the total amount of carbon dioxide (the sum of the carbon dioxide, carbonic acid, bicarbonate, and carbonate concentrations). Under physiological conditions, more than 99 percent of the carbon dioxide is in the form of HCO_3^-. When strong concentrated acid (HCl) is added to a measured volume of blood or plasma, all the carbonic acid, bicarbonate, and carbonate are released as carbon dioxide. This carbon dioxide can then be measured with a carbon dioxide electrode. The resulting total carbon dioxide, minus the concentration of carbon dioxide measured before adding the HCl, is very close to being equal to the concentration of bicarbonate.

While there are a number of indirect methods for measuring intracellular pH, the best results are achieved directly with micro pH electrodes which can be used to impale single cells. These techniques must be done on isolated tissues or anesthetized, restrained animals, whose acid-base status does not, in the least, resemble that of resting, undisturbed animals. The techniques are useful for studying pH maintenance at the cellular level and useful extrapolations can be made to the whole animal.

The usual approach to the study of pH maintenance is to induce a disturbance in the pH of an experimental animal and then to follow changes in the pH, partial pressure of CO_2, and the concentration of HCO_3^- in the blood. Breathing rates and volumes of air exchanged by the lungs can be measured to determine the contribution of the lungs to pH maintenance. For example, increasing the rate of breathing will increase the rate of elimination of carbon dioxide and lower the partial pressure of this gas in the blood. Urine collection can be done to assess the contribution of the kidneys to pH maintenance, and in aquatic animals that do not rely heavily on their kidneys for pH maintenance, changes in the composition of the water in which the animals are situated can be used to assess the contribution of the gills (fish) or skin (amphibians) to pH maintenance.

Examples of experimental manipulations that can produce pH disturbances in animals include infusion of acids and bases into the blood to produce metabolic acidosis and alkalosis respectively. Alternatively, animals can be exercised to elevate lactic acid in their blood to produce a metabolic acidosis. Respiratory acidosis and alkalosis can be induced by artificially altering breathing rates. Alternatively, elevation of the partial pressure of carbon dioxide in the air or water of an experimental animal will produce a respiratory acidosis.

pH Maintenance and Homeostasis

Maintenance of pH is part of the overall phenomenon of homeostasis or constancy of the internal environment in a changing external environment. The primary reason for maintaining consistent acid-base balance is the need to keep constant the conformation (shape) of proteins. Proteins are very sensitive to changes in pH, and the usual consequence of such changes is a change in the shapes of proteins. Protein conformation is critically important to proper protein function.

All enzymes (proteins that cause necessary biochemical reactions to occur) have very specific shapes that must be maintained in order for them to function properly. Changes in pH cause alterations in those shapes and compromise the effectiveness of the enzymes. Many other proteins require certain conformational states. Cell membranes contain proteins that determine which chemicals can enter and leave the cell's interior. These proteins are called carriers if they actively transport specific substances into or out of the cell interior. Alternatively, membrane proteins can act as specific channels that allow passive movement of only specific molecules or ions across the membrane. Both carriers and channels are sensitive to pH-induced conformational changes. There are other membrane-bound proteins called receptors that respond best to hormones and other chemical messengers when the pH is maintained at the norm. Proteins in the immune system, such as antibodies, also require proper pH in order to function optimally.

Cold-blooded animals do not regulate their pH to a specific point, but instead regulate the difference between their pH and the pH of pure water at the same temperature. This relative alkalinity ensures constant protein conformation in the same way that constant pH at constant temperature ensures constant protein conformation in warm-blooded animals, because the weak acid nature of proteins causes them to be partially dissociated from H^+ in the pH and temperature range experienced by animals. The conformation of protein is determined, to a large extent, by the net electrical charge (distribution of + and − charges over the protein). The net charge is influenced by the degree of dissociation of the protein and H^+. As discussed above, the dissociation in protein that is important to pH maintenance is the dissociation of H^+ from histidine. As long as the dissociated fraction of histidine remains constant, the protein conformation will remain constant. As long as the relative alkalinity is maintained, the fractional dissociation of histidine remains constant and thus protein conformation does not change. In reality, the regulation of mammalian pH at 7.4 (at 37 degrees Celsius) is one specific example of this general rule. The relative alkalinity of mammalian blood at 37 degrees is the same as the relative alkalinity of fish blood at 5 degrees or of reptile blood at 42 degrees.

—Daniel F. Stiffler

See also: Cell types; Gas exchange; Osmoregulation; Respiration; Water balance in mammals.

Bibliography

Berne, R. M., and M. N. Levey. *Principles of Physiology*. 3d ed. St. Louis: C. V. Mosby, 2000. An introductory-level text on systemic physiology that presents the physiology of the lungs and kidneys in a straightforward, nonmathematical way. There is also a section on acid-base balance that will be useful to the beginning student. The major emphasis of the book is regulation.

Davenport, Horace W. *The ABC of Acid-Base Chemistry*. 6th ed. Chicago: University of Chicago Press, 1974. Although the title refers to chemistry, the subject matter is the chemistry of body fluids pertaining to pH regulation. This is an excellent quantitative approach to the understanding of pH maintenance in detail. It is the introductory text of choice for the student wishing to explore further the intricacies of acid-base balance.

Heisler, N., ed. *Acid-Base Regulation in Animals*. Amsterdam: Elsevier, 1986. This is a very detailed survey of acid-base balance in all groups of animals (as opposed to treatments dealing with mammalian-human species). It is extremely comprehensive and contains the most complete bibliographic information available on acid-base balance. It is appropriate for the advanced student.

Jacquez, John A. *Respiratory Physiology*. New York: McGraw-Hill, 1979. This is an advanced text on respiratory physiology. It is included here to provide a detailed reference to respiratory physiology topics which are central to acid-base balance. Of particular interest to the reader is the treatment of control of breathing rate by neural and chemical means.

Keyes, Jack L. *Fluid, Electrolyte, and Acid-Base Regulation*. Boston: Jones and Bartlett, 1999. An intermediate-level text dealing with pH maintenance and with an emphasis on medical aspects of fluid and acid-base balance and imbalance. There is a very clear explanation of the processes of ion-ion transport, which is useful in understanding the role of ion exchange in pH maintenance. Presents clinical case histories in a clear, understandable manner.

Lowenstein, Jerome. *Acid and Basics: A Guide to Understanding Acid-Base Disorders*. New York: Oxford University Press, 1993. A guide aimed at medical students and young physicians, but provides a good, basic guide to the concepts for anyone with a background in chemistry and physiology.

Stewart, P. A. *How to Understand Acid-Base: A Quantitative Acid-Base Primer for Biology and Medicine*. New York: Elsevier, 1981. An advanced text that derives the basis of acid-base balance in terms of ion balance in general. This is a controversial book that looks at pH maintenance from a very different perspective. In this treatment, H+ transport is considered not to exist and all changes in pH result from changes in the distribution of so-called strong ions such as sodium, potassium, calcium, magnesium, and chloride.

Valtin, Heinz. *Renal Function: Mechanisms Preserving Fluid and Solute Balance in Health*. 3d ed. Boston: Little, Brown, 1995. This book is primarily about the function of the kidneys. It gives one of the clearest presentations of their role in pH maintenance available. The section on pH maintenance in general will also be useful to the intermediate reader.

PHEROMONES

Type of animal science: Physiology
Fields of study: Ecology, ethology, evolutionary science, invertebrate biology

Pheromones are chemicals or mixtures of chemicals that are used as messages between members of a species. They are integral parts of the social communication within most species. They may prove to be of great value in pest control and in enhancing agricultural production.

Principal Terms

ALLELOCHEMIC: a general term for a chemical used as a messenger between members of different species; allomones and kairomones are allelochemics, but hormones and pheromones are not

ALLOMONE: a chemical messenger that passes information between members of different species, resulting in an advantage to the sender

KAIROMONE: a chemical messenger that passes information between members of different species, resulting in an advantage to the receiver

PRIMER PHEROMONE: a chemical that generates an often subtle physiological response in another member of the same species; a behavioral response may follow

SEMIOCHEMICAL: a chemical messenger that carries information between individual organisms of the same species or of different species; pheromones and allelochemics are semiochemicals, but hormones are not

VOLATILES: chemical compounds that are vapor or gas at environmental temperatures or that readily release many of their molecules to the vapor phase

Pheromones are chemical signals. Originally defined to include only signals between individuals of the same animal species, the term has been generalized to designate any chemical or chemical mixture that, when released by one member of any species, affects the physiology and/or behavior of another member of the same species. "Pheromone" is one of a set of terms developed to express the chemical interactions in ecological communities, and it can best be understood in the context of that set of terms.

In this context, pheromones are semiochemicals that carry information between members of a single species. To do so, the pheromone must be released into the atmosphere or placed on some structure in the organism's environment. It is thus made available to other members of the species for interpretation and response. It is also available to members of other species, however, so it is a potential allelochemic.

For example, klipspringer antelope mark vegetation in their environment with a chemical secreted from a special gland. Other klipspringer investigate the marks to gather information on the marking individual. Ticks that parasitize the klipspringer, however, are also attracted to the chemical marks and thus increase their chance of attaching to a host when the mark is renewed or when another klipspringer investigates it. The tick is using the pheromone as a kairomone. Pheromones can act as allomones as well, though the interaction is sometimes less direct. Bolas spiders produce the sex-attractant pheromone of a female moth and use it to lure male moths to a trap. The spider uses the moth pheromone as an allomone.

To appreciate fully the complexity of the interactions under consideration, it is important to remember that a pheromone may also be acting as a kairomone, allomone, or hormone. There are two general types of pheromone: those that elicit an immediate and predictable behavioral response, called releaser or signal pheromones, and those that bring about a less obvious physiological response, called primer pheromones (because they prime the system for a possible behavioral response). Pheromones are also categorized according to the messages they carry. There are trail, marker, aggregation, attractant, repellant, arrestant (deterrent), stimulant, alarm, and other pheromones. Their functions are suggested by the terms used to name them.

Pheromone Composition and Function
The chemical compounds that act as pheromones are numerous and diverse. Most are lipids or chemical relatives of the lipids, including many steroids. Even a single pheromonal message may require a number of different compounds, each present in the proper proportion, so that the active pheromone is actually a mixture of chemical compounds.

Different physical and chemical characteristics are required for pheromones with different functions. Attractant pheromones must generally be volatile to permit atmospheric dispersal to their targets. Many female insects emit sex-attractant pheromones to advertise their readiness to mate. The more widely these can be dispersed, the more males the advertisement will reach. On the other hand, many marking pheromones need not be especially volatile because they are placed at stations which are checked periodically by the target individuals. The klipspringer marking pheromone is an example. Some pheromones are exchanged by direct contact, and these need not have any appreciable volatile component. Many mammals rub, lick, and otherwise contact one another in social contexts and exchange pheromones at these times.

Specificity also varies for pheromones with different functions. Sex attractants usually need to be very specific, directed only to members of the opposite sex and the same species. Alarm phero-

mones, on the other hand, need not be so specific. These pheromones simply alert other members of the same species to a disturbance. It is usually harmless, and sometimes even helpful, to alert members of other species as well. In keeping with this argument, related groups of ant species produce species-specific sex-attractant pheromones: Each female attracts only males of its own species. In contrast, alarm pheromones of any species in the group will stimulate defensive reactions in individuals in many of the species.

Pheromonal systems are not organized in any standard way in different species. Many mites and ticks also have nonspecific alarm pheromones. Surprisingly, some groups also have nonspecific sex-attractant pheromones. In these cases, the specificity necessary for reproductive efficiency is generated by species-specific mating stimulant pheromones. These pheromones are produced by a female after males have been attracted to her. They stimulate mating behavior, but only in males of the same species. Thus the required specificity is achieved by a different mechanism. This is only one of many examples of the diversity of pheromonal schemes among organisms.

Pheromone Sources and Receptors
The sources of pheromones are also diverse. Some pheromones are produced by specialized glands; many insect species have glands specialized for the production of pheromones. One example is the harvester ant's alarm pheromone, which is produced in the mandibular gland at the base of the jaws. Other pheromones seem to be byproducts of other bodily functions. The lipids of mammalian skin are probably primarily important in waterproofing and in maintaining the outer layer of the skin, but many also function as pheromones.

The reproductive tract is an important source of pheromones in many species. These usually act as sex-attractant or sex-stimulant pheromones or as signal pheromones that give information on the sexual state of the emitter. The urine and feces of many species also contain pheromones which are used to mark territory boundaries and to transmit other information about the marking individual. Many pheromones seem to be produced not by the sending organism alone but by microorganisms living on the skin or in the glands or cavities of the sender's body. These microbes convert products of their host into the actual signal molecules, or pheromones, used by the host.

The receptors for pheromones are also of many different types, and the chemical receptors for taste and smell are often involved. In vertebrates, the vomeronasal organ (Jacobson's organ) seems to be an important receptor for many pheromones. It is a pouch off the mouth or nasal passages, and it contains receptors similar to those for smell. It is nonfunctional in humans, but it functions in more primitive mammals and seems to be of great importance to snakes and other reptiles. Insects and other invertebrates have many specialized structures for receiving pheromonal messages. Perhaps the best-known example is the feathery antennae of many male moths, which are receptors for the female moth's sex-attractant pheromone. Some pheromones seem to be absorbed through the skin or internal body linings and to bring about their effects by attaching to some unknown internal receptor.

Pheromones are widespread in nature, occurring in most, if not all, species. Most are poorly understood. The best-known are those found in insects, partly because of their potential use in the control of pest populations and partly because the relative simplicity of insect behavior allowed for rapid progress in the identification of pheromones and their actions. Despite these advantages, much remains to be learned even about insect pheromones. Mammalian pheromones are not as well known, although they may also be of economic importance. The more complex behavior of mammals makes the study of their responses to pheromones much more difficult.

Behavioral and Chemical Research

Both behavioral and chemical techniques are required in order to study pheromones and other semiochemicals. The observation of behavior, either in nature or in captivity, often suggests pheromonal functions. These hypothesized functions are then tested by presenting the pheromone to a potentially responsive organism and observing the response. Situations may be arranged which demand the subject's response to a particular pheromone under otherwise natural conditions. Alternatively, the organisms may be observed in enclosures to help control the experimental context. The presentation of the hypothetical pheromone may be in the form of another organism of the same species or some structure to which the presumed pheromone has been applied. The observed response (or lack of response) gives information on the status of the presented chemical as a pheromone in that behavioral context.

While the pheromonal function of secretions from a gland or other source can be determined from these behavioral tests, the tests can give information on specific chemical compounds only if the compounds can be isolated and identified. The isolation and identification of pheromonal compounds are challenging because of the great complexity of the secretions in which they are found and the exceptionally small amounts that are required to elicit a response. Many separation and identification techniques are used. One of the most powerful is a combination of gas chromatography and mass spectrometry.

Gas chromatography is used to separate and sometimes to identify chemicals that are volatile or can be made volatile. The unknown chemical is mixed with an inert gas, called the mobile phase of the gas chromatography system. This mixture is passed through a tube containing a solid, called the stationary phase. The inert gas does not interact with the solid; however, many of the compounds mixed with it do, each to an extent determined by the characteristics of the compound and the characteristics of the stationary phase. Some members of the mixture will interact very strongly with the solid and so move slowly through the tube, whereas others may not interact with the solid at all and so pass through rapidly. Other members of the mixture interact at intermediate strengths and so spend intermediate amounts of

Attempts have been made to use sexual attractant pheromones for controlling insect pests, such as gypsy moths. (Rob and Ann Simpson/Photo Agora)

time in the tube. The different compounds are recorded and collected separately as they exit from the tube.

For identification, the compounds are often passed on to a mass spectrometer. In mass spectrometry the compound is broken up into electrically charged particles. The particles are then separated according to their mass-to-charge ratio, and the relative number of particles of each mass-to-charge ratio is recorded and plotted. The original compound can usually be identified by the pattern produced under the specific conditions used. After separation and identification, the individual chemicals may be subjected to behavioral studies.

Uses for Pheromones

Pheromones and other semiochemicals are of interest simply from the standpoint of understanding communication between living things. In addition, they have the potential to provide effective, safe agents for pest control. The possibilities include sex-attractant pheromones to draw pest insects of a particular species to a trap (or to confuse the males and keep them from finding females) and repellant pheromones to drive a species of insect away from a valuable crop species. One reason for the enthusiasm generated by pheromones in this role is their specificity. Whereas insecticides generally kill valuable insects as well as pests, pheromones will often be specific for one or a few species.

These chemicals were presented as a panacea for insect and other pest problems in the 1970's, but most actual attempts to control pest populations failed. Many people in the field have suggested that lack of understanding of a particular pest and its ecological context was the most common cause of failure. They maintain that pest-control applications must be made with extensive knowledge and careful consideration of pest characteristics and the ecological system. In this context, pheromones have become a part of integrated pest management (IPM) strategies, in which they are used along with the pest's parasites and predators, resistant crop varieties, insecticides, and other weapons to control pests. In this role, pheromones have shown great promise.

Some consideration has been given to the control of mammalian pests with pheromones, though this field is not as well developed as that of insect control. Pheromonal control of mammalian reproduction has received considerable attention for other reasons: Domestic mammals are of great economic importance, and many wild mammalian species are endangered to the point that captive breeding has been attempted. The manipulation of reproductive pheromones may be used to enhance reproductive potentials in both cases. The complexity of mammalian behavioral and reproductive systems, however, and the subtle changes brought about by mammalian phero-

mones present a particular challenge. As with insect pest control, the key to progress is a complete understanding of the entire system being manipulated.

Pheromones and other semiochemicals are of great potential economic importance as substitutes for or adjuncts to toxic pesticides in pest management. Mammalian reproductive pheromones are being explored as tools to enhance reproductive efficiency in domestic and endangered mammals. A complete understanding of the complex roles of pheromones in each of the systems being managed is necessary for success in all these endeavors.

—*Carl W. Hoagstrom*

See also: Communication; Courtship; Ecology; Endocrine systems of invertebrates; Endocrine systems of vertebrates; Ethology; Hormones and behavior; Insect societies; Insects; Mammalian social systems; Mating; Sexual development; Territoriality and aggression.

Bibliography

Agosta, William C. *Chemical Communication: The Language of Pheromones*. New York: Scientific American Library, 1992. An easy-to-read overview of pheromones and their functions in the animal world. Requires some background in physiology and zoology.

Albone, Eric S. *Mammalian Semiochemistry: The Investigation of Chemical Signals Between Mammals*. New York: John Wiley & Sons, 1984. Written as an introduction for scientists, but accessible to nonspecialists for the most part. Covers terminology; most known mammalian pheromones; their chemistry, origins, and effects; mammalian chemical receptors; and the chemical methods used in semiochemistry. Extensive bibliography. Well indexed.

Booth, William. "Revenge of the 'Nozzleheads.'" *Science* 239 (January 8, 1988): 135-137. A news article written for a general audience. Describes the history of pheromones as pest-control devices, providing basic information on pheromones, examples of successes, and continuing problems.

Carde, Ring T., and Albert K. Minks, eds. *Insect Pheromone Research: New Directions*. New York: Chapman and Hall, 1997. Proceedings of a symposium, with articles covering control of pheromone production, sensory processing of pheromone signals, neuroethology of pheromone-mediated responses, use of pheromones in direct control, and the evolution of pheromone communication.

Mayer, Marion S., and John R. McLaughlin. *Handbook of Insect Pheromones and Sex Attractants*. Boca Raton, Fla.: CRC Press, 1991. Encyclopedic coverage of the chemical composition of insect mating pheromones, written from the perspective of agricultural pest control.

Mitchell, Everett R., ed. *Management of Insect Pests With Semiochemicals: Concepts and Practice*. New York: Plenum, 1981. Addressed to pest management professionals, but much is accessible to nonspecialists. Contains forty articles on insect pest-control efforts and experiments. Boll weevil, gypsy moth, spruce budworm, bark beetles, southern pine beetles, and other pests are considered. Population monitoring, mass trapping, and mating disruption are the management techniques featured. Articles are generally good, and many references are given after each article. Cumulative index.

Mittler, Thomas E., Frank J. Radovsky, and Vincent H. Resh, eds. *Annual Review of Entomology*. Vol. 46. Palo Alto, Calif.: Annual Reviews, 2000. Written for entomologists, but most articles are accessible to nonspecialists. Each article lists many references, and each volume is well indexed.

Nordlund, Donald A., Richard L. Jones, and W. Joe Lewis, eds. *Semiochemicals: Their Role in Pest Control*. New York: John Wiley & Sons, 1981. Addressed to scientists, but accessible to nonspecialists. Fourteen review articles cover history, terminology, allelochemics, pheromones, chemistry, and evolution. Each chapter lists references. Well indexed.

Vandenbergh, John G., ed. *Pheromones and Reproduction in Mammals*. New York: Academic Press, 1983. Written for both the general reader and the specialist. Nine articles review mammalian pheromones, including mother-young interaction, regulation of puberty and the ovarian cycle, pregnancy block, effect of pheromones on hormones, detection of pheromones, and use of pheromones to control reproduction in domestic animals. References after each chapter. Good index.

Wilson, Edward O. "Pheromones." *Scientific American* 206 (May, 1963): 100-114. An early summary article on the concept of pheromones. Pheromones of the fire ant are given most consideration, but human, mouse, cockroach, gypsy moth, and other pheromones are considered. Explores differences in volatility, longevity, and specificity among pheromones. Excellent.

PHYLOGENY

Type of animal science: Evolution
Fields of study: Anthropology, archaeology, evolutionary science, population biology, systematics (taxonomy), zoology

Phylogeny is the study of the evolutionary history of an animal or of groups of animals. It is the methodology that traces the lines of descent among living things and tries to classify them in some systematic order based on ancestry.

Principal Terms

DEOXYRIBONUCLEIC ACID (DNA): the genetic material of most living organisms

EVOLUTION: the process by which the variety of plant and animal life has developed over time from the most primitive to the most complex life-forms

KINGDOM: the highest category into which organisms are classified; there are now believed to be five kingdoms

MOLECULE: the smallest part of a chemical compound

PHYLUM: a group that consists of several closely related classes

SPECIES: a group of organisms that can produce offspring with each other

---

Phylogeny traces the history of life on earth through the study of how animals and plants have developed over time and how they are related to one another. It is similar to taxonomy, the science of classifying organisms based on their structure and functions. Taxonomists create family trees of living and extinct species in order to discover the origins and lines of descent of various forms of life. Very few family trees are complete to their fossil origins, however, because of gaps in the fossil record. The first system of classification was devised by the eighteenth century Swedish scientist Carolus Linnaeus. Linnaeus classified life-forms based on their appearance; the more they resembled each other in size, shape, and form, he believed, the more closely they were related.

The theory of evolution developed by nineteenth century naturalist Charles Darwin was based on accumulated changes over time through natural selection and led to a new way of looking at the history of life and to the development of phylogeny as a method of classification. The new system classified the living world by similarities in ancestry rather than appearance. Life histories of species were derived from the study of comparative anatomy (the search for common features in different species), embryology (the study of the development of life from the egg to birth), and biochemistry (the study of invisible chemical characteristics of cells that link species that can look very dissimilar). Modern technology allows scientists to measure differences in deoxyribonucleic acid (DNA) molecules among species. DNA carries the genetic material of an organism and plays a central role in heredity. The degree of difference between two species helps scientists determine how much modern species have changed from their ancestors and to estimate when important events in evolution occurred. If a sufficient fossil record exists, biochemists can determine the timing of a major change that led to the development of a new characteristic such as a longer neck in giraffes or a different size eyeball in a bat.

Classifying Earth's Species

Studies of evolutionary change suggest that anywhere from three million to twenty million species now exist, with millions more having become extinct. The earth began about 5 billion years ago, and it has been occupied by living organisms for about 3.5 billion years. For about 2.5 billion years, the planet was populated only by single-celled bacteria. The earliest forms of life evolved by chance during a long period of chemical reactions. Random encounters between chemicals in the seas and the atmosphere produced amino acids and proteins. Small drops of proteins, made up of carbon, hydrogen, oxygen, nitrogen, and sulfur, somehow bound together and became organisms; that is, they began to reproduce themselves, which is the quality that makes them life-forms.

Every living thing is given a two-part Latin name under a system of binomial nomenclature. This name is usually based on an evolutionary, phylogenetic relationship. The first part of the name indicates the organism's genus; the second part identifies the specific species. For example, *Acer saccharum*, *Acer nigrum*, and *Acer rubrum* are all kinds of maple trees. They share the genus *Acer*, and their species names mean sugar, black, and red, respectively. No rules govern species names, though frequently they refer to a prominent feature such as color or characteristic, such as in the case of *Homo sapiens*, the Latin name for human beings, which literally means "the thinking one."

Of the five kingdoms of life-forms, one (containing the bacteria) is made up of prokaryotes; the other four are eukaryotes. Prokaryotes do not have cell nuclei; eukaryotes do. In prokaryotes, cells are smaller and simpler in structure, and the DNA is not organized into chromosomes. Chromosomes, the threadlike structures found in the cell nucleus of all eukaryotes, determine the characteristics of individual organisms, with offspring receiving an equal number of chromosomes from each parent. In bacteria, however, reproduction takes place by a simple division of the parent into two masses that eventually separate. Bacteria do not share chromosomes with other members of the species; in other words, reproduction is asexual.

The Nonanimal Kingdoms

All bacteria belong to the kingdom Prokaryotae. They are in terms of total numbers the most suc-cessful form of life in the universe. Some bacteria, which are germs, can cause disease, while some bacteria, such as antibiotics, can be used to cure disease. Bacterium-like fossils that were found in rocks from an Australian gold mine are about 3.5 billion years old. Bacteria are found in every climate and habitat, and are the first to invade and populate new habitats. Seventeen phyla of bacteria are known to exist, and none can be seen without a microscope.

The four other kingdoms of living things and the common names given to their most important phyla are Protoctista (algae, protozoa, slime molds), Fungi (mushrooms, molds, lichens), Animalia (sponges, jellyfish, flatworms, ribbon worms, rotifers, spiny-headed worms, parasitic nematodes, horsehair worms, mollusks, priapulid worms, spoon worms, earthworms, tongue worms, velvet worms, insects, beard worms, starfish, arrowworms, and chordates), and Plantae (mosses, ferns, and pine-bearing and flowering plants). Phyla are based on similarities in evolutionary development. Life-forms are divided into ninety-two phyla: There are seventeen phyla in the kingdom Prokaryotae; twenty-seven in Protoctista; five in Fungi; thirty-three in Animalia, and ten in Plantae. Each phylum is divided into classes, then orders, families, genera (the singular form is genus), and finally, species. The last two divisions are based on the most recent evolutionary differences. Some phyla have only a few genera and species, while others, such as those in Animalia, have millions. A species consists of a group of similar individual organisms that can breed and reproduce offspring. A genus contains similar species, but members of a genus cannot reproduce; only members of a species can.

Protoctista (also called Protista) means the "very first to establish." It is the kingdom made up of living forms that are neither animals, plants, nor fungi. It includes red, green, and brown algae, seaweeds, water and slime molds, and most protozoa (single-celled life-forms including amoebas). The twenty-seven phyla in this kingdom are found mostly in water habitats such as rivers, freshwater lakes, and oceans.

The kingdom Fungi (from a Greek word meaning "sponge") may have descended from the kingdom Protoctista. The oldest fossil Fungi are about 300 million years old. There are more than one hundred thousand identified species, including bread molds, yeasts, and mushrooms. Thousands more species are believed to exist but have yet to be identified.

The Animal Kingdom

All plants, from mosses to giant redwoods, belong to the kingdom Plantae (from a Latin word for "plant"), and most animals belong to the kingdom Animalia. This kingdom contains all multicellular, heterotrophic, diploid organisms that develop from an egg and sperm. A heterotrophic animal takes food into the body and digests it, or breaks it down, into energy for use. A diploid organism has two sets of chromosomes, one derived from the female parent and the other from the male. The characteristic that defines all Animalia, however, is that they develop from a blastula, a hollow ball of cells found in animal embryos (the earliest stage of development, when the egg just begins to divide) but not in plants. The blastula forms a layer around a central open space or cavity in the embryo.

Animalia (from a Latin word meaning "soul" or "breath") is the kingdom with the largest number of life-forms, from Placozoa, a phylum with a single species usually found growing on aquarium walls, to the phylum Arthropoda (from a Greek word meaning "jointed foot") that contains more than a half million identified species, including spiders, scorpions, beetles, shrimp, lobsters, crayfish, crabs, flies, centipedes, millipedes, butterflies, moths, and all other species of insects. Seventeen of the thirty-three phyla in Animalia are various kinds of worms found in the shallow and deep water of lakes, rivers, and oceans and in the ground. One worm phyla, Pentastoma, consists of more than seventy species that live exclusively in the tongues, lungs, nostrils, and nasal sinuses of dogs, foxes, goats, horses, snakes, lizards, and crocodiles. Another phyla, Platyhelminthes, consists of worms that live in bat dung and other

equally unusual environments. Only two of the phyla, Arthropoda and Chordata (animals with nerve cords), live entirely on land. Among the phyla, Arthropoda has the largest number of identified species, more than a million, and possibly millions more not yet identified. Most of the species of Animalia that have ever lived are now extinct. Thirty-two of the thirty-three phyla are invertebrates, which means they lack backbones.

The History of Life-Forms

The phylogenetic method of classifying life, using animals as an example, begins with the earliest fossil records available. The earliest members of the Animalia kingdom evolved from members of the Protoctista kingdom about 565 million years ago. Exactly when and which members were the first is subject to debate; however, the first large, multicellular fossils date to this time. About 530 million years ago, there was a gigantic explosion of life-forms, including all kinds of clams, snails, arthropods, crabs, and trilobites. After a few tens of millions more years came echinoderms (starfish and sea urchins) and eventually chordates, out of which emerged fish and mammals.

The periods of explosive growth and massive extinction of life-forms are believed to have been caused by some major environmental changes, the exact nature and cause of which have not been determined. Scientists believe an explosion of new life could have been caused by an increase in atmospheric oxygen, which would allow life-forms to live out of water, and a mass extinction could have been the result of gigantic dust clouds, large enough to darken the sun's light for millions of years, stirred up by huge meteors crashing into earth. Such a critical darkening of the earth is believed to have taken place about sixty-five million years ago, plunging temperatures to near freezing and killing off thousands of species including all the dinosaurs.

The Phylum Chordata

Despite the tremendous losses resulting from the extinction of as much as 95 percent of the existing species, life-forms continued to survive and evolve. The most successful of the new forms emerged from the ancestors of the phylum Chordata (from a Latin word for "cord"). This phylum includes all mammals, birds, amphibians, reptiles, and species with backbones (vertebrates). More than forty-five thousand species of chordates exist. Three key features are used to classify members of the phylum. A member must have a single nerve cord along its back. In mammals, this cord has developed into the spinal cord and brain. It must also have a notochord, a bony rod located between the nerve cord and the digestive tract, which supports the body and the muscles and is found in the embryo and the adult. The last requirement is that members have gill slits in the throat at some stage of development, whether in the embryo stage as is true with land animals or throughout their entire life as in fish. Gill slits show that land animals developed from sea creatures.

The phylum Chordata includes two superclasses, Pisces (fish) and Tetrapoda (four-limbed forms). There are two classes of living Pisces, Chondrichthyes (boneless creatures), which includes sharks, skates, and rays, and Osteichthyes (bony fish), a class that contains about twenty-five thousand species. The earliest fish fossils are about 500 million years old. There are four classes in Tetrapoda: Amphibia, Reptilia, Aves, and Mammalia.

The class Amphibia contains about two thousand species of frogs, toads, and salamanders. Members of this class evolved about 370 million years ago and were the first vertebrates to live on land. Amphibia must lay their eggs in water and must live close to water, or their soft skin will dry out and cause their deaths.

The class Reptilia has more than five thousand species. It includes turtles, lizards, snakes, and crocodiles. Reptiles develop from an egg, live on land, and have dry, scaly skin. The largest reptiles, the dinosaurs, died out more than sixty million years ago. Reptiles are cold-blooded, which means they cannot control the temperature of their blood, they breathe air through lungs, and they are not required to lay their eggs in water.

They probably evolved out of an amphibian species between 300 million and 320 million years ago.

The class Aves contains nine thousand species of living birds, with thousands more extinct. Aves evolved from dinosaur species perhaps 200 million years ago. Unlike reptiles, Aves class members have the ability to regulate their internal temperature. They have feathers rather than scales, but they lack teeth.

The fourth class is Mammalia, which has more than forty-five hundred living species, including human beings. Mammals are classed into Metatheria (marsupials) and Eutheria (placentals). Metatheria, such as kangaroos, have external pouches in which their young are born live. Eutheria have vaginas through which the fully developed young pass during birth. Most mammals are members of Eutheria. This includes the orders of Insectivora (hedgehogs, shrews, and moles), Primates (lemurs, monkeys, apes, and human beings), Carnivora (dogs, cats, and bears), and Pinnipedea (seals and sea lions.)

The Benefits of Classification
Life on earth includes millions of different types of organisms. Phylogeny is one method of making sense out of so many different life-forms. It provides a system that links organisms together on the basis of their evolutionary history of development. By grouping vast numbers of forms into related groups, phylogenetics helps bring some order to what seems like chaos. Attempts to classify life-forms have always been controversial, and there still is much disagreement in the scientific community over various types of classification. However, phylogeny is the most widely agreed upon system.

Developments in techniques used to analyze molecules and DNA in the 1980's have shed light on the evolutionary development of species and have demonstrated new links among living plants and animals. The information gleaned using these techniques can also be used to trace the ancestry of modern species back to the earliest fossil evidence available from 575 million years ago. Molecular comparisons have shown that the development of species from spores and eggs into embryos and adult stages is very similar across a wide range of phyla.

These techniques produce data that consist of long sequences of the four nucleic acids that make up the information contained in DNA. The patterns formed by the acids are very similar among related organisms. The more closely related the sequence of nucleic acids, the more closely the species are related. Closely related species differ only very slightly in the way their DNA is structured. The more species differ in their DNA structure, the more distant they are in evolutionary terms. By measuring and comparing the differences in a gene that controls basically the same function in various species, scientists can con-struct an evolutionary tree. Species can be placed on the tree at the points where they begin to diverge from other genus members. In this way, examinations of the phylogenetic development of organisms can be used to create a tree of life showing the close relationships among all organisms. The tree also shows that all forms of life are related because they originally came from the same source in some ancient pool where the right chemicals just happened to bump into one another, mix, and begin to produce offspring.

—Leslie V. Tischauser

See also: Amphibians; Evolution: Historical perspective; Extinction; Fish; Genetics; Human evolution analysis; Insects; Mammals; Multicellularity; Primates; Reptiles; Sex differences: Evolutionary origin; Systematics.

Bibliography

Copeland, Herbert F. *The Classification of Lower Organisms*. Palo Alto, Calif.: Pacific Books, 1956. The classic work by a biologist given credit for developing the current system of classification. It has many drawings of microscopic life-forms, and though old, is still useful because of the tremendous effort made by the author to revise the phylogenetic history of many species.

Gould, Stephen Jay. *Wonderful Life: The Burgess Shale and the Nature of History*. London: Vintage, 2000. A history of life on earth, that pays particular attention to the role of catastrophe in evolution. Tells how a major disaster destroyed 95 percent of life-forms some 65 million years ago. An interesting discussion of the problems of classification is included in this work by one of the best science writers and naturalists in the field. Contains an index and a useful bibliography.

Margulis, Lynn, and Karlene V. Schwartz. *Five Kingdoms: An Illustrated Guide to the Phyla of Life on Earth*. 3d ed. New York: W. H. Freeman, 1999. A very good source for the history of life. It contains a picture or drawing of each phylum of plants and animals in the five kingdoms. The illustrations and descriptions are clear and easy to understand. The discussion of each phylum includes an evolutionary history of its development and phylogeny. Includes a bibliography for each kingdom and phylum.

Romer, A. S. *The Vertebrate Story*. Chicago: University of Chicago Press, 1971. The best, most detailed evolutionary history of a subphylum of the phylum Chordata, out of which *Homo sapiens* (humans) developed. Good illustrations, a well-constructed family tree, and a good history of the classification debate.

Wilson, Edward O. *The Diversity of Life*. New York: W. W. Norton, 1992. A highly regarded zoologist discusses the history of life on earth and the many forms of life that have developed. Well-written with many drawings and charts that display the richness of animal and plant life. Describes how plants and animals evolved and discusses the current threat from environmental pollution to the future of living things. Contains a detailed index.

PHYSIOLOGY

Types of animal science: Anatomy, fields of study, physiology
Fields of study: Anatomy, developmental biology, embryology, evolutionary science, physiology, zoology

Physiology is a fundamental branch of biology, predating all disciplines except perhaps anatomy. Physiologists study the functions of living organisms. These functions are studied at many levels of organization: the molecular level, the cellular level, the tissue-organ level, and even the organismic (whole animal) level. The goal is to explain living functions in quantifiable terms that obey the laws of chemistry and physics.

Principal Terms

ARTERY: a blood vessel that carries blood from the heart to the body tissues

CAPILLARY: a small vessel that connects arteries to veins; this is where respiratory gases, nutrients, and wastes are exchanged between blood and tissues

GLOMERULUS: a specialized capillary in the kidneys that filters blood

NEPHRON: a tubular structure in the kidneys that extracts filtrate, and reabsorbs nutrients and other valuable substances and secretes wastes

NEURON: an individual nerve cell; nerves are made up of many neurons bundled together

VEIN: a blood vessel that returns blood to the heart

How animal bodies work is the subject of physiology. Early Greek philosophers such as Aristotle used deductive reasoning (the use of observations, logic, and intuition) to explain function. William Harvey (1578-1657) is generally considered the father of modern physiology, as he was among the first and most successful in employing inductive reasoning (experimentation) to explain body function. Inductive reasoning uses the scientific method to study unsolved problems. This approach starts with a hypothesis. Experiments are designed to test the hypothesis. Data are collected using quantified measurements; physiologists never simply say heart rate increased in response to the experimental treatment. The result is always expressed in quantified terms; for example, one might say that heart rate increased from thirty to forty-five beats per minute. Once the data are collected, they are analyzed and interpreted in terms of the original hypothesis. The interpretation usually leads to new hypotheses which are then tested with new experiments.

The Study of Systems

Classically, physiology has been divided into several organ systems. These include the neuromuscular system, the cardiovascular system, the pulmonary system (lungs), the renal system (kidneys), the digestive system, and the endocrine system. This nomenclature is based on mammalian systems and has had to be modified to include such alternative systems as gills (branchial system) in aquatic animals.

Physiology has been characterized as a synthetic science because it involves a synthesis of biology, chemistry, mathematics, and physics to describe body functions. The objective is to explain a given physiological function in terms which obey the laws of chemistry and physics. This often involves model building. A physiological model is a construct of hypotheses, which can be qualitative or quantitative. Harvey proposed a model in which the blood circulates from the heart to arteries, and then to capillaries which connect to veins to return the blood to the heart. He then did experiments to prove this. Quantitative or mathematical models are used to describe many functions using mathematical constructs and equations.

Physiology is studied at the molecular, cellular, tissue, organ, system, and organismic levels. There are several branches of physiology. Mammalian, fish, and insect physiologists restrict their studies to a certain group. General physiology seeks to describe functions that are common to all life forms, such as cell membrane function. Comparative physiology seeks to examine how different groups of animals accomplish similar goals while living in completely different circumstances, for example, in aquatic environments exchanging oxygen and carbon dioxide with water versus living on land and exchanging these respiratory gases with air.

Comparative studies are important for several reasons. First, the acquisition of each new piece of specific knowledge raises questions about its broader applicability. Once it is known that land animals need oxygen in air to breathe, the question arises: If a fish spends its entire life under water, why does it not drown? The answer is that it uses gills to extract oxygen efficiently from water (with a notoriously scarce oxygen supply) and to excrete carbon dioxide. Comparative physiology also helps to better understand evolution. The evolution of the vertebrate cardiovascular system from fish with two-chambered hearts, to amphibians and most reptiles with three-chambered hearts, to birds and mammals with four-chambered hearts, has attracted much interest in how the circulation of oxygen-rich blood is kept separate from oxygen-poor blood. Finally, comparative physiology can be used to study simple physiological systems in primitive animals such as invertebrates to help explain more complex systems in more advanced animals such as mammals.

Gaining Knowledge Through Experimentation
Our knowledge of how information is carried by neurons (the individual fibers of nerves) began with studies of squid neurons. Squid have giant neurons; they are so large that physiologists were able to insert electrodes inside them to discover the electrical events that produce nervous impulses and thus information transfer. Beyond this, the neurons are so large that it is possible to remove their contents and substitute artificial solutions to see how this alters neural impulse production. This is how it was discovered that the ions sodium and potassium are responsible for neural impulses. This research started in the 1920's, and developing technology has shown that mammalian neural function operates essentially the same way as squid neural function. Another example of such use of comparative physiology has led to the understanding of kidney function. In the early twentieth century, it was observed that kidney tubules in amphibians such as frogs and salamanders are large enough to allow samples of nephric tubular fluid to be removed and analyzed. Such studies led to the discovery that the initial event in urine formation is filtration of blood plasma in glomerular capillaries and that the final urine composition results from selective reabsorption and filtration of water and solutes in nephrons.

Physiology is a very interesting field that can prepare one for careers in teaching at all levels, basic research, or commercial research and development.

—*Daniel F. Stiffler*

See also: Adaptations and their mechanisms; Anatomy; Antennae; Beaks and bills; Biology; Bone and cartilage; Brain; Cell types; Circulatory systems of invertebrates; Circulatory systems of vertebrates; Convergent and divergent evolution; Digestive tract; Ears; Embryology; Endoskeletons; Exoskeletons; Eyes; Feathers; Fossils; Genetics; Heart; Horns and antlers; Hydrostatic skeletons; Kidneys and other excretory structures; Lungs, gills, and tracheas; Muscles in invertebrates; Muscles in vertebrates; Nervous systems of vertebrates; Noses; Reproductive system of female mammals; Reproductive system of male mammals; Respiratory system; Scales; Sense organs; Shells; Tails; Teeth, fangs, and tusks; Tentacles; Wings.

Bibliography

Boron, W., and E. Boulpaep. *Medical Physiology*. St. Louis: C. V. Mosby, 2001. Designed for medical students, but offers the most up-to-date coverage of molecular physiology.

Dukes, H. H., M. J. Swenson, and W. O. Reese, eds. *Dukes' Physiology of Domestic Animals*. 11th ed. Ithaca, N.Y.: Comstock, 1993. The leading authority on veterinary physiology.

Prosser, C. L. *Comparative Animal Physiology*. 2 vols. 4th ed. New York: John Wiley & Sons, 1991. The leading reference text in comparative physiology.

Randall, D., W. Burggren, and K. French. *Eckert Animal Physiology*. 4th ed. New York: W. H. Freeman, 1997. One of the most popular physiology texts in America, this book provides a comprehensive treatment of our knowledge of animal physiology.

Schmidt-Nielsen, K. *Animal Physiology*. 5th ed. New York: Cambridge University Press, 1997. This text stresses adaptations to various environments.

PIGS AND HOGS

Type of animal science: Classification
Fields of study: Anatomy, zoology

Pigs are social animals of the genus Sus, *living in family groups led by an older female and her offspring. The male likes to be alone except during the breeding season. The female will have between six and twelve babies at a time.*

Principal Terms

BARROW: a castrated male pig that has not produced any secondary sexual characteristics

BOAR: male pig

GILT: a young female pig that has not produced a litter

HOG: the general name applied to a pig over three months old

PORK: the meat of the pig used for food

SHOAT: an immature pig of either sex that weighs between 35 and 160 pounds

SOW: female pig that has produced a litter of babies

SWINE: the general name applied to the domesticated pig

Pigs are estimated to have evolved about 38 million years ago in central Europe, and about 25 million years ago to have become established in Africa and Asia. Pigs became domesticated in China about 5000 B.C.E. The reason for the delay in agricultural domestication is believed to be that pigs were unwilling to adapt to a nomadic lifestyle, as dogs, sheep, and goats did. The domesticated pig is thought to have descended from two different lines of wild hogs. It is believed that the East Indian pig (*Sus vittatus*) has had major influence on the swine of China and the surrounding region. The European wild boar has probably had major influence on the rest of the various breeds of pigs in the world.

The domestic pig is a compact, solid-looking animal with a large head, short, fast-moving legs, a rough coat, and a small tail. The color of a pig can vary greatly, from white or black to brownish-red, and any combination thereof, including spotted, solid colored, and banded. Some may reach a height of only twelve inches at the shoulder, others may reach all the way to four feet. The weight

---

Pig Facts

Classification:
Kingdom: Animalia
Phylum: Chordata
Subphylum: Vertebrata
Class: Mammalia
Subclass: Theria
Order: Artiodactyla
Family: Suidae
Subfamily: Suinae
Genus: Sus

Geographical location: There are no true wild pigs in the Western Hemisphere, but there are feral hogs, except in Antarctica; the javelina, native to North and South America, is not a true pig; wild pigs are found in Southeast Asia and Eurasia

Habitat: Wide range, from swamps and rain forests to dry lands and mountains

Gestational period: A little less than four months (111-115 days)

Life span: Domestic pigs generally have a life span of from six to ten years, with an occasional occurrence of over twenty years

Special anatomy: Pigs have an excellent sense of smell and exceptional hearing, but their eyesight is weak

---

Pigs are very useful domesticated animals because they are exceptionally omnivorous and produce a large quantity of meat with relatively little care. (Robert Maust/Photo Agora)

of a pig also can vary, with some weighing only about 60 pounds, where others can go over 900. The average meat pig falls in the range from 225 to 300 pounds.

Humans use pigs in many different ways. From the point of view of the percentage of the carcass used, the pig is the most utilized of all domesticated animals. The hair is used for brushes because it is very strong, yet flexible. The hide is used for numerous products, from shoes to purses; in general it is lightweight and durable. The blood and offal are used for fertilizers, soaps, and medical supplies. Domestic pigs have been used by mankind for centuries as a main supplier of dietary protein.

Types of Pigs

Pigs are cloven-hoofed ungulates, and are closely related to the hippopotamus. As mentioned, the domestic pig (*Sus scrofa*) is now found worldwide. They are omnivorous, eating almost anything. However, pigs generally prefer to eat soft tissue plants, especially roots and tubers, which they dig

up with their noses. They will also eat leaves, seeds, bugs, and anything that is found on the ground, including bird eggs, baby birds, snakes, and carrion.

There are hundreds of different breeds of pigs in the world. Most of the breeds that the various departments of agriculture recognize as major contributors to current swine production come from the United States or Europe. The pigs can be identified as being either meat, lard, or bacon types. Lard types have lost a great deal in numbers over the past fifty years, and have been bred into a style more like meat. The pig population in the world is large, and is approaching almost one billion in number, with China having the largest number, 250 million. Russia and the United States are tied in second place with about 60 million each. Brazil would be fourth with 35 million.

The largest consumer of pork per capita is Denmark (105 pounds annually), followed by Hungary, Germany, Austria, Luxembourg-Belgium, and the United States. In the United States, the state of

Iowa leads in production of hogs, followed by Illinois, Missouri, Indiana, and Minnesota.

Pigs are very smart domesticated animals. Their abilities are thought of as being greater than those of the domestic dog. Indeed, pigs make fine pets, as is evidenced by the pot-bellied pig of Southeast Asia.

—Earl R. Andresen

See also: Domestication; Claws, nails, and hooves; Omnivores; Ungulates.

Bibliography

Briggs, Hilton M., and Dinus M. Briggs. *Modern Breeds of Livestock*. 4th ed. New York: Macmillan, 1980. Reviews the various breeds of modern pigs and gives a brief history of each major breed.

Ensminger, E. M., and R. O. Parker. *Swine Science*. 5th ed. Danville, Ill.: Interstate Printing, 1984. Explains in detail various aspects of swine production. Gives a brief explanation of breeds.

Porter, Valerie. *Pigs: A Handbook to the Breeds of the World*. Ithaca, N.Y.: Comstock Press, 1993. Review of the various swine, pigs, hogs of the world.

Taylor, Robert E., and Thomas G. Field. *Scientific Farm Animal Production: An Introduction to Animal Science*. 6th ed. Upper Saddle River, N.J.: Prentice Hall, 1998. A review of the various breeds of modern pigs in America.

PLACENTAL MAMMALS

Types of animal science: Development, reproduction
Fields of study: Anatomy, biochemistry, cell biology, embryology, physiology, reproductive science

Placental reproduction is the most common form of mammalian reproduction. All mammals except the marsupials (such as the kangaroo and opossum) and the monotremes (platypus and echidna) form placentas.

Principal Terms

CHORION: the outer cellular layer of the embryo sac of reptiles, birds, and mammals; the term was coined by Aristotle

EMBRYO: a young animal that is developing from a fertilized or activated ovum and that is contained within egg membranes or within the maternal body

ENDOMETRIUM: an inner, thin layer of cells overlying the muscle layer of the uterus

FETUS: a mammalian embryo from the stage of its development where its main adult features can be recognized, until birth

MATERNAL: referring to the female parent

OVUM: an unfertilized egg cell

UTERUS: in female mammals, the organ in which the embryo develops

VIVIPAROUS: producing young that are active upon birth (often referred to as live birth); the embryo is nurtured within the uterus

Monotremes are oviparous, or egg-laying mammals. Marsupials are ovoviviparous, meaning that the egg is large and has a yolk adequate to nourish the embryo during its early development, but it remains unattached to the wall of the uterus. Gestation in marsupials is necessarily short, therefore, and the young are born in an immature, fetal stage. They make their way to the mother's pouch and continue to grow, nourished by the mother's milk. All other mammals, termed placental, are viviparous. The small egg, lacking food substance, becomes attached to the uterine wall, and the developing embryo is nourished by the mother's blood passing through a placenta. This process allows longer gestation, and as a result the young are born in a more advanced (precocial) state of development.

The placental mammals form a diverse and successful group that includes the insectivores (such as shrews, hedgehogs, and moles), bats, sloths, anteaters, armadillos, primates (to which humans belong), rodents, rabbits, whales, dolphins and porpoises, carnivores (such as cats, dogs, and bears), seals, aardvarks, elephants, hyraxes, manatees, uneven-toed mammals (such as tapirs, horses, and rhinoceroses), and even-toed (cloven-hoofed) mammals such as pigs, camels, deer, sheep, cattle, and goats.

Placenta Structure

The term "placenta" comes from the Latin meaning "flat cake," and is used to describe the flat structure (in most animals) which attaches the developing embryo and fetus to the wall of the uterus. The term was first used in 1559 C.E., although knowledge of the placenta, at least in humans, goes far back into antiquity. Reference to it may be found in many ancient texts and drawings, including the Old Testament books of the Bible. Early Egyptians considered the placenta to be the seat of the external soul.

The placenta is the organ responsible for the transmission of materials between mother and fetus prior to birth—the only bridge between them. In many species, it has important endocrine functions, producing hormones necessary for development of the fetus or for maintenance of the pregnant state. It is essentially a product of both the developing ovum and the mother. The placenta is commonly referred to as the afterbirth, extruded from the mother's uterus at the end of the birth process.

When the ovum is released by the ovary and fertilized by sperm, it eventually comes to rest in the hollow cavity of the uterine horn (one of two chambers) or the uterus (single chamber). While it moves toward that area it begins to divide, forming a ball of cells known as a morula. As development continues, the ball becomes hollow, and it is then referred to as a blastocyst. Within this hollow blastocyst a few cells protrude into the cavity, forming a knob. These are the only cells that will eventually develop into the embryo and fetus. The rest of the cells are responsible for forming the supportive structures, including the chorion, the placenta, and the umbilical cord (which attaches the embryo to the placenta).

Among the eutherian (placental) mammals, a variety of placental configurations occurs. In many species the entire chorionic sac becomes connected to the uterine wall, and transfer of materials between the maternal and fetal compartments occurs over the whole surface. In other species, a much more specialized system develops. Here parts of the chorion (a membrane equivalent to the one that lines the shell of reptile and bird eggs) become highly specialized, establishing an intimate relationship with the uterine tissues. Thus, transfer of materials occurs only in one select region of the chorion, referred to as the placenta. It is these tissues, with their flattened, cakelike appearance, from which the name derives.

The most important feature of the placenta is the close contact between the fetal blood vessels in the placenta and the maternal blood vessels in the uterine wall. While it is a common misconception that the fetal and maternal blood mix or flow together, this is not a correct picture. What actually occurs is that the two blood systems come close to each other, at which point materials that can diffuse out of one vessel may diffuse into another.

Thus, transfer of materials from mother to fetus, and vice versa, can occur. The closer the two blood pools come to each other, the better it is for transfer. Nutritional, respiratory, and excretory products are transferred.

Placenta Types

In the epithelio-chorial placenta, as found in the hoofed mammals (such as Artiodactyla and Perissodactyla), whales, and lemurs, the wall of the uterus retains its surface epithelium. The minimum separation of bloods is four cells thick (two epithelial cell layers and the endothelium of the blood vessels). Thus, it is vital to have a very large surface area to allow for adequate movement of materials. While this is a large improvement over the marsupial system, it is nonetheless 250 times less efficient at salt transfer than the placenta used by humans.

The separation between mother and fetus is reduced in carnivores and sloths. Here the chorion invades the uterine epithelium and comes in direct contact with the epithelium of the maternal blood vessels, allowing a more uniform transfer of materials. The most advanced form, showing minimal separation, is the hemo-chorial placenta, found in humans, rodents, bats, and most insectivores. Here the maternal blood vessel walls are chemically broken down and the invading chorion is now in direct contact with the maternal blood stream. Because this is so much more efficient for materials exchange, the size of the interacting surfaces can be much reduced.

In addition to the exchange of materials, the placenta plays an important role in immunology, and without the placenta, the mother's body would reject the developing embryo like any other foreign body. It is this tolerance of the embryo that separates the placental mammals from the marsupials, and allows for gestation periods to be extended. Fetuses in placental mammals receive antibodies from their mothers, thus enhancing their early immunity to disease.

The epithelio-chorial placenta, being only in contact with the uterine wall and not being invasive, is readily shed by the uterus when the fetus is born. There is no damage to the maternal tissue. The more invasive types of placenta, including the hemo-chorial type, can only be lost by separation through the uterine tissues. Thus, birth in species with hemo-chorial placentas is of necessity associated with some degree of maternal bleeding. In fact, in many hemo-chorial placentas, the blastocyst actually digests (chemically) the endometrial lining of the uterus and comes to lie completely within it. The endometrium then heals over the blastocyst, which then grows, fully surrounded by endometrium. The true placenta forms on the deep pocket of the endometrium in which the blastocyst lies. When the fetus is expelled from the uterus (that is, at birth), it ruptures the now very thin layer of stretched endometrium that covers the chorion. The placenta separates from the uterus as a result of rupture of the uterine blood vessels and tissues when the uterus contracts down after expulsion of the fetus. Bleeding between uterus and placenta produces a clot which eventually seals off the broken blood vessels and forms the basis for endometrial repair.

Most mammalian placentas, regardless of type, have some type of endocrine (hormonal) function. While the specific hormones may vary from species to species, two in particular are found in most placental mammals. Chorionic gonadotropin is a hormone secreted by the placenta which acts upon the ovary to increase progesterone synthesis. Progesterone, in return, is responsible for maintenance of pregnancy during the early phases. Placental lactogen, another hormone secreted by most placentas, acts on the mother to stimulate mammary gland development. This occurs throughout gestation, so that the mammary glands are ready for suckling by the time the offspring is born.

Gestation Periods

The length of gestation varies tremendously among placental mammals. In elephants the gestation period is as long as twenty-two months. However, size alone is not the determining factor. The giant among all mammals, the blue whale, has a gestation period of only eleven months, not appreciably longer than the human (nine to ten months).

Many bats and other mammals have a delayed implantation, in which the fertilized ovum remains dormant or its development is retarded at first, thus considerably extending the gestation period and delaying birth until the optimal season of warm weather or abundant food is present. Often a placenta is not found during this period of delay, with arrest of development occurring in the blastula stage. Thus, the gestation period of the fisher, a small North American carnivore with delayed implantation, is forty-eight to fifty-one weeks, or about the same as that of the blue whale. Gestation varies from twenty-two to forty-five days in squirrels, twenty to forty days in rats and mice, two to seven months in porcupines, six months in bears, and fourteen to fifteen months in giraffes.

Gestation length is ultimately constrained by the size of skull which will fit through the maternal pelvis. Where agility, speed, or long distances of travel put a premium on the mother's athleticism, gestation length is often shortened, and the birth weight of the offspring will be low.

Animals having long gestation periods, or whose young mature slowly and are suckled for a long period of time, generally do not breed as often as others. Many species of mice breed repeatedly throughout the spring, summer, and fall, having a gestation period of about twenty days, and being mature and ready to breed by twenty-one days of age. Many others, such as bears, coyotes, and weasels, breed only once a year. Environmental conditions, and the adaptability of various species to these conditions, play a large role in breeding cycles. It is clearly advantageous for young to be born during the season of least severe weather and to be weaned when food is most abundant. Many tropical mammals breed and give birth throughout the year, whereas in temperate or cold climates young are usually born in the spring or summer.

Similar factors also influence the number of young born in each litter among different species. Their rate of growth until weaned, mortality rates, adult activity cycles, and other factors no doubt help determine the litter size as well. Many rodents have three to six young per litter; a few species of mice can have as many as eighteen. Seals, whales, and most species of bats and primates bear only a single young at one time.

—*Kerry L. Cheesman*

See also: Asexual reproduction; Birth; Breeding programs; Cleavage, gastrulation, and neurulation; Cloning of extinct or endangered species; Copulation; Courtship; Determination and differentiation; Development: Evolutionary perspective; Estrus; Fertilization; Gametogenesis; Hermaphrodites; Hydrostatic skeletons; Mating; Parthenogenesis; Pregnancy and prenatal development; Reproduction; Reproductive strategies; Reproductive system of female mammals; Reproductive system of male mammals; Sexual development; Vertebrates.

Bibliography

Gilbert, Scott. *Developmental Biology*. 6th ed. Sunderland, Mass.: Sinauer Associates, 2000. Beautiful artwork showing placental formation and the process of implantation. Easy-to-read text.

Harris, C. Leon. *Concepts in Zoology*. 2d ed. New York: HarperCollins, 1996. A college text. Excellent chapters on reproduction in mammals, including photographs of developing blastocysts and placentas.

Macdonald, David, ed. *The Encyclopedia of Mammals*. New York: Facts on File, 1984. An excellent chapter on "what is a mammal," along with in-depth reviews of each family or group of mammals. A lot of attention to social aspects of mammals.

Randall, David, Warren Burggren, and Kathleen French. *Eckert Animal Physiology*. 4th ed. New York: W. H. Freeman, 1997. An in-depth college text that is not easy to read, but well written for the advanced student.

Silverthorn, Dee Unglaub. *Human Physiology: An Integrated Approach*. 2d ed. Upper Saddle River, N.J.: Prentice Hall, 2001. Chapters on reproduction illustrate the placenta well, but the book covers only the human.

PLANT AND ANIMAL INTERACTIONS

Type of animal science: Ecology
Fields of study: Physiology, zoology

Interactions between plants and animals in natural environments often revolve around either food acquisition or pollination and seed dispersal. By studying such processes, scientists have discovered strategies to increase agricultural production and to duplicate naturally produced pesticides.

Principal Terms

CELLULAR RESPIRATION: the release of energy in organisms at the cell level, primarily through the use of oxygen

CHLOROPHYLL: one of several forms of photoactive green pigments in plant cells that is necessary for photosynthesis to occur

COEVOLUTION: a mutualistic relationship between two different organisms in which, as a result of natural selection, the organisms become interdependent

CROSS-POLLINATION: the transfer of pollen grains and their enclosed sperm cells from the male portion of a flower to a female portion of another flower within the same species

FOOD CHAIN: a diagram illustrating the movement of food materials from green plants (producers) through various levels of animals (consumers) within natural environments

NATURAL SELECTION: the survival of variant types of organisms as a result of adaptability to environmental stresses

PISTIL: a female portion of a flower that produces unfertilized egg cells

STAMEN: a male portion of a flower that produces pollen grains and their enclosed sperm cells

Ecology represents the organized body of knowledge that deals with the interrelationships between living organisms and their nonliving environments. Increasingly, the realm of ecology involves a systematic analysis of plant-animal interactions through the considerations of nutrient flow in food chains and food webs, exchange of such important gases as oxygen and carbon dioxide between plants and animals, and strategies of mutual survival between plant and animal species through the processes of pollination and seed dispersal.

Ecologists study both abiotic and biotic features of such plant and animal interactions. The abiotic aspects of any environment consist of nonliving, physical variables, such as temperature and moisture, that determine where species can survive and reproduce. The biotic (living) environment includes all other plants, animals, and microorganisms with which a particular species interacts. Certainly, two examples of plant and animal interactions involve the continual processes of photosynthesis and cellular respiration. Green plants are often classified as ecological producers and have the unique ability to carry out both these important chemical reactions. Animals, for the most part, can act only as consumers, taking the products of photosynthesis and chemically releasing them at the cellular level to produce energy for all life activities.

Plant-Animal Mutualism

One topic that has captured the attention of ecologists involves the phenomenon of mutualism, in which two different species of organisms beneficially reside together in close association, usually revolving around nutritional needs. One such example demonstrating a plant and animal association is a certain species of small aquatic flatworm that absorbs microscopic green plants called algae into its tissues. The benefit to the animal is one of added food supply; the adaptation to this alga has been so complete that the flatworm does not actively feed as an adult. The algae, in turn, receive adequate supplies of nitrogen and carbon dioxide and are literally transported throughout tidal flats in marine habitats as the flatworm migrates, thus exposing the algae favorably to increased sunlight. A similar example of mutualism has been reported by ecologists studying various types of reef-building corals, which are actually marine, colonial animals that grow single-celled green algae called zooxanthellae within their bodies. The coral organisms use the nutrients produced by these algae as additional energy supplies, enabling them to build more easily the massive coral reefs associated with tropical waters. In 1987, William B. Rudman reported a similar situation while researching the formation of such coral reefs in East African coastal waters. He discovered a type of sea slug called a nudibranch that absorbs green algae into its transparent digestive tract, producing an excellent camouflage as it moves about on the coral reefs in search of prey. In turn, the algae growing within both the coral and sea slugs receive important gases from these organisms for their own life necessities.

An example of plant-animal mutualism that has been documented as a classic example of coevolution involves the yucca plant and a species of small, white moth common throughout the southwestern United States. The concept of coevolution builds upon Charles Darwin's theories of natural selection, reported in 1859, and describes situations in which two decidedly different organisms have evolved into a close ecological relationship characterized by compatible structures in both. Thus, coevolution is a mutualistic relationship between two different species that, as a result of natural selection, have become intimately interdependent. The yucca plant and yucca moth reflect such a relationship. The female moth collects pollen grains bearing sperm cells from the stamens of one flower on the plant and transports these pollen loads to the pistil of another flower, thereby ensuring cross-pollination and fertilization. During this process, the moth will lay her own fertilized eggs in the flower's undeveloped seed pods. The developing moth larvae have a secure residence for growth and a steady food supply. These larvae will rarely consume all the developing plant seeds; thus, both species (plant and animal) benefit.

Defensive Mutualism

Although these examples demonstrate the evolution of structures and secretions that reflect mutual associations between plants and animals, other interactions are not so self-supportive. Plant-eating animals, called herbivores, have always been able to consume large quantities of green plants with little fear of reprisal. Yet, some types of carnivorous plants have evolved that capture and digest small insects and crustaceans as nutritional supplements to their normal photosynthetic activities. Many of these plants grow abundantly in marshlike environments, such as bogs and swamps, where many insects congregate to reproduce. Such well-known plants as the Venus's-flytrap, sundews, butterworts, and pitcher plants have modified stems and leaves to capture and consume insects and spiders rich in protein. On a smaller scale, in freshwater ponds and lakes, a submerged green plant commonly known as the bladderwort partially satisfies its protein requirements by snaring and digesting small crustaceans, such as the water flea, within its modified leaves.

A form of ecological interaction commonly classified as mimicry can be found worldwide in diverse environments. In such situations, an animal or plant has evolved structures or behavior patterns that allow it to mimic either its surroundings or another organism as a defensive or offensive strategy. Certain types of insects, such as leafhoppers, walking sticks, praying mantids, and katydids (a type of grasshopper), often duplicate plant structures in environments ranging from the tropical rain forests to the northern coniferous forests of the United States. Such exact mimicry of their plant hosts affords these insects protection from their predators as well as camouflage that enables them to capture their own prey readily. In other examples of mimicry, some insects will absorb unpalatable plant substances in their larval stages and retain these chemicals in their adult forms, making them undesirable to birds as food sources. The monarch butterfly demonstrates this type of interaction with the milkweed plant. The viceroy butterfly has evolved colorations and markings similar to those of the monarch, thereby ensuring its own survival against bird predators. Certain species of ambush bugs and crab spiders have evolved coloration patterns that allow them to hide within flower heads of such common plants as goldenrod, enabling them to grasp more securely the bees and flies that visit these flowers.

Nonsymbiotic Mutualism

Many ecologists have been studying the phenomenon known as nonsymbiotic mutualism: different plants and animals that have coevolved morphological structures and behavior patterns by which they benefit each other without necessarily living together physically. This type of mutualism can be demonstrated in the often unusual and bizarre shapes, patterns, and colorations that more advanced flowering plants have developed to attract various insects, birds, and small mammals for pollination and seed dispersal purposes. Pollination essentially is the transfer of pollen grains (and their enclosed sperm cells) from the male portion of a flower to the egg cells within the female portion of a flower. Pollination can be accomplished by the wind, by heavy dew or rains, or by animals, and it results in the plant's sexual production of seeds that represent the next generation of new embryo plants. Accessory structures, called fruits, often form around seeds and are usually tasty and brightly marked to attract animals for seed dispersal. Although the fruits themselves become biological bribes for animals to consume,

often the seeds within these fruits are not easily digested and thus pass through the animals' digestive tracts unharmed, sometimes great distances from the original plant. Other types of seed dispersal mechanisms involve the evolution of hooks, barbs, and sticky substances on seeds that enable them to be easily transported by animals' fur, feet, feathers, or beaks to new regions for possible plant colonization. Such strategies of dispersal reduce competition between the parent plant and its new seedlings for moisture, living space, and nutritional requirements.

The evolution of flowering plants and their resulting use of animals in pollination and seed dispersal probably began in dense, tropical rain forests, where pollination by the wind would be cumbersome. Because insects are the most abundant form of animal life in rain forests, strategies based upon insect transport of pollen probably originated there. Because structural specialization increases the possibility that a flower's pollen will be transferred to a plant of the same species, many plants have evolved a vast array of scents, colors, and nutritional products to attract many insects, some birds, and a few mammals. Not only does pollen include the plant's sperm cells, but it is also rich in food for these animals. Another source of animal nutrition is a substance called nectar, a sugar-rich fluid often produced in specialized structures called nectaries within the flower itself or on adjacent stems and leaves. Assorted waxes and oils are also produced by plants to ensure plant-animal interactions. As species of bees, flies, wasps, butterflies, and hawkmoths are attracted to flower heads for these nutritional rewards, they unwittingly become agents of pollination by transferring pollen from male portions of flowers (stamens) to the appropriate female portions (pistils). Some flowers have evolved distinctive, unpleasant odors reminiscent of rotting flesh or feces, thereby attracting carrion beetles and flesh flies in search of places to reproduce and deposit their own fertilized eggs. As these animals consummate their own relationships, they often become agents of pollination for the plant itself. Some tropical plants such as orchids even mimic a

Butterflies, as well as bees, help pollinate plants while obtaining nectar for themselves. (Rob and Ann Simpson/Photo Agora)

female bee, wasp, or beetle, so that its male counterpart will attempt to mate with it, thereby ensuring precise pollination.

Among the bird species, probably the hummingbirds are the best examples of plant pollinators. Various types of flowers with bright, red colors, tubular shapes, and strong, sweet odors have evolved in tropical and temperate regions to take advantage of the hummingbirds' long beaks and tongues as an aid to pollination.

Because most mammals, such as small rodents and bats, do not detect colors as well as bees and butterflies do, flowers that use them as pollinators do not rely upon color cues in their petals but instead focus upon the production of strong, fermenting or fruitlike odors and abundant pollen rich in protein to attract them. In certain environments, bats and mice that are primarily nocturnal have replaced day-flying insects and birds in these important interactions between plants and animals.

Experiments with Plant-Animal Interaction

Contemporary ecologists have gone beyond the purely descriptive observations of plant-animal interactions (initially within the realm of natural history) and have designed controlled experiments that are crucial to the development of such basic concepts as coevolution. For example, the use of radioactive isotopes and the marking of pollen with dye and fluorescent material in field settings have allowed ecologists to demonstrate precise distances and patterns of pollen dispersal. Ecologists and insect physiologists have cooperatively studied how certain insects, such as bees, are sensitive to ultraviolet light. When some flowers are viewed under ultraviolet light, distinct floral patterns become evident to guide these insects to nectar pollen sources. Through basic research, Carolyn Dickerman reported in 1986 that animal color preferences vary throughout the season. Insect pollinators, who must feed every day, will adapt to these changes by shifting their foraging behavior. Research in the field has demonstrated that some species of flowers, such as the scarlet gilia, will produce differently colored flowers to accommodate shifts in pollinator species. Early in the growing season, this plant will produce long, red, tubular-shaped flowers to attract hummingbirds. As the hummingbirds migrate, the flowers will later become lighter in hue and be pollinated primarily by nocturnal hawkmoths.

In the laboratory, ecologists and biochemists have cooperatively analyzed the chemical composition of plant secretions and products. The chemical analysis of nectar indicates great variation in composition, correlating with the type of pollinator. Flowers pollinated by beetles generally have high amino acid content. The nectar associated with hummingbird-pollinated flowers is rich in sugar. Pollen also varies widely in chemical composition within plant species. Oils and waxes are major chemical products in the pollen of plants visited primarily by bees and flies. For bat-pollinated flowers, the protein content is quite high.

Research has also successfully analyzed how certain plants have been able to develop toxins as chemical defenses against animals. These protective devices include such poisons as nicotine and rotenone that help prevent insect and small mammal attacks. A more remarkable group of protective compounds recently isolated from some plants are known as juvocimines. These chemicals actually mimic juvenile insect hormones. Insect larvae feeding on leaves containing juvocimines are prevented from undergoing their normal development into functional, breeding adults. Thus, a specific insect population that could cause extensive plant damage is locally reduced.

Ecological interactions between plants and animals are diverse and varied. These plant-animal interactions can be viewed as absolute necessities for developing food chains and food webs and for maintaining the global balances of such important gases as oxygen and carbon dioxide. The interactions can also be very precise, limited, and crucial for determining species survival or extinction. By analyzing varied plant-animal interactions, from the microscopic level to the global perspective, one can more fully appreciate all the ecological relationships that exist on the earth.

The Ecology of Interaction
The ecological importance of plant-animal interactions cannot be stressed enough. Modern-day agriculture owes its existence to the activities of such insect pollinators as honeybees in regard to the production of domestic fruits, vegetables, and honey. It is becoming increasingly evident to many ecologists and forestry scientists how important certain bird species, such as blue jays and cedar waxwings, are in natural reforestation of burned and blighted areas through their seed dispersal strategies. The plant horticulture and floral industries also are developing an appreciation of specific plant-animal interactions that produce more viable natural strains of flowers and ornamental shrubbery. The study of natural chemical defenses produced by some plants against animal invasions is most promising. The renewed interest in earlier efforts to extract such plant products as nicotine, rotenone, pyrethrum, and caffeine may produce natural compounds that can be effective insecticides without the long-term, environmental hazards associated with such human-made pesticides as malathion, chlordane, and dichlorodiphenyl-trichloroethane (DDT).

Finally, humankind is realizing that it is important to understand and protect certain plant-animal interactions associated with the tropical regions of the earth; otherwise, the global balance of oxygen and carbon dioxide could be seriously disrupted. Also, these tropical areas represent the last natural environments for the continuation of important plant species that produce secretions and products that have favorable medicinal qualities for humans and domestic livestock. By maintaining these populations and understanding how certain animals interact with them, humans can be guaranteed a viable supply of beneficial plant species whose medicinal values can be duplicated within the laboratory.

—Thomas C. Moon

See also: Adaptations and their mechanisms; Coevolution; Digestion; Ecological niches; Ecology; Ecosystems; Food chains and food webs; Habitats and biomes; Herbivores; Insect societies; Predation; Rhythms and behavior; Symbiosis.

Bibliography
Abrahamson, Warren G., ed. *Plant-Animal Interactions*. New York: McGraw-Hill, 1989. Covers herbivore-plant interactions, carnivorous plant ecology and evolution, pollination and population dispersal agents, plant communities as habitats for animals, interactions in agroecosystems, and coevolution.
Barth, Frederick G. *Insects and Flowers: The Biology of a Partnership*. Princeton, N.J.: Princeton University Press, 1991. This beautifully illustrated book, rich in both color

photographs and electron micrographs, describes all aspects of flower-insect interactions. Includes bibliography and index. The writing style is appropriate for both general readers and introductory-level biology students.

Dickerman, Carolyn. "Pollination: Strategies for Survival." *Ward's Natural Science Bulletin*, Summer, 1986, 1-4. This article traces the coevolution of flowering plants with their animal pollinators, especially the beetles, butterflies, and hawkmoths. An excellent introduction to both strategies of pollination and seed dispersal. The author also outlines some current research techniques and includes a simplified biological key that focuses upon flower structures that will help determine which animals pollinate them.

Howe, Henry F., and Lynne C. Westley. *Ecological Relationships of Plants and Animals*. New York: Oxford University Press, 1990. Offers introductory students in botany, zoology, and ecology a comprehensive summary of both field and experimental studies on the ecology and evolution of plant and animal interactions. Excellent black-and-white photographs, illustrations, charts, and tables are found throughout. This textbook is written with clarity but is not only for the casual reader.

John, D. M., S. J. Hawkins, and J. H. Price, eds. *Plant-Animal Interactions in the Marine Benthos*. Oxford, England: Clarendon Press, 1992. The proceedings of a symposium on shallow-water marine plant-animal interactions.

Lanner, Ronald M. *Made for Each Other: A Symbiosis of Birds and Pines*. New York: Oxford University Press, 1996. Based on the author's research into the mutualism between Clark's nutcracker and several species of western white pines with wingless seeds. Accessible to the lay reader.

Meeuse, Bastian, and Sean Morris. *The Sex Life of Flowers*. New York: Facts on File, 1984. This well-written text includes beautiful photographs and illustrations, and it appeals to readers with varied scientific backgrounds. It describes the evolution of the flower as an organ of sexual reproduction and portrays selected animal adaptations in structure and behavior to accommodate pollination and seed dispersal. Includes a chapter-by-chapter listing of further reading sources. The photographs and writing are based upon the Public Broadcasting Service series entitled "Sexual Encounters of the Floral Kind."

Norstag, Knut, and Andrew J. Meyerriecks. *Biology*. 2d ed. Columbus, Ohio: Charles E. Merrill, 1985. A textbook written with exceptional clarity and organization. Discusses ecology and coevolution of plants and animals. Includes a glossary and index. Excellent photographs and diagrams. This is a college-level textbook but is also suitable for advanced high school students.

Rudman, William B. "Solar-Powered Animals." *Natural History* 96 (October, 1987): 50-53. A well-written article for all readers interested in plant-animal interactions, especially focusing upon mutualism involving marine corals, sea slugs, and their green algae allies. The undersea photography is excellent.

PLATYPUSES

Type of animal science: Classification
Fields of study: Conservation biology, embryology, reproduction science, wildlife ecology, zoology

The platypus is one of only three egg-laying mammals and is the only member of its family, Ornithorhynchidae. Its evolution in the relative isolation of Australia has led to such unique characteristics that the first specimen collected was believed to be a hoax.

Principal Terms

DORSAL: at or near the back

ELECTRORECEPTORS: sensors in the bill of the platypus that detect the weak electric field given off by animals; the electroreceptors help the platypus locate prey in murky water

METAZOA: organisms which are multicellular

MONOTREME: an animal that has a single, common, outer body opening for the excretory and reproductive systems; includes platypuses and echidnas

NOTOCHORD: longitudinal, flexible rod located between the gut and nerve cord

---

When the first platypus pelt arrived at London's Natural History Museum in the late 1790's, it was thought to be a fake made from bits of animals sewn together. This unusual mammal has a leathery bill, webbed feet, and fur, and it is one of only three mammals that lay eggs. Its body length is about eighteen inches, and its broad, flat tail is about seven inches long. The reclusive platypus spends most of its time in streams, rivers, and some lakes, foraging for food in the evening and sleeping during the day in burrows dug into the river banks.

Platypus Life

The unusual anatomical features of the platypus provide perfect adaptations for its life in water. The webbed feet are efficient paddles for swimming through the water. Claws on the feet help the platypus to dig burrows. Dense, waterproof fur covers the entire body except the feet and bill. The eyes and ear holes of the animal lie in folds that close when the animal is submerged, and the nostrils are located toward the end of the beak and also close under water. The bill is highly sensitive to touch, and is equipped with electrosensors that detect weak electrical fields produced by prey. Thus, the platypus can locate and capture prey in murky river bottoms without relying on vision, hearing, or smell.

Bottom-dwelling invertebrates, especially crustaceans, aquatic insects, and insect larvae, compose the majority of the platypus diet. Behind the bill are located two internal cheek pouches containing horny ridges that substitute for teeth, which are lost early in the life of the platypus. The pouches are used to store food while it is being chewed and sorted by the animal.

A male platypus has a spur on each rear ankle that is connected to a venom gland in the thigh. The spur is used against attackers, but also against competing males during the mating season. The venom is not fatal to humans, but can cause a great deal of pain. This feature makes the platypus one of very few mammals that are venomous.

From Egg to Adult

Courtship and mating occur in the water. After initial approaches by the female, the male chases and grasps her by the tail and inseminates her. After mating, a female will lay two to three eggs

Platypus Facts

Classification:
Kingdom: Animalia
Subkingdom: Bilateria
Phylum: Chordata
Subphylum: Vertebra
Class: Mammalia
Subclass: Prototheria
Order: Monotremata
Family: Ornithorhynchidae
Genus and species: Ornithorhynchus anatinus
Geographical location: Eastern Australia, Tasmania, Kangaroo Island in southern Australia
Habitat: Streams, rivers, and some lakes; must have permanent water and banks suitable for burrows
Gestational period: Seven to fourteen days
Life span: Ten or more years in the wild; seventeen years and longer in captivity
Special anatomy: Rubbery beak equipped with electrosensors used to detect prey; webbed feet; females have slits in abdominal walls from which milk expresses during lactation; males have spurs on rear feet used primarily to inject venom into rivals during mating season; reproductive and excretory tracts have one common opening to the exterior of the body

the water. The female blocks the entry to the nesting burrow with soil plugs to protect the eggs and young from predators and flooding. She removes and replaces the plug each time she leaves to forage for food. When the eggs hatch, in seven to fourteen days, the young are about one inch long and totally dependent on the mother. Platypuses do not have nipples, but milk is produced in the mammary glands and expressed through openings in the abdominal wall. The young suck the milk directly from the fur. At about five months of age, the young emerge from the nesting burrow and begin learning to search for prey themselves. A typical platypus in the wild will live about ten years.

—*Karen E. Kalumuck*

See also: Beaks and bills; Fauna: Australia; Lactation; Monotremes; Poisonous animals.

and incubate them in a special nesting burrow, which may extend one hundred feet away from

Bibliography

Brown, Malcolm W. "If a Platypus Had a Dream, What Would It Mean?" In *The Science Times Book of Mammals*, edited by Nicholas Wade. New York: Lyons Press, 1999. This highly readable article discusses scientific studies of brain activity during sleeping in platypuses, and how the study results have shed light on the evolution of monotremes.

Macdonald, David, ed. *The Encyclopedia of Mammals*. 2d ed. New York: Facts on File, 1995. This lavishly illustrated compendium of mammals gives detailed, accessible information on hundreds of wild mammals, organized by phylogenetic order. Of particular interest are the introductory chapter on mammals, and the chapter on monotremes, including the platypus.

National Geographic Society. *National Geographic Book of Mammals*. Washington, D.C.: Author, 1998. This excellent book for children and adults alike features detailed essays and marvelous photographs of over five hundred animals in their wild habitats.

Serena, Melody. "Duck-Billed Platypus: Australia's Urban Oddity." *National Geographic* 197, no. 4 (April, 2000): 118-129. Article chronicles the movement of platypuses to urban areas near Melbourne and the research study that tracks them; includes superb photographs.

Whitfield, Philip, ed. *The Simon and Schuster Encyclopedia of Animals*. New York: Simon & Schuster, 1998. This book, a who's who of the world's creatures, gives concise information on the platypus and other monotremes early in the work; its organization helps the reader compare animals across groups.

POISONOUS ANIMALS

Types of animal science: Classification, ecology, geography, physiology
Fields of study: Anatomy, biochemistry, ecology, invertebrate biology, zoology

Poisons disrupt life processes or kill. Animal poisons are venoms, delivered by biting, stinging, or body contact. Poisonous species occur throughout the animal kingdom, and include snakes, insects, spiders, other arachnids, mammals, lizards, and fish.

Principal Terms

ANAPHYLAXIS: hypersensitivity to a foreign substance, such as a venom, that causes discomfort and can even kill

ARACHNID: an arthropod having eight legs; a spider

ARTHROPOD: an organism with a horny, segmented external covering and jointed limbs

HEMOTOXIN: a substance that causes blood vessel damage and hemorrhage

NEUROTOXIN: a substance that damages the nervous system, most often nerves that control breathing and heart action

VENOM: a poison made by an animal

Substances that cause disease symptoms, injure tissues, or disrupt life processes on entering the body are poisons. When ingested in large quantities, most poisons kill. Poisons may be contacted from minerals, in vegetable foods, or in animal attack. Any poison of animal origin is a venom. Venoms are delivered by biting, stinging, or other body contacts. These animal poisons are used to capture prey or in self-defense. Often, it seems that ability to make venom arose in animals that were too small, too slow, or too weak to otherwise maintain an ecological niche. The mechanism for development of the ability to make venom is not clear.

The most familiar poisonous animals are snakes, insects, spiders, and some other arachnids. Poisonous species, however, occur throughout the animal kingdom, including a few mammals and lizards, and some fish. The severity of venom effects depends on its chemical nature, the nature of the contact mechanisms, the amount of venom delivered, and victim size. For example, all spiders are poisonous. However, their venom is usually dispensed in small amounts that do not affect humans. Hence, few spiders kill humans, though they kill prey and use venom in self-defense very effectively.

Chemically, venoms vary greatly. Snake venoms are mixtures of enzymes and toxins. Study of their effects led to the identification of hemotoxins, which cause blood vessel damage and hemorrhage; neurotoxins, which paralyze nerves controlling heart action and respiration; and clotting agents, which excessively promote or prevent blood clotting. Cobras, coral snakes, and arachnids all have neurotoxic venoms.

Poisonous Lizards, Arachnids, and Insects

The poisonous lizards are useful to explore first, because only two species are known: Gila monsters and beaded lizards (both holoderms). They inhabit the southwestern deserts of the United States and Mexico. They do not strike like snakes; rather, they bite, hold on, and chew to poison. Holoderm bites kill prey, but rarely kill humans. Beaded lizards grow to three feet long and Gila monsters grow to two feet long.

Most poisonous arthropods are spiders and scorpions. Both use venom to subdue and/or kill prey. As stated earlier, few spiders endanger humans because their venom is weak and is not injected in large quantities, but some species have very potent venom and harm or even kill humans. Best known of these are black widow spiders. Though rarely lethal to humans, black widow bites cause cramps and paralysis.

There are approximately six hundred scorpion species, of sizes between one and ten inches. All have tail-end stingers. Large, tropical scorpions can kill humans, while American scorpions are smaller and less dangerous. Scorpions are more dangerous than spiders because they crawl into shoes and other places where their habitat overlaps that of humans.

Many insects, such as caterpillars, bees, wasps, hornets, and ants, use venom in self-defense or to paralyze prey to feed themselves or offspring. Caterpillars use poison spines for protection. Bees, wasps, hornets, and ants use stingers for the same purpose. The venom of insects also kills many organisms that seek to prey on them. Humans, however, are rarely killed by insect bites. Such bites are usually mildly to severely painful for a period from a few minutes to several days. However, severe anaphylaxis occurs in some cases, followed by death.

Poisonous Snakes

Poisonous snakes are colubrids, elapids, or vipers, depending on their anatomic characteristics. All have paired, hollow fangs in the front upper jaw. The fangs fold back against the upper palate when not used, and when a snake strikes they swing forward to inject a venom that attacks the victim's blood and tissues. The heads of poisonous snakes are scale-covered and triangular. Such snakes are found worldwide and among them are pit vipers, named for the pits on each side of the head that contain heat receptors. The pits detect warm-blooded prey, mostly rodents, in the dark. Pit vipers include rattlesnakes, moccasins, copperheads, fer-de-lance, and bushmasters.

The population and species of American and European poisonous snakes differ. In North America, twenty such snake types occur: elapid coral snakes and copperheads, sixteen rattler

---

Bee Stings

Most American honeybees came from Europe. They were brought to North America by early colonists to make honey for human consumption. Due to hive escapees, honeybees are now found wild throughout the country. Familiar worker honeybees are often seen collecting nectar and pollen from flowers.

When alarmed, worker bees sting. This is unfortunate for them, as their barbed stingers stay in the skin of victims, and in pulling away after stinging, essential bee body tissue is ripped away. This causes the bee to die quickly. Workers also defend hives in this way, committing suicide to protect the hive. Bee stings often disable or kill predators who seek to eat the bees or their honey.

Honeybees are so widespread that most people are stung at least once during their lives. Some individuals are severely allergic to bee stings, and occasionally the severe anaphylactic shock kills. Sensitive individuals should carry sting kits. A kit includes epinephrine and antihistamine, a tourniquet, and alcohol swabs. A stinger works its way deep into the skin and can cause infection. Thus, it should scraped off after a sting.

---

types, and cottonmouths (all vipers). Vipers are found everywhere but Alaska. Rattlers have the widest habitat, as shown by their abundance in the snake-rich Great Plains, Mississippi Valley, and southern Appalachia. In contrast, copperheads and cottonmouths are abundant in Appalachia and the Mississippi valley, respectively. Mexican poisonous snakes are divided into two ranges, the northern from the U.S.-Mexican border to Mexico City, and the southern, south of Mexico City. In the north, snakes are mostly rattlers, as in the contiguous United States. Coral snakes and pit vipers are plentiful in the south. Most perilous are the five- to eight-foot fer-de-lance, whose venom kills many humans. In South America, all vipers but rattlers are tropical. Bushmasters, the largest South American vipers, and elapid coral snakes are nocturnal and rarely endanger humans. Tropical rattlers and lance-headed vipers, somewhat less nocturnal, kill many. Europe has few snakes, due to its cool climates and scarce suitable habitats. Its few vipers range almost to the Arctic Circle. Eastern Mediterranean regions hold most of the European vipers.

There are many poisonous snakes in Africa and Asia. North Africa, mostly desert, has few snakes. Central Africa's diverse poisonous snakes are colubrid, elapid, and viper types. Elapids include dangerous black mambas, twelve to fourteen feet long, and smaller cobras, which also occur in South Africa. Among diverse vipers, the most perilous are Gaboon vipers and puff adders. The Middle East, mostly desert, has few poisonous snakes. Southeast Asia has the most poisonous snakes in the world—elapids, colubrids, and vipers. This is due to snake habitats that range from semiarid areas to rain forests. The huge human population explains why this area has the world's highest incidence of snakebite and related death, due to vipers, cobras, elapids, and sea snakes. Vipers bite most often, but elapids cause a larger portion of deaths. The Far East snake population is complex and its snakebite incidence is also high. Its important poisonous snakes are pit vipers.

Australia and New Guinea have large numbers of poisonous snakes. Australia has 65 percent of the world's snakes, while New Guinea has 25 percent. Also, sea snakes occur offshore and in some rivers and lakes. However, these countries have few snakebite deaths, due to the small size and nocturnal nature of most of indigenous snakes.

Poisonous Fish and Amphibians

Venomous fish are dangerous to those who enter the oceans, especially fisherman who take them from their nets. The geographical distribution of these fish is like that of all other fish. The highest population density is in warm temperate or tropical waters. Numbers and varieties of poisonous fish decrease with proximity to the North and South Poles and they are most abundant in Indo-Pacific and West Indian waters.

A well-known group of poisonous fishes is the stingrays (dasyatids). They inhabit warm, shallow, sandy-to-muddy ocean waters. Dasyatids lurk almost completely buried, awaiting prey that they sting to death with barbed, venomous teeth in their tails. The tail poison is made in glands at the bases of the teeth. Small, freshwater dasyatids are found in South American rivers, such as the Amazon, hundreds of miles from the river mouths. Stingrays near Australia grow to fifteen-foot lengths. Emphasizing wide distribution of stingrays and their danger to humans is their mention in Aristotle's third century B.C.E. writings, and the death of John Smith in 1608, killed by a stingray while exploring Chesapeake Bay.

Also well known are the venomous Scorpaenidae fish family, many members of which cause very painful stings. Zebra fish and stonefish are good examples. Both, like all scorpaenids, have sharp spines supporting dorsal fins. The spines, used in self-defense, have venom glands. The

---

Sea Snakes

The name sea snake identifies members of about fifty species of poisonous water snakes, living in tropical oceans from the Persian Gulf to the southwest Pacific. They are very abundant in the Indian Ocean and Australian coastal waters. Sea snakes are five to nine feet long, depending on species. They have strong, flat, paddlelike tails used in swimming. Most species have nostrils atop the head. Sea snakes have lungs, not gills, and must surface for air. However, they remain underwater for several hours and can dive down twenty to thirty feet.

Most species live in warm, shallow waters, especially river mouths, bays, and swamps. Most rarely leave the water, so they lack the enlarged abdominal scales that terrestrial snakes use to move along the ground. A few types have abdominal scales like terrestrial snakes because they spend part of the time crawling in marshes. The eggs of sea snakes hatch internally and young are born alive. Sea snakes eat fish, especially eels, prawns, and fish eggs. They paralyze prey with their venom. They usually bite humans only if stepped on in shallow water or when removed from nets.

most deadly fish venom is that of the stonefish, which, when stepped on, can kill humans.

Poisonous animals which endanger by contact are exemplified by zebra fish and stonefish, just mentioned, and by poisonous frogs or toads. Most such frogs and toads, such as poison dart frogs, live in Africa and South America. They secrete poisons through the skin. Contact with the poisons causes effects which range from severe irritation to death in humans. The poisons frighten away or kill most predators that attempt eat them. The ecological function of poisonous animals is seen as helping to keep down the population of insects, rodents, arachnids, and small fishes. They thus contribute to maintaining the balance of nature. Poisonous land animals, such as scorpions and many poisonous snakes, are often nocturnal and add another dimension to pest control by nighttime predation.

—*Sanford S. Singer*

See also: Ants; Arthropods; Bees; Defense mechanisms; Fish; Frogs and toads; Insects; Jellyfish; Lizards; Predation; Scorpions; Snakes; Spiders; Wasps and hornets.

Bibliography

Aaseng, Nathan. *Poisonous Creatures*. New York: Twenty-first Century Books, 1997. This valuable book describes species from each animal family that uses venom for protection or to acquire food. It is well illustrated and has a useful bibliography.

Edström, Anders. *Venomous and Poisonous Animals*. Malabar, Fla.: Krieger, 1992. An interesting book that covers many types of poisonous animals, their venoms, and the basis for their toxicity.

Foster, Steven, and Roger Caras. *Venomous Animals and Poisonous Plants*. Boston: Houghton Mifflin, 1994. A very detailed book covering a great many poisonous animals, their habitat ranges, and their characteristics. It also includes many illustrations and numerous references.

Grice, Gordon D. *The Red Hourglass: Lives of the Predators*. New York: Delacorte, 1998. A fascinating book which describes predatory and poisonous animals, their characteristics, and their lives.

POLAR BEARS

Type of animal science: Classification
Fields of study: Anatomy, conservation biology, evolutionary science, wildlife ecology, zoology

The polar bear lives in one of the harshest environments on earth: the Arctic. Yet it succeeds through a combination of hunting skill and the ability to store and conserve energy in the form of fat.

Principal Terms

BLUBBER: thick layer of fat under the skin
CARNIVOROUS: eating meat
GEOGRAPHICALLY ISOLATED: living in different habitats
PLANTIGRADE: walking on the entire foot, not just the toes
TERRESTRIAL: living on land

Protected from cold by a thick fur coat and a layer of blubber, the polar bear hunts seals in open areas in the Arctic sea ice. A strong swimmer, the polar bear uses its large front paws as paddles. Its white fur blends in with the ice and snow as it stalks or still-hunts seals. Ringed seals are the polar bear's primary food, but it also consumes bearded seals, and occasionally walruses, belugas, narwhals, musk oxen, and carrion (dead terrestrial and marine mammals). Although largely carnivorous, when on land the polar bear may eat grasses, kelp, berries, and lichens. Males and nonpregnant females do not make dens or hibernate, but spend the winter hunting on the sea ice.

A thick layer of blubber and a heavy fur coat provide insulation that allows polar bears to swim in icy Arctic waters. (PhotoDisc)

The polar bear evolved from the terrestrial brown bear about 100,000 to 200,000 years ago. In captivity, polar bears and brown bears have interbred and produced fertile offspring. This shows a high degree of genetic relatedness. In nature, however, these two species are geographically isolated from one another and would rarely meet.

Swimming in icy water among the ice floes of the Arctic, quietly stalking a resting seal, then killing it with a crushing blow of its forepaw, the polar bear is an impressive example of an animal's ability to adapt to and live in one of the harshest environments on earth.

Physical Characteristics of Polar Bears

Male polar bears are up to five feet high at the shoulder while on all fours and up to ten feet long. When standing on its hind legs, a male can be eleven feet tall. Adult males are generally much bigger than females: 1,100 to 1,770 pounds for males, 330 to 770 pounds for females. Like other bears, polar bears are plantigrade.

Mating takes place in late March to late May. This is the only time that the male is with the female. Other than family groups of females with their cubs, polar bears are solitary. Pregnant females dig snow or earth dens in which they will give birth to one to three cubs in late November to early January. The mother's milk is very rich, with an average fat content of 33 percent. The cubs, which weigh 1 to 1.5 pounds at birth, grow quickly and weigh 22 to 33 pounds when they emerge from the den with their mother in late February to early May. Polar bear cubs usually leave their mother at 2.5 years of age, at which time the mother is ready to breed again. Therefore, the female usually gives birth every third year.

Conservation

By the 1960's, polar bear populations were in serious decline due to sport hunting. In 1967, the five "polar bear nations" (the United States, Canada, Denmark, Norway, and the Soviet Union) limited hunting to the Inuit (Eskimo) people. By 2001, polar bears had recovered, and there were twenty thousand to forty thousand in the world.

A 1999 study by longtime polar bear biologist Ian Stirling and his colleagues shows that polar bears at Hudson Bay are 10 percent thinner and have 10 percent fewer cubs than they did twenty years ago. In 1999, the ice on Hudson Bay melted three weeks earlier than it did twenty-five years earlier. Polar bears must wait for ice to form each

---

Polar Bear Facts

Classification:
Kingdom: Animalia
Subkingdom: Bilateria
Phylum: Chordata
Subphylum: Vertebrata
Class: Mammalia
Subclass: Eutheria
Order: Carnivora
Suborder: Fissipedia
Family: Ursidae
Subfamily: Ursinae
Genus and species: Ursus maritimus

Geographical location: Northern marine areas of Alaska, Canada, Greenland, Norway's Svalbard Archipelago, and Russia

Habitat: The sea ice and adjacent land areas of the circumpolar Arctic

Gestational period: About eight months

Life span: Up to thirty-two years in the wild, over forty years in captivity

Special anatomy: Fur, hide, and blubber providing effective insulation from the extreme arctic cold; white fur providing camouflage against the backdrop of ice and snow and greatly aiding in stalking seals; large paws used for paddling in water and which act like snowshoes on thin ice; pads of the feet covered with small, soft papillae that improve traction on ice; small ears and tail to reduce heat loss and a large body mass to conserve heat

fall to hunt ringed seals, their main food source. Hudson Bay polar bears fast six to eight months each year and then hunt seals intensively during the ice season. The three-week reduction of hunting time has not yet resulted in significant decline of the polar bear population, but is expected to do so if the climate trend continues.

—Thomas Coffield

See also: Bears; Carnivores; Fauna: Arctic; Grizzly bears; Pandas.

Bibliography

Lynch, Wayne. *Bears: Monarchs of the Northern Wilderness*. Seattle: The Mountaineers, 1993. Well-researched, readable book for the general public. Excellent photographs by the author.

Matthews, Downs. *Polar Bear*. San Francisco: Chronicle, 1993. Clear, informative text. Excellent photos by Dan Guravich.

Stirling, Ian. *Polar Bears*. Ann Arbor: University of Michigan Press, 1988. Written for the general reader by the world's foremost polar bear researcher. Richly illustrated with photographs by Dan Guravich.

Stirling, Ian, Nicholas J. Lunn, and John Iacozza. "Long Term Trends in Population Ecology of Polar Bears in Western Hudson Bay in Relation to Climate Change." *Arctic* 52, no. 3 (1999): 294-306. Details the effect of climate change on polar bear weight and hunting patterns.

POLLUTION EFFECTS

Types of animal science: Ecology, geography, physiology, scientific methods
Fields of study: Conservation biology, ecology, environmental science, pathology, physiology

Pollutants in soil, water, and atmosphere have created enormous problems for the living world. Destroyed habitats and polluted food sources and drinking water for animals have caused deformations in animal growth, development, and reproduction, as well as a shortening of life span, all of which contribute to an accelerated decrease in biodiversity and the extinction of more species. Realization of the severity of pollution is the first step toward seeking a long-term remedy.

Principal Terms

ACID RAIN: the burning of primarily fossil fuels leads to an excessive release of nitrogen oxides or sulfur dioxide, which in turn combines with water to form nitric or sulfuric acid; when acids dissolve into rainfall, it is called acid rain

CHEMICAL POLLUTANTS: harmful chemicals manufactured and released to the environment; normally referring to those that contaminate ecosystems

CHLOROFLUOROCARBONS (CFCs): a group of very stable compounds used widely since their development in 1928 for refrigeration, coolants, aerosol spray propellants, and other purposes; once risen into the stratosphere, they cause ozone depletion

GREENHOUSE EFFECT: the process by which certain gases, such as carbon dioxide and methane, trap sunlight energy in the atmosphere as heat, resulting in global warming as more gases are released to the atmosphere by human activities

OZONE LAYER: the ozone-enriched layer of the upper atmosphere that filters out some of the sun's ultraviolet radiation, which causes skin and other types of cancer

During the last decade of the twentieth century, the environmental problems predicted by environmental scientists decades previously began to be aggravated in a variety of ways. These included population explosion, food imbalances, inflation brought about by energy resource scarcity, acid rain, toxic and hazardous wastes, water shortages, major soil erosion, a punctuated ozone layer, and greenhouse effects. The list goes on. As a result of pollution, decreases in biodiversity and the extinction of both plant and animal species have accelerated. The burning and cutting of thousands of square miles of rain forest not only destroyed habitats for numerous animal species, but also caused irreversible damage to ecosystems and climate. The recurring drought and famine in Africa are testimony to human mischief toward Mother Nature. The well-being of animals as well as humans will not be protected against the ecological consequences of human actions by remaining ignorant of those actions. Effective measures taken to reduce pollution and protect natural resources and the environment first come with a recognition of these problems. The ignorance and inaction of ordinary citizens will lead to disastrous consequences for the environment, threatening humanity's very existence.

Source and Types of Pollution

Industrialization and the expansion of the human population have left relatively few places on earth undisturbed. In simple terms, human interference in natural ecosystems is the single most important source of pollution. First, heavy dependence upon fossil fuels for energy, and on synthetic chemicals and materials helped to dump millions of metric tons of nonnatural compounds and chemicals into the environment. Among them are agrichemicals such as fertilizers, insecticides, fungicides, and herbicides, and home products. Application of excess chemical fertilizers to soil hampers natural cycling of nutrients, depletes the soil's own fertility, and destroys the habitat for thousands of small animals residing in the soil. Farm runoff carries priceless topsoil, expensive fertilizer, and animal manure into rivers and lakes, where these potential resources become pollutants. In the city, water pours from sidewalks, rooftops, and streets, picking up soot, silt, oil, heavy metals, and garbage. It races down gutters into storm sewers, and a weakly toxic soup gushes into the nearest stream or river. Many of these chemicals also seep through into the ground, contaminating groundwater.

Plants and factories manufacturing these chemical products are another source of pollutants and contamination. Burning fossil fuels releases greenhouse gases, carbon oxides, and methane. Coupled with deforestation in many regions of the world, carbon dioxide concentration in the atmosphere has steadily climbed, from 290 parts per million in 1860 to 370 parts per million in 1990, more than a 25 percent rise due to industrialization. The resultant global warming will have far-reaching effects on plants, animals, and humans in ways still not understood. Acid rain, a result of overcharging the atmosphere with nitric oxides and sulfur dioxide (two gases also released by burning of fossil fuels), has increased the acidity of soil and lakes to levels many organisms cannot survive. The most acidic rain is concentrated in the northeastern United States. In New York's Adirondack Mountains, for instance, acid rain has made about a third of all the lakes and ponds too acidic to support fish. First, much of the food web that sustains the fish is destroyed. Clams, snails, crayfish, and insect larvae die first, then amphibians, and finally fish. The detrimental effect is not limited to aquatic animals. The loss of insects and their larvae and small aquatic animals has contributed to a dramatic decline in the population

The Freshwater Crisis

Toxic chemicals are contaminating groundwater on every inhabited continent, endangering the world's most valuable supplies of fresh water, reports a study published in 2000 from the Worldwatch Institute, a Washington, D.C.-based research organization on the environment. This worldwide survey of groundwater quality shows that pesticides, nitrogen fertilizers, industrial chemicals, and heavy metals are contaminating groundwater everywhere, and that the damage is often worst in places where people most need water.

There are at least three essential roles of groundwater: providing drinking water, irrigating farmland, and replenishing rivers, streams, and wetlands. About one third of the world population relies almost exclusively on groundwater for drinking. Groundwater provides irrigation for some of the world's most productive farmland. Over 50 percent of irrigated croplands in India, and 40 percent in the United States, is watered by groundwater. Groundwater plays a crucial role in replenishing rivers, streams, and wetlands. It provides much of the flow for great rivers such as the Mississippi, the Niger, the Yangtze, and many more.

The range of groundwater contamination is stunning. Groundwater in all twenty-two major industrial zones surveyed by the Indian government in the late 1990's was unfit for drinking. In a 1995 study of four northern Chinese provinces, groundwater was contaminated in more than 50 percent of surveyed locations. One third of the wells tested in a California region in 1988 contained the pesticide 1,2-dibromo-3-chloropropane (DBCP) at levels ten times higher than the maximum allowed for drinking. The list goes on. There is a compelling urgency to prevent groundwater contamination.

Pollutants in water, from pesticides to acid rain, cause disruption and death all the way up the food chain. (PhotoDisc)

ozone to oxygen gas. As a result of the decline of ozone and the punctuation of the ozone layer, UV radiation has risen by an average of 8 percent per decade since the 1970's. This depletion of the ozone layer poses a threat to humans, animals, plants, and even microorganisms.

Pollution Effects of Chemicals
The degradation of air, land, and water as a result of the release of chemical and biological wastes has wide-ranging effects on animals. On a large scale, pollution destroys habitats and produces population crashes and even the extinction of species. Hazardous chemicals introduced into an environment sometime render it unfit for life (as at Love Canal, New York, or Times Beach, Missouri). At the individual level, pollution causes abnormalities in growth, development, and reproduction. Hazardous chemicals, introduced either intentionally (such as fertilizers, herbicides, and pesticides) or through neglect (as with industrial wastes), have a variety of detrimental, sometimes devastating effects on animals. They affect the metabolism, growth and development, reproduction, and average life span of many species.

A few examples will illustrate the effects of chemical pollution on animals. In the 1940's, the new insecticide dichloro-diphenyl-trichloroethane (DDT) was regarded as a miracle. It saved millions of lives in the tropics by killing the mosquitoes that spread deadly malaria. DDT saved millions more lives with increased crop yields resulting from DDT's destruction of insect pests. This miraculous pesticide, however, turned out to be a long-lasting nemesis to many species of wildlife and to the environment. In the United States, ecologists and wildlife biologists during the 1950's and 1960's witnessed a stunning decline in the populations of several predatory birds, especially fish-eaters, such as bald eagles, cormorants, ospreys, and brown pelicans. The population de-

of black ducks that feed on them. The result is a crystal-clear lake, beautiful but dead.

Another serious problem created by the chemical industry is ozone depletion. CFC compounds contain chlorine, fluorine, and carbon. Since their development in the 1930's, these compounds have been widely used as coolants in refrigerators and air conditioners, as aerosol spray propellants, as agents for producing Styrofoam, and as cleansers for electronic parts. These chemicals are very stable and for decades were considered to be safe. Their stability, however, turned out to be a real problem. In gaseous form they rise into the atmosphere. There, the high energy level of ultraviolet (UV) light breaks them down, releasing chlorine atoms, which in turn catalyzes the breakdown of

cline drove the brown pelican and bald eagle close to extinction. In 1973, the U.S. Congress passed the Endangered Species Act, which banned the use of DDT. The once-threatened species have somewhat recovered since. In the mid 1950's, the World Health Organization used DDT on the island of Borneo to control malaria. DDT entered food webs through a caterpillar. Wasps that fed on the caterpillar were first destroyed. Gecko lizards that ate the poisoned insects accumulated high levels of DDT in their bodies. Both geckos and the village cats that ate the geckos died of DDT poisoning. The rat population exploded with its natural enemy, cats, eliminated. The village was then threatened with an outbreak of plague, carried by the uncontrolled rats.

Although DDT has been banned in much of the world, there is a growing concern over the effects of a number of chlorinated compounds. These chemicals, described as "environmental estrogens," interfere with normal sex hormone functions by mimicking the effects of the hormone estrogen or enhancing estrogen's potency. High levels of chlorinated compounds, such as dioxin and polychlorinated biphenyls (PCBs), in the Great Lakes have led to a sharp decline in populations of river otters and a variety of fish-eating birds, including the newly returned bald eagles. These chemicals are also the cause of deformed offspring, or eggs that never hatch. In Florida's Lake Apopka, a spill of chlorinated chemicals in 1980 led to a 90 percent drop in the birth rate of the lake's alligators. These are only a few examples of the detrimental effects of synthetic chemicals on various animals.

Effects of Air Pollution

Air pollution leads to acid rain and the greenhouse effect, as well as damage to the ozone layer. Acid rain drops out of the skies onto areas at great distances from the source of the acids, and destroys forests and lakes in sensitive regions. As a result, fish populations are dwindling or being eliminated from lakes and streams by a lower pH caused by acid deposition. The strongest evidence comes from data collected from the past twenty-

Endangered Amphibians

Frogs and toads have been amazingly resilient throughout their evolution and hence have a wide distribution in ponds and swamps all over the planet. Their evolutionary longevity, however, appears to offer inadequate defense against the pollution to the environment brought about by human activities. During the 1990's, biologists around the world have documented an alarming decline in amphibian populations. Thousands of species of frogs, toads, and salamanders are experiencing dramatic decreases in numbers, and many have gone extinct or become endangered.

The specific causes of the worldwide decline in amphibians are not fully understood. All the likely factors, however, point at human modification of the biosphere, the portion of the earth that sustains life. The draining of wetland habitats, one type of habitat destruction, is a major cause of the decline. Amphibian bodies at all developmental stages are protected only by thin, permeable skins through which pollutants can easily penetrate. The double life on land and in water exposes their permeable skin to a wide range of aquatic and terrestrial habitats and hence a correspondingly wide range of environmental toxins.

Many amphibians' eggs and larvae develop in ponds and streams during spring, a time when acid rain causes an acidity increase in freshwater ecosystems. Increased ultraviolet (UV) light damages eggs and causes deformity among offspring. In addition, many amphibians are herbivores as larvae and insectivores as adults. Thus, they are vulnerable to both herbicides and insecticides in their food. Dramatic decline of amphibian populations provides an early warning of environmental degradation. It also depletes the food source for large carnivores that feed on them and may risk keeping insect populations in check.

five to forty-five years in Adirondack lakes and in Nova Scotia rivers. Studies during this period clearly show declines in acid-sensitive species. Similar results were obtained from analyzing fish population and water acidity in Maine, Massachusetts, Pennsylvania, and Vermont. The consensus is that fish populations would be eliminated if the surface waters were to acidify to between pH 5.0 and pH 5.5. The effects of acid rain on other animals are indirect, either through the dwindling fish population (as a food source for other animals) or stunted forest growth (disturbance to habitats).

The effect of global warming on the animal kingdom is also a serious and complex issue. As global temperature rises, ice caps in polar regions and glaciers melt, ocean waters expand in response to atmospheric warming, and thus the sea level elevates. The expected sea level rise will flood coastal cities and coastal wetlands. These threatened ecosystems are habitats and breeding grounds for numerous species of birds, fish, shrimp, and crabs, whose populations could be severely diminished. The Florida Everglades will virtually disappear if the sea level rises two feet. The impact of global warming on forests could be profound. The distribution of tree species is exquisitely sensitive to average annual temperature, and small changes could dramatically alter the extent and species composition of forests. This in turn could dramatically alter the population distribution of animals. A rise may make temperatures unsuitable for some species, hence reducing biodiversity.

The effect of the punctuated ozone layer on animals is yet to be fully understood. It is known that the high energy level of UV radiation can damage biological molecules, including the genetic material deoxyribonucleic acid (DNA), causing mutation. In small quantities, UV light helps the skin of humans and many animals produce vitamin D, and causes tanning. However, in large doses, UV light causes sunburn and premature aging of skin, skin cancer, and cataracts, a condition in which the lens of the eye becomes cloudy. Due to UV radiation's ability to penetrate, even animals covered by hair and thick fur cannot escape from these detrimental effects. Ozone damage costs U.S. farmers over $2 billion annually in reduced crop yields. All who depend on forestry and agriculture may bear a much higher cost if the emission of pollutants that destroy ozone is not regulated soon.

Possible Remedies

The various types of pollution all have serious effects on the plant and animal species that share this planet. It is all too easy to document the impacts of pollution on human health and ignore its effects on the rest of the living world. Any possible remedies to alleviate these problems should start with education, the realization of these problems at an individual as well as a global level. The tasks seem to be insurmountable, and no organization, no country can do it alone. It takes willingness to accept short-term inconvenience or economic sacrifice for the benefit of the long run. A couple of examples serve to illustrate what can be done to alleviate the problems of pollution.

Synthetic chemical pollutants that are poisoning both people and wildlife could be largely eliminated without disrupting the economy, as reported in a study published in 2000 by the Worldwatch Institute, a Washington, D.C.-based environmental organization. The report presents strong evidence from three sectors that are major sources of these pollutants—paper manufacturing, pesticides, and PVC plastics—to show that nontoxic options are available at competitive prices. Agricultural pollution can be mitigated, significantly reduced, or virtually eliminated through the use of proper regulation and economic incentives. Farmers from Indonesia to Kenya are learning how to use less of various chemicals while boosting yields. Since 1998, all farmers in China's Yunnan Province have eliminated their use of fungicides while doubling rice yields, by planting more diverse varieties of the grain. In most, if not all, cases, the question is not whether it is possible to alleviate pollution of the environment; rather, it is whether we realize the urgency and/or are willing to take a high road to do it. For the common well-being of generations to come, better approaches have to be taken to preserve the environment and biodiversity.

—*Ming Y. Zheng*

See also: Chaparral; Death and dying; Diseases; Ecology; Ecosystems; Endangered species; Extinction; Food chains and food webs; Forests, coniferous; Forests, deciduous; Grasslands and prairies; Habitats and biomes; Lakes and rivers; Marine biology; Mountains; Mutations; Rain forests; Savannas; Tidepools and beaches; Tundra; Urban and suburban wildlife; Wildlife management; Zoology.

Bibliography

Brown, Lester. *State of the World 2000*. New York: W. W. Norton, 2000. Provides concerned citizens with a comprehensive framework for the global debate about our future in the new century. Makes invaluable analysis of negative environmental trends and a guide to emerging solutions.

Hill, Julia B. *The Legacy of Luna: The Story of a Tree, a Woman, and the Struggle to Save the Redwoods*. San Francisco: HarperSanFrancisco, 2000. Perched in a thousand-year-old redwood named Luna, the author tells the amazing story of how she is battling the elements in a fight against a major corporation to save the ancient redwood forests. A must read for anyone who is concerned about future generations and the living legacy we will leave them.

Johnson, Arthur H. "Acid Deposition: Trends, Relationships, and Effects." *Environment* 28, no. 4 (May, 1986): 6-11, 34-39. The rain in many parts of the world has taken on a new and threatening complexity in the form of acid rain, which kills fish and other wildlife, corrodes buildings, and destroys forests.

Lippmann, Morton, ed. *Environmental Toxicants: Human Exposures and Their Health Effects*. 2d ed. New York: John Wiley & Sons, 2000. This book, written by a number of authors, describes all major environmental toxicants and their possible short- and long-term effects on human health. Well-written and very informative.

Sampat, Payal. *Deep Trouble: The Hidden Threat of Groundwater Pollution*. Washington, D.C.: Worldwatch Institute, 2000. Describes the essential roles of groundwater, the status of contamination by various chemicals including fertilizers, pesticides, and industrial wastes. Outlines possible measures for reducing the risk of groundwater contamination.

POPULATION ANALYSIS

Type of animal science: Ecology
Fields of study: Ecology, wildlife ecology, zoology

Many animal populations are becoming threatened or endangered, primarily due to loss of suitable habitat. Population analysis enables biologists to examine the factors which lead to declines in animal populations and thus is important in the management of wild species.

Principal Terms

CONTINUOUS GROWTH: growth in a population in which reproduction takes place at any time during the year rather than during specific time intervals

DENSITY-DEPENDENT GROWTH: growth in a population in which the per capita rates of birth and death are scaled by the total number of individuals in the population

DISCRETE GROWTH: growth in a population that undergoes reproduction at specific time intervals

POPULATION ANALYSIS: the study of factors that influence growth of biological populations

---

A population is a group of organisms belonging to the same species that occur together in the same time and place. For example, a wildlife biologist might be interested in studying the population of porcupines that inhabits a hemlock forest or the population of bark beetles that lives on a particular tree. Populations can change over time. They increase or decrease in size, and their change in size can depend on a wide variety of factors. Population analysis is the study of biological populations, with the specific intent of understanding which factors are most important in determining population size.

In order to conduct a population analysis, one must first determine whether the population of interest is best understood as discrete or continuous.

A discrete population is one in which important events such as birth and death happen during specific intervals of time. A continuous population is one in which births, deaths, and other events take place continuously through time. Many discrete populations are those with nonoverlapping generations. For example, in many insect populations, the adults mate and lay eggs, after which the adults die. When the juveniles achieve adulthood, their parental generation is no longer living. In contrast, most continuous populations have overlapping generations. For instance, in antelope jackrabbits (*Lepus alleni*), females may give birth at any time during the year, and members of several generations occur together in space and time.

Modeling Animal Populations

The dynamics of animal populations is affected by a wide variety of demographic factors, including the population birth rate, death rate, sex ratio, age structure, and rates of immigration and emigration. In order to understand the effects of these factors on a population, biologists use population models. A model is an abstract representation of a concrete idea. The representation created by the model boils the concrete idea into a few critical components. By building and examining population models, population analysts investigate the relative importance of different factors on the dynamics of a given population.

A basic mathematical model of population size is as follows: $N_{t+1} = N_t + B - D + I - E$ (equation 1), where N_{t+1} equals the population size after one time interval, N_t equals the total number of individuals in the population at the initial time, B equals the number of births, D equals the number of deaths, I equals the number of immigrants into the population, and E equals the number of emigrants leaving the population. This simple model boils population size down to just four factors, B, D, I, and E. This model is not meant to be a true or precise representation of the population; rather, it is meant to clarify the importance of the factors of birth, death, immigration, and emigration on population size. To use the same model to examine the rate of growth of a population through time, it can be rearranged as follows: $N_{t+1} - N_t = B - D + I - E$ (equation 2). That is, the increase or decrease in the population size between time intervals t and $t+1$ is based on the number of births, deaths, immigrants, and emigrants.

When population biologists choose to focus specifically on the importance of birth and death in population dynamics, population models are simplified by temporarily ignoring the effects of immigration and emigration. In this case, the degree of change in the population between time intervals t and $t+1$ becomes: $N_{t+1} - N_t = B - D$ (equation 3). It is usually safe to assume that the total number of births (B) and deaths (D) in a population is a function of the total number of individuals in the population at the time, N_t. For example, if there are only ten females in a population at time t, it would be impossible to have more than ten births in the population. More births and deaths are possible in larger populations. If B equals the total number of births in the population, then B is equal to the rate at which each individual in the population gives birth, times the total number of individuals in the population. Likewise, the total number of deaths, D, will be equal to the rate at which the individuals in the population might die times the total number of individuals in the population. In other words, $B = bN_t$ and $D = dN_t$ (equation 4), where b and d represent the per capita rates of birth and death, respectively.

Given this understanding of B and D, the original model becomes $N_{t+1} - N_t = (bN_t) - (dN_t)$ or $N_{t+1} - N_t = (b - d)N_t$ (equation 5). It would be useful to find a variable which can represent per capita

births and deaths at the same time. Biologists define r as the per capita rate of increase in a population, which is equal to the difference between per capita births and per capita deaths: $r = b - d$ (equation 6). Thus, the equation which examines the changes in population size between time intervals t and $t+1$ becomes: $N_{t+1} - N_t = rN_t$ (equation 7).

A numerical example works as follows. In a population that originally had 1000 individuals, a per capita birth rate of 0.1 births per year and a per capita death rate of 0.04 deaths per year, the net change in the population size between the year t and $t+1$ would be:

$$r = 0.1 - 0.04 = 0.06$$
$$N_{t+1} - N_t = 0.06(1000) = 60$$

In other words, the population would increase by sixty individuals over the course of one year.

This model works for populations in which events take place during discrete units of time, such as a population of squirrels in which reproduction takes place at only two specific times in a single year. In contrast, many populations are continuously reproductive. That is, at any given time, any female in the population is capable of reproducing. When these conditions are met, time is viewed as being of a more fluid than discrete nature, and the population exhibits continuous growth. Models of population growth are slightly different when births and deaths are continuous rather than discrete. One way to imagine the difference between a population with continuous rather than discrete growth is to imagine a population in which each time interval is infinitesimally small. When these conditions are met, the model for population growth becomes: $\delta N / \delta t = rN$ (equation 8), where $\delta N / \delta t$ represents the changes in numbers in the population over very short time intervals. The per capita rate of increase (r) can now also be called the instantaneous rate of increase because the population is one with minute time intervals.

How does a population biologist select the best model? Which model is best depends on exactly what it is that a scientist is trying to understand about a population. In the first model presented above (equation 1), the different effects of birth, death, immigration, and emigration can be compared relative to one another. In the second model, the effects of immigration and emigration are ignored and the effects of birth and death are summarized into one constant called the per capita rate of increase (equations 7 and 8). If the scientist is trying to understand the cumulative effects of B, D, I, and E on the population, then equation 1 would represent a good model. On the other hand, if the scientist is trying to understand how births and deaths influence the net changes in population size, equation 7 or 8 would be a better model.

Effects of Density on Population Growth

When dealing with a continuous rather than a discrete population, equation 8 represents the rate of population growth as a function of per capita births and per capita deaths in the population. Equation 8 represents a population that is growing exponentially without bound. In other words, regardless of the population size at any given

time, the per capita rate of increase remains the same. It would be reasonable to assume that per capita rates of increase can actually change with changes in overall population size. For example, in a population of bark beetles inhabiting the trunk of a tree, many more resources are available to individual beetles when the population is small. Resources must be shared between more and more individuals as the population size increases, which can result in changes to the per capita rate of increase. A model of population growth that incorporates the effect of overall population density on the per capita rate of increase might look like: $\delta N/\delta t = r(1-N/K)N$ (equation 9), where K is equal to the carrying capacity, the maximum number of individuals in the population that there are adequate resources to support. The per capita rate of increase in equation 9 is not simply r by itself, but becomes $r(1-N/K)$. The per capita rate of increase is a function of rates of birth and death scaled by the population size and the carrying capacity of the habitat. If the population is very large relative to the number of individuals that the habitat can support, then $N \approx K$, and the expression $(1-N/K)$ becomes approximately equal to 0. When so, equation 9 takes the form $\delta N/\delta t = r(0)N = 0$ (equation 10) and the rate of population growth is zero. In other words, the population has ceased growing. On the other hand, if the population is very small relative to the number of individuals the habitat can support, then $N \ll K$ and the expression $(1-N/K)$ becomes approximately equal to 1. When so, equation 9 takes the form $\delta N/\delta t = r(1)N = rN$ (equation 11) and the rate of population growth remains a function of the rates of birth and death, but not the population size or carrying capacity. Thus, equa-

Declining Songbird Populations

Population biologists are interested in monitoring the status of migratory songbirds in the United States. The warbling vireo is a neotropical migrant that overwinters in Mexico and flies northward into the United States during the summer to breed. Population analysis has shown a decline in the coastal California population of warbling vireos over the last twenty-two years. Biologists examined the level of nest productivity and adult survivorship as indices of the per capita birth and death rates in the population, and found that declines in the population were most likely related to low reproductive success and not low survival rates. In terms of population models, this finding would indicate a population decline resulting from the low birth rate leading to an overall decline in numbers.

This example illustrates the importance of population analysis for conservation biology. By understanding the relative effects of birth and death on the warbling vireo population size, conservation biologists can develop the best strategies to sustain vireo populations. By determining that it is nest failure and not adult survivorship that is causing population declines, biologists can focus on eradicating factors that limit nest success and thereby increase the per capita birth rate in the population.

tion 9 represents what is called density-dependent growth.

Effects of Sex Ratio and Age Structure on Population Growth

The model set forth in equation 9 only takes into account the ways in which births, deaths, and population density relative to carrying capacity influence population growth. Sometimes it is helpful to understand how other factors such as the sex ratio and age structure in a population influence rates of growth. For example, deer hunters are not always allowed to take equal numbers of bucks and does from a population. Similarly, fishermen are often restricted in the size of fish they are allowed to keep when fishing. These wildlife management restrictions on the sex and size of animals that can be hunted arise from the fact that both age and sex can influence population growth rates. Models that incorporate the effects of age structure and population sex ratios will not be covered here. Suffice it to say that a population that consists mostly of young individuals yet to reproduce will grow more quickly than a popula-

tion equal in size but consisting of mostly older individuals who have finished reproducing. Similarly, a population with a highly skewed sex ratio that has many more males than females will not grow as quickly as a population of equal size in which the numbers of males and females are equal.

Population analysis is the study of biological populations, with the specific intent of understanding which factors are most important in determining population size. Factors such as the per capita rates of birth and death, the population density, age structure, and sex ratio all contribute to population size. Understanding how these factors interact to influence population size is critical if biologists hope to manage populations of organisms at sustainable levels for hunting or fishing and if conservation biologists hope to prevent populations from going extinct.

—*Erika L. Barthelmess*

See also: Birth; Communities; Death and dying; Demographics; Groups; Herds; Migration; Packs; Population fluctuations; Population genetics; Population growth; Wildlife management.

Bibliography

Gardali, Thomas, et al. "Demography of a Declining Population of Warbling Vireos in Coastal California." *The Condor* 102, no. 3 (August, 2000): 601-609. Primary research article presenting nineteen years worth of data on a California warbling vireo population. Valuable for a specific example of population analysis used in a conservation context.

Hastings, Alan. *Population Biology: Concepts and Models*. New York: Springer-Verlag, 1997. Graduate-level textbook with several chapters on population analysis. Detailed mathematical examination of different analytical models.

Hedrick, Philip W. *Population Biology: The Evolution and Ecology of Populations*. Boston: Jones and Bartlett, 1984. Undergraduate-level textbook on population biology. Includes several useful chapters about population analysis, including models of growth and methods used in population demography.

Johnson, Douglas H. "Population Analysis." In *Research and Management Techniques for Wildlife and Habitats*, edited by Theodore A. Bookhout. 5th ed. Bethesda, Md.: The Wildlife Society, 1994. Applied and detailed chapter dealing with analysis of wildlife populations. Many specific examples and a detailed discussion of several different population models.

POPULATION FLUCTUATIONS

Type of animal science: Ecology
Fields of study: Genetics, population biology, reproduction science

The simplest realistic models of population growth produce populations that rise to some level and then stay there. These models cannot produce the complicated array of fluctuations observed in natural populations. Fluctuations vary in period from a few weeks to many decades and reach sufficient amplitude to threaten populations (and entire species) with extinction.

Principal Terms

ABUNDANCE: in ecology, the number of organisms living in a particular environment

ANTHROPOGENIC DISTURBANCE: a change (usually a reduction) in population size caused by human activities

DENSITY-DEPENDENT POPULATION REGULATION: the regulation of population size by factors or interactions intrinsic to the population; the strength of regulation increases as population size increases

IRRUPTION: a sudden increase in the size of a population, usually attributed to a particularly favorable set of environmental conditions

POPULATION: all the individuals of the same species in the same place at the same time

POPULATION DENSITY: the number of individuals in a population per unit of area or volume

QUADRAT: a sample plot of a specific size and shape used in one method of determining population size or species diversity

SESSILE: not free to move around; mollusks are sessile organisms

Every species on earth is composed of one or more populations. The number of organisms making up a population is never constant; it always changes over time. The populations of some species change in predictable or cyclical ways, whereas populations of other species frequently exhibit seemingly unpredictable and noncyclic changes. Fluctuations in population size may be caused by changes in the population's environment; for example, seasonal changes in temperature or moisture produce seasonal fluctuations in population size. Resource limitations may produce density-dependent reductions in the growth rate of a population, which, if the reduction is not instantaneous, can result in oscillations in population size. Interactions with other species also produce population fluctuations; mathematical models of predator-prey systems typically produce oscillations in the abundance of both predators and prey. Finally, natural or anthropogenic disturbances often reduce the size of a population, which then either recovers its former abundance over time or declines further to local (or global) extinction.

The Time Scales of Population Fluctuations

Population fluctuations occur over many different time scales. On a geological time scale (occurring over millions of years), species arise, increase to some level of abundance, and finally become extinct. These long-term patterns of species abundance provide a background for understanding population fluctuations that occur over ecological time (over days, weeks, years, or centuries). Fluctuations at these shorter time scales draw most of the attention of ecologists interested in population dynamics.

Many species of animals, including numerous insects and several small vertebrates, exhibit a more or less annual life cycle, characterized by increasing numbers and higher levels of activity during the summer (or wet season) and by dormancy or decreasing numbers during the winter (or dry season). Even highly mobile animals, such as birds, exhibit a strong seasonal pattern of abundance, if viewed from a local perspective; in North America, for example, most songbirds migrate to more tropical latitudes in the fall and to temperate latitudes in the spring, thereby producing a yearly cycle of abundance in each location. Yearly cycles of abundance are predictable and easily explainable in terms of seasonal patterns of temperature, moisture, and sunlight. Of more interest to ecologists are population fluctuations that appear to be random or unpredictable from year to year or those fluctuations that occur out of synchrony with climatic cycles.

Regular Fluctuations

Nonseasonal fluctuations are of two main types: those that exhibit more or less regular cycles of abundance over several years and those that seem to fluctuate irregularly or noncyclically. A three-to-four-year cycle of abundance is characteristic of several species of mice, voles, and other rodents found in far northern latitudes. Probably the best-known example of this type of cycle is that observed in lemming species in the northern tundra of Europe and North America. Lemming populations exhibit very high densities every three to four years, with such low densities in the intervening years that they are difficult to locate and study. This boom-or-bust cycle is apparently caused by alternating selection regimes. When lemmings are rare, high reproductive capacity and nonaggressive social behavior are favored, and the population grows rapidly. As the growing population becomes more crowded, aggressive individuals are favored, because they can hold territories, secure mates, and protect offspring better than passive individuals. The aggressive in-

teractions, however, inhibit reproductive capacity, increase mortality attributable to fighting and infanticide, and expose more lemmings to predation as subordinate individuals are forced by dominants to occupy more marginal habitats. The behavioral changes that occur in response to crowding apparently persist for some time even as the density declines, so that aggressive interactions and a depressed birthrate continue until the lemming population reaches very low levels. Finally, passive individuals with high reproductive rates are again favored, and the cycle repeats.

Although the breeding cycles of many predators, including snowy owls, weasels, and foxes, are tied to lemming abundance, it appears that the regular fluctuation of lemming populations is a product of crowding and resource limitation rather than of a classical predator-prey cycle; that is, there is no tight coupling between the population fluctuations of lemmings and those of their predators. There is, however, a tight coupling between the population cycles of the snowshoe hare and the Canadian lynx. Since about 1800, the Hudson's Bay Company has kept records of furs produced each year. Both the hare and the lynx show a regular ten-year cycle, with the peaks in lynx abundance occurring about a year behind the hare's peak abundances. Since the hare is a major food source for the lynx in northern Canada, it is logical to assume that this is a coupled oscillation of population sizes, precisely as predicted by classical predator-prey theory.

Some regular cycles of abundance appear to have evolved as a means of avoiding predation rather than being a direct reduction caused by predation. There is a periodicity in the populations of cicadas and locusts. The hypothesized explanation is that predators cannot reproduce rapidly enough to increase their population sizes quickly in response to the sudden availability of a large food supply. When millions of adult cicadas appear above ground for a few weeks after surviving for seventeen years as nymphs in the soil, predators cannot possibly consume them all: No predator could specialize on adult cicadas unless it also had a seventeen-year cycle.

Several northern bird populations (such as crossbills, grosbeaks, and waxwings) fluctuate dramatically, in some years rising to several times their usual levels. This fluctuation may be a response to changing habitat quality. These bird populations always produce as many eggs as food availability and their natural fecundity allow, even though many offspring will not survive. In a good year, a higher proportion of the offspring survives, and the population experiences an irruption, often leading to intense competition and consequent expansion of the range of the population. In subsequent years, population size returns to preirruption levels. Thus, these fluctuations are entirely consistent with normal density-dependent processes responding to a fluctuating environment.

Irregular Fluctuations

Population fluctuations that occur irregularly or noncyclically often appear to be responses to natural disturbances rather than to density-dependent processes or predator-prey relationships. For example, blue grouse persist at a relatively low level of abundance in coniferous forests until a fire occurs. The species rapidly increases in number following a fire and gradually diminishes again as the forest regenerates over the next several decades.

The population fluctuations of some species are not easily attributed to disturbance or to any other single cause. For example, swarming locusts typically remain at low abundance in a restricted area for several years; then, apparently without warning, they may increase more than a hundred-fold and swarm over large areas, consuming large amounts of vegetation. The locust outbreak lasts for several years, then the population declines as rapidly as it initially increased. In the early part of the twentieth century, it was discovered that locusts exhibit two phases: a solitary phase, corresponding to low abundance, and a gregarious phase, corresponding to an outbreak. While it is still not known how locusts transform from one phase to another, it is clear that several stages are involved and that weather conditions seem to initiate a transformation. Moisture seems to be the most important determinant, because of its influence on nymph development and survival, on egg development, and on predator abundance, but wind and plant nitrogen levels have also been implicated. Furthermore, it appears that environmental conditions are only effective in inducing a phase transformation if a certain concentration of locusts already exists and if the existing locusts are adequately sensitive to crowding.

Measuring Fluctuations

There are two parts to the study of population fluctuation: detecting and measuring the pattern of the fluctuation and identifying the underlying causes of the fluctuation. In general, any method designed for measuring population size can be used repeatedly over time to detect fluctuations in the population. Reference to a specialized textbook on ecological sampling techniques is strongly recommended when using any of these methods, in order to assure validity of the sampling for subsequent statistical analysis. The mark-recapture method is commonly used with animal populations. There are many variants of this technique, but they all involve capturing and marking some number of individuals, then releasing them; after some time period appropriate to the study, a second sample is captured and the proportion of marked individuals in the second sample (those that are "recaptured") is recorded. This proportion is used to estimate the size of the population at the time when the individuals were originally marked.

The quadrat method is used primarily with plants and other sessile organisms. Plots (called quadrats) are laid out, either randomly or in some pattern; all individuals within the plots constitute a sample. Quadrats are usually square, but any regular shape may be used. The appropriate size of each quadrat depends on the sizes of organisms to be sampled and on their spatial distribution. If nondestructive sampling techniques are used, the same quadrats may be sampled repeatedly; otherwise, new quadrats must be established for each sampling episode.

A variety of plotless techniques is available for sessile organisms, in lieu of the quadrat method. These techniques were developed to eliminate some of the uncertainties associated with selecting proper quadrat size and location. Most plotless methods locate points on the ground, then measure distances to nearby organisms; each plotless technique identifies the individuals to be measured in a slightly different way.

None of these techniques is adequate by itself to identify the origin or cause of any fluctuation in population size. Experimental manipulation of a population is necessary to elucidate the underlying mechanisms and determining factors. Populations of small, rapidly reproducing species (such as species of *Paramecium* or *Daphnia*) can be manipulated in the laboratory, and hypothesized causes of fluctuation can be tested under controlled conditions. This has been done primarily to develop theoretical predictions regarding environmental conditions (such as temperature, moisture, and humidity), resource limitations and fluctuations, and the effects of predators and competitors.

Grosbeak populations fluctuate dramatically; the birds produce as many offspring as the habitat will bear, and in a good year the population will grow enormously, immediately creating a situation of food scarcity for the next year. (Rob and Ann Simpson/Photo Agora)

Identifying Causes of Fluctuation

The most interesting examples of population fluctuation, however, occur over spatial and temporal scales too large to handle in the laboratory. Their underlying mechanisms must be elucidated in the field. Because suites of factors typically produce the complex patterns of population fluctuation observed in nature, an effective field study must include all relevant factors. Generally, the most effective studies have been those that have sought to understand the complete life history of a species. Superb examples include the long-term studies of the wolves of Isle Royal National Park, by David Mech and colleagues, and the equally ambitious studies of the grizzly bears of Yellowstone National Park and surrounding areas by Frank and John Craighead and their many coworkers.

Most equilibrium models of population dynamics are capable of producing regular oscillations that mimic the patterns observed in nature. If the model parameters are properly manipulated, many of these models can produce apparently random fluctuation. More sophisticated models have been constructed that incorporate a mathematical equivalent of random environmental fluctuation, although they usually still assume that a population has a tendency to stabilize and that environmental change simply prevents stabilization. The underlying assumption of almost all these models is that species normally exist at equilibrium. This assumption is consistent with the long-held belief that there is a "balance of nature"—that species exist in harmony with their environments.

If an entire species is considered, perhaps the assumption of equilibrium is warranted, at least for extended periods; yet at the level of the population, fluctuation is the rule—indeed, it may be that extreme fluctuation is the rule. As noted earlier, many populations fluctuate so markedly that they often disappear; they are reestablished only by colonization from large populations within dispersal range. If populations become too small or too isolated from one another, this colonization cannot occur. Additionally, because small populations are more subject to extinction associated with fluctuation, there is additional risk of species extinction if only small populations remain.

The problem of extinction is severe, since habitat destruction is occurring at an unprecedented rate on a global scale. The fragments of intact habitat that remain because of inaccessibility or preservation efforts contain populations that are smaller and more isolated than in the past. If an isolated population fluctuates markedly, resulting in its extinction from a habitat fragment, its replacement by recolonization is unlikely. Furthermore, the genetic variation maintained by a complex population structure within a species is reduced. As the genetic variation within a species is lost, the ability of a species to respond to environmental change is reduced, and extinction of the species is more likely.

Ultimately, if populations normally fluctuate severely enough that they can be expected to become extinct at frequent intervals, then effective conservation requires the maintenance of pathways for exchange and dispersal of individuals among populations within a species. It also requires the preservation of the largest population size possible to allow for normal fluctuation without extinction.

—Alan D. Copsey

See also: Competition; Demographics; Ecology; Endangered species; Extinction; Extinctions and evolutionary explosions; Habitats and biomes; Mark, release, and recapture methods; Migration; Population analysis; Population Growth; Predation; Punctuated equilibrium and continuous evolution; Rhythms and behavior.

Bibliography

Begon, Michael, Martin Mortimer, and David J. Thompson. *Population Ecology: A Unified Study of Animals and Plants.* 3d ed. Cambridge, Mass.: Blackwell Science, 1996. An up-to-date and readable population ecology textbook aimed at undergraduates. Sections cover single-species interactions, interspecies interactions, and population ecology. Extensive bibliography.

Craighead, Frank C., Jr. *Track of the Grizzly.* San Francisco: Sierra Club Books, 1982. An exciting and accurate book about grizzly bear behavior, the history and prospects of the grizzly, and the methods and people involved in studying it. Craighead has spent his career studying the grizzly and is singularly qualified to present its life history.

Gotelli, Nicholas J. *A Primer of Ecology.* 3d ed. Sunderland, Mass.: Sinauer Associates, 2001. Focuses on ecological problem solving and concepts in mathematical ecology.

Krebs, Charles J. *Ecology: The Experimental Analysis of Distribution and Abundance.* 4th ed. New York: HarperCollins College Publishers, 1994. One of the most widely used college-level ecology texts, this book has excellent clarity and scope. Provides good coverage of population regulation, and gives three examples of the kinds of factors that influence population abundance and produce population fluctuation. Emphasizes the importance of population biology in ecology; consequently, its bibliography regarding population biology is quite good.

Krebs, Charles J., and J. H. Myers. "Population Cycles in Small Mammals." *Advances in Ecological Research* 8 (1974): 267-399. The authors review the evidence and hypotheses regarding these cycles and present a synthesis that has provided the framework for much subsequent research. They provide a particularly good summary of the proposed mechanisms causing the cyclic fluctuations.

Mech, L. David. *The Wolf: The Ecology and Behavior of an Endangered Species*. Garden City, N.Y: Doubleday, 1970. The result of many years of dedicated work, this is a fascinating and detailed account of the wolf, including its population dynamics and the dynamics of its prey. More recent articles about the wolf, including work by Mech, are cited in Robert Smith's book; nevertheless, this book is a masterpiece of natural history.

Smith, Robert Leo. *Elements of Ecology*. 4th ed. San Francisco: Benjamin/Cummings, 2000. A text widely recognized for its field orientation. Provides a very readable discussion of population biology, population regulation, and an interesting example of extreme population fluctuation.

Wohrmann, K., and S. K. Jain, eds. *Population Biology: Ecological and Evolutionary Viewpoints*. New York: Springer-Verlag, 1990. A collection of essays aimed at graduate level students and researchers. Focuses mostly on the interface between population biology and population genetics, from a theoretical perspective.

POPULATION GENETICS

Type of animal science: Evolution
Fields of study: Ecology, genetics, systematics (taxonomy)

Population genetics is the analysis of genes and genetic traits in populations to determine how much variability exists, what maintains the variability, how selection (natural or controlled) affects a population, and what the mechanisms of evolution are.

Principal Terms

DEOXYRIBONUCLEIC ACID (DNA): the chemical basis of genes

DOMINANT: requiring only one copy of a gene for expression of the trait

GENE: the unit of heredity; a short stretch of DNA encoding a specific product, usually protein

GENOTYPE: the gene makeup of an individual

INBREEDING: the mating of individuals more closely related than the population average

MEIOSIS: the two cell divisions leading to egg or sperm, during which the genes from the two parents are mixed

MUTATION: a sudden, unpredictable change in a gene

PHENOTYPE: a trait or combination of traits, the result of the genotype and environment

RANDOM GENETIC DRIFT: the random change of gene frequencies because of chance, especially in small populations

RECESSIVE: requiring two copies of a gene for the trait to be expressed

SELECTION: differential survival and reproduction rates of different genotypes

---

Population genetics is the description and analysis of genetic traits and their causative genes at the population level. Classical genetics deals with the rules of genetic transmission from parents to offspring, developmental genetics deals with the role of genes in development, and molecular genetics looks at the molecular basis of genetic phenomena. Population genetics uses information from all three fields and helps explain why populations are so variable, why some harmful traits are common, why most animals and plants reproduce sexually, how evolution works, why some animals are altruistic in a cutthroat world, and how new species arise.

The Sources of Variability

Simple observation tells us that animals are highly variable. Some dogs are big, others small; some wiry, others big boned; some long haired, others curly; some with special talents such as herding or retrieving, others with none; and some with diseases or defects, others normal. All of these are the results of various genes combined with environmental influences. Unless an animal is an identical twin, no one else shares that individual's genotype and no one ever will. Population genetics looks at variability in a population and examines its sources and the forces that maintain it.

Variability can come from genetic mutations. For example, about one child in ten thousand is born with dominant achondroplasia (short-legged dwarfism). Some children with the trait inherit the condition from an affected parent, but most have normal parents. They are therefore the result of a new mutation. Many mutations are deleterious and are eventually eliminated from the population by the lowered survival or fertility rates of those who have the mutation, but while they remain in the population, they add to its variability. Occasionally, a seemingly harmful mutation persists, for example, the gene that causes sickle-cell anemia, a severe disease characterized by red blood cells that become sickle shaped in certain laboratory tests. The causative gene is recessive, meaning that two copies are needed to produce the anemia, but the disease is very common in some parts of Africa. The harmful, anemia-causing gene persists in the population because if a person has only one gene with the trait rather than two, that gene confers resistance to malaria, the major cause of debility in that part of the world. Although the genes in these two examples have large and conspicuous effects, the great majority of mutations and the great bulk of genetic variability in the population are the result of a large number of genes with individually small effects, often detected only through statistics. The variability of quantitative traits such as size is due mainly to the cumulative action of many individual genes, each of which produces its small effect. The average size stays roughly constant from generation to generation because individuals who are too large or too small are at a disadvantage. However, such individuals continuously arise from new mutations.

The driving force in evolution is natural selection, that is the differential survival and fertility of different genotypes. New mutations occur continuously. Most of these are harmful, although usually only mildly so, but a small minority are beneficial. The rules of Mendelian inheritance ensure that the genes are thoroughly scrambled every generation. Natural selection acts like a sieve, retaining those genes that produce favorable phenotypes in the various combinations and rejecting others. Such a process, acting over eons of time, has produced the variety and specific adaptations that can be found throughout the animal kingdom.

The Forces Against Change

Although evolutionary progress is the result of natural selection, most selection does not accomplish any systematic change. Most selection is directed at maintaining the status quo—eliminating harmful mutations, keeping up with transitory changes in the environment, and eliminating statistical outliers (extremes of variation). Most of the time, evolutionary change is very slow.

In most populations, mating is essentially random in that mates do not choose each other because of the genes they carry. There are exceptions, of course, but for the most part, random mating can be assumed. This permits a great simplification known as the Hardy-Weinberg rule. This rule says that if the proportion of a certain gene, say A, in the population is p and of another, say a, is q, then the three genotypes AA, Aa, and aa appear in the proportions p_2, $2pq$, and q_2, respectively. (Remember that p and q are fractions between zero and one.) This is a simple application of elementary probability and the binomial theorem. Furthermore, after a few generations of random mating, genotypes at different loci also equilibrate, which means that the frequency of a composite genotype is the product of the frequencies at the constituent loci. The reason that this is so useful is that the number of genotypes is enormous, but a population can be characterized by a much smaller number of gene frequencies.

Genotypes are transient, but genes may persist unchanged for many generations. This has led the great theorists of population genetics, J. B. S. Haldane and R. A. Fisher in England and Sewall Wright in the United States, to make the primary units the frequency of individual genes and develop theories around this concept, making free use of the simple consequences of random mating. Such a gene-centered view has been described by scholar Richard Dawkins as the "selfish gene." A population can be thought of as a collection of genes, each of which is maximizing its chance of being passed on to future generations. This causes the population to become better adapted because those genes that improve adaptation have the best chance of being perpetuated.

An extension of this notion is kin selection. The concept holds that, to the extent that behavior is determined by genes, individuals should be protective of close relatives because relatives share genes. The fact that brothers and sisters share half their genes should lead a brother to be half as concerned with his sister's survival and reproduction as with his own. Evolutionists believe that altruistic behavior in various animals, including humans, is the result of kin selection. The degree of self-sacrifice to protect a close relative is proportional to the fraction of shared genes. Parents regularly make sacrifices for their children, and this is what evolutionary theory would predict.

One way in which populations depart from random mating is inbreeding, the mating of individuals more closely related than if they were randomly chosen. Related individuals share one or more ancestors; hence an inbred individual may get two copies of an ancestral gene, one through each parent. In this way, inbreeding increases the proportion of homozygotes. Because many deleterious recessive genes are hidden in the population, inbreeding can have a harmful effect by making genes homozygous. Similarly, if the population is subdivided into local units, mating mostly within themselves, these local units will be more homozygous than if the entire population mated at random. Small subpopulations will be more subject to purely random fluctuations in gene frequencies known as random genetic drifts. Therefore, subdivisions of a population often differ significantly, particularly with respect to unimportant genes.

The Argument Against Average Effects

The gene-centered view of evolution is not always accepted. Some evolutionists believe that it is simplistic to view an individual as a bag of genes, each trying to perpetuate itself. They emphasize that genes often interact in complicated ways, and that a theory that deals with only average gene effects is incomplete. Modern theories of evolution take such complications into account.

This different viewpoint has led to a major controversy in evolution, one that has not yet been settled. Wright emphasized that many well-adapted phenotypes depend on genes that interact in very specific ways; two or more genes may be individually harmful but when combined produce a beneficial effect. He argued that selecting genes on the basis of average effects cannot produce such combined effects. He believed that a population subdivided into many partially isolated units provides an opportunity for such interactions. An individual subpopulation, by random drift, might chance upon such a happy gene combination, in which case, the whole population can be upgraded by migrants from this subpopulation. Whether evolutionary advance results from gene interactions in subpopulations, from mass selection in largely unstructured populations, or from a combination of both is a question that remains unresolved.

Population genetics theory, along with the techniques of molecular genetics, has greatly deepened our knowledge of historical evolution. Everyone is familiar with tree diagrams of common ancestry that show, for example, birds and mammals branching off from early ancestors. In the past these had to be constructed using external phenotypes and fossils. These techniques for measuring the relatedness of different species and determining their ancestral relations have been replaced by DNA sequencing, which produces much surer results. It has long been suspected that genes can persist for very long evolutionary periods, changing slightly to perform new, often related but sometimes quite different functions. This belief has been confirmed repeatedly by molecular analysis. The similarity of the DNA sequences between some plant and animal genes is so great as to leave no doubt that they were both derived from a common ancestral gene a billion or more years ago.

Neutral Mutation and the Benefits of Sexual Reproduction

Most gene mutations have very small effects, and the smaller the effect, the less likely it is to be noticed. Molecular techniques have enabled scientists to detect changes in DNA without regard to the traits they cause or whether they have any effect at all. The Japanese geneticist Motoo Kimura has advanced the idea that most evolution at the DNA level is not the result of natural selection but simply the result of mutation and chance, a concept termed neutral mutation. In vertebrates, especially in mammals, most of the DNA has no known function. The functional genes make up a very small fraction of the total DNA. Many scientists believe that most DNA evolution outside the genes—and some within—is the result of changes that are so nearly neutral as to be determined by chance. How large a role random drift plays in the evolution of changes in functional proteins is still not certain.

A few animals and a large number of plants reproduce asexually. Instead of reproducing by using eggs and sperm, the progeny are carbon copies of the parent. Asexual reproduction has obvious advantages. If females could reproduce without males, producing only female offspring like themselves, reproduction would be twice as efficient. However, despite its inherent inefficiency, sexual reproduction is the rule, undoubtedly because of the gene-scrambling process that sex produces. The ability of a species to produce and try out countless gene combinations confers an evolutionary advantage that outweighs the cost of males. Another advantage of gene scrambling is that it permits harmful mutations to be eliminated from the population in groups rather than individually.

Population genetics is also concerned with the processes by which new species arise. Scientists believe that a population somehow becomes divided into two or more isolated groups, separated perhaps by a river, mountain range, or other geographical barrier. Each group then follows its own separate evolutionary course, and the groups' dissimilar environments accentuate their differences. Eventually so many differences between the two groups evolve that they are no longer compatible. The products of interspecies crosses, or hybrids, often do not develop normally or are sterile (like the mule). Sometimes the two species do not mate because they are so different.

Theory, Observation, and Experiment

Population genetics involves theory, observation, and experiment. Population genetics examines how genes are influenced by mutation, selection, population size, migration, and chance. Scientists develop mathematical models that embody these theories and compare the results obtained using the models with data from laboratory experiments or field observations. These genetic models have become more and more sophisticated to take into account complex gene interactions and increasingly realistic population structures. The models are further complicated by efforts to account for random processes. Often the mathematical geneticist relies on computers to perform complex analyses and computations.

One of the simpler models, which makes the assumption that mating is random, is the Hardy-Weinberg principle. If the proportion of gene A in the population is p and that of gene a is q, then the three genotypes AA, Aa, and aa appear in the proportions p_2, $2pq$, and q_2, respectively. The proportion of Aa is $2pq$ rather than simply pq because this genotype represents two combinations, maternal A with paternal a and paternal A with maternal a. This principle can be used to predict the frequency of persons with malaria resistance from the incidence of sickle-cell anemia. If one-tenth of the genes are sickle-cell genes and the other nine-tenths are normal, the frequency of two genes coming together to produce an anemic child is 0.1×0.1, or 0.01. The frequency of those resistant to malaria, who have one normal and one sickle-cell gene, is $2 \times 0.1 \times 0.9$, or 0.18. A slight extension of the calculation (using the rates of malaria infection and death from the disease) can be used to estimate the death rate from malaria. Another mathematical model can be formed based on the molecular genetics theory of neutral mutation. A neutral mutation, because it is not influenced by natural selection, has an expected rate of evolution that is equal to the mutation rate. Mathematical models embodying this theory are used to quantitatively predict what will happen in an experiment or what an observational study will find and act as a test of the theory. Neutral mutation theory is quite complicated and requires advanced mathematics.

Observational population genetics consists of studying animals and plants in nature. Evolution rates are inferred from the fossil record. Field observations can determine the frequency of genes in different geographical areas or environments. The frequency of self- and cross-pollination can often be observed directly. Increasingly, DNA analysis, which can detect relationships or alterations that are not visible, is being used to support field observations. For example, molecular markers have been used to determine parentage and relationship. DNA analysis revealed that certain birds that do not reproduce but care for the progeny of others are in fact close relatives, consistent with kin selection theory.

Increasingly, population genetics has begun to rely on experimentation. Plants and animals can be used to study the process of selection, but to save time and reduce costs, most laboratory experiments involve small, rapidly reproducing organisms such as the fruit fly, *Drosophila*. Some of the most sensitive selection experiments have involved the use of a chemostat, a container in which a steady inflow of nutrients and steady outflow of wastes and excess population permit a population to maintain a stable number of rapidly growing organisms, usually bacteria. These permit very sensitive measurements of the effects of mutation. Evolutionary studies that would require eons if studied in large animals or even mice can be completed in a very short time.

Explaining, Quantifying, and Predicting Evolution

The greatest intellectual value of population genetics has been to provide a theory of evolution that is explanatory, quantitative, and predictive. Population genetics places knowledge of mutation, gene action, selection, inbreeding, and population structure in a unified framework. It brings together Charles Darwin's theory of evolution by natural selection, Gregor Mendel's laws of inheritance, and molecular genetics to create a coherent picture of how evolution took place and is still occurring.

Population genetics has provided explanations for variability in a population, the prevalence of sexual rather than asexual reproduction, the origin of new species, and behavioral traits such as altruism. It has also provided an understanding of why some harmful diseases are found in the population. Population genetics has been used in animal and plant breeding to create rational selection programs. Using quantitative models, the results of various selection schemes can be compared and the best one chosen.

A particular telling example of a situation in which population genetics predicted an outcome that has become painfully obvious is the development of resistance to insecticides, herbicides, and antibiotics. As people used these products more and more, the insects, weeds, and bacteria they were trying to eliminate developed resistance, and new products had to be developed to replace those rendered ineffective. The development of resistance represents evolution by natural selection that took place not over hundreds or thousands of years but in just a few years. Probably the most problematic area of resistance is antibiotics because some treatable diseases are again threatening to move beyond the ability of medicine to cure. A major challenge to ecologists, microbiologists, physicians, and population geneticists is how to deal with the increasingly difficult problem of disease-producing microorganisms that are resistant to antibiotics.

—James F. Crow

See also: Aging; Asexual reproduction; Convergent and divergent evolution; Demographics; Development: Evolutionary perspective; Evolution: Historical perspective; Genetics; Hardy-Weinberg law of genetic equilibrium; Isolating mechanisms in evolution; Mutation; Natural selection; Neutral mutations and evolutionary clocks; Population analysis; Reproduction; Sex differences: Evolutionary origin.

Bibliography

Crow, James F. *Basic Concepts in Population, Quantitative, and Evolutionary Genetics*. New York: W. H. Freeman, 1986. College level. A knowledge of elementary calculus is helpful but not necessary. Illustrated with diagrams. Problems and answers are included.

Dawkins, Richard. *The Blind Watchmaker*. New York: W. W. Norton, 1986. This gifted writer has produced a delightfully interesting book about evolution that is also scientifically sound.

_____. *The Selfish Gene*. Rev. ed. Oxford, England: Oxford University Press, 1999. This book is one of the most enjoyable ways to learn about the gene-centered view of evolution.

Falconer, Douglas S. *Introduction to Quantitative Genetics*. 4th ed. New York: John Wiley & Sons, 1996. A clearly written, elementary exposition of population and quantitative genetics, with particular emphasis on animal breeding.

Fisher, R. A. *The Genetical Theory of Natural Selection*. Rev. ed. New York: Dover, 1958. Originally published in 1930, this has been called the greatest book on evolution since Darwin. Tough reading, but well worth the effort.

Haldane, J. B. S. *The Causes of Evolution*. Reprint. Ithaca, N.Y.: Cornell University Press, 1993. This classic, first published in 1931, is still well worth reading. Haldane's clear explanation of complex subjects is legendary. Includes an appendix for the mathematically minded.

Hartl, Daniel. *A Primer of Population Genetics*. 3d ed. Sunderland, Mass.: Sinauer Associates, 2000. A short, elementary, clearly written textbook. College level, but suitable for high-school students. Illustrated with diagrams; includes problems with answers and a useful list of references.

Hartl, Daniel, and Andrew Clark. *Principles of Population Genetics*. 3d ed. Sunderland, Mass.: Sinauer Associates, 1997. A standard textbook with broad coverage, examples, and problems.

Kimura, Motoo. *The Neutral Theory of Molecular Evolution*. Cambridge, England: Cambridge University Press, 1985. The originator and chief protagonist of the neutral theory has consolidated his thoughts and produced a book that, despite its difficult subject matter, is quite easy to understand.

Maynard Smith, John. *The Theory of Evolution*. Cambridge, England: Cambridge University Press, 1993. An elementary description by one of the leaders in contemporary evolutionary thought. It may be the best nontechnical account of modern evolutionary thought.

Wright, Sewall. *Evolution and the Genetics of Populations*. 4 vols. Chicago: University of Chicago Press, 1968-1978. This set, by one of America's greatest geneticists, is too long and complicated to read as a textbook but is excellent as a reference source.

POPULATION GROWTH

Type of animal science: Ecology
Fields of study: Anthropology, ethology, genetics, population biology, reproduction science

Populations typically grow when they are found on sites with abundant resources, and biologists have developed two models to describe growth. In exponential growth, the population is exposed to ideal conditions, and new individuals are added at an ever-increasing rate. Logistic growth recognizes that resources are eventually depleted, however, and that the population density ultimately stabilizes at some level, which is defined as the carrying capacity.

Principal Terms

CARRYING CAPACITY: the number of individuals of a given species that a site can support

EMIGRATION: the process whereby individuals leave a site and move elsewhere, leading to a decrease in the size of the population

EXPONENTIAL GROWTH: a pattern of population growth in which the rate of increase becomes progressively larger over time

IMMIGRATION: the process by which new individuals arrive at a site from elsewhere, leading to an increase in size of the population

INTRINSIC RATE OF INCREASE: the growth rate of a population under ideal conditions, expressed on a per individual basis

LOGISTIC GROWTH: a pattern of population growth that involves a rapid increase in numbers when the density is low but slows as the density approaches the carrying capacity

POPULATION: a group of individuals of the same species that live in the same location at the same time

In nature, organisms of a particular species rarely occur by themselves. Instead, they usually exist with other individuals of the same species. Biologists use the term "population" to refer to an aggregation of organisms of a given species that live in the same general location at the same time. In some cases, populations can be well defined, such as herds of cattle or flocks of geese. In other cases, a population is not well defined, often because several species may be found in the same location. For example, a meadow may contain intermingled populations of several species, including daisies, timothy grass, earthworms, and grasshoppers.

Biologists have studied populations for many years. Many of those studies have been conducted to answer three separate but interrelated questions. First, how many individuals are there in a given population at a particular time? Second, how do those numbers change from one time to another? Third, what environmental factors are responsible for any population increases or decreases? Studies have shown that for most species of plants, animals, and microbes, the number of individuals in the population changes over time. Some populations increase steadily, other populations decrease, while still others fluctuate. Thus, populations are generally dynamic, rather than static, when viewed over time.

Population Behaviors

Most populations change so much through time because there is constantly turnover among individuals. That is, new individuals are constantly being born, hatched, or germinated, while others die. Moreover, animals are also able to enter a population by immigration and leave by emigration. Since the number of births and new immigrants hardly ever exactly matches the number of deaths and emigrants, the dynamic nature of populations should not be a surprise. Because changes in population size are common in nature, biologists have tried to understand the changes that are observed. One approach has been to model populations; the model is a simplified graphical or mathematical summary of the actual changes that are occurring in the species of interest. The relationship between a model and the actual population that it represents is similar to that between a map and the area of land that it represents. Because modeling is such an important aspect of population biology, a biologist who studies population must often have a good background in mathematics.

Perhaps the simplest model of population behavior is the difference equation, which states that the number of individuals in a population at some specified time in the future is equal to the number at present, plus the number of births, minus the number of deaths, plus the number of new immigrants, minus the number of emigrants. Thus, by knowing how many individuals are on a site at a given time and knowing the usual number of births, deaths, immigrants, and emigrants, one can predict the number of individuals on the site at some future time.

Obviously, the number of births, deaths, immigrants, and emigrants will vary from one place to another and from one time to another. For example, on a site with abundant food and space and with favorable physical conditions for growth and development, births and immigration will be much greater than deaths and emigration. Thus, the population will increase. Conversely, if food or space is limiting or if the physical conditions are more severe, losses to the population through death and emigration will equal or exceed gains through birth and immigration. Thus, the population will remain constant or decline.

Biologists often are concerned about what happens in extreme conditions, because such conditions define the limits within which the population normally operates. When conditions are very bad, a population normally declines rapidly, often to the point of local extinction; when conditions are very good, a population will increase. That increase is attributable to the fact that each individual normally has the capacity to produce many offspring during its lifetime. For example, a woman could produce more than forty children if she conceived every time that she was fertile. Other individual organisms, particularly many invertebrates and plants, can produce hundreds of thousands of offspring in their lifetimes.

Influences on Birth and Death Rates
At least three different traits influence the reproductive output of a given species. The first is the number of offspring per reproductive period (elephants produce only one child at a time, whereas flies can lay thousands of eggs). The second is the age at first reproduction (most dogs can reproduce when less than three years old, whereas humans do not usually become fertile until they reach an age of thirteen or fourteen). The third is the number of times that an individual reproduces in its lifetime (salmon spawn, that is, lay eggs, only once before they die, whereas chickens lay eggs repeatedly). Even under ideal conditions, death must also be considered when examining population growth. Most all organisms have a maximum life span that is determined by their innate physiology and cannot be exceeded, even if they are supplied with abundant food and kept free from disease.

Population biologists frequently express birth and death in the form of rates. This can be done by counting the number of new births and deaths in a population during a predetermined period of time and then dividing by the number of individuals in the population. That will give the per capita (per individual) birth and death rates. For example, suppose that during the course of a year there were thirty births and fifteen deaths in a population of one thousand individuals. That per capita birthrate would be 0.030 and the per capita death rate would be 0.015.

Next, one can subtract the death rate from the

birthrate to find the per capita rate of population growth. That rate should be greatest under ideal conditions, when the birthrate is greatest and the death rate is least. That per capita rate of population growth is called the "maximal intrinsic rate of increase" or the "biotic potential" by population biologists, and it is a very important attribute. It is often symbolized as r_{max} or referred to as "little r." Normally, r_{max} is considered an inherent feature of a species. As one might expect, it varies greatly among different types of organisms. For example, r_{max}, expressed per year, is 0.02-1.5 for birds and large mammals, 4-50 for insects and small invertebrates, and as high as 20,000 for bacteria.

Exponential Growth
By knowing the intrinsic rate of increase and the number of organisms in a population, one can predict much about the behavior of a population under ideal conditions. The rate at which the population grows is merely the intrinsic rate of increase (r_{max}) multiplied by the number of individuals in the population. For example, suppose that there are ten individuals in a population whose annual r_{max} is 2. That population would increase by an annual rate of twenty (which would be a healthy increase). Next, suppose that one returned to that population at some later time when the population was fifty individuals. At that point, the annual rate of population increase would be one hundred new individuals (which would be an even healthier increase). If the rate of increase were measured when the population reached five hundred, the annual rate of increase would be one thousand individuals.

Under such circumstances, the population would keep on growing at an ever-increasing rate. That type of growth is called "exponential growth" by population biologists, and it typifies the behavior of many populations in ideal conditions. If the number of individuals in a population undergoing exponential growth is plotted as a function of time, the curve would resemble the letter J. That is, it would be somewhat flat initially, but it would curve upward, and at some point it would be almost vertical. Exponential population growth has

been observed to a limited extent in many different kinds of organisms, both in the laboratory and under field conditions. Examples include protozoans, small insects, and birds. It should be obvious, however, that no species could behave in this manner for long. If it did, it would overrun the earth (and indeed the universe) in time. Instead, population growth is slowed by limited resources, accumulated wastes, behavioral stresses, and/or periodic catastrophes caused by the environment.

Logistic Growth
Biologists have created a second model to account for the behavior of populations under finite resources and have called it logistic growth. If the number of individuals in a population undergoing logistic growth is plotted as a function of time, the curve resembles a flattened S shape. In other words, the curve is initially flat, but then curves upward at a progressively faster rate, much like exponential growth. At some point (called the inflection point), however, the curve begins to turn to the right and flatten out. Ultimately, the curve becomes horizontal, indicating a constant population over time.

An important aspect of pure logistic growth is that the population approaches, but does not exceed, a certain level. That level is called the carrying capacity, and is represented by the symbol K in most mathematical treatments of logistic growth. The carrying capacity is the maximum number of individuals that the environment can support, based on the space, food, and other resources available. When the number of individuals is much fewer than the carrying capacity, the population grows rapidly, much as in exponential growth. As the number increases, however, the rate of population growth becomes much less than the exponential rate. When the number approaches the carrying capacity, new population growth virtually ceases. If the population were to increase above the carrying capacity for some reason, there would be a net loss of organisms from the population.

There are few studies that have documented logistic growth in nature. It would be necessary to

The study of population growth must consider how often an animal reproduces and how many offspring it produces at one time. Some animals, such as the elephant, have only one offspring with a long gestation period, while others, such as the fruit fly, lay thousands of eggs at a time. Significantly, elephants are an endangered species while fruit flies are not. (Corbis)

fects may be attributable to reduced space within which the organism can operate, to less food and other resources, to physiological and behavioral stress caused by crowding, and to increased incidence of disease. Those factors are commonly designated as being density-dependent. They are considered much different from the density-independent factors that typically arise from environmental catastrophes such as flooding, drought, fire, or extreme temperatures. For many years, biologists argued about the relative importance of density-dependent versus density-independent factors in controlling population size. It is now recognized that some species are controlled by density-independent factors, whereas others are controlled by density-dependent factors.

Classically, when a species undergoes logistic growth, the population is ultimately supposed to stabilize at the carrying capacity. Most studies that track populations over the course of time, however, find that numbers actually fluctuate. How can such variability be reconciled with the logistic model? On the one hand, the fluctuations may be caused by density-independent factors, and the logistic equation therefore does not apply. On the other hand, the population may be under density-dependent control, and the logistic model can still hold despite the fluctuations. One explanation for the fluctuations could be that the carrying capacity itself changes over time. For example, a sudden increase in the amount of food available would increase the carrying capacity and allow the population to grow. A second explanation relates to the presence of time lags; that is, a population might not respond immediately to a given resource level. For example, two animals in a rapidly expanding population might mate when the number of individuals is less than the carrying capacity. The progeny, however, might be born sev-

watch a species in a habitat from the time of its first introduction until its population stabilized. Such studies are necessarily of a very long duration and thus are not normally conducted. Logistic growth has been found in a number of experimental studies, however, particularly on small organisms, including protozoans, fruit flies, and beetles.

An important aspect of logistic growth is that, as the population increases, the birthrate decreases and the mortality rate increases. Such ef-

eral weeks or months later, into a situation in which the population has exceeded the carrying capacity. Thus, there would have to be a decline, leading to the fluctuation.

Approaches to Studying Population Growth

Two main approaches can be used to investigate logistic and exponential population growth among organisms. One approach involves following natural populations in the field; the other involves setting up experimental populations. Each approach has its benefits and drawbacks, and, ideally, both should be employed. To study population growth in the field, it is important to study a species from the time that it first arrives on a site until its population stabilizes. Thus, any species already present are automatically eliminated from consideration unless they are brought to local extinction and a new population is then allowed to recolonize. Population growth studies can be profitably done on sites that are very disturbed and are beginning to fill up with organisms. Examples would be an abandoned farm field or strip mine, a newly created volcanic island, or a new body of water. Moreover, studies could also be done on a species that is purposely introduced to a new site.

In either case, one needs to survey the population periodically to assess the number of individuals that it contains. The size of the population can be determined directly or by employing sampling techniques such as mark-recapture methodologies. The number of individuals can then be plotted on a graph (on the y axis) as a function of time (on the x axis).

To study population growth in experimental conditions, one sets up an artificial habitat according to the needs of the species in question. For example, investigators have examined population growth in protozoans (unicellular animals) by growing populations in test tubes filled with food dissolved in known volumes of water. Others have grown fruit flies in stoppered flasks. Still others have grown beetles in containers filled with oatmeal or other crushed grain. In those cases, it was typically necessary to replenish the food to keep the population going. Whenever the population was placed into an artificial habitat with a nonrenewable food source, it would generally consume all the food and then die out.

More detailed experiments can be performed to test whether density-dependent mortality is occurring. Such experiments would involve setting up a series of containers with different densities of organisms and then following the mortality of those organisms. In theory, mortality rates should be highest in containers that have the greatest densities of organisms and lowest in containers with the sparsest populations. One could also examine the birthrate in those containers, with the expectation that birthrates should be highest in the sparsest containers and lowest in those that have the most organisms. The investigation should be long enough to allow the population to reach equilibrium at the carrying capacity. For short-lived organisms such as protozoans or insects, that could take days, weeks, or months. For longer-lived organisms such as fish or small mammals, one to several years may be required. For long-lived animals, a truly adequate study may take decades.

Another consideration in studying exponential and logistic population growth is that immigration and emigration should be kept to a minimum. Thus, organisms that are highly active, such as birds, large mammals, and most flying insects would be extremely difficult to study. Finally, one can set up numerous populations and expose each to a slightly different set of conditions. That would enable the researcher to ascertain which environmental factors are most important in determining the carrying capacity. For example, populations of aquatic invertebrates could be monitored under a range of temperature, salinity, pH, and nutrient conditions.

Implications of Logistic Growth

Since exponential growth is unrealistic in practical terms for almost all populations, its scientific usefulness is limited. The concepts derived from logistic growth, however, have important implications to biologists and nonbiologists alike. One important aspect of logistic growth is that the maximum rate of population growth occurs when the population is about half of the environment's carrying capacity. When populations are very sparse, there are simply too few individuals to produce many progeny. When populations are dense, near the carrying capacity, there is not enough room or other resources to allow for rapid population growth. Based on that relationship, those who must harvest organisms can do so at a rate that allows the population to reestablish itself quickly. Those who can apply this concept in their everyday work include wildlife managers, ranchers, and fishermen. Indeed, quotas for hunting and fishing are often set in a way that allows for the population to be thinned sufficiently without depleting it too severely. Unfortunately, there are two problems that biologists must confront when they try to use the logistic model to help manage populations. The first is that it is often difficult to establish the carrying capacity for a given species on a particular site. One reason is that the populations of many species are profoundly affected by density-independent factors, as well as by other species, in highly complex and variable ways. Further, for reasons unclear to most biologists, some species have a maximum rate of population growth at levels well above or well below the level (one half of the carrying capacity) that is normally assumed. Thus, the logistic model typically gives only a very rough approximation for the ideal size of a population. However, the logistic model is useful because it emphasizes that all species have natural limits to the sizes of their populations.

—*Kenneth M. Klemow*

See also: Biodiversity; Competition; Demographics; Ecology; Ecosystems; Mark, release, and recapture methods; Population analysis; Population fluctuations; Reproduction; Reproductive strategies.

Bibliography

Brewer, Richard. *The Science of Ecology*. 2d ed. Philadelphia: Saunders College Publishing, 1994. This clearly written textbook is aimed at upper-level undergraduates and has the benefit of being written by an author who has experience researching both plants and animals. Contains a well-presented discussion of exponential and logistic growth, including ample information concerning the effects of environmental factors on the intrinsic rate of increase. The models are treated in a way that explains them first in a nonmathematical context, then mathematically for those who desire to understand them in that way.

Elseth, Gerald D., and Kandy D. Baumgardner. *Population Biology*. New York: Van Nostrand, 1981. Intended for graduate students and advanced undergraduates, this book is recommended for anyone who desires a rigorous mathematical treatment of population biology. Reviews the theory and the experimental evidence pertaining to the exponential and logistic models. It also discusses some methodological aspects, some factors that contribute to deviations of the logistic model, and the effects of different attributes of organisms on their population biology.

Gotelli, Nicholas J. *A Primer of Ecology*. 3d ed. Sunderland, Mass.: Sinauer Associates, 2001. Focuses on ecological problem solving and concepts in mathematical ecology.

Hutchinson, G. Evelyn. *An Introduction to Population Ecology*. New Haven, Conn.: Yale University Press, 1978. This book presents an engaging account of the study of populations by a well-respected ecologist. The first chapter outlines the historical basis for the development of logistic theory, focusing on the personal lives and contributions of several people who significantly advanced knowledge in the area.

Kormondy, Edward J. *Concepts of Ecology*. 4th ed. Englewood Cliffs, N.J.: Prentice-Hall, 1996. This book is intended for students who have had general biology, but can be understood by readers without it. In chapter 4, "The Ecology of Populations," the reader can find a highly informative discussion of the exponential and logistic models, complete with useful examples, historical perspectives, and explanations of many of the pertinent studies.

Krebs, Charles J. *Ecology: The Experimental Analysis of Distribution and Abundance*. 5th ed. San Francisco: Benjamin/Cummings, 2001. A basic introduction to the basic issues and theories of ecology, emphasizing problem solving and critical analysis.

Raven, Peter, H., and George B. Johnson. *Biology*. 5th ed. St. Louis: Times Mirror/Mosby, 1999. An unusually well-written and profusely illustrated textbook covering all phases of modern biology. Chapter 23 provides a cogent summary of population biology, including the logistic and exponential models. Chapter 27 provides an overview of the human population as it relates to resource availability and pollution.

Wilson, Edward O., and William Bossert. *A Primer of Population Biology*. Sunderland, Mass.: Sinauer Associates, 1977. This classic handbook has been used extensively by students wishing to master the basics of population biology. Chapter 1 presents an extremely well-thought-out introduction to the field of population biology and includes a section on how to construct a mathematical model. The exponential and logistic models are discussed in chapter 3. That chapter includes a well-paced presentation of the mathematical aspects of the two growth models, reinforced by problems for the reader to solve (solutions are given).

PORCUPINES

Types of animal science: Behavior, classification, ecology
Fields of study: Ecology, wildlife ecology, zoology

North American porcupines are large rodents who are vegetarians. These rather clumsy mammals are best known for their prickly quills that can inflict pain and suffering, and even result in death if the quill finds its way to a vital organ.

Principal Terms

ARBOREAL: living in or spending time in trees
DORSAL: the back portion of an animal
HERBIVORE: an animal that consumes vegetation for its diet
NOTOCHORD: a dorsal, flexible, rodlike structure extending the length of a vertebrate's body; serves as an axis for muscle attachment
PELAGE: a mammal's fur coat
PREDATOR: any animals that preys on another animal
QUILLS: sometimes referred to as spines; modified guard hairs; quills have barbed tips which can work themselves deeper into flesh once they have penetrated

---

Agile and armed, although they may seem slow, a porcupine is quick to make its point with its defensive quills. Porcupines do not throw or cast their quills into a potential predator; instead, quills penetrate a predator's body on contact with the porcupine's prickly body. The more than thirty thousand quills on a porcupine's back and sides are actually modified hairs (one of the characteristics of mammals). Other common names of porcupines are quillpig and pricklepig.

North American porcupines are arboreal or semiarboreal, spending much of their day climbing trees and consuming tree bark. These herbivores ingest a variety of plant materials, from buds to roots. On occasion, porcupines may eat shed antlers of deer or elk for the various minerals, such as calcium, that they contain.

Second in size only to the beavers in the class Rodentia, adults porcupines weigh between four and six kilograms, although much larger ones have been reported. The length attained by adults ranges from about sixty to one hundred centimeters. While color variations occur, most individuals have dark colored pelage. Porcupines are mostly nocturnal, but may be observed during the day either on the ground or in trees.

Porcupine Life Cycle
Adult porcupines are solitary mammals for most of the year, except during the breeding season, between September and November. Female porcupines begin reproductive activities at about 1.5 years of age. It is common to find several males around a female during her brief (eight- to twelve-hour) time of receptivity. Mating is brief and occurs on the ground, with the female raising her tail over her back. After the male has inseminated her, each porcupine goes its separate way.

Usually only one porcupette, as the young are sometimes called, is born after the lengthy gestation period. Weighing between four hundred and five hundred grams at birth, newborn porcupines are quite precocial. Their eyes are open and their quills are present, as are their incisors and premolar teeth. Although capable of consuming vegetation within a week of birth, the young are nursed by their mother through the summer months.

Porcupines consume the inner bark of trees and shrubs, especially in the fall and winter when the plants on the ground are becoming dormant or

Porcupine Facts

Classification:

Kingdom: Animalia

Phylum: Chordata

Class: Mammalia

Order: Rodentia

Family: Erethizontidae

Genus and species: Erethizon dorstaum

Geographical location: North American porcupines are found throughout Canada, extending into the northeastern and western United States; South American porcupines live in the tropical rain forests of South and Central America; various other porcupine species live in Africa and Asia

Habitat: Ranges from tropical rain forests to deserts; some inhabit coniferous and deciduous forests, while others live in grasslands

Gestational period: Averages 211 days

Life span: Five to six years in the wild, ten to twelve years in captivity

Special anatomy: Quills

dying. It is easy to observe porcupine feeding sites in the forests by observing the limbs and trunks of trees. If the outer bark has been stripped away, the whitish colored areas beneath are quite apparent. During the spring and early summer, porcupines spend more time on the ground feeding on tender shoots and buds of emerging plants.

While their vision is not acute, their olfactory (smell) and auditory (hearing) senses are well developed. Some researchers have reported observing porcupines standing up on their hind legs and sniffing their surroundings. If a porcupine detects a potential predator, it will form a defensive posture of lowered head and back, at the same time raising the tail for swinging. The heavy muscular tail can drive quills deep into a predator's face and head.

—*Sylvester Allred*

See also: Beavers; Defense mechanisms; Fur and hair; Mice and rats; Rodents; Squirrels.

Porcupines spend much of their time in trees, eating bark. (Jack Milchanowski/Photo Agora)

Bibliography

Buchsbaum, Ralph, et al. *The Audubon Society Encyclopedia of Animal Life*. New York: Crown, 1982. An old but useful reference with many colorful photos of mammals and other animal life from around the world.

Chapman, J. A., and G. A. Feldhamer, eds. *Wild Mammals of North America: Biology, Management, and Economics*. Baltimore: The John Hopkins University Press, 1982. An excellent and extensive reference covering all aspects of mammals living in North America.

Feldhamer, G. A., L. C. Drickamer, S. H. Vessey, and J. F. Merritt. *Mammalogy: Adaptation, Diversity, and Ecology*. Boston: WCB/McGraw-Hill, 1999. This is an excellent mammalogy textbook with many examples, illustrations, taxonomy, and references.

PRAYING MANTIS

Type of animal science: Classification
Fields of study: Ecology, entomology, ethology, invertebrate biology, physiology, population biology, systematics (taxonomy), zoology

All species of praying mantids (or mantises) are predators on other arthropods, but their importance in natural food webs is poorly understood. They have been models for a wide variety of experimental studies on behavior, neurophysiology, and ecology.

Principal Terms

ECOSYSTEM: the collection of all species and their abiotic environment in an area (habitat)

FOOD WEB: the feeding relationships among species in an ecosystem

HEMIMETABOLOUS: incomplete metamorphosis in which juveniles resemble wingless adults

NYMPH: the term for any of the juvenile stages of a hemimetabolous insect

SEX PHEROMONE: a volatile chemical released into the air by females to attract males

TROPHIC LEVEL: feeding position in the food web of an ecosystem

---

Praying mantis is the common name for any insect in the order Dictyoptera, suborder Mantodea. All of these insects are predators. The most important family in this group is Mantidae; hence, the general name for these insects often is given as "mantid," rather than "mantis." Mantids are closely related to cockroaches, termites, and grasshoppers. Approximately 1,800 species have been identified worldwide, most of which are tropical. The most common native species in the United States is the Carolina mantid, *Stagmomantis carolina*, which is found from New Jersey to the Gulf Coast in the eastern half of the United States. The most abundant and widespread species, the Chinese mantid, *Tenodera aridifolia sinensis*, was introduced from Asia at the end of the nineteenth century and can be found from South Carolina to Long Island east of the Mississippi, as well as in portions of the Midwest and West Coast. Another well-known imported species is *Mantis religiosa*, the distribution of which is limited to northern states and Canada because its eggs require cold winters before they will develop and hatch in the spring.

Mantid Physiology

The praying mantis has the most highly mobile head of any insect, attached to the front of an elongated prothorax (foremost midbody segment), and spiny front legs also attached to the prothorax. These specialized forelegs are folded when the animal waits motionless in ambush for its prey, giving it an attitude of prayer (hence, the common name for the group). Although most mantids are sit-and-wait ambush predators, some species actively hunt and chase their prey over short distances.

The sensory systems of mantids have been well studied. Mantids can integrate detailed information from their environments, and have exhibited a highly sophisticated array of responses to external stimuli, such as light, chemicals, and sound. They are able to use binocular vision to accurately estimate the striking distance to their prey, or the distance between perching sites in vegetation. Female mantids produce sex pheromones to attract males during mating season. Some species can hear ultrasound emitted by bats, and thus avoid predation when they fly at night.

Sexual behavior and cannibalism in mantids has been the subject of much folklore and scientific speculation. Females of many species sometimes eat males before, after, or even during copulation. The outcome of male-female encounters is mainly a function of hunger level in females, and males often escape to mate with other females later on. The reputedly suicidal behavior of males is not really a sacrifice, because a male cannot tell in advance the hunger state of his prospective mate, or whether her eggs have already been fertilized by an earlier encounter with another male.

Mantid Life Cycle

Mantids inhabiting temperate geographic zones generally live from spring to autumn, and the adults all die with the onset of cold weather, leaving the eggs to winter over. Development of mantids is hemimetabolous: Eggs hatch out into nymphs that are miniatures of the adults, except without wings. Nymphs grow through as many as seven stages (depending on species), increasing body length tenfold before developing wings as adults near the end of the season. The largest species in the United States, *Tenodera aridifolia sinensis*,

---

Mantid Facts

Classification:
Kingdom: Animalia
Subkingdom: Bilateria
Phylum: Arthropoda
Subphylum: Uniramia
Class: Insecta
Order: Dictyoptera
Suborder: Mantodea (praying mantids)
Geographical location: Europe, Africa, Asia, the Americas, mainly in the tropics
Habitat: Open fields, forests, and deserts
Gestational period: Highly variable among species; most temperate zone species have a single generation each year, with eggs overwintering
Life span: Variable among species; temperate zone species die at the end of a single growing season; females of some tropical species survive until their eggs hatch
Special anatomy: Highly mobile head; elongated prothorax (first segment of thorax); raptorial front legs for grasping prey

---

may grow as large as ten centimeters (four inches), and often appears even larger to startled human observers. Very few hatchling nymphs survive the growing season to reach adulthood, most of them dying of starvation and the rest from predators such as spiders. Adults are large enough to escape predation by most other invertebrates, but vertebrate predators such as birds and lizards actively prey on them. Cannibalism between same-sized mantids is relatively rare except under crowded conditions in captivity where they cannot avoid one another, but larger nymphs will readily eat smaller ones. The variable feeding opportunities in natural ecosystems cause variable growth rates among nymphs within a season,

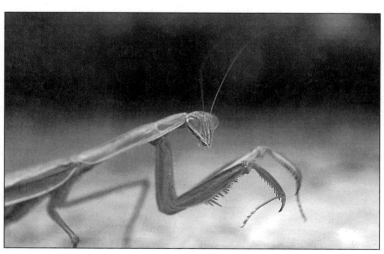

The praying mantis gets its name from the "prayerful" posture of its forearms and claws. In fact, "preying" mantis would be a better name for this avid carnivore. (Richard Martin/Photo Agora)

so cannibalism among different-sized individuals may be common in nature.

Praying mantids simultaneously occupy two trophic levels, feeding on both herbivorous and carnivorous arthropods. This makes their use in pest control problematic: If they eat grasshoppers they may be beneficial, but if they eat spiders they may be harmful. There is evidence that mantids can enhance plant growth by eating herbivorous insects, but their impact is likely to vary depending on the plants and other arthropod species present.

—Lawrence E. Hurd

See also: Arthropods; Cannibalism; Copulation; Insects; Mating; Metamorphosis; Reproductive strategies.

Bibliography

Helfer, Jacques. *How to Know the Grasshoppers, Cockroaches, and Their Allies*. New York: Dover, 1987. A taxonomic guide that includes all the common species of praying mantids in the United States.

Moran, M. D., T. P. Rooney, and L. E. Hurd. "Top-Down Cascade from a Bitrophic Predator in an Old-Field Community." *Ecology* 77 (1996): 2219-2227. This paper presents the first scientific evidence that praying mantids can influence plant growth by preying on herbivorous insects in a natural ecosystem.

Prete, F. R., H. Wells, P. H. Wells, and L. E. Hurd, eds. *The Praying Mantids*. Baltimore: The Johns Hopkins University Press, 1999. This is currently the only book on all aspects of research involving praying mantids, including taxonomy, behavior, physiology, ecology, and evolution.

Speight, M. R., M. D. Hunter, and A. D. Watt. *Ecology of Insects: Concepts and Applications*. Malden, Mass.: Blackwell Science, 1999. A highly readable book on how insects interact with their environments.

PREDATION

Type of animal science: Ecology
Fields of study: Environmental science, evolutionary science, wildlife ecology

The relationships among predators and their prey in natural communities are varied and complex. These interactions provide clues as to how natural populations regulate one another, as well as to how to preserve and manage exploited populations more successfully.

Principal Terms

COMPETITION: the interaction among two or more organisms of the same or different species that results when they share a limited resource

FOOD WEB: the sum total of the feeding relationships (links) between trophic levels in ecosystems

FUNCTIONAL RESPONSE: the rate at which an individual predator consumes prey, dependent upon the abundance of that prey in a habitat

MIMICRY: the resemblance of one species (the model) to one or more other species (mimics), such that a predator cannot distinguish among them

NUMERICAL RESPONSE: the abundance of predators dependent upon the abundance of prey in a habitat

POPULATION REGULATION: stabilization of population size by factors such as predation and competition, the relative impact of which depends on abundance of the population in a habitat

TROPHIC LEVEL: a level at which energy is acquired in a food web—the herbivore level obtains energy from plants; the carnivore level from herbivores and other carnivores

Predation is an interaction between two organisms in which one of them, the predator, derives nutrition by killing and eating the other, the prey. Obvious examples include lions feeding on zebras and hawks eating rodents, but predation is not limited to interactions among animals. Birds that feed on seeds are legitimate predators since they are killing individual organisms (embryonic plants) to derive energy. There are a number of species of carnivorous plants, such as sundews and pitcher plants, that capture and consume small animals to obtain nitrogen in habitats lacking sufficient quantities of that nutrient. Most animals that feed on plants (herbivores) do not kill the entire plant and therefore are not really predators. Exceptions to this generalization are some insects that reach infestation levels, such as gypsy moths or locusts, and can kill the plants upon which they feed. The majority of herbivore-plant associations, however, are more properly described as parasite-host interactions in which the host plant may suffer damage but does not die.

There are special cases in which parasitism and predation may be combined. One of these is the interaction between parasitoid wasps and their hosts, usually flies. Adult female parasitoids attack and inject eggs into fly pupae (the resting stage, during which fly larvae metamorphose into adults), and the larvae of the wasp consume the fly. The adult parasitoid is therefore a parasite, while the larval wasp acts as a predator.

Predator-Prey Interactions

Predator-prey interactions can be divided into two considerations: the effects of prey on predators, and the effects of predators on their prey. Predators respond to changes in prey density (the number of prey organisms in the habitat) in two principal ways. The first is called numerical response, which means that predators change their numbers in response to changes in prey density. This may be accomplished by increasing or decreasing reproduction or by immigrating to or emigrating from a habitat. If prey density increases, predators may immigrate from other habitats to take advantage of this increased resource, or those predators already present may produce more offspring. When prey density decreases, the opposite will occur. Some predators, which are known as fugitive species, are specialized at finding habitats with abundant prey, migrating to them, and reproducing rapidly once they are established.

Cape May warblers are good at finding high densities of spruce budworms (a serious pest of conifers) and then converting the energy from their prey into offspring. This strategy allows the birds to persist only because the budworms are never completely wiped out; they are better at dispersing to new habitats than are the birds.

The second response of predators to changing prey density is called functional response. The rate at which predators capture and consume prey depends upon the rate at which they encounter prey, which is a function of prey density. If the predator has a choice of several prey species, it may learn to prefer one of them. If that prey is sufficiently abundant, this situation results in a phenomenon known as switching, the concentration by the predator on the preferred prey. It may entail a change in searching behavior on the part of the predator, such that former prey items will no longer be encountered as frequently.

Animals have evolved a number of defense mechanisms that reduce their probability of being eaten by predators. Spines on horned lizards, threatening displays by harmless snakes, camouflage of many cryptic animals, toxic or distasteful chemicals in insects and amphibians, and simply rapid movement—all are adaptations that may have evolved in response to natural selection by predators. A predator that can learn to prefer one prey item over another is smart enough to learn to avoid less desirable prey. That capability is the basis for a phenomenon—known as aposematism—among potential prey species which are toxic and/ or distasteful to their predators. Aposematic organisms advertise their toxicity by bright coloration, making it easy for predators to learn to avoid them, which in turn saves the prey population from frequent taste-testing. Many species of insects are aposematic. Monarch butterflies are bright orange with black stripes, an easy signal to recognize. They owe their toxicity to the milkweed plant, which they eat as caterpillars. The plant contains cardiac glycosides, which are very toxic. The monarch caterpillar is immune to the poison and stores it in its body so that the adult has a high concentration of it in its wings. If a bird grabs the butterfly in flight, it is likely to get a piece of wing first, and this will teach it not to try orange butterflies in the future.

Some potential prey species that are not themselves toxic have evolved to resemble those that are; these are called Batesian mimics. Viceroy butterflies, which are not toxic, mimic monarchs very closely, so that birds cannot tell them apart. One limit to Batesian mimicry is that mimics can never get very numerous, or their predators will not get a strong enough message to leave them alone. Another kind of mimicry involves mimics that are as toxic as their models. The advantage with this type, Müllerian mimicry, is that the predator has to learn only one coloration signal, which reduces risk for both prey populations. In this relationship, the mimic population does not have to remain at low levels relative to the model population. A third type of mimicry is more insidious— aggressive mimicry, in which a predator resembles a prey or the resource of that prey in order to lure it close enough to capture. There are tropical praying mantids that closely resemble orchid flowers, thus attracting the bees upon which they prey. There are some species of fireflies that eat other species of fireflies, using the flashing light signal of their prey to lure them within range.

The Choice of Prey

What determines predator preference for prey? Since prey are a source of energy for the predator, it might be expected that predators would simply attack the largest prey they could handle. To an extent, this choice holds true for many predators, but there is a cost to be considered. The cost involves the energy a predator must expend to search for, capture, handle, and consume prey. In order to be profitable, a prey item must yield much more energy than it costs. Natural selection should favor reduction in energetic cost relative to energetic gain, the basis for optimal foraging theory. According to this theory, many predators have evolved hunting strategies to optimize the time and energy spent in searching for and capturing prey. Some predators, such as web-building spiders and boa constrictors, ambush their prey. The low energetic cost of sit-and-wait is an advantage in environments that provide plentiful prey. If encounters with prey become less predictably reliable, however, an ambush predator may experience starvation. Spiders can lower their metabolic energy requirements when prey is unavailable, whereas more mobile predators, such as boa constrictors, can simply shift to active searching. Probably because of the likelihood of facing starvation for extended periods of time, ambush predation is more common among animals that do not expend metabolic energy to regulate their body temperatures (ectotherms) than among those that do (endotherms). Some predators, such as wolves and lions, hunt in groups. This allows them to tackle larger (more profitable) prey than if they hunted alone. Solitary hunters generally have to hunt smaller prey.

Natural communities consist of food webs, constructed of links (feeding relationships) among trophic levels. Each prey species is linked to one or more predators. Most predators in nature are generalists with respect to their prey. Spiders, snakes, hawks, lions, and wolves all feed on a variety of prey. Some of these prey are herbivores, but some are themselves predators. Praying mantids eat grasshoppers (herbivores), but they also eat spiders (carnivores) and each other.

Thus, generalist predators have a bitrophic niche, in that they occupy two trophic levels at the same time.

Predatory Relationships and Population Fluctuation

It is an open question whether predators and prey commonly regulate each other's numbers in nature. There are many examples of cyclic changes in abundance over time, in which an increase in prey density is followed by an increase in the numbers of predators, and then the availability of prey decreases, also followed by a decrease in predators. Are predators causing their prey to fluctuate, or are prey responding to some other environmental factor, such as their own food supply? In the second case, prey may be regulated by food, and in turn, may be regulating predators, but not the reverse.

Predators can sometimes determine the number of prey species that can coexist in a habitat. If a predator feeds on a prey species that could outcompete (competitively exclude) other prey species in a habitat, it may free more resources for those other species. This relationship is known as the keystone effect. Empirical studies have indicated that the number of prey species in some communities is directly related to the intensity of predation (numerical and functional responses of predators) such that at low intensity, few species coexist because of competitive exclusion; at intermediate intensity, the diversity of the prey community is greatest; and at high intensity, diversity decreases because overgrazing begins to eliminate species. This intermediate predation hypothesis depends upon competition among prey species, which is not always the case.

Studying Predation

The central question in the study of predation is: To what extend do predators and their prey regulate one another? Most studies suggest that predators are usually food limited, but the extent to which they regulate their prey is uncertain. It is one thing to observe predators in nature and another to assess their importance to the dynamics

In the course of evolution, predatory species must choose whether to expend small amounts of energy to capture large numbers of small prey or to expend a relatively large amount of energy to capture a few large victims, as this lion has done. (Digital Stock)

of natural communities. Like other aspects of ecology, studies of predation can be descriptive, experimental, and/or mathematical.

At the descriptive level, characteristics of both predator and prey populations are assessed: rates of birth and mortality, age structure, environmental requirements, and behavioral traits. Qualitative and quantitative information of this type is necessary before predictions can be made about the interactions between predator and prey populations. General lack of such information in natural ecosystems is largely responsible for failures at biological control of pests and management of exploited populations.

Experimental studies of predation involve manipulation of predator and/or prey populations. A powerful method of testing the importance of predation is to exclude a predator from portions of its accustomed habitat, leaving other portions intact as experimental controls. In one such experiment, excluding starfish from marine intertidal communities of sessile invertebrates resulted in domination by mussels and exclusion of barnacles and other attached species; in the absence of the predator, one prey species was capable of competitively excluding others. This keystone effect depends upon two factors—that the prey assemblage structure is determined by competition and that the predator preferentially feeds on the species that is the best competitor in the assemblage. Clearly, not all food webs are likely to be structured in this way.

Another method of experimental manipulation is to enhance the numbers of predators in a community. For complex natural communities, both additions and exclusions of predators have revealed direct (depression of prey) and indirect (enhancement) effects. Since generalist predators are bitrophic in nature, they may interact with other carnivores in such a way as to enhance the survival of herbivores that normally would fall victim. In one experiment, adding praying mantids to an insect community resulted in a decrease in spiders and a consequent increase in aphids, normally eaten by these spiders. Such results are not uncommon and contribute to the uncertainty of prediction. Mathematical models have been constructed to depict predator-prey interactions in terms of how each population affects the growth of the other. The simplest of these models, known as the Lotka-Volterra model for the mathematicians who developed it, describes a situation in which prey and predator populations are assumed to be mutually regulating. This model, which was developed for a single prey population and single predator species, has been modified by many workers to provide more realism, but it is far from predicting many competitive situations in complex natural communities.

As with the rest of modern ecology, these different approaches must be blended in order to build a robust picture of how important predators are in natural ecosystems. This knowledge would allow more successful prediction of the outcomes of human intervention and more intelligent management of exploited populations. Predation is a key interaction in natural ecosystems; understanding the nature of this interaction is central to any understanding of nature itself.

—*Lawrence E. Hurd*

See also: Altruism; Cannibalism; Carnivores; Communities; Competition; Ecological niches; Ecology; Groups; Herbivores; Herds; Omnivores; Packs; Population fluctuations; Symbiosis; Territoriality and aggression.

Bibliography

Allen, K. R. *Conservation and Management of Whales*. Seattle: University of Washington Press, 1980. A fascinating account of the history and methods of human exploitation of a poorly managed resource. Requires no special training to read and appreciate. Diagrams and references provided.

Crawley, M. J., ed. *Natural Enemies: The Population Biology of Predators, Parasites, and Diseases*. Boston: Blackwell Scientific Publications, 1992. Covers evolution, morphology, and population dynamics, the biology of predator-prey interactions, and predator-prey coevolution.

Ehrlich, Paul R., and Anne H. Ehrlich. *Extinction: The Causes and Consequences of the Disappearance of Species*. New York: Random House, 1981. This is a thoughtful and lucid treatment of one of the most crucial of environmental topics. Paul Ehrlich is one of the best known American environmentalists and is also a highly respected researcher in evolutionary ecology. As such, his point of view is not to be taken lightly. Illustrated with references.

Fabre, Jean Henri. *The Life of the Spider*. Translated by Alexander Teixeira de Mattos. New York: Dodd, Mead, 1916. The famous French naturalist provides some interesting insights into the natural history of this important group of predators. Written for lay readers and amateur naturalists, this work is both informative and entertaining. Considered to be a classic.

Gause, G. F. *The Struggle for Existence*. Reprint. New York: Dover, 1971. This is a reprint of the classic 1934 edition. Gause blends mathematical insights with simple laboratory experiments in competition and predation. This book can be read by those with a basic understanding of algebra. Diagrams and references.

Hassell, Michael P. *Arthropod Predator-Prey Systems*. Princeton, N.J.: Princeton University Press, 1978. This is a book for serious students of mathematical ecology who have had some college calculus. Among topics covered are basic models of predator-prey dynamics, functional responses, and theoretical considerations of biological control. Diagrams and references provided.

Krebs, J. R., and N. B. Davies, eds. *Behavioral Ecology: An Evolutionary Approach*. 4th ed. Boston: Blackwell Scientific Publications, 1997. A textbook for upper level undergraduates or graduate students.

Levy, Charles Kingston. *Evolutionary Wars: A Three-Billion Years Arms Race*. New York: Basingstoke, 2000. Traces the coevolution of predators and prey from the earliest days of animal life on earth. Focuses on animals' natural "weapons systems."

Mowat, Farley. *Never Cry Wolf*. Boston: Atlantic-Little, Brown, 1963. An intimate natural history of wolves of the North from the standpoint of a man who lived in close proximity to them. Mowat reveals details about the natural history of these diminishing predators in an entertaining and sympathetic manner. Written in narrative form for a general readership.

PREGNANCY AND PRENATAL DEVELOPMENT

Types of animal science: Development, reproduction
Fields of study: Developmental biology, embryology, physiology, reproduction science

After conception, the life and normal development of the new individual depend on establishing a mutual relationship with its mother. Both mother and embryo must contribute to the success of the pregnancy.

Principal Terms

CORPUS LUTEUM (pl. CORPORA LUTEA): the structure on the ovary that is formed from the follicle after the egg has been released; it secretes progesterone

EMBRYO: a fertilized egg as it undergoes divisions from one cell to several thousand cells, but before the individual is completely differentiated into a fetus

ESTROGEN: a hormone secreted by the ovary and placenta for development of the uterus

FETUS: a differentiated but undeveloped individual with organ systems usually identifiable as a member of a species

OVARY: the female gonad that is the source of eggs to be fertilized and hormones to maintain pregnancy

OVIDUCT: a narrow, hollow tube which takes the newly ovulated egg from the ovary, provides the site for fertilization, and transports the new embryo to the uterus.

PLACENTA: the tissue providing contact for exchange of nutrients between the lining of the mother's uterus and the blood supply from the fetus

PROGESTERONE: the hormone, essential for maintenance of pregnancy, that is secreted by the corpus luteum

UTERUS: the hollow internal female organ that accommodates the embryo as it grows to term

Gestation is a period of intrauterine development in animals that are "viviparous," meaning "bearing live young." "Pregnant" comes from a Latin word meaning "to possess important contents." Intrauterine development is important for several reasons. The uterus provides a unique environment that not only permits development, but promotes it. A recently fertilized egg has little capability of living if not surrounded by the uterus. The pregnant female sustains a completely undifferentiated single cell in the form of a fertilized egg through the stages of embryonic differentiation to that of a fetus capable of living outside the uterus. Proper temperature is maintained, the necessary fluids and nutrients are provided, wastes are removed, attacks by microorganisms are prevented, oxygen is supplied in a form useful to the embryo, and other compounds essential for cell growth are present. The uterus is normally a safe haven with all the accommodations needed for the new potential animal. Because gestation takes place concealed inside the body, much myth and mystery have arisen over the process. Recent advances have dispelled some of the mystery, but much is still not completely understood. The present knowledge has been obtained from research and observations on laboratory and domestic animals, supplemented by observations on humans. The mechanisms of maintenance of pregnancy, number of young, length of gestation, anatomy of the uterus, and hormones vary considerably between species, thus preventing generalizations that might apply to more than a few species.

The Course of Pregnancy

Pregnancy begins when the recently ovulated egg is fertilized by a sperm in the upper segment of the oviduct. Depending on the species, the fertilized egg resides in the oviduct for forty to one hundred twenty hours before being transported to the uterus. At a particular time during that interval, upon the laying down of certain fundamental tissues, it is referred to as an embryo. The rate of passage is influenced by the concentration of hormones in the mother. Normally, estrogen production from the ovary is great before fertilization, when progesterone is not produced. After ovulation, the corpus luteum forms on the ovary, estrogen declines, and progesterone rises. This sequence permits the embryo to enter the uterus when the uterus is ready to support it. Progesterone will cause the oviduct to relax and allow the embryo to enter the uterus. Administration of large amounts of progesterone by injection will cause the oviduct to relax prematurely, and embryos will enter the uterus in eight or ten hours. High concentrations of estrogen, on the other hand, will cause the embryos to be retained in the oviduct for many days. Estrogen and progesterone also have a profound influence on the nature and function of the lining of the uterus. Estrogen causes rapid growth and, in synergism with progesterone, causes proliferation of glands lining the uterus. These glands secrete a complex of the compounds necessary for embryonic growth and development, called "uterine milk," for the nourishment of the young embryo before it becomes attached to the uterus.

The embryo must in some way make its presence known to the mother's constitution so that the uterus continues to be a hospitable environment for the entire gestation. There are a great variety of ways that have evolved in different species to maintain pregnancy. The act of mating will cause a female rabbit to ovulate and maintain a pseudopregnancy for about eighteen days. Signals from the fetus take over during the next thirteen days to maintain the corpora lutea. If the mating was not a fertile one, the rabbit will mate again about eighteen days after the first mating. Rats normally have recurrent periods of estrus every four or five days. Mating will delay the next estrus for about twelve days. When the mating is fertile, the fetus provides signals for maintenance for the remaining nine days of gestation. Even without mating, dogs will often become pseudopregnant. This pseudopregnancy can extend as long as the normal gestation period of about sixty-five days. In these cases of pseudopregnancy, the hormones present, changes in the uterus, and maternal behavior are very similar to a normal pregnancy that would produce live young.

In many animal species, the interval between ovulations is somewhat prolonged by the spontaneous formation of one or more corpora lutea if there is no pregnancy. Examples are: sheep, seventeen days; horse, pig, goat, cow, twenty-one days; and human, twenty-eight days. In each of these cases, the embryo must signal the mother to establish the pregnancy several days before the next expected ovulation. Embryos of some species produce estrogens that are thought not only to cause uterine growth and enlargement, so as to accommodate the growing fetus, but also to maintain the corpora lutea. In all species, the corpus luteum is necessary to produce progesterone during at least the first one-third of gestation. In the goat, pig, rabbit, and rat, the ovaries and the corpora lutea on them are essential throughout the gestation period. Progesterone is essential throughout gestation in all species, but in the sheep and human, the fetus and placenta supply progesterone for maintenance of pregnancy after about fifty-five days, even when the ovaries are removed. In summary, estrogen and progesterone are both essential for pregnancy for a number of functions. The source of these hormones can be either or both the ovaries and placenta, with variation among species.

The Uterine Environment

Embryos arriving in the uterus must have not only sustenance in the form of an ever-changing, compatible, fluid environment, but also a place and space to develop. In species with only one young born, this may not seem to be a problem, but in litter bearers (with as many as ten per birth), it is important that each embryo have sufficient space. In the pig, the uterus is V-shaped, with the ovaries at the tips. The two uterine horns form the sides of the V, and the body of the uterus and the birth canal consisting of the cervix and vagina are at the bottom. One uterine horn may be as long as 150 to 200 centimeters. The fetuses each occupy a segment of the uterus about 30 centimeters long; thus they are arranged like peas in pods in the two horns. The embryos enter the uterus forty-eight hours after ovulation at the four-cell stage. Embryos cannot move by themselves, so the motility of the uterus helps intrauterine migration. Embryos gradually move down each uterine horn during the period from day two to day nine when they reach the body of the uterus. From day nine to day twelve, some of the embryos continue to move into the other uterine horn. Some embryos arise from one ovary implant in the uterine horn on the side of the ovary of origin while others migrate through the body of the uterus to the other side. Movement stops at day twelve, when the embryos cannot move farther.

During this period of distribution, the embryo is dividing and growing. At day six, the embryo is about thirty-two cells and is still compact, resembling a tiny raspberry or morula about two hundred microns in diameter. The embryo forms a hollow ball of cells called a blastocyst, at days seven and eight. Some of the cells of the blastocyst are destined to become the new fetus and some will develop into the placenta. At around day twelve, the embryo is twenty to thirty centimeters in length—a very thin, fragile bit of tissue. It is at this stage that each embryo occupies the place and space in the uterus that it will grow within during the remaining 104 days of gestation. At this stage also the embryo histologically signals the mother of its presence so that the pregnancy may continue. The corpora lutea form and produce progesterone for about 14 days after ovulation, regardless of the presence of embryos. There is no need for a signal before day twelve. When embryos are present at day twelve and provide an adequate signal, the corpora lutea persist, maintaining the secretion of

progesterone that maintains the pregnancy and prevents recurring ovulation and a new estrus cycle. In the event of an inadequate signal (as the result of no embryos or too few), the corpora lutea regress and a new ovulation occurs.

The pig is unique in that the proportion of the uterus that is occupied determines whether the signal is adequate. If half of the uterus is not occupied by embryos at day thirteen, the signal is inadequate and the corpora lutea will regress, even though there may have been some live embryos in some segment of the uterus. The anatomy of the uterus of the rat and rabbit is such that each horn is entirely separate and embryos cannot migrate between horns. In cows, sheep, and mice, intrauterine migration occurs infrequently. The embryos of the sheep and the cow signal the corpora lutea on the ovaries adjacent to the horns in which they are implanted. If an embryo is removed and placed in the opposite horn, the signal rarely goes from the uterine horn on one side to the ovary on the opposite side. The pregnancy will cease in that case in spite of the presence of an embryo. The uterus of the horse consists primarily of the body with no pronounced horns. Embryos of the horse do not migrate great distances and have only a local effect, influencing only the ovary on one side of the uterus.

The heart of the pig embryos begins to beat at about day sixteen. The uterus begins to expand at day eighteen and by day thirty the fetuses are about two centimeters long. The fetus is now surrounded by a round ball of fluid, the amnion. It will be floating in a zero-gravity environment for the next fifty days. The pig fetus grows roughly two centimeters every ten days until term. The first thirty days of gestation are critical for establishing systems and forming organs. It is during these embryonic stages that the embryo is very sensitive to toxins. The mother, the placenta, and the fetus all strive to protect the developing fetus from the harsh external environment.

Fetal Loss

Not all embryos and fetuses survive to term. For most species, only about 55 percent of the eggs finally end up as live fetuses. There are many potential causes for this prenatal loss. Some eggs are not fertilized because either the sperm or egg is faulty. Because the lives of an unfertilized egg and a sperm are both finite, limited to a matter of hours, the timing of the meeting of the sperm and egg at fertilization must be synchronized precisely, or a nonviable embryo will result. Even if fertilization occurs, the egg or sperm may have inherent chromosomal defects that produce a nonviable embryo. Any disruption of the rate of development of the embryo or in the rate of change in the uterus or composition of uterine milk will cause loss of the embryo. Synchronous, parallel development of mother's uterus and the embryo are absolutely essential for survival of the embryo. Factors such as toxic chemicals, estrogens, and severe dietary changes can greatly affect chances of survival. In some cases, the amounts of progesterone and estrogen and other hormones essential for pregnancy are slightly abnormal. In species that bear large litters, the uterine space available to each fetus is not always equal, with the result that some fetuses do not have sufficient uterine resources for survival. Considering these and other hazards, it is remarkable that the proportion of individuals that are normal at birth is as great as it is.

Not all embryos grow and implant within a few days of entering the uterus. The embryos of badgers, mink, lactating rats, seals, kangaroos, armadillos, and roe deer are among those that have a delayed implantation. The embryo develops to the early blastocyst, stops for several months until uterine conditions are favorable, and then completes the normal development. When a lactating rat with delayed implantation weans the pups, the embryos resume growth. In mink and some other species, the ratio of light to dark each day seems to influence the time of resumption of growth. The knowledge of this phenomenon goes back more than one hundred years, as indicated by the records of roe deer collected by the physician of a noble on a large estate. He had an opportunity to examine the uteri of roe deer killed for meat. He had noted the dates of mating but found no fetuses even after several months. This led to the discovery that the blastocyst's development had been arrested for a prolonged period before resumption.

The Study of Gestation

The accumulation of knowledge on pregnancy and prenatal development has taken place over a long period. From the early superficial, gross observations of dead animals to the recent sophisticated and detailed techniques of observation of both mother and embryo or fetus, methods of exploring the changes that take place have developed. One procedure in current practice has its origin in antiquity. A cesarean section is a surgical procedure in which the fetus is brought out of the uterus through an incision in the abdominal wall and the uterus. (The term derives the name from the belief that Julius Caesar was delivered by that procedure.) Currently, embryos can be removed very early and still survive. The ability to recover embryos and keep them alive outside the body for several days during the first six or seven days after fertilization has been a major step in study of embryos. The culture fluid must have many characteristics similar to uterine milk to be compatible with living embryos. Temperature, the ratio of gases, the osmotic pressure of the fluid, and chemical composition must all be near those of the normal uterine environment. Each component may be modified slightly to achieve optimum livability at different stages of development and in different species. Embryos from several species may now be frozen and preserved for many years. This creates many possibilities for studies but also creates ethical and moral dilemmas.

The transfer of embryos between mothers of the same species has provided a means of separating the maternal genetic influence from the effect of intrauterine environment on the fetus. Embryos from smaller breeds of rabbits grow larger than normal when transferred to a larger breed. Zebra foals have been born to horse mares as the result of transferring zebra embryos to the horse. The use of ultrasound to obtain an image of the fetus, the uterus, and surrounding tissue by noninvasive means has been very helpful. Now it is possible to get a frequent picture of the course of events in the uterus to monitor development of the fetus. Nuclear magnetic resonance imaging can generate a picture of the uterus and contents noninvasively. By nuclear magnetic resonance spectroscopy, chemical reactions can be measured in very small segments of the uterus and placenta. Optical systems inserted into the amniotic cavity with the fetus permit direct visualization of the fetus. Perfection of surgical techniques has permitted insertion of cannulas into blood vessels of experimental fetuses to monitor the concentration of hormones. The development of the brain of fetuses has been measured by use of minute electrodes and cannulas placed in specific points.

The concentration of hormones circulating in the mother and fetus can be measured, often and with great precision and specificity, by radioimmunoassay of very small quantities of blood. These hormones change with different stages of gestation. The normal changes and concentrations have been determined for a wide variety of animals. An interesting aspect of the emerging picture of the various species studied is the great variety of mechanisms by which pregnancy exists. No two species have the same mechanism even though they may appear to be very similar. As this rapidly growing area of knowledge expands, humans will be able to shed light on the inside of the uterus, which has been called one of the darkest places on earth.

Fostering Optimal Pregnancies

Humankind has entered into an era of unprecedented ability to understand the complexities of pregnancy in animals. The efficiency of the animals related to human consumption—in production of food, fiber, recreation, and companionship—depends largely on the efficiency of pregnancy. Among cows, horses, pigs, sheep, and goats, the proportion of mated females that conceive and bear young is less than 100 percent. The proportion of the fertilized eggs that develop into differentiated embryos is less than 100 percent. The proportion of embryos that develop into full-term fetuses born alive is less than 100 percent. Of those fetuses born alive, less than 100 percent survive more than a few days because of problems that had occurred prenatally. As an example, the runt of the litter often cannot compete successfully and perishes. Often, these animals were stunted in the uterus because of inadequate space or other resources, such as energy and gas exchange. Some of the inadequacies of resources may be avoided by proper prenatal care.

Avoiding adverse effects on embryos and fetuses of all species should be humanity's aim; it may be possible to correct both maternal and fetal problems by judicious prenatal monitoring and treatment. A problem anticipated is half-solved. A sheep with twins needs different feeding and management from one with a single lamb. Prenatal monitoring can anticipate a problem, and treatment may prevent it.

—*Philip J. Dziuk*

See also: Asexual reproduction; Cleavage, gastrulation, and neurulation; Determination and differentiation; Fertilization; Marsupials; Parthenogenesis; Placental mammals; Reproduction; Reproductive system in female mammals; Reproductive system in male mammals.

Bibliography

Cole, H. H., and P. T. Cupps. *Reproduction in Domestic Animals*. 3d ed. New York: Academic Press, 1977. A reference and classroom text covering most aspects of reproduction. Subjects include basic endocrinology and gamete biology in addition to events after conception. There are many figures, graphs, and illustrations to supplement the text. Each chapter is well-referenced. Index.

Foxcroft, G. R., D. J. A. Cole, and B. J. Weir. *Control of Pig Reproduction II*. Cambridge, England: Journals of Reproduction and Fertility, 1985. The proceedings of a conference on pig reproduction. The fetal pig is often a model for studies on maternal effects and teratology. The seventeen chapters give a review of pig reproduction. This book is a place to start for the person seeking a background knowledge in pig reproduction. Bibliography. Indexed by author and subject.

Gilbert, Scott F. *Developmental Biology*. 6th ed. Sunderland, Mass.: Sinauer Associates, 2000. An upper-level college textbook on developmental biology, covering all the basics of animal development from conception to birth.

Hennig, Wolfgang, ed. *Early Embryonic Development of Animals*. New York: Springer-Verlag, 1992. An upper-level college textbook. Covers early embryonic development in *Drosophila, Caenorhabditis*, mice, and zebra fish.

Mossman, H. W. *Vertebrate Fetal Membranes*. New Brunswick, N.J.: Rutgers University Press, 1987. This text is primarily a histological and gross description of vertebrate fetal-placental units. There is some textual material, but the figures are probably the most useful aspect of the work. There is a glossary, a bibliography, and an index that covers a wide variety of species and forms. A fine reference.

Novy, M. J., and J. A. Resko. *Fetal Endocrinology*. New York: Academic Press, 1981. Focuses on the endocrinology of primate fetuses. Most of the chapters describe normal fetal hormone production. The effects of the hormones on growth, development, behavior, and subsequent function of the fetus are also covered. The book is the published result of a conference, and is valuable as a reference to students of primate pregnancy. There is a subject index for the book, and each chapter has a reference list.

Pasqualini, J. R., and F. A. Kincl. *Hormones and the Fetus*. Elmsford, N.Y.: Pergamon Press, 1986. Emphasizes the hormones of pregnancy from before conception to birth. The peculiarities of several different species are brought to the readers' attention. The details of hormone production and metabolism by both mother and fetus are described thoroughly. Each of the five chapters has a complete list of references. The book is indexed by subject, and is especially useful to the more advanced student of fetal-maternal endocrinology.

Pierrepoint, Colin G., ed. *The Endocrinology of Pregnancy and Parturition*. Cardiff, Wales: Alpha Omega Alpha, 1973. The proceedings of a conference on experiments using pregnant sheep. Emphasizes signals for maintenance of pregnancy, hormone secretion, and metabolism. Each section is referenced extensively. The coverage of the subjects is broad; good background material is provided; there is no index.

Shostak, Stanley. *Embryology: An Introduction to Developmental Biology*. New York: HarperCollins, 1991. A detailed introduction to embryology. Includes definitions of terms, topic summaries, discussion questions, recommended readings and extensive bibliography, and numerous tables, diagrams, and illustrations.

Wilson, J. G., and F. C. Fraser. *Handbook of Teratology*. New York: Plenum, 1977. The book consists of nine chapters, each with a bibliography. Cumulative subject index. Topics include the effects of drugs and their rates of absorption by the mother, possible effects of drugs on the fetus, spontaneous fetal abnormalities, and placental pathologies and the incidence of prenatal loss. Students investigating the frequency and causes of fetal abnormalities, including those leading to fetal death, will find this book informative.

PREHISTORIC ANIMALS

Types of animal science: Anatomy, classification, ecology, evolution
Fields of study: Anatomy, evolutionary science, paleontology, zoology

Prehistoric animals are those ancient animals that lived before modern humans described them in writing. The vast majority of prehistoric animals are now extinct.

Principal Terms

ARTHROPODS: the phylum of invertebrate animals having clearly segmented bodies and appendages (legs, antennae, and mouth parts) with many segments and joints

CHORDATES: the phylum of animals that have a stiff, rodlike structure called the notochord running their length; chordates include the boneless fishlike animals similar to amphioxus, as well as fish, amphibians, reptiles, birds, and mammals

EVOLUTION: the science that studies the changes in astronomical bodies, the earth, and living organisms and creates explanations of how changes occur and how entities and events are related

GEOLOGICAL PERIODS: the twelve divisions in successive layers of sedimentary and volcanic rocks, which are differentiated by the distinctive fossils present within each division; because of many recently discovered fossils and careful dating, the beginning and ending dates for these periods tend to vary somewhat among different references

PALEONTOLOGY: the branch of geology that deals with prehistoric forms of life through the study of fossil animals, plants, and microorganisms

---

The fossil record indicates that eubacteria evolved as long as 3,800 million years ago, whereas a comparison of ribosomal ribonucleic acid (RNA) suggests that cells with nuclei probably diverged from archaebacteria as long as 2,800 million years ago. Although cells with respiratory and photosynthetic organelles (mitochondria and chloroplasts) evolved much later, about 1,500 million years ago, no trace of animals has been found in the fossil record before 1,000 million years ago. The fossil record of microscopic animals is nearly nonexistent between 1,000 and 600 million years ago. All of our knowledge of prehistoric animals comes from the fossil record that formed during the last 600 million years.

Near the end of the Precambrian period (540 million years ago), a few sponges, jellyfish, colonial filter feeders, wormlike animals, early mollusks, and primitive arthropods are found in the fossil record. One or more minor extinction events that eliminated up to 50 percent of all species brought the Precambrian period to an end. Animals whose body plan was based on segmentation quickly diversified into a myriad of forms.

How Did So Many Animals Develop So Quickly?

The Cambrian period (540 to 500 million years ago) is characterized by the "sudden" appearance (in 1 to 10 million years) of a large number of morphologically different animals. One of the most innovative body plans that evolved near the end of the Precambrian period was that associated with the many wormlike animals that developed segmented bodies. Segmentation allowed for rapid evolution of all sorts of animals simply by changing the number of segments, their size and shape, as well as the associated appendages. Evolution of homeotic genes and the genes they regulate altered not only segment number and segment characteristics but also appendages. For example, legs could be easily changed into preening, mating, and sensory appendages. During the Cambrian period, all kinds of marine worms, caterpillar-like animals, arthropods, and chordates made their appearance. Some of these caterpillar-like animals had broad legs (*Aysheaia*) and resembled velvet worms of today, whereas others had long, sharp spikes for legs (*Hallucigenia*) and are unrelated to any other known animals.

Many of the early arthropods vaguely resemble today's sow bugs, shrimp, centipedes, and horseshoe crabs. A very successful group of marine arthropods that superficially resemble sow bugs were the trilobites. A trilobite had a prominent, centrally located, segmented, concave, dorsal ridge running the full length of its body. The trilobites differentiated and evolved for over 300 million years until the few remaining forms went extinct during the Permian-Triassic mass extinction (250 million years ago) that wiped out nearly 90 percent of all species. A trilobite had a head, thorax-abdomen, and tail (pygidium). The segments of the head were fused but the thoracic-abdominal and tail segments were varied. Depending upon the trilobite, thoracic-abdominal segments ranged from nine (low of six) to thirteen (high of twenty-two) in number, whereas tail segments varied from six to sixteen (high of twenty-two). Legs and other specialized appendages protruded ventrally from the segments. Gills were also associated with a number of the segments. Although some of the first trilobites had "eyes," their compound nature is not clear because of their size. Most of the later trilobites developed compound eyes that closely resemble the compound eyes seen in extant insects.

One of the most unusual arthropod-like creatures of the Cambrian period was the enormous predator known as *Anomalocaris*, which ranged up to three feet in length. It had a long, oval, dorsoventrally flattened head with compound eyes on short stalks near the back of the head. Protruding from the ventral region at the front of the head were two grasping feeding appendages that for many years were mistaken for separate shrimplike organisms. Just behind the feeding appendages was a circular mouth resembling a cored pineapple slice. This mouth had been misinterpreted as a jellyfish by early discoverers. Saw-shaped "teeth" lined the inside of the mouth and most likely contracted around animals that were positioned in the mouth by the feeding appendages. A number of fossil trilobites have been found with missing chunks that appear to have been torn away by *Anamalocaris*. No arthropods of today have any sort of mouth like that of *Anamalocaris*. This creature was a specialized swimmer, propelled by winglike flaps that appear to consist of numerous overlapping flaps originating from most of the body segments.

The Evolution of Fish
The oldest fossils of jawless fish, referred to as the Agnatha, are about 470 million years old. The jawless lampreys and hagfish of today are most closely related to the ancient Agnatha of the Ordovician period (500 to 440 million years ago). These prehistoric Agnatha attached to larger animals through their whorls of "teeth" and rasped the flesh of the host. Lamprey teeth are horny, sharp structures devoid of calcium, derived from the skin. Because the teeth contained enamel-like proteins, they are considered to be precursors to modern teeth, which consist of enamel over dentine.

The more advanced Agnatha that developed in the Silurian and Devonian periods did not survive the Devonian-Carboniferous mass extinction. Yet they are important because they suggest when the evolution of mineralized teeth, bone, body armor, eyes, and paired limbs occurred. The jawed fish first evolved in the Devonian period (410 to 360 million years ago) but did not dominate until after the Devonian-Carboniferous mass extinction. The earliest jawed fishes were sharks. The discovery of fossilized shark teeth dated to the Devonian period suggest that they evolved during this period.

The earliest fossil skeletons of sharks discovered to date, however, are from the Carboniferous period (360 to 290 million years ago).

The First Terrestrial Vertebrates

The first land-dwelling vertebrates were amphibians that evolved from freshwater, lobe-finned fishes closely related to *Eusthenopteron*, a crossopterygian that lived midway through the Devonian period (410 to 360 million years ago). Most amphibians are tetrapods that begin life in a watery environment but turn to a terrestrial life as adults. Extant amphibians include frogs, toads, salamanders, and the legless, snakelike caecilians.

An unusual prehistoric amphibian was *Ichthyostega*, a three-foot-long crocodile-like animal that lived somewhere between 350 and 370 million years ago. Its jaws were lined with two rows of teeth. Along the front was a row of large teeth and behind them was a row of densely packed, small, sharp teeth bounded by a couple of very large canine teeth. The skull was similar in some respects to the lobe-finned fish that lived during the Devonian. Its neck was only two to three vertebrae long and the pectoral bones were attached to the skull and backbone by muscle.

External gills were a prominent feature of the juvenile or larva. The adult however, lost the external gills when it underwent metamorphosis and developed primitive lungs. The aquatic juvenile amphibians depended on gills to breath, but when they metamorphosed into terrestrial adults they developed primitive lungs out of part of the swim-bladder. Air was swallowed into the intestine-swim bladder tract. A high concentration of capillaries in the wall of these simple lungs carried oxygen to the heart.

The pelvic bones were attached to the spinal chord through muscles. *Ichthyostega* had a heavy rib cage to support its weight out of water, in contrast to other contemporaneous vertebrates. In general, the early amphibians had very short leg bones compared to the primitive reptiles that evolved from them. *Ichthyostega* had seven digits on both its front and back feet, whereas *Acanthostega* had eight digits. Other ancient amphibians had variations in the number of digits. Living amphibians generally have four to five digits on the front feet and five digits on the back feet. *Ichthyostega*'s diet was mainly fish, but on land it may have consumed arthropods and smaller amphibians.

Mastodonsaurus giganteus, another crocodile-like animal, lived 230 million years ago, during the early Triassic period (250 to 205 million years ago). It was the largest amphibian that ever existed, growing to nearly thirteen feet in length. Like *Ichthyostega*, two rows of pointed teeth lined the jaws of this amphibian, making it a formidable match for the early reptiles.

A small group of amphibians lost their legs and took on the appearance of snakelike animals. Today these legless amphibians are highly segmented, making them look more like giant earthworms than snakes. One of the largest extant caecilians is *Caecilia thompsoni*, which lives in Colombia and grows to a length of five feet.

Reptile Diversity

The first reptiles evolved from amphibians during the early Carboniferous period (360 to 290 million years ago). Early reptile skeletons can be distinguished from amphibian remains by the skull and jaw characteristics, the number of neck vertebrae, the absence of external gills, the length of the leg bones, and the number of foot and toe bones.

Prehistoric amphibians had dorsoventrally flattened skulls and skull bones similar to their crossopterygian fish relatives. Amphibians had no nasal holes in their skulls because they breathed through their gills or through their mouths. Reptiles, however, had anterior nasal openings in the snout. The diapsid reptiles had two additional holes behind the eye openings for muscle attachment, whereas synapsid reptiles had one hole, and anapsid reptiles had none. The early amphibians had necks with two to three vertebrae, whereas reptiles generally had necks with more than six vertebrae. The early amphibians generally had very short front and rear leg bones and feet with six to eight digits, whereas early reptiles had front and rear feet with only five digits. Later amphibians reduced the number of digits to four in the front and five in the rear, similar to what is found in extant frogs. Early amphibians had digits that varied considerably, whereas early reptiles usually had digits with two, three, four, five, and three bones (moving from thumb to little finger).

The evolution of reptiles required the development of a new type of egg. The new egg provided a watery environment in which the embryo and juvenile could develop. An amniotic sac evolved to contain the developing reptile and its watery environment, the yolk sac developed to provide the nutrients needed for growth, and the allantoic sac was acquired to help eliminate wastes. The evolution of a tough, membranous egg covering allowed oxygen and carbon dioxide exchange but prevented water loss. Fish and amphibians have to lay their eggs in water because they never evolved the adaptations needed to lay their eggs on land. Reptiles, dinosaurs, birds, and mammals are all amniotes and can lay their eggs on land because their eggs have amniotic sacs, yolk sacs, or allantoic sacs.

Although amphibians' skin was protected by thick layers of secreted mucous, it was always in danger of drying or burning under the sun if the animal strayed too far from water. Reptiles and dinosaurs evolved thick scales to prevent drying and burning, whereas birds and mammals evolved feathers and hair to protect their skin.

Flying Reptiles
The oldest flying reptiles are dated to the late Triassic period (250 to 205 million years ago). The first of these pterosaurs (flying lizards) had long, thin tails, often more than half the length of their bodies. The wings consisted of a leathery skin that was supported along its anterior margin by the pterosaurs' front legs and a highly elongated fourth digit. Digits one, two, and three were tipped with claws. The wings were attached to the body and hind legs, and the wingspan ranged from one to six feet. Their elongated jaws were lined with numerous sharp teeth that they used to grab surface fish, like some present day birds. In the late Jurassic and Cretaceous period, ptero-

saurs lost their long tails. Some became quite large. *Pteranodon* had a wingspan of up to twenty-five feet, whereas the largest *Quetzalcoatlus* had a wingspan of nearly forty feet. Toothless *Pteranodon* fished on the move, scooping up surface fish, whereas *Quetzalcoatlus* hunted much like the vultures of today, but swooping down on a dead dinosaur instead of the carcass of a wildebeest. Both these large flying reptiles were consummate soaring reptiles, able to take advantage of the slightest updraft of air because of their long, narrow wings.

Reptiles Returned to the Sea
Some reptiles became adapted to living most of their lives in the sea. *Nothosaurus* lived near the beginning of the Triassic period (250 to 205 million years ago) and grew up to ten feet long. Its neck and tail each accounted for a third of its length. *Nothosaurus* used its legs and webbed feet to swim and to drag itself onto beaches, where it laid its eggs.

By the end of the Triassic period, marine reptiles such as *Liopleurodon* and *Plesiosaurus* had flippers instead of legs and feet. *Liopleurodon* sometimes grew to lengths of seventy-five feet and achieved weights of nearly one hundred tons. It was twenty times heavier than *Tyrannosaurus rex* and had teeth twice as long. Long necks and tails were characteristic of the plesiosaurs, which had anywhere from twenty-eight to seventy neck vertebrae, depending upon the genus. The small mouth and teeth of *Plesiosaurus* were adapted to capturing small fish and squid.

Over a period of millions of years, ichthyosaurs evolved an extremely long, thin snout. The leg bones of these reptiles shortened and the digits were remodeled to form flippers. A dorsal and tail fin appeared. Some of these reptiles have the distinction of having the largest eyes of any animal that ever lived. Fully grown *Temnodontosaurus* had eyes with diameters of nearly eleven inches. (In comparison, adult elephants have eyes that are two inches in diameter, and those of blue whales are six inches wide.) These reptiles became so adapted to their marine lives that they could not return to land to lay eggs. Ichthyosaurs solved the

---

An Early Chordate Related to the Vertebrate Ancestor

A large number of soft-bodied chordates from the Cambrian period have been studied. *Haikouella* is a 530-million-year-old fishlike animal up to 1.25 inches long. The bulbous head had a ventral mouth cavity surrounded by short tentacles. "Teeth" in the pharyngeal cavity of *Haikouella* represent one of the earliest known biomineralizations in chordates. A single ventral artery and paired dorsal arteries were connected by six branchial arches, each with gill filaments. A heart was located at the end of the ventral aorta, more than a third of the way along the body. The narrow tuberous body was strengthened by a notochord that began less than a third of the way along the body and extended nearly to the tail. A small brain headed the spinal chord, which stretched to the tail just above the notochord. Some of these chordates may have had primitive eyes. A gut ran from the mouth cavity ventrally along the animal to the anus, very near the tail. The dorsal rear two-thirds of the body consisted of a muscular, finlike structure. The vertical muscle groups of this vertical fin allowed the animal to undulate from side to side through its watery environment. The extant lancelet, *Branchiostoma*, appears to be most closely related to *Haikouella*.

---

problem of producing the next generation by giving birth to live babies. A fossil of *Stenopterygius* illustrates a baby exiting the reptile's birth canal much like the marine mammals of today.

All these marine reptiles had nasal openings and lungs that had to be filled with air from time to time. This need for oxygen required them to surface just like extant marine mammals. Even though the marine reptiles were highly adapted to the sea, none survived the Cretaceous-Tertiary extinction, 65 million years ago.

Dinosaurs

Dinosaurs evolved from reptiles sometime after the Permian extinction, 250 million years ago. Dinosaurs developed legs and a skeleton that supported them from beneath, in contrast to most reptiles, which supported themselves on legs that protruded from the sides of their bodies. Dinosaurs were slow to diversify during the Triassic (250 to 205 million years ago) and early Jurassic (205 to 145 million years ago) periods. Even so, many distinctive animals evolved from four major groups present near the beginning of the Jurassic.

Evolution of animals in the Eurypoda led to tetrapods such as *Stegosaurus*. These huge vegetarians up to thirty feet in length had large, diamond-shaped plates that protected their necks and backs from predators. It is believed that these large dinosaurs were cold-blooded and also used the plates to rapidly cool or heat their bodies.

Animals in the Euornithopoda were mostly tetrapods. *Iguanosaurus* was a large vegetarian with big, hornlike claws on its thumb and toes that ended in "hoofs." *Hadrosaurus* was a large vegetarian with a long, hollow, head crest along the top of its skull that protruded up to 1.5 feet behind the head. *Parasaurolophus* grew up to thirty feet in length and had a head crest over 5 feet long. These animals could blow air through the head crest to create various trumpeting sounds. *Triceratops* was another large vegetarian that grew up to nine feet long. It had a sharp, horny beak, a rhinoceros-like horn above and somewhat posterior to its nostrils, two long horns above its eyes, and a bony neck frill protruding from the back of its skull to protect its neck from predators. *Styracosaurus*, a relative of *Triceratops*, grew to lengths of seventeen feet. A number of smaller horns on either side of two long horns extended the neck frill to make *Styracosaurus* appear larger and to protect its neck.

Certain Sauropoda evolved into some of the largest land animals that ever existed. The vegetarian tetrapods *Titanosaurus* weighed up to seventy tons, while *Supersaurus* ranged up to fifty-five tons, *Brachiosaurus* forty-five tons, and *Seismosaurus* thirty tons. Some scientists argue that these reptiles must have been cold-blooded, because they could not possibly process enough vegetation to maintain a constant temperature. In comparison, warm-blood mammals such as elephants often reach weights of six to eight tons, whereas warm-blooded blue whales may attain weights of up to ninety tons.

Archaeopteryx, the Dinosaur That Flew

Theropod dinosaurs were warm-blooded, bipedal animals that came in a myriad of sizes and superficially resembled *Tyranosaurus rex*. Approximately 155 million years ago, a number of theropod dinosaurs were developing feathered coats. Most of the feathers were similar to down feathers that superficially resemble a tuft of hair. A tuft of hair, however, consists of hairs derived from many hair follicles, whereas a down feather is a single unit in which each hairlike barb attaches to the base of the feather. Some theropod dinosaurs also produced vaned feathers having a distinct shaft and barbs attached along the length of the shaft. Like scales, feathers develop from the differentiation of certain epithelial cells. The earliest stages of feather development resemble the first stages of reptilian scale formation. This suggests that one or more genes evolved that blocked scale development in favor of feather development.

During the late Jurassic period, many of the small, bipedal theropod dinosaurs related to the ancestor of birds evolved long arms, three-fingered hands with claws, fused clavicles (the beginnings of a wishbone), feet with three toes for running, first toes protruding from the back of the foot, and hollow bones.

The first primitive bird, *Archaeopteryx*, is dated to the late Jurassic, about 150 million years ago. Fossil remains suggest that *Archaeopteryx* had primary, secondary, and tertiary flight feathers originating from its palm, forearm, and upper arm, respectively. These flight feathers were covered at their base by other feathers known as main coverts and lesser coverts. The three clawed digits of the hand appear not to have supported primary feathers; instead they protruded three-quarters of the way along the wing.

In modern birds, the palm and first digit bones have fused and greatly diminished so that a single, small thumb bone protrudes near the wrist. This first finger supports feathers that run parallel to the front of the wing. The palm bones associated with the second and third digits have partially fused at their ends to become a bone that roughly resembles a fused miniature radius and ulna. The second digit that supports the primary feathers that extend the wing and compose the wing tip has shortened and thickened. The third finger has been reduced to a single, tiny bone.

Modern flying birds have large, keel-like sterna for attaching the muscles that move the wings up and down. Although *Archaeopteryx* had wings that suggest it was capable of performing intricate maneuvers in flight, its tiny sternum indicates it probably lacked the power to take off from the ground. It probably launched itself by running rapidly until it developed sufficient lift. Its feet are similar to those on fast-running dinosaurs. The first digit is high on the foot, suggesting that *Archaeopteryx* did not depend upon trees or high bushes to launch itself.

Archaeopteryx inherited a long, bony tail from its dinosaur ancestors. The tail supported numerous vaned feathers along its length. In modern birds, the bony tail has been greatly reduced by the loss of numerous segments at the end and the fusion of segments at the base. The short, bony tail in modern birds is called the pygostyle. Most of the tail in modern birds consists of long feathers that extend toward the rear from the pygostyle.

Archaeopteryx also inherited a head that closely resembled that of its dinosaur ancestors. The skull and mandible are light because of cavities in the bones, whereas its elongated snout, not yet a true beak, has a mouth lined with teeth. Although *Archaeopteryx* and many primitive birds of the Cretaceous period did not survive the Cretaceous-Tertiary extinction, 65 million years ago, a few did. These survivors gave rise to the myriad bird species of today.

—Jaime Stanley Colomé

See also: *Allosaurus; Apatosaurus; Archaeopteryx; Brachiosaurus;* Dinosaurs; Evolution: Animal life; Extinction; Fossils; Hadrosaurs; Ichthyosaurs; Mammoths; Neanderthals; Paleoecology; Paleontology; Pterosaurs; Sauropods; Stegosaurs; *Triceratops; Tyrannosaurus;* Velociraptors.

Bibliography

Chen, Ju-Yuan, Di-Ying Huang, and Chia-Wei Li. "An Early Cambrian Craniate-like Chordate." *Nature* 402 (December 2, 1999): 518-522. Particularly clear fossil remains of a 530-million-year-old, soft-bodied chordate show interior structures of this type of animal.

Dowswell, Paul, John Malam, Paul Mason, and Steve Parker. *The Ultimate Book of Dinosaurs*. Bath, England: Parragon, 2000. A very nicely illustrated book containing a brief description of a few interesting reptiles (such as pterosaurs, plesiosaurs, ichthyosaurs) and numerous dinosaurs.

Gould, Stephen Jay. *Wonderful Life: The Burgess Shale and the Nature of History*. New York: W. W. Norton, 1989. An engrossing account of the unusual animals of the Burgess Shale in Canada.

Hofrichter, Robert, ed. *Amphibians: The World of Frogs, Toads, Salamanders, and Newts*. Willowdale, Ontario: Firefly Books, 2000. A beautifully illustrated book with very useful information on amphibians and how to distinguish them from reptiles.

Motani, Ryosuke. "Rulers of the Jurassic Seas." *Scientific American* 283, no. 6 (2000): 52-59. A beautifully illustrated article explains the adaptations that had to be made for reptiles to return to the sea.

Padian, Kevin, and Luis M. Chiappe. "The Origin of Birds and Their Flight." *Scientific American* 278, no. 2 (1998): 38-47. A good outline of how birds evolved from dinosaurs.

Sereno, Paul C. "The Evolution of Dinosaurs." *Science* 284, no. 5423 (June 25, 1999): 2137-2146. An outstanding and very informative review of phylogeny. The figures are particularly informative, illustrating when most of the important lineages arose and what typical animals were like. There is a detailed figure illustrating the evolution of birds and when important clads arose.

PRIMATES

Type of animal science: Classification
Fields of study: Anthropology, human origins, systematics (taxonomy)

The primates are an order of mammals that includes monkeys, apes, lemurs, tarsiers, and humans. Many primates live in trees and have adaptations which include freely movable limbs, grasping hands and feet with opposable thumbs, reliance on vision, a high degree of intelligence, and complex, learned behavior patterns.

Principal Terms

ARBOREAL: tree-dwelling, or pertaining to life in the trees

BINOCULAR VISION: vision using two eyes at once, with overlapping visual fields

BRACHIATION: a form of locomotion, also called arm-swinging, in which the body is held suspended by the arms from above

CATARRHINI: a primate group including Old World monkeys, apes, and humans, with reduced tails and only two pairs of premolar teeth

CLAVICLE: the collarbone, connecting the top of the breastbone to the shoulder

OPPOSABLE: capable of rotating so that the fingerprint surface of the thumb or big toe approaches the corresponding surfaces of other fingers or toes

PLACENTALS: mammals whose unborn young are nourished within the mothers' uteri

STEREOSCOPIC VISION: vision with good depth perception

VISUAL CORTEX: the part of the cerebral cortex concerned with vision

VISUAL PREDATION: catching prey (such as insects) by sighting them visually, judging their exact position and distance, and pouncing on them

The primates are an order of mammals that includes monkeys, apes, lemurs, tarsiers, and humans. The other primates are thus the nearest relatives to human beings. Anatomically, primates have many similarities to the earliest placental mammals. Primitive features retained by primates, but lost in many other mammals, include five-fingered hands and feet with individually mobile fingers and toes, a collarbone (clavicle), a relatively simple cusp pattern in the molar teeth, and the freedom of movement within the forearm that allows the wrist to rotate without moving the elbow.

Many primates live in trees; those that do not, have anatomical features showing that their ancestors were tree-dwellers. These include features directly concerned with arboreal locomotion, those concerned with vision and intelligence, and those concerned with reproduction. Arboreal locomotion, the ability to climb trees, has many direct consequences in primate anatomy. Most primates, for example, possess long, agile arms and legs. Grasping hands and feet give them the ability to hold objects and to climb by wrapping the fingers around branches. This is different from the many other arboreal animals that dig their claws into the bark. The primate grasp is aided by the development of opposable thumbs and big toes. If a primate grasps an object, the fingerprint surface of the thumb faces the corresponding surfaces of the other fingers. This is possible because the bone supporting the thumb can rotate out of the plane of the other fingers. Most primate thumbs and big toes are opposable; human feet are unusual. Individual mobility of all fingers and toes is a primitive mammalian ability that primates have kept but which many other mammals have lost through evolution. Combined with the grasping ability of hands and feet, this characteristic allows primates to manipulate objects of all sizes rather skillfully.

Primates also have hairless friction-skin on their palms and soles, which are supplied with a series of parallel ridges in complex patterns (fingerprints). These ridges provide a high-friction surface, which helps in grasping objects—or in holding branches without slipping. Claws, which could get in the way and cause injury, have generally been reduced to fingernails that rub against things and thus stay short. The clavicle (collarbone), which strengthens the shoulder region, is present in primitive mammals. Many modern mammals have lost this bone, but in primates, it has been retained and often strengthened.

Primate Vision
Primates are primarily visual animals. Whereas many other animals rely strongly on smell, primates rely on sight, and particularly on color vision. One reason for this visual orientation of primates is that the exact position of the next branch can best be judged visually, and good depth perception is essential for avoiding a fall. Primates often hunt by visual predation, stalking an insect or other small prey, judging its exact position and distance, and then pouncing on it.

Vision in depth (stereoscopic vision) requires two eyes with overlapping visual fields (binocular vision). Each eye sees the same objects from slightly different angles; the brain combines the two images into a single three-dimensional image. Binocular vision is made possible by the forward position of the eyes. The eyes need protection in this position, and it is furnished in most primates by a bony structure called the postorbital bar. A complex organization of the brain is required for binocular vision. The cerebral cortex of the brain is therefore expanded, especially the part known as the visual cortex. The surface of the brain develops a large number of folds—primate brains are more heavily folded than any others.

One characteristic fold is called the calcarine fissure, a characteristic feature of all primates.

Primates can pick up unfamiliar objects with the fingers and bring them in front of the face for closer visual inspection. This behavior requires coordination of eye, brain, and hand in a very complex interaction. It is through this behavior that young primates learn to manipulate and understand their environments and to investigate them with a characteristic curiosity. Primates rely heavily on learned behavior, including exploratory play. Primates are the most intelligent animals, and the great expansion of the cerebral cortex reflects this. Most primates grow up in social groups, where they interact in complex ways with other individuals. They have a curiosity about the world, especially about other primates.

Primate behavior is complex, and most of it is learned. The price paid for reliance on learned behavior is youthful inexperience. Infant primates have much to learn, and they have not yet had much time to learn it; they therefore depend significantly on their parents. A long and extended period of intensive parental care of offspring characterizes nearly all primates. This would not be possible if primates were born in large litters, because their mothers would be too busy caring for so many babies at once. Furthermore, infant primates are carried frequently, placing a burden on their mothers, some of whom need both arms free for climbing. Accordingly, primates are usually born one at a time. Twins and other multiple births are uncommon. There are several related features in primate reproductive systems. Female primates need only one uterus to carry their young and one pair of nipples to nourish them after birth. The right and left uteri, paired in other mammals, are fused into a single uterus called a uterus simplex. Unlike female dogs or pigs, which have many nipples in parallel rows, female primates have only a single pair, located high in the chest region.

Primate Classification
There are many different types of primates and several ways of classifying them. The classifica-

tion followed here divides primates into the suborders Paromomyiformes, Haplorhini, and Strepsirhini. The suborder Paromomyiformes contains the earliest primates, all now extinct. They lived from about fifty to eighty million years ago (from the Cretaceous period to the Eocene epoch). They are known mostly from fossilized teeth, but the genus *Plesiadapis* is known from nearly complete skeletons. All these primates were small, from roughly the size of a mouse to that of a cat. Their limbs were long and agile, with hands and feet showing the grasping ability typical of arboreal primates. The teeth, which differed among the three families of Paromomyiformes, were modified for a diet of mostly fruits but also some insects. A postorbital bar was not present; the eyes had not yet shifted forward. The visual fields overlapped much less than they do in modern primates, showing that these animals were not efficient visual predators. Some paleontologists

Primates are visually oriented animals, with forward-facing eyes to optimize binocular vison. (PhotoDisc)

have suggested excluding them from the order Primates for this reason.

Suborder Haplorhini

The majority of primates belong to the suborder Haplorhini and possess relatively large brains and relatively short faces. The noses of all Haplorhini are dry and have no external connection to the mouth; the upper lip goes straight across the mouth and is not divided by a groove, as it is in the Strepsirhini. The structures of the ear region, the placenta, and certain other anatomical features show that the Haplorhini are closer to the ancestral mammals than are the Strepsirhini. The geologic record of haplorhines extends from the Paleocene to the Recent (or Holocene) epoch.

There are three subgroups of Haplorhini: Tarsioidea, Platyrrhini, and Catarrhini. Of these, the Tarsioidea are the oldest and most primitive. The tarsioids are small primates with rounded heads. The eyes are rotated forward and protected from behind by a postorbital bar; visual fields overlap significantly. Tarsioids flourished during the Eocene epoch, about forty to fifty million years ago. They lived across Europe, Asia, and North America, all of which then had subtropical climates. Tarsioids disappeared from Europe and North America as climates became colder, but they survived in Asia. The living genus *Tarsius* now inhabits the East Indies from Borneo to Celebes. *Tarsius* is a big-eyed, nocturnal primate with an elongated ankle. It has a peculiar form of locomotion called vertical clinging and leaping. It clings to vertical bamboo stalks much of the time, then uses its hind legs to jump to the next perch. At least one extinct tarsioid had similar adaptations.

The Platyrrhini include the South and Central American primates. They are characterized by nostrils that open directly forward, three pairs of premolar teeth, and tails that are strong and sometimes assist in locomotion. Platyrrhines probably evolved from tarsioid primates that reached South America prior to Miocene times. Included in the Platyrrhini are the small forest primates called marmosets (*Callithricidae*). Male marmosets often have striking white or yellowish facial markings (such as tufted ears, eyebrows, and mustaches) which may be useful in mate recognition. The other Platyrrhini are the New World monkeys (*Cebidae*). Familiar ones include squirrel monkeys (*Saimiri*), capuchin monkeys (*Cebus*), and howler monkeys (*Alouatta*). All these monkeys are arboreal, and several of them are skilled acrobats. A few species, including the squirrel monkeys, can hang by their tails.

The Catarrhini include the Old World monkeys, apes, and humans. All Catarrhini are characterized by noses that protrude from the face with the nostrils opening downward, as human noses do. There are only two pairs of premolars. The tails are weak or absent and are useless in locomotion. Two very early catarrhine species occurred in Burma, but they are poorly known and of uncertain relationships. The oldest undoubted Catarrhini occur as fossils in the Fayum deposits of Egypt, which are Oligocene in age—about thirty-five million years old. Fayum primates are thought to be of tarsioid ancestry. One type, *Apidium*, has definite tarsioid resemblances. Other Fayum primates include several small apes, such as an early gibbon, a possible ancestor of Old World monkeys, and *Aegyptopithecus*.

Living Catarrhini include the Old World monkeys, or cercopithecoids, along with apes and humans. Familiar cercopithecoids include baboons (*Papio*), guenons and vervet monkeys (*Cercopithecus*), macaques (*Macaca*), langurs (*Presbytis* and *Pygathrix*), and colobus monkeys (*Colobus*); there are many others. The large and complex social groups of macaques and baboons have been studied repeatedly. There are dominance relationships within these primate societies, but they are usually expressed by gestures and displays instead of by fighting.

The family Pongidae includes the apes. The smaller, more lightly built apes are called gibbons (Hylobates and Symphalangus). These skillful acrobats are often placed in a family by themselves, the Hylobatidae. The more typical "great" apes include the orangutans (*Pongo*) of Asia, the gorillas and chimpanzees (*Pan*) of Africa, and several fossil apes, such as *Dryopithecus*. Modern apes have

long arms and shorter legs. They exhibit a variety of locomotor patterns: arm-swinging (brachiation) in gibbons, knuckle-walking in gorillas and chimpanzees, and a four-handed type of clambering in the orangutan.

The family Hominidae includes humans, who walk upright and communicate using language. All living humans belong to the single species *Homo sapiens*.

Suborder Strepsirhini

The Strepsirhini, or Lemuroidea, include lemurs and their relatives. Once thought to be "lower" on the evolutionary scale, the lemuroids are now known to be separately specialized in many ways. Their placentas, for example, are of a peculiar type not found among other primates or among primitive mammals. The ear regions of their skulls show certain anatomical peculiarities; the same is true of the brain and the facial muscles.

All true lemurs live on the island of Madagascar, off the eastern coast of Africa. Most lemurs have the lower front teeth modified to form a comblike structure used in cleaning the fur. Other lemuroids include the small, agile galagos of mainland Africa and four types of slower-moving lorises in Africa and southern Asia. Many fossil lemuroids are known from the Eocene epoch.

Genetic Comparisons

For many years, the classification of primates was based largely on anatomy, physiology, and paleontology. Now, however, the molecular structure of proteins and other types of biochemical evidence are used to make additional comparisons. Closely related primates have proteins with similar or identical sequences. For the most part, family trees based on protein sequences confirm the traditional classifications, with some exceptions; the separateness of gibbons from other apes is based largely on such biochemical evidence.

Geneticists have compared the chromosome sequences of humans and apes. They have found that the sequences of banding patterns in chimpanzee and gorilla chromosomes are nearly identical and that fewer than twenty chromosomal inversions (end-to-end changes) separate chimpanzees and humans.

Types of Primate Study

Primates are studied by different specialists, each using different methodologies. Primate anatomists and functional morphologists dissect primates and compare their structures to those of other species. Morphologists may also take numerous measurements and analyze the results statistically, often with the aid of a computer.

Primate ethologists and sociobiologists study the behavior of living primates, both in the field and in the laboratory. Field studies have been conducted among the chimpanzees in the Gombe Stream Reserve in Tanzania, the rhesus macaques of India, the Japanese macaques of Japan, the howler monkeys of Barro Colorado Island in the Panama Canal Zone, and the baboons of the East African savannas. Studies have confirmed that these primates have diverse locomotor patterns, varied diets, and complex and flexible social organizations that help them find food, avoid predators, and survive under difficult circumstances.

Molecular biologists and geneticists analyze the structures of important proteins and of deoxyribonucleic acid (DNA). This information provides important evidence of the relatedness of one group of primates to another. It was through such studies that molecular biologists were able to show that chimpanzees and gorillas are nearly identical, that humans are very closely related to these African apes, and that the gibbons of Asia are much more distantly related. In general, the findings of molecular biology and genetics have tended to confirm the earlier findings based on comparative anatomy and paleontology, but with occasional exceptions.

Observations on living species are supplemented by studies of fossil primates. Paleontologists are always trying to connect living and extinct species into common family trees. Some molecular biologists have made estimates, based on biochemical differences, of the time of evolutionary divergence between apes (Pongidae) and humans (Hominidae). These estimates are somewhat controversial because they place the hominid-pongid split at less than five million years ago—in contrast to earlier estimates of ten million years or more made by paleontologists in the 1960's. Reinterpretations of the fossil primate *Ramapithecus* and other fossils from Asia support the biochemists' view that the hominid-pongid split took place less than five million years ago.

Many scientists with other fields of interest use nonhuman primates as experimental subjects for medical or biochemical research, including cancer research, space exploration, and the testing of new drugs. These scientists also make important additions to knowledge of how similar these primates are to humans. In particular, monkeys and apes are subject to nearly all the same diseases as are humans. Many drugs and surgical procedures designed to aid in the fight against these diseases are first tested in monkeys or apes because the physiological responses of these primates to the drugs or other medical procedures are nearly always the same as they are in humans. One case of particular importance is acquired immunodeficiency syndrome (AIDS). This disease, which attacks the immune system, does not occur in rats and mice, but it does occur in monkeys and apes. For these reasons, monkeys and apes are used extensively in tests on the AIDS virus and on possible treatments or cures for this disease.

Humans and Other Primates

Since the nonhuman primates are *Homo sapiens'* closest relatives, most studies of primates involve inevitable comparisons with humans. Most of the knowledge that is gained in studying the primates helps scientists understand the human species better. Primates, especially macaques, are commonly used in medical and behavioral research. New drugs, for example, are often tested in monkeys because the physiological reactions of these primates are likely to be more similar to those of humans than would be the case in rats or mice.

Chimpanzees are used when an even closer approximation to human anatomy or human intelligence is important to the experiment. From the use of these and other primates in medical research, knowledge of human health and human diseases has been greatly enhanced.

Scientific understanding of human behavior and other aspects of human biology is similarly enhanced by studies of other primates. Studies of sign language among apes, for example, have greatly increased the understanding of human language and the ways in which it is learned. Both gorillas and chimpanzees have successfully been taught American Sign Language (ASL), a gesture-language commonly used by deaf people. Many researchers who have worked with these apes believe that they show true language skills in their use of ASL, although some linguists disagree. Apes who use sign language can converse about past, present, and future events, faraway places, hypothetical ("what if?") situations, pictures in books, and individual preferences. These apes apparently can use language to lie, to play games or make puns, or to create definitions, such as "a banana is a long yellow fruit that tastes better than grapes."

Studies on the social organization of baboons, langurs, and other nonhuman primates have greatly increased understanding of how human social organization might have originated. Most nonhuman primates have social groups based on friendships and dominance relationships. Larger and stronger individuals tend to get their way more often, but usually by gesturing or threatening rather than by actually fighting. Even this observation, however, must be qualified, because encounters involving three or more individuals are generally very complex, and less dominant individuals can often manipulate these complex situations to their advantage.

—Eli C. Minkoff

See also: Apes to hominids; Baboons; Chimpanzees; Evolution: Historical perspective; Gorillas; Hominids; *Homo sapiens* and the diversification of humans; Human evolution analysis; Lemurs; Mammalian social systems; Mammals; Monkeys; Neanderthals; Orangutans.

Bibliography

Ciochon, Russell L., and John G. Fleagle, eds. *Primate Evolution and Human Origins.* Menlo Park, Calif.: Benjamin/Cummings, 1985. This book is a collection of papers reprinted from various journals. It contains an extensive bibliography of papers on fossil primates and primate evolution. The individual articles vary greatly, but most are excellent. The level of technical detail is high, but this should not deter the determined reader. Most of the papers are very well illustrated.

Fleagle, John G. *Primate Adaptation and Evolution.* 2d ed. New York: Academic Press, 1999. A good and very readable summary of knowledge of primates, including their behavior, classification, anatomy, and evolution. The living and fossil members of each group are artificially separated, but the treatment is in every other way direct, readable, and to the point. Many illustrations and lengthy bibliographies for each chapter.

Jolly, Alison. *Evolution of Primate Behavior.* New York: Macmillan, 1972. This book discusses the ecology, behavior, intelligence, and social organization of many kinds of primates. All kinds of primates are treated, but emphasis is on Old World monkeys and apes. The book is very readable and nontechnical. Illustrations are well used, and a good bibliography is provided.

Kinzey, Warren G., ed. *New World Primates: Ecology, Evolution, and Behavior.* New York: Walter de Gruyter, 1997. A collection of papers from a symposium on primate anthropology.

Napier, J. R., and P. H. Napier. *A Handbook of Living Primates*. New York: Academic Press, 1967. This a good photographic atlas of all living primate genera, arranged alphabetically by their Latin names. Short summaries of the geographic range, ecology, structure, and habits of each genus are provided. Also contains short descriptions of primate anatomy and some useful tables of such data as chromosome numbers and gestation periods. Useful particularly as a reference.

_____. *The Natural History of the Primates*. Cambridge, Mass.: MIT Press, 1985. A nontechnical account of primates, with emphasis on the habits of the living primates; anatomy and social behavior receive much less treatment. Fossils are not treated at all, except for a short section on human fossils. Many good pictures and a medium-length bibliography. A good introduction to the subject.

Rowe, Noel. *The Pictorial Guide to the Living Primates*. East Hampton, N.Y.: Pogonias Press, 1996. Written and photographed by the director of Primate Conservation, Inc., based on his ten years of research around the world.

Sleeper, Barbara. *Primates: The Amazing World of Lemurs, Monkeys, and Apes*. San Francisco: Chronicle Books, 1997. Covers one hundred primate species by geographic region. Lavishly illustrated with color photographs.

Smuts, Barbara B., et al., eds. *Primate Societies*. Chicago: University of Chicago Press, 1987. An excellent and compendious summary of knowledge of primate societies, including all species for which reliable field studies exist. The style is quite readable, and technical jargon is avoided. Good illustrations and a thorough bibliography make this an essential reference for anyone interested in primate behavior and social organization in the wild.

PROTOZOA

Type of animal science: Classification
Fields of study: Cell biology, evolutionary science, pathology, systematics (taxonomy)

The protozoa are a heterogeneous group of organisms the only common characteristic of which is that they are all unicellular. Otherwise, they are extremely diverse in their structures and habits: Some are animal-like and others plantlike; some are free-living, while others are parasitic and cause pernicious diseases.

Principal Terms

AUTOTROPH: any organism capable of synthesizing its own food by using either solar or chemical energy

CILIATE: a protozoan that uses short, hairlike structures called cilia for locomotion

FLAGELLATE: a protozoan that uses long, whiplike structures called flagella for locomotion

HETEROTROPH: any organism that must consume other organisms or organic substances to obtain nutrition

NUCLEUS: a cellular organelle that coordinates the cell's activities and contains the genetic information

ORGANELLE: any of several structures found inside an individual cell, analogous to the organs of multicellular organisms

PHAGOCYTOSIS: obtaining food by engulfing it

PSEUDOPODS: cytoplasmic extensions of a protozoan's body, used for locomotion and the engulfing of food

UNICELLULAR: consisting of only one cell

The protozoa are a vast assemblage of extremely diverse organisms whose only common characteristic is that they are all unicellular. They were previously classified as members of the animal kingdom but are now recognized as a distinct group and occupy a kingdom of their own, Protista.

Protozoa have been known only since the seventeenth century, when Dutch naturalist Antoni van Leeuwenhoek observed them with simple microscopes of his own making. Leeuwenhoek's discovery was accidental; he was simply curious about the microscopic nature of such common substances as rainwater and scrapings from his mouth and teeth. His observations of small, active creatures associated with these substances were the first recorded descriptions of microscopic life. Although Leeuwenhoek made no effort to classify protozoa (which he called "little beasties"), his observations were later confirmed by scientists who recognized the importance of his discoveries.

Protozoan Diversity

The diversity of protozoan types cannot be overemphasized. Though some have a close evolutionary relationship, others have followed very different paths. Some protozoa are animal-like, obtaining energy by consuming food from external sources, and others resemble plants in that they contain chlorophyll and use the energy of the sun to manufacture food.

Protozoa occur in salt water and freshwater, in the soil, and inside larger organisms. In short, they are wherever moisture is present. Most are motile, although some are sessile (attached), and others, like volvox, form colonies. The great majority of protozoa are microscopic, although the very largest are just visible to the naked eye. The blood parasite *Anaplasma* is only one-sixth to one-tenth the size of a human red blood cell, while the ciliate *Spirostomum* (a freshwater protozoan that uses tiny hairlike cilia for locomotion) is up to three millimeters long and can be seen without a microscope.

An individual protozoan, because it consists of only one cell, does not contain any organs. Instead, it harbors analogous structures called organelles, which enable it to carry out its life functions. Central among these organelles is the nucleus, which contains the genes and serves as a sort of control center for cell activities. The cells of protozoa contain from one to many nuclei, depending on the species. Other organelles that distinguish the protozoa are those used for locomotion. These include whiplike flagella, hairlike cilia, and flowing extensions of the cell called pseudopodia, or false feet, which are generated in the direction of movement. Locomotory organelles are important in the classification of the protozoa. Many protozoa, especially freshwater species, possess an organelle called the contractile vacuole, which serves to pump excess water out of the cell. This organelle is especially important in freshwater protozoa, which tend to accumulate water by osmosis because their bodies contain more dissolved substances than the water in which they live. As water flows into these protozoa, the contractile vacuole absorbs it, expands, and finally contracts to flush the water through the cell membrane and into the surrounding environment.

Protozoa can be divided into two groups based on how they get their food: autotrophic and heterotrophic. Autotrophic protozoa contain chlorophyll and are able to convert solar energy into usable energy. The freshwater *Euglena* are autotrophs. Heterotrophic protozoa must capture their food, either funneling it through mouthlike openings or, like the amoeba, engulfing food particles with their pseudopods. This latter method of obtaining food is called phagocytosis.

Gas exchange in protozoa occurs through simple diffusion of gases across the cell membrane. Some protozoa live in environments such as stagnant water or the guts of animals where little oxygen is present. These protozoa are called facultative anaerobes, which means they can live with or without oxygen.

Protozoan Reproduction

Most protozoa reproduce asexually: a single parent produces offspring that are identical genetic copies of itself. Asexual reproduction in protozoa is accomplished in two ways, through fission and budding. In fission, the parent cell simply divides into two offspring cells of equal size. In budding, the offspring is a smaller cell that grows off the larger parent cell. Some protozoa carry out sexual reproduction, in which two parent cells release smaller sex cells called gametes. Two gametes fuse and grow into a new protozoan that is genetically different from either parent.

Conjugation is an interesting process that occurs in ciliate protozoa and may be linked to the reproductive process. It involves the temporary linking of two cells, followed by an exchange of nuclei, after which the cells separate. However, no offspring are formed. It is thought that conjugation is a means of renewing the nuclei of certain protozoa so that they may continue to reproduce asexually. Another process that many protozoa undergo is called encystment, in which the protozoa secrete a thickened sheath around themselves and enter a period of dormancy. Encystment enables protozoa to survive adverse environmental conditions such as drought or cold. When conditions become favorable again, the protozoa emerge from their cysts and resume their normal activities.

Protozoan Phyla

The kingdom Protista is subdivided into five groups, or phyla: Mastigophora, Sarcodia, Sporozoa, Cnidospora, and Ciliophora. The phylum Mastigophora contains the flagellates: protozoa that move by means of one or more whiplike flagella. The mastigophorans are of two types: phytoflagellates, which contain chlorophyll and are primarily autotrophic; and zooflagellates, which are primarily heterotrophic. Many zooflagellates are parasites of insects and vertebrates. The flagellates of the phylum Mastigophora show such a high degree of structural diversity that it is impossible to describe a typical representative. However, among the more interesting of the phylum members are the dinoflagellates, which occur in both marine and freshwater environments. They have two flagella, one of which points away from the cell and the other of which lies transversely around the cell. Many dinoflagellates contain a golden-brown photosynthetic pigment called xanthophyll and are therefore autotrophic, but others are unpigmented and are heterotrophic. These protozoa are encased in a cellulose sheath, or pellicle, which is often highly ornamented with spikes, hooks, or other prominences. Certain marine species are responsible for red tides, a population explosion that is responsible for massive fish kills and causes paralytic poisoning in humans who consume shellfish taken from red tide waters.

The protozoa in the phylum Sarcodina move about and engulf their food by means of pseudopodia. Amoebas belong to this phylum, as well as many marine, freshwater, and terrestrial species. The members of this phylum are the simplest protozoa and have few organelles. Some have exquisitely crafted external skeletons. The radiolarians, which are restricted to the marine environment, have beautifully sculptured exoskeletons made of silicon dioxide or strontium sulfate. The skeleton is riddled with tiny pores through which the living organism within is able to extrude thin filaments of cytoplasm for purposes of feeding, locomotion, and anchorage. Radiolarians exist in great numbers in the surface waters of the ocean and, upon dying, sink to the bottom where in some areas their skeletons form a large proportion of ocean bottom sediments.

Two phyla are closely related: the Sporozoa and Cnidospora. These are parasitic protozoa collectively referred to as sporozoans because some members of both groups go through sporelike infective stages. The phylum Sporozoa contains parasites that infect cells of the intestine and blood and are responsible for causing malaria in humans. These sporozoans belong to the genus *Plasmodium*, a group containing more than fifty species, four of which infect humans. All require a mosquito for transmission of the disease-causing protozoan. When the mosquito bites a person, the parasite is introduced into the blood with the saliva of the insect. From the blood, the protozoan travels to the individual's liver, where it reproduces, then escapes to invade red blood cells. Once in the red blood cells, it consumes hemoglobin, reproduces, and escapes to invade additional red blood cells. This cycle of invasion and escape is responsible for the debilitating chills and fever characteristic of malaria. The phylum Cnidospora contains amoeboid parasites of fishes and insects. This phylum's members are separated from those in Sporozoa based on differences in the nature of the infective spores.

The largest phylum is Ciliophora, which has more than eight thousand species. Of all the phyla, this one has the most homogeneous members as far as their evolutionary relationships are concerned. They all appear to have arisen from a common ancestor. Members of this phylum are commonly referred to as ciliates because they use cilia for locomotion and for sweeping food particles into the cytostome (cell mouth). They are also characterized by the presence of two nuclei, a small micronucleus and a larger macronucleus whose sole purpose is the synthesis of deoxyribonucleic acid (DNA, the genetic material). Although some ciliophorans are sessile, or attached, many are nimble, quick, and carnivorous and are therefore perhaps the most animal-like of all the protozoa. They reproduce either asexually by fission or through sexual conjugation, but they never release free gametes into the surrounding environment. Ciliophorans are common in fresh and salt water. Some are parasitic.

Protozoology

Although Antoni van Leeuwenhoek had little sense of what he was looking at when he first observed protozoa in the seventeenth century, he used the same tool that later scientists did, the microscope. Protozoology—the study of organisms too small to be observed with the naked eye—has become an area of specialization within microbiology.

Protozoa are best examined while alive, as specimens that have been killed, fixed, and stained often become distorted. Some of the smaller species can be viewed under the cover slip of a microscope slide; others, especially the larger specimens, must be viewed under less restrictive conditions: in a petri dish with a dissecting microscope or under a raised cover slip with corners that have been supported with paraffin or some other material.

One of the challenges of studying protozoa with the microscope is the speed with which they move and the tedium of keeping track of them by constantly adjusting the position of the slide. For this reason, a solution of methyl cellulose is often added to the slide preparation to increase the viscosity of the water and slow down the organisms.

As many protozoa are almost transparent, it is necessary to highlight their features by using dyes or stains that will not kill them. Such substances, called vital stains, render various structures in the protozoa visible under the microscope. For example, the vital stain janus green b selectively stains mitochondria, the energy-producing organelles, while neutral red has an affinity for vacuoles, the organelles that contain engulfed food particles. In studying protozoan structure, scientists use an assortment of vital stains and regular stains, which are used when protozoa are killed and preserved on a slide. Electron microscopy, used to examine the extremely fine structure of protozoa, has special stains and fixatives of its own for maintaining the structure of these organisms under the sustained electron bombardment required to illuminate the specimen.

Protozoa are ideal subjects of study for cytologists, geneticists, and developmental biologists because they reproduce asexually and therefore provide the researcher with an endless succession of clones, or genetically identical individuals. Scientists can subject these clones to various environmental conditions and observe the effects in their offspring.

The primary method of maintaining protozoa in a laboratory is a culture, which usually consists of one species of protozoa and the bacterium on which it feeds. The object is to exclude other organisms that might harm the protozoan. As toxic waste products build up in the culture dish, a small number of protozoa are transferred to a clean culture medium. This procedure is called subculturing and continues as long as the culture is maintained. There are two types of cultures: monoxenic and axenic. In monoxenic cultures, a single species of protozoa is maintained with a single species of bacteria. In axenic cultures, a single species of protozoa lives in isolation from any other organisms. The axenic culture is of value in certain types of nutritional studies.

Parasitic protozoa can be examined separate from their hosts in a laboratory but only when kept under conditions that provide them with a chemical and physical environment that will keep them alive for as long as necessary. For example, some parasitic protozoa of warm-blooded animals must be kept in a warm culture medium to prevent their being distorted by cooling.

The Base of the Food Chain

Protozoa are of first and foremost importance to ecology. These organisms make up the lowest links of the food chain that eventually leads to humans. Autotrophic protozoa are called producers because, like plants, they are able to use the energy of sunlight to manufacture food substances. Other protozoa are heterotrophs, or consumers, and must take food from their environments. Dinoflagellates are a group of autotrophs with a notorious role in the marine environment. These organisms, which are responsible for red tides, can pass their potent toxins up the food chain to cause paralytic shellfish poisoning in humans. The toxic materials they release into the ocean during periods of intense reproductive activity also lead to massive fish kills, which mean losses for the fishing industry.

Foraminifers are a type of shelled amoeba in the phylum Sarcodina. Some foraminifers existed millions of years ago, when the earth's fossil fuel deposits were being generated. Therefore, the presence of their fossil remains in drill cores helps indicate where oil is likely to be found.

An interesting relationship exists between protozoa and termites. Cellulolytic (cellulose-breaking) protozoa inhabit the guts of termites and are the organisms actually responsible for digestion of the wood that termites eat. Without these protozoa, the termites would not be able to feed on wood and would no longer be able to damage homes and other structures.

Protozoan Infections

In people, protozoa cause numerous infections, generally limited to four sites in the body: the intestines, the genital tract, the bloodstream and tissues, and the central nervous system (brain and spinal cord). Not all parasitic protozoa cause death in humans, but they are responsible for untold misery in wide areas of the world, especially underdeveloped countries where preventive medicine is unknown and too expensive for these societies to employ.

The two most common protozoan infections are malaria and amoebic dysentery. The malarial parasite is transmitted through the bite of the *Anopheles* mosquito and eventually destroys the body's red blood cells, causing anemia and recurrent debilitating chills and fever. Malaria is a major health problem in parts of Africa, Asia, and Central and South America. More than one hundred million people are afflicted at any given time, and about one million die of the disease each year. Amoebic dysentery flourishes in overcrowded, unsanitary conditions. The disease gradually erodes the inner lining of the intestines, causing ulcers. Amoebic dysentery is entirely avoidable when food and water are handled carefully to prevent their contamination or are purified if contamination is suspected.

Other protozoa of medical importance are *Trichomonas vaginalis*, which parasitizes the genital tract; *Giardia lamblia*, a parasite of the small intestine; and *Toxoplasma gondii*, which infects the fetus and can cause congenital abnormalities.

Protozoa play an important role in the degradation of both human and industrial wastes in the environment. In sewage treatment, for example, anaerobic protozoa (those that flourish in the absence of oxygen) and bacteria play a role in the degradation of raw sewage. After they have done their work, the effluent is passed to a set of tanks with aerobic (oxygen-requiring) protozoa for further processing. Autotrophic protozoa have been used to deal with industrial wastes containing high levels of nitrates and phosphates. The settling tanks are illuminated to promote the growth of the autotrophs, which absorb and metabolize the industrial chemicals as part of their normal life processes.

—*Robert T. Klose*

See also: Deep-sea animals; Ecosystems; Food chains and food webs; Histology; Marine animals; Marine biology.

Bibliography

Barnes, Robert D. *Invertebrate Zoology*. 6th ed. Philadelphia: Saunders College Publishing, 1993. This general invertebrate zoology textbook contains a well-organized chapter on protozoa that is detailed yet accessible and has numerous illustrations. Indispensable for those seeking a comprehensive yet concise understanding of protozoa.

Hegner, Robert W. *Big Fleas Have Little Fleas: Or, Who's Who Among the Protozoa*. New York: Dover, 1968. A lively and lighthearted look at protozoa. This book is a rarity: fact-based science that also functions as literature. Highly recommended as an introduction to protozoa.

McKane, L., and J. Kandel. *Microbiology: Essentials and Applications*. New York: McGraw-Hill, 1996. This microbiology text presents a comprehensive picture of protozoa, from their evolution through their classification to the human diseases they cause. Contains some beautiful photomicrographs of protozoa feeding, regulating water balance, and reproducing.

Margulis, L., et al. *Illustrated Glossary of Protoctista*. Boston: Jones and Bartlett, 1993. A handsome reference volume, nicely organized, easy to use, and illustrated with line drawings. Contains a detailed glossary of protozoology terms, making it a good companion to other books on the subject. For the advanced student.

Morholt, E., and Paul F. Brandwein. *A Sourcebook for the Biological Sciences*. San Diego, Calif.: Harcourt Brace, 1980. The information in this extremely useful reference book is arranged cookbook-style. The entries are brief, practical, and readable, describing procedures for the collection, staining, observation, and preservation of specific protozoa. What is unique about this book is that it prefaces the practical how-to entries with well-written background information.

Roberts, Larry S., et al. *Foundations of Parasitology*. 6th ed. Boston: McGraw-Hill, 2000. The sections on protozoa are limited to those species that parasitize human hosts. The text is densely written, emphasizing scientific terminology, classification, and the fine structure of these organisms. For the reader with a serious interest in this topic.

Whitten, R., and W. Pendergrass. *Carolina Protozoa and Invertebrates Manual*. Burlington, N.C.: Carolina Biological Supply, 1980. A short, attractive, ready-reference guide to the culture and observation of protozoa by a biology supply house. Entirely accessible to the interested middle school or high-school student. Has large, colorful illustrations, recipes for media, and suggestions for further reading.

PTEROSAURS

Type of animal science: Classification
Fields of study: Evolutionary science, paleontology, physiology

Pterosaurs were the flying reptiles of the Mesozoic. Pterosaurs are part of the group Archosauria, and the first of only three vertebrate groups to have evolved active flight. Pterosaurs were contemporaries of the Mesozoic dinosaurs, and quite closely related to them.

Principal Terms

ARCHOSAURS: diapsid reptiles whose skeletal evolution distinguishes them from the diapsid lepidosaurs

MESOZOIC ERA: the middle era of the Phanerozoic eon, 250 million to 65 million years ago

PATAGIUM: a soft, flexible membrane

Pterosaurs ("winged reptiles") were flying reptiles of the Mesozoic Era, and the first of only three vertebrate groups known to have evolved active flight. Pterosaurs first appear in the paleontological record about 225 million years ago, during the late Triassic Period, and persist to the terminal Cretaceous extinction event, about 65 million years ago. Pterosaurs are recognized as being members of the group Archosauria, a group including thecodonts, crocodiles, birds, and dinosaurs. While pterosaurs were contemporaries to the great Mesozoic dinosaurs, and often confused with them, pterosaurs were not dinosaurs.

The origin of pterosaurs is a subject of great debate. Pterosaurs share some physical characteristics in the hips and legs with early dinosaurs. It has also been suggested they could run bipedally and took to the air by the energy-consuming method of running and flapping their wings. Another view suggests pterosaurs pursued a tree-dwelling way of life, and developed parachuting, gliding, and eventually flight as an energy-saving byproduct of their lifestyle. Another debate surrounds whether pterosaurs were warm-blooded or cold-blooded creatures. Some fossil remains suggest certain pterosaurs may have been covered with short, thin fur, or even very fine feathers. This, combined with the energy-consuming activity of flying, leads many researchers to favor warm-blooded pterosaurs.

The first pterosaur fossils were discovered in the Solnhofen Limestone Formation in Germany, in 1784. Pterosaur fossils range in size from those of *Pterodactylus*, whose forty-centimeter wingspan was about that of a modern song bird, to *Quetzalcoatlus*, whose fifteen-meter wingspan was as big as a modern private aircraft wing. Pterosaurs had light, bony skeletons made of hollow, tubular bones, and compact bodies to improve wing support. The pelvis and hind limbs were small, yet long and slender when compared to the fore limbs and shoulders. The fore limbs were exaggerated in comparison to the hind limbs, with their great length derived from the extended fourth finger, which alone supported the flight patagium. The pterosaur patagium was a soft membrane thought to have been tougher and thicker than a bat's wing, and divided into three anatomical segments. One short segment stretched from the torso to the elbow end of the humerus; a second, longer segment stretched from the radius and ulna and the bones of the wrist and palm; and the third, longest segment was an elongated fourth finger supporting the wing membrane to its tip. The segments of the wing were progressively longer as they got farther from the trunk. The first, second, and third fingers were not involved in membrane support

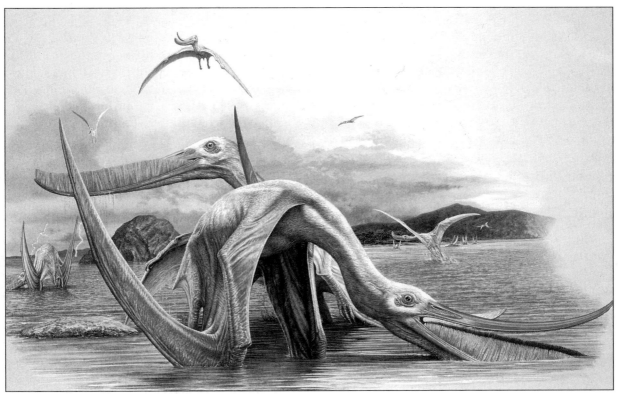

Pterodaustro *had an array of teeth much like the baleen of whales for straining out small animals scooped from the* *water.* (©John Sibbick)

but were free to grasp and cling; there was no fifth finger. This "finger-wing" configuration is quite different from the wing support structure seen in bats and birds.

Types of Pterosaurs

Pterosaurs are generally divided into two groups: The rhamphorhynchoids first appear during the Triassic, and the pterodactyloids appear some 108 million years later. The Rhamphorhynchoidea had teeth, long tails, short metacarpals, long fifth toes, and a head held in front of and slightly above the torso. The Pterodactyloidea had no tail or fifth toe, an elongated metacarpus forming the largest support structure of its wing, a more "birdlike" neck posture with the neck curving in an S-shape entering the cranium from below rather than from behind, holding the head higher above the torso, and many had evolved toothless beaks. Ptero-

dactyloidea, such as *Pteranodon*, also evolved extreme head crests, and like many later pterosaurs, lacked well- developed teeth. Many other pterosaur fossils exhibit well-developed and specialized teeth: *Eudimorphodon* and *Dorygnathus* had teeth designed to spear and hold prey; *Dsungaripterus* had bony jaws and broad, flattened teeth to winkle out shellfish and snails; and *Pterodaustro* had a comblike array of teeth, ideal for sieving plankton.

The majority of pterosaur bones have been recovered from marine sedimentary rock deposits, suggesting these flying reptiles took advantage of the atmospheric updraft conditions of gentle and constant breezes along shorelines, as well as the abundant food of the marine environment. Pterosaur aerodynamics and flight characteristics have been studied in great detail, and they suggest pterosaurs were active, wing-flapping fliers, but also better gliders than are most modern bird species. Pterosaur anatomy suggests they were able to glide and soar over great distances and for extended periods of time under calm conditions, but that they were not well adapted for staying aloft under turbulent conditions. It is estimated that the pterosaur *Pteranodon*, who represents one of the high points in pterosaur evolution with a wingspan of over seven meters, could soar at speeds in excess of 30 kilometers per hour, continue gliding aloft for nearly twenty hours, and possibly cover distances of over 750 kilometers without landing.

The greatest number of pterosaur bones have been recovered from the Kansas marine chalk deposits of North America. Over eight hundred skeletal fragments have been found there. While many of the remains suggest these Kansas pterosaurs died nonviolently, other bones suggest the pterosaurs were eaten by marine reptiles. Apparently, pterosaurs swooping low to snatch prey from the ocean's surface were often in turn preyed upon by larger marine reptiles or sharks.

—*Randall L. Milstein*

See also: *Allosaurus*; *Apatosaurus*; *Archaeopteryx*; Dinosaurs; Evolution: Animal life; Extinction; Fossils; Hadrosaurs; Ichthyosaurs; Paleoecology; Paleontology; Prehistoric animals; Sauropods; Stegosaurs; *Triceratops*; *Tyrannosaurus*; Velociraptors.

Bibliography

Norman, David. *Prehistoric Life: The Rise of the Vertebrates*. New York: Macmillan, 1994. A richly illustrated book and good general reference, outlining evolutionary development, anatomy, and ongoing scientific debates concerning the life history and ecological function of pterosaurs within the Mesozoic environment.

Padian, Kevin. "Pterosaurs: Were They Functional Birds or Functional Bats?" In *Biomechanics in Evolution*, edited by J. M. V. Rayner and R. J. Wootton. New York: Cambridge University Press, 1991. Suggests the course of vertebrate flight evolution is best supported by phylogenetic analysis of biomechanics and functional morphology.

Russell, Dale A. *An Odyssey in Time: The Dinosaurs of North America*. Toronto, Ontario: University of Toronto Press, 1989. This book provides a general overview of pterosaur fossil discoveries in North America, and gives some general information on pterosaur behavior based on paleoecology of the rock formations in which the fossils were found.

Shipman, Pat. *Taking Wing: Archaeopteryx and the Evolution of Bird Flight*. New York: Simon & Schuster, 1998. The chapter "Pathways to the Skies" gives a very good overview of the current theories surrounding the biomechanics and physiology of pterosaurs. The chapter also provides a brief history of the study of pterosaurs, and the historical use of comparative anatomy to define pterosaur morphology and behavior.

PUNCTUATED EQUILIBRIUM AND CONTINUOUS EVOLUTION

Type of animal science: Evolution
Fields of study: Evolutionary science, genetics, systematics (taxonomy)

According to classical evolutionary theory, new species arise by gradual transformation of ancestral ones. Speciation theory of the 1950's and 1960's, however, predicted that new species arise from small populations isolated from the main population, where they diverge rapidly. In 1972, Niles Eldredge and Stephen Jay Gould applied this concept to the fossil record, predicting that species should arise suddenly ("punctuated" by a speciation event) rather than gradually, and then persist virtually unchanged for millions of years in "equilibrium" before becoming extinct or speciating again.

Principal Terms

ALLOPATRIC: populations of organisms living in different places and separated by a barrier that prevents interbreeding

GRADUALISM: the idea that transformation from ancestor to descendant species is a process spanning millions of years

MACROEVOLUTION: large-scale evolutionary processes that result in major changes in organisms

MICROEVOLUTION: small-scale evolutionary processes resulting from gradual substitution of genes and resulting in very subtle changes in organisms

SPECIATION: the process by which new species arise from old species

SPECIES SELECTION: a higher level of selection above that of natural selection is postulated to take place on the species level

STASIS: the long-term stability and lack of change in fossil species, often spanning millions of years of geologic time

SYMPATRIC: populations of organisms living in the same place, not separated by a barrier that would prevent interbreeding

Although Charles Darwin's most influential work was entitled *On the Origin of Species* (1859), in fact it did not address the problem in the title. Darwin was concerned with showing that evolution had occurred and that species could change, but he did not deal with the problem of how new species formed. For nearly a century, no other biologists addressed this problem either. Darwin (and many of his successors) believed that species formed by gradual transformation of existing ancestral species, and this viewpoint (known as gradualism) was deeply entrenched in the biology and paleontology books for a century. In this view, species are not real entities but merely arbitrary segments of continuously evolving lineages that are always in the process of change through time. Paleontologists tried to document examples of this kind of gradual evolution in fossils, but remarkably few examples were found.

The Allopatric Speciation Model

By the 1950's and 1960's, however, systematists (led by Ernst Mayr) began to study species in the wild and therefore saw them in a different light. They noticed that most species do not gradually transform into new ones in the wild but instead have fairly sharp boundaries. These limits are established by their ability and willingness to interbreed with each other. Those individuals that can interbreed are members of the same species, and those that cannot are of different species. When a population is divided and separated so that formerly interbreeding individuals develop differences that prevent interbreeding, then a new species is formed. Mayr showed that, in nature, large populations of individuals living together (sympatric conditions) interbreed freely, so that evolutionary novelties are swamped out and new species cannot arise. When a large population becomes split by some sort of barrier so that there are two different populations (allopatric conditions), however, the smaller populations become isolated and prevented from interbreeding with the main population. If these allopatric, isolated populations have some sort of unusual gene, their numbers may be small enough that this gene can spread through the whole population in a few generations, giving rise to a new species. Then, when the isolated population is reintroduced to the main population, it has developed a barrier to interbreeding, and a new species becomes established. This concept is known as the allopatric speciation model.

The allopatric speciation model was well known and accepted by most biologists by the 1960's. It predicted that species arise in a few generations from small populations on the fringe of the range of the species, not in the main body of the population. It also predicted that the new species, once it arises on the periphery, will appear suddenly in the main area as a new species in competition with its ancestor. These models of speciation also treated species as real entities, which recognize one another in nature and are stable over long periods of time once they become established. Yet, these ideas did not penetrate the thought of paleontologists for more than a decade after biologists had accepted them. In 1972, Niles Eldredge and Stephen Jay Gould proposed that the allopatric speciation model would make very different predictions about species in the fossil record than the prevailing dogma that they must change gradually and continuously through time. In their paper, they described a model of "punctuated equilibrium." Species should arise suddenly

in the fossil record (punctuation), followed by long periods of no change (equilibrium, or stasis) until they went extinct or speciated again. They challenged paleontologists to examine their biases about the fossil record and to see if in fact most fossils evolved gradually or rapidly, followed by long periods of stasis.

In the years since that paper, hundreds of studies have been done on many different groups of fossil organisms. Although some of the data were inadequate to test the hypotheses, many good studies have shown quite clearly that punctuated equilibrium describes the evolution of many multicellular organisms. The few exceptions are in the gradual evolution of size (which was specifically exempted by Eldredge and Gould) and in unicellular organisms, which have both sexual and asexual modes of reproduction. Many of the classic studies of gradualism in oysters, heart urchins, horses, and even humans have even been shown to support a model of stasis punctuated by rapid change. The model is still controversial, however, and there are still many who dispute both the model and the data that support it.

Implications of Punctuated Equilibrium

One of the more surprising implications of the model is that long periods of stasis are not predicted by classical evolutionary theory. In neo-Darwinian theory, species are highly flexible, capable of changing in response to environmental changes. Yet, the fossil record clearly shows that most species persist unchanged for millions of years, even when other evidence clearly shows climatic changes taking place. Instead of passively changing in response to the environment, most species stubbornly persist unchanged until they either go extinct, disappear locally, or change rapidly to some new species. They are not infinitely flexible, and no adequate mechanism has yet been proposed to explain the ability of species to maintain themselves in homeostasis in spite of environmental changes and apparent strong natural selection. Naturally, this idea intrigues paleontologists, since it suggests processes that can only be observed in the fossil record and were not predicted from studies of living organisms.

The punctuated equilibrium model has led to even more interesting ideas. If species are real, stable entities that form by speciation events and split into multiple lineages, then multiple species will be formed and compete with one another. Perhaps some species have properties (such as the ability to speciate rapidly, disperse widely, or survive extinction events) that give them advantages over other species. In this case, there might be competition and selection between species, which was called species selection by Steven Stanley in 1975. Some evolutionary biologists are convinced that species selection is a fundamentally different process from that of simple natural selection that operates on individuals. In species selection, the fundamental unit is the species; in natural selection, the fundamental unit is the individual. In species selection, new diversity is created by speciation and pruned by extinction; in natural selection, new diversity is created by mutation and eliminated by death of individuals. There are many other such parallels, but many evolutionary biologists believe that the processes are distinct. Indeed, since species are composed of populations of individuals, species selection operates on a higher level than natural selection.

If species selection is a valid description of processes occurring in nature, then it may be one of the most important elements of evolution. Most evolutionary studies in the past have concentrated on small-scale, or microevolutionary, change, such as the gradual, minute changes in fruit flies or bacteria after generations of breeding. Many evolutionary biologists are convinced, however, that microevolutionary processes are insufficient to explain the large-scale, or macroevolutionary, processes in the evolution of entirely new body plans, such as birds evolving from dinosaurs. In other words, traditional neo-Darwinism says that all evolution is merely microevolution on a larger scale, whereas some evolutionary biologists consider some changes too large for microevolution. They require different kinds of processes for macroevolution to take place. If there is a differ-

ence between natural selection (a microevolutionary process) and species selection (a macroevolutionary process), then species selection might be a mechanism for the large-scale changes in the earth's history, such as great adaptive radiations or mass extinctions. Naturally, such radical ideas are still controversial, but they are taken seriously by a growing number of paleontologists and evolutionary biologists. If they are supported by further research, then there may be some radical changes in evolutionary biology.

Patterns of Evolution

Determining patterns of evolution requires a very careful, detailed study of the fossil record. To establish whether organisms evolve in a punctuated or gradual mode, many criteria must be met. The taxonomy of the fossils must be well understood, and there must be large enough samples at many successive stratigraphic levels. To estimate the time spanned by the study, there must be some form of dating that allows the numerical age of each sample to be estimated. It is also important to have multiple sequences of these fossils in a number of different areas to rule out the effects of migration of different animals across a given study area. Once the appropriate samples have been selected, then the investigator should measure as many different features as possible. Too many studies in the past have looked at only one feature and therefore established very little. In particular, changes in size alone are not sufficient to establish gradualism, since these phenomena can be explained by many other means. Finally, many studies in the past have failed because they picked one particular lineage or group and selectively ignored all the rest of the fossils in a given area. The question is no longer whether one or more cases of gradualism or punctuation occurs (they both do) but which is predominant among all the organisms in a given study area. Thus, the best studies look at the entire assemblage of fossils in a given area over a long stratigraphic interval before they try to answer the question of which tempo and mode of evolution are prevalent.

Since the 1940's, evolutionary biology has been

dominated by the neo-Darwinian synthesis of genetics, systematics, and paleontology. In more recent years, many of the accepted neo-Darwinian mechanisms of evolution have been challenged from many sides. Punctuated equilibrium and species selection represent the challenge of the fossil record to neo-Darwinian gradualism and overemphasis on the power of natural selection. If fossils show rapid change and long-term stasis over millions of years, then there is no currently understood evolutionary mechanism for this sort of stability in the face of environmental selection. A more general theory of evolution may be called for, and, in more recent years, paleontologists, molecular biologists, and systematists have all been indicating that such a radical rethinking of evolutionary biology is on the way.

—*Donald R. Prothero*

See also: Evolution: Animal life; Evolution: Historical perspective; Convergent and divergent evolution; Extinction; Gene flow; Isolating mechanisms in evolution; Natural selection.

Bibliography

Bennett, K. D. *Evolution and Ecology: The Pace of Life.* New York: Cambridge University Press, 1997. Provides a history of evolutionary theory and the data supporting current approaches to the topic.

Eldredge, Niles. *Time Frames: The Rethinking of Darwinian Evolution and the Theory of Punctuated Equilibria.* New York: Simon & Schuster, 1985. A general introduction to the development of punctuated equilibria by one of its original authors, with an excellent discussion of its implications for evolutionary biology. This text is the best book with which to begin reading about this topic. Includes the classic article by Eldredge and Stephen J. Gould, "Punctuated Equilibria: An Alternative to Phyletic Gradualism," that started the whole revolution in evolutionary paleobiology.

Gerhart, John. *Cells, Embryos, and Evolution: Toward a Cellular and Developmental Understanding of Phenotypic Variation and Evolutionary Adaptability.* Malden, Mass.: Blackwell Science, 1997. Explores phenotypic variation and evolutionary variability, attempting to explain how phenotype and genotype are connected, drawing upon the explosion of new data over the 1990's.

Gould, Stephen J. "The Meaning of Punctuated Equilibria and Its Role in Validating a Hierarchical Approach to Macroevolution." In *Perspectives on Evolution,* edited by Roger Milkman. Sunderland, Mass.: Sinauer Associates, 1982. Gould's clearest defense and discussion of the macroevolutionary implications of punctuated equilibria and species selection.

Gould, Stephen J., and Niles Eldredge. "Punctuated Equilibrium: The Tempo and Mode of Evolution Reconsidered." *Paleobiology* 3 (1977): 115-151. Five years after the original article, the same authors evaluate the criticisms and new research that the paper generated and suggest an agenda for future research.

Hoffman, Antoni. *Arguments on Evolution: A Paleontologist's Perspective.* New York: Oxford University Press, 1988. One of the harshest and most cogent of the critics of punctuated equilibria. In this book, Hoffman details his objections to the whole model. Although some of his criticisms are well taken, others miss some of the most important points or overlook key evidence.

Levinton, Jeffrey S. *Genetics, Paleontology, and Macroevolution.* 2d ed. New York: Cambridge University Press, 2000. A thorough discussion of the topic by one of the leading critics of punctuated equilibria; unfortunately, he misses some of the central points of the species selection argument.

Mayr, Ernst. *Animal Species and Evolution*. Cambridge, Mass.: Harvard University Press, 1963. The classic study of speciation theory by the foremost proponent of the allopatric speciation model. This book dominated speciation theory for more than forty years from its first edition, *Systematics and the Origin of Species*, in 1942.

Moller, A. P. *Asymmetry, Developmental Stability, and Evolution*. New York: Oxford University Press, 1997. Argues that the prevalence of symmetry in animal physiology is related to genetic stability and fitness.

RABBITS, HARES, AND PIKAS

Type of animal science: Classification
Fields of study: Anatomy, ecology, evolutionary science, physiology, reproduction science, systematics, wildlife ecology, zoology

At least eighty species of rabbits, hares, and pikas have been identified in the order Lagomorpha. Lagomorphs constitute approximately 2 percent of the known four thousand mammalian species.

Principal Terms

ALTRICIAL: born naked, with eyes and ears closed for several days (rabbits)
DIGITIGRADE: walking on toes
FENESTRATION: a latticework of openings on the sides of the skull
PRECOCIAL: born fully haired, with functional eyes and ears (hares)
RUDIMENTARY: short or small

Lagomorphs have two families: Leporidae (rabbits and hares) and Ochotonidae (pikas). Lagomorphs range in size from pikas, which are six inches in length and 3.5 ounces in weight, to hares which are almost twenty-eight inches in length and 10 pounds in weight. Most lagomorphs have coats that are brownish or reddish brown above and lighter brown to white below. There are color differences according to species, location, and season. Lagomorphs are well adapted to a herbivorous diet. Rabbits and hares are not known to store food, while pikas not only store food but also dry or cure vegetation for winter. Rabbits and pikas burrow or inhabit abandoned burrows, and hares shelter in natural depressions.

Physical Characteristics of Rabbits, Hares, and Pikas

Lagomorph incisor teeth are long and grow throughout life. They are extremely effective for severing plant stems and for gnawing on bark. Behind the long incisors is a tiny peglike incisor. There are no canine teeth. Cheek teeth, located further back in the jaw, also grow throughout the animal's life, wearing away as they grind abrasive vegetation. The upper tooth rows are more widely separated than the lower rows, and chewing is done with a transverse movement. Vegetation

Lagomorphs have fused bones in their hind limbs that allow them to hop with great rapidity. (Corbis)

passes through the small intestine, which has a spiral valve, providing greater surface area for digestion. A large cecum is located at the point of attachment of the large intestine, which contains bacteria that aid in digestion. Lagomorphs have the ability to produce two types of fecal material, one that is wet and eaten again for further nutrient absorption, and one that is dry and discarded.

The bones of the hind limb are fused where they move against the calcaneuni, resulting in increased leverage in locomotion. Lagomorphs are digitigrade, with five digits on the forefoot and five on the hind foot.

Rabbits and hares have a rudimentary tail, while pikas have none at all. Folds of skin on the lips can meet behind the incisors so that gnawing can take place with the mouth cavity closed. Other flaps of skin are able to close the nostrils. The skull is peculiarly fenestrated. The ears are usually long. The testes are located in front of the penis rather than behind.

Rabbits and hares usually vocalize only when frightened or injured. Pikas express themselves with a whistle or bark and a chattering call.

Lagomorphs can yield two or more litters during each breeding season, with two or three litters common among hares and pikas and three to six among rabbits. Litter size ranges between two and eight. Young rabbits are altricial and are cared for in a nest. Hares are born in the open and are precocial, able to run soon after birth.

Destructive and Beneficial Lagomorphs

Lagomorphs can be vectors for disease, as well as pests to human agriculturists. Rabbits eat flowers and vegetables in spring and summer, causing problems in flowerbeds, gardens, and fields. In the fall and winter, they may damage and kill valuable woody plants. Lagomorphs are the dietary staple of many carnivorous mammals and birds, among them wolves, foxes, bobcats, weasels, predatory hawks, and owls. They are therefore an important link in the food chain.

—*Jason A. Hubbart*

See also: Home building; Mammals; Reproductive strategies; Teeth.

Rabbit, Hare, and Pika Facts

Classification:
Kingdom: Animalia
Subkingdom: Deuterostomia
Phylum: Chordata
Subphylum: Vertebrata
Class: Mammalia
Subclass: Eutheria
Order: Lagomorpha
Family: Leporidae (hares and rabbits), Ochotonidae (pikas)
Genera: Bunolagus (bushman hares), *Caprolagus* (hispid hares), *Lepus* (hares, twenty-one species), *Oryctolagus* (European rabbits), *Nesolagus* (Sumatran hares), *Pentalagus* (amami rabbits), *Poelagus* (Bunyoro rabbits), *Pronolagus* (red rock hares, three species), *Romerolagus* (volcano rabbits), *Sylvilagus* (cottontails, thirteen species); *Ochontona* (pikas, fourteen species)
Geographical location: All continents except Australia and Antarctica; absent from southern South America and most islands; especially diverse in North America and Eurasia, the only continents where pikas are found
Habitat: Range from tropical forest to arctic tundra
Gestational period: Rabbits, approximately twenty-eight days; hares, up to forty-seven days; pikas, approximately thirty days
Life span: In the wild, rarely more than nine months; up to fifteen years in captivity
Special anatomy: Flaps of skin able to close the nostrils; peculiarly fenestrated skull; long incisors that grow throughout life; generally long hind feet, with the hind legs strong and positioned for leaping; usually long ears; testes located in front of the penis rather than behind; lagomorphs produce two types of fecal material, one that is wet and eaten again for further nutrient absorption and one that is dry and discarded

Bibliography

Chapman, J. A., and J. E. C. Flux. *Rabbits, Hares and Pikas: Status Survey and Conservation Action Plan*. Gland, Switzerland: IUCN/SSC Lagomorph Specialist Group, 1990. A thorough and concise piece of literature for those who have some experience in this field.

Feldhamer, G. A., L. C. Drickamer, S. H. Vessey, and J. F. Merritt. *Mammalogy: Adaptation, Diversity, and Ecology*. Boston: WCB/McGraw-Hill, 1999. These authors have done a superior job describing and explaining mammalian diversity.

Hoffmann, R. S. "Order Lagomorpha." In *Mammal Species of the World: A Taxonomic and Geographic Reference*, edited by D. E. Wilson and D. M. Reeder. Washington, D.C.: Smithsonian Institution Press, 1993. An excellent account of distribution and phylogeny of the mammals.

Vaughan, T. A. *Mammalogy*. 3d ed. Fort Worth, Tex.: Saunders College Publishing, 1996. Very thorough and easily understood. A great place to start for those unfamiliar with this discipline.

Wilson, D. E., and D. M. Reeder. *Mammal Species of the World: A Taxonomic and Geographic Reference*. 2d ed. Washington, D.C.: Smithsonian Institution Press, 1993. A detailed account of the mammalian species. Nicely organized and easy to follow.

RACCOONS AND RELATED MAMMALS

Types of animal science: Anatomy, classification, reproduction
Fields of study: Anatomy, zoology

The raccoon family—raccoons, coatimundis, kinkajous, and pandas—are agile arboreal mammals. Most are nocturnal omnivores.

Principal Terms
ARBOREAL: living in trees
CARRION: dead animals
DIURNAL: active during the day
GESTATION: duration of pregnancy
NOCTURNAL: active at night
OMNIVORE: an animal that eats plant and animal foods
PREHENSILE: able to grip things
SOLITARY: living alone

The raccoon family, Procyonidae, includes raccoons, coatimundis, kinkajous, and red pandas. (Giant pandas are part of the bear family, rather than the raccoon family.) All are arboreal. Except for pandas, found in the southeastern Himalayas at elevations up to twelve thousand feet, procyonids occur only in the New World. Most are mammals living in the lower elevations of temperate and tropical regions, in areas rich in water and trees.

Procyonids are omnivores, eating insects, crayfish, crabs, fishes, amphibians, reptiles, birds, small mammals, nuts, fruits, roots, and young plants. Procyonids eat what is available, including carrion, depending on season, locale, and availability. Red pandas eat bamboo, honey, grass, vines, other plants, and some meat.

The raccoon family is nocturnal, except coatis and red pandas. They mate at different times of the year. Gestation is about three months. Life spans of procyonids are up to thirty years in captivity but only ten years in the wild.

Raccoons

All members of the raccoon family have small bodies, long tails, ringed tails, and facial markings. Kinkajous lack markings, but have prehensile tails that aid arboreal movement. Raccoons are carnivores of the genus *Procyon*. They are common throughout the United States and also inhabit southern Canada, Central America, and South America. They are foxlike in

Raccoons have distinctive facial markings that look like masks. (Corbis)

appearance, with a broad head, a pointy muzzle, and short, erect ears. Raccoons have long fur and bushy tails. They are gray to brown above and lighter beneath. Black cheek patches and white whiskers adorn their heads. Their tails are marked with dark rings. Each paw has five toes, and most raccoons are about 3.5 feet long, including a 1-foot tail. The crab-eating raccoon, a species of Central and South America, is larger than common raccoons and has dark gray fur with yellow patches.

Common U.S. raccoons inhabit trees near ponds and streams, or near human homes. They hunt at night for poultry, rodents, bird eggs, insects, fish, frogs, carrion, nuts, and fruit. Those that live near human habitations are particularly fond of scavenging in garbage cans and raiding bird feeders. In northern areas, raccoons winter in dens, rarely emerging. Males are solitary except for mating, while females and young live in groups. Raccoons mate in winter. The following spring, they give birth to up to six young, born in tree dens. Young raccoons depend on their mothers for five months. Mothers and offspring stay together for a year.

Coatimundis

Coatimundis inhabit South American lowland forests and grasslands, and dry, high-altitude forests. They have long, furry, ringed tails, are fine climbers, and live in trees and on the ground. Unlike most Procyonidae, they are diurnal. Their bodies look like those of other raccoons, and their front limbs have long claws. Males grow to 4 feet long and weigh fourteen pounds, while females are 3.5 feet long and weigh ten pounds. They are a yellow-red color on their bodies, while their faces are black, with white spots around each eye and on each cheek. Their throats and bellies are also

---

Raccoon Family Facts

Classification:
Kingdom: Animalia
Subkingdom: Bilateria
Phylum: Chordata
Subphylum: Vertebrata
Class: Mammalia
Order: Carnivora
Family: Procyonidae (raccoons and related animals)
Genus and species: Ailurus fulgens (lesser or red panda); *Procyon lotor* (common raccoon), *P. cancrivorous* (crab-eating raccoon), *P. insularis* (Tres Marías raccoon), *P. gloveranni* (Barbados raccoon), *P. pygmaeus* (Cozumel Island raccoon), *P. minor* (Guadalupe raccoon); *Nasua nasua* (ringtailed coatimundi), *N. narica* (white-nosed coati), *N. nelsoni* (island coati); *Nasuella olivacea* (mountain coati)
Geographical location: North, Central, and South America, and the Himalayas
Habitat: Mostly lower elevations of temperate and tropical regions, in areas well supplied with water and trees, although some inhabit mountains twelve thousand feet high
Gestational period: Raccoons, 2 months; coatis, 2.5 months; red pandas, 3 months
Life span: Ten to thirty years, depending on species
Special anatomy: Dexterous five-toed front paws; sixth, thumb-like toe; prehensile tail

---

white. Coatis eat insects and other arthropods, crabs, frogs, lizards, mice, and reptile eggs.

Adult males are solitary but females and young live in bands of up to twelve members. Mating occurs in February or March, when a visiting male impregnates each adult female in a band. Gestation is 2.5 months, and about a month before birthing, each female leaves the band and builds a tree nest. There, she gives birth to up to five babies, each weighing six ounces. Babies are nestbound for six weeks. Then mother and young rejoin the band. Life spans of coatis are seven to nine years in the wild and up to fourteen years in captivity.

—Sanford S. Singer

See also: Fauna: North America; Fauna: South America; Omnivores; Pandas; Urban and suburban wildlife.

Bibliography

Laidler, Keith, and Liz Laidler. *Pandas: Giants of the Bamboo Forest*. London: BBC Books, 1992. Describes habitats, lives, and behavior of giant pandas and red pandas.

MacClintock, Dorcas, and J. Sharkey Thomas. *A Natural History of Raccoons*. New York: Scribner's, 1981. Discusses raccoons thoroughly, and has a brief section on their relatives.

Patent, Dorothy Hinshaw. *Raccoons, Coatimundis, and Their Family*. New York: Holiday House, 1979. Introduces highly adaptable raccoon family members, including North American raccoons, coatimundis, and pandas.

Ritchie, Rita, Jeff Fair, Alan Carey, Sandy Carey, and John F. McGee. *The Wonder of Raccoons*. Milwaukee: Gareth Stevens, 1996. Provides information on physical characteristics, habits, and behavior of raccoons.

RAIN FORESTS

Types of animal science: Ecology, geography
Fields of study: Conservation biology, ecology, environmental science

A forest growing in a region that receives over one hundred inches of rain annually is considered to be a rain forest. Rain forests can be found in both tropical and temperate climates, and are noted for their remarkable biodiversity. Thousands of different animal and plant species can be found within only an acre or two of a rain forest.

Principal Terms

HERBIVOROUS: a plant-eating animal

TEMPERATE ZONE: the regions of the earth north of the Tropic of Cancer but south of the Arctic Circle, or south of the Tropic of Capricorn but north of the Antarctic Circle

TOLERANCE: ability of a tree to grow in the shade of other trees

TROPICS: the region of the earth lying between the Tropic of Cancer and the Tropic of Capricorn

---

Rain forests are forests found in regions of the world that receive large amounts of precipitation annually. Rain forests present an incredibly diverse range of habitats, as they exist both at low elevations and high in mountain ranges. Many unusual and seldom-seen creatures inhabit the world's rain forests, including spiders so large they eat small birds, and colorful but highly poisonous tree frogs. The enigmatic sloth, an animal that spends its entire life hanging upside down from tree limbs and moving so slowly that moss grows on its fur, is found in the rain forests of South America.

Although tropical rain forests such as those in the Amazon River drainage system of South America are perhaps the best known, rain forests do exist in temperate regions as well. Olympic National Park in the state of Washington preserves a temperate climate rain forest, while much of the coast of British Columbia and southeastern Alaska also receives well over one hundred inches of rain annually. The primary difference between temperate and tropical rain forests is that in a temperate rain forest, often one or two species of trees will become dominant. In the coniferous rain forest of the Pacific Northwest, for example, Douglas fir and western red cedar are the dominant species, while other trees are found in much smaller numbers. In a tropical rain forest, in contrast, several hundred species of trees may grow side by side within a very small geographic area. The majority of trees found in tropical rain forests tend to be broad-leaved, such as the rubber tree, while temperate rain forests are dominated by conifers. The leaves of many plants in rain forests have a waxy texture or come to a point to help shed water more quickly and prevent the growth of fungi or mold.

Characteristics of Rain Forests

Although rain forests are remarkably diverse, they do share a few characteristics in common. The abundant moisture in a rain forest gives the woodland a lush, fertile appearance. This is particularly true in tropical regions. Even in the understory, close to ground level where light is limited, vegetation may be dense. This appearance of fertility is often deceptive. Dead plant matter decays rapidly in a tropical forest, but the nutrients are used quickly by the numerous competing plants. In addition, the trees in tropical forests are evergreen, which means the litter that does fall to the forest floor does so irregularly, unlike temperate broadleaf forests, where trees lose

The rain forests of the Pacific Northwest are largely composed of coniferous trees. (Corbis)

their leaves annually as the seasons change. Leaves will remain indefinitely on tropical species, such as fig and rubber trees, which is one reason small specimens of these trees are popular as houseplants. As a consequence of this lack of mulch, topsoil is often thin and the root systems of the trees are quite shallow.

One reason tropical rain forests are evergreen is that in the tropics there is little seasonal variation in the hours of daylight. The closer to the equator a forest lies, the less change there is from season to season. In temperate climates, many plants have evolved to bloom, set seeds, or lose their leaves based on the number of hours of sunlight available each day. As the seasons change annually, plants bloom in the spring or early sum-

mer; fruit ripens in the fall, photosynthesis slows, and leaves change color and die. In the tropics, where the number of hours of daily sunlight never varies, plants follow a different schedule. Many tropical plants bear new flowers and mature fruit simultaneously. The evergreen foliage and continuous supply of certain fruits has led to the adaptation of some animals to a very restricted diet: koalas, for example, which feed exclusively on eucalyptus leaves, or parakeets that eat only figs. Exceptions to this pattern are the forests where rainfall is seasonal, such as regions of the world like southeast Asia, where much of the rain comes in the form of annual monsoon storms. In those cases, flowering and setting fruit will coincide with the seasonal rains.

Forest Zones

A rain forest can be divided into four zones, each of which has its own distinct characteristics. The lowest level, the forest floor, is often dark and gloomy. Little sunlight penetrates to this level, and there is little air movement. Numerous insects, such as beetles, cockroaches, and termites, live in the decaying litter and provide food for larger animals and birds. Many of the insects, birds, reptiles, and amphibians that live in the lower levels of the rain forest are brightly colored. Scientists speculate that the animals have evolved in this fashion to more easily attract potential mates. Other scientists believe that colors warn potential predators to stay away. In either case, the vivid colors make the animals more easily seen in what is otherwise a dark environment.

Just above the forest floor is the understory. Many of the plants in the understory have large, dark leaves to maximize their light-collecting ability. Because there is little natural air movement within the lower levels of a rain forest due to the canopy blocking any natural breezes, the flowering plants in the understory often have strongly scented or vividly colored flowers to help attract insects or birds to assist with pollination. Lizards, snakes, amphibians such as tree frogs and salamanders, small birds, and mammals as large as the jaguar all call the understory home. The plants found only in the understory seldom exceed fifteen to twenty feet in height. The coffee shrub is an example of a small, shade-tolerant, tropical tree. Until horticulturalists developed strains of coffee for use in plantations where the coffee bushes are the only plants grown, coffee grew naturally in the understory of tropical forests.

The densest layer of plant life is the canopy. High above the rain forest floor, the branches of mature trees form a dense intertwined zone of vegetation extending up as much as 150 feet above the ground. Numerous plants sprout in the crotches of trees, where debris may collect. Tree limbs are festooned with vines and mosses, and bromeliads and orchids grow on the rough bark of tree trunks. Even other trees may start their life cycles a hundred feet above the ground: The strangler figs of Borneo are a relatively shade-intolerant species. A fig seed that lands and sprouts on the ground will probably not survive due to low light levels on the forest floor. Strangler figs have adapted so that their seedlings do best high in the canopy. The figs begin life in the crotches of other trees. The roots of a young fig will gradually creep down the trunk of the tree on which it sprouted. Over time, the strangler fig's roots will completely encircle the host tree, as well as penetrating the forest floor. The fig thrives, but the host tree dies, choked off by the strangler fig. Primates such as gibbons, orangutans, and lemurs spend much of their lives in the canopy, feeding on the fruit of trees such as the strangler fig, as do the sloth and other herbivorous mammals.

The emergent layer of the rain forest consists of the tallest trees, some of which exceed two hundred feet in height. The tops of these trees provide a habitat for large, predatory birds, such as eagles, as well as being home to assorted snakes, monkeys, and other animals. Every layer of the rain forest teems with life, and often what can be found at ground level gives no hint of the diversity that exists two hundred feet above in the tree tops.

Rain Forest Conservation

Many of the trees found in rain forests are valued for their commercial use as lumber, while others have been exploited for their fruits or other products, causing much habitat loss. Tropical hardwoods, such as teak and mahogany, for example, have long been used in construction and in furniture. Teak resists rotting and as a result is often used for products that are going to be exposed to the weather, such as garden furniture. Due to teak's desirability as lumber, timber companies are increasingly planting it in plantations for a sustainable yield rather than relying solely on natural forests as a source.

Activists hoping to preserve the tropical rain forest have encouraged indigenous peoples to collect forest products, such as nuts or sap, as a way to create a viable economy while at the same time discouraging industrial clear-cutting of the forest. Native people tap rubber trees in Amazonia, for example, to collect latex. Rubber trees are native to the rain forests of South America, although they are also grown in plantations in other tropical regions of the world, such as Southeast Asia.

The biggest threat to the world's rain forests may not come from commercial logging, however. In many regions of the world, rain forests have fallen victim to population pressures. Forests continue to be clear-cut for agricultural use, even when the farmers and ranchers know the exposed soil's fertility will be quickly exhausted. In some cases, the cleared land becomes an arid wasteland as the tropical sun bakes the soil too hard to absorb rain water. In others, the land is farmed for a year or two and then abandoned. Given enough time, the rain forest may regenerate, but the process will take hundreds of years.

—*Nancy Farm Männikkö*

See also: Ecosystems; Food chains and food webs; Forests, coniferous; Forests, deciduous; Habitats and biomes; Lakes and rivers; Mountains; Tidepools and beaches.

Bibliography

Bowman, David M. J. S. *Australian Rainforests: Islands of Green in a Land of Fire.* New York: Cambridge University Press, 2000. Looks at Australia's rain forests, which exist as patches separated from each other by vegetation from other environments, and discusses the ecosystems unique to Australia.

Durbin, Kathie. *Pulp Politics and the Fight for the Alaska Rain Forest*. Corvallis: Oregon State University Press, 1999. Fascinating study of the Tongass National Forest in southeast Alaska.

Goldsmith, Frank Barrie. *Tropical Rain Forest: A Wider Perspective*. New York: Chapman & Hall, 1998. A good overview of tropical rain forests globally. This book helps the reader gain an appreciation of the incredible diversity of tropical forests.

Killman, Wolf, and Lay Thong Hong. "Rubberwood: The Success of an Agricultural By-Product." *Unasylva* 51 (2000): 66-72. Looks at the rubber tree and its uses once the tree no longer produces harvestable latex. An interesting, focused discussion of one specific type of tropical tree.

Maser, Chris. *Forest Primeval: The Natural History of an Ancient Forest*. Reprint. Corvallis: Oregon State University Press, 2001. An in-depth study of the old-growth forests of the Pacific Northwest. Provides insight into temperate rain forests and how they differ from tropical.

Mysteries of the Rain Forest. Pleasantville, N.Y.: Reader's Digest, 1998. A lavishly illustrated book meant for the general reader. Deals with the rain forests as ecosystems and provides information on plants, animals, and indigenous peoples of the rain forests of the world.

Tricart, Jean. *The Landforms of the Humid Tropics, Forests, and Savannas*. Translated by Conrad J. Kiewiet de Jonge. New York: St. Martin's Press, 1972. Discusses the different ecosystems found in the tropics and helps the reader to understand what distinguishes one type of forest from another.

REEFS

Type of animal science: Geography
Fields of study: Ecology, invertebrate biology, marine biology, oceanography

Coral reefs are the largest structures constructed by living organisms and have been around for hundreds of millions of years. The importance of these marine habitats extends far beyond their aesthetic beauty.

Principal Terms

ATOLL: a remnant horseshoe or ring-shaped barrier reef surrounding a sunken island

BARRIER REEF: a reef that is separated from land by a lagoon

FRINGING REEF: a reef that is directly adjacent to land, lacking a lagoon

HERMATYPIC CORAL: reef-building coral species, belonging to the Cnidarian order of Scleractinia

LAGOON: shallow, sandy, protected areas located behind the reef flat of barrier reefs and atolls

PATCH REEF: a small, isolated reef that is typically found in lagoons

---

Reefs are hard, rocky habitats typically found in shallow seas or along coasts. When people speak of reefs, they are typically referring to coral reefs, which are the largest structures ever built by living organisms. Coral reefs have been around for hundreds of millions of years. Hermatypic, or reef-building, corals are small animals that excrete a calcium carbonate shell that forms most of the structure of reefs. Calcareous seaweeds and some other animals also contribute calcium carbonate to reefs. Carbon dioxide is removed from water (and eventually the atmosphere) when reef organisms form calcium carbonate. The formation of calcium carbonate removes seven hundred billion kilograms of carbon every year, which is tied up as rock and not easily released back into the atmosphere. Coral reefs are typically found between thirty degrees north and south latitude in warm, shallow, clear, nutrient-poor seas, where hermatypic corals grow best. Some species of ahermatypic (nonreef building) corals can live outside these latitudes, but their growth rates are too slow to build reef structures.

Rain Forests of the Sea

Coral reefs are beautiful, but they are important for environmental reasons as well. Despite occupying just 0.17 percent of earth's surface, coral reefs house an estimated 5 percent of all known species. Nearly every phylum of organism can be found on coral reefs. This high degree of biodiversity found on coral reefs is why they have been called "the rainforests of the sea."

Most of the species that contribute to the beauty of reefs are poorly studied because of the difficulty of visiting marine habitats. However, with the use of scuba gear, underwater laboratories, and ships, scientists are beginning to explore the rich array of reef organisms. Because predation is so intense on reefs, many reef creatures produce unique chemical defenses. Scientists study these chemicals with hopes of finding ones that can be useful to humans. Compounds with anti-inflammatory, antibiotic, antifungal, and anticancer properties have been isolated from reef organisms and are in various stages of clinical testing. These compounds might also be useful as shark repellents, as antifoulants, or as agrochemicals. Protecting species because they might someday benefit humans is one of many good reasons for the conservation of biodiversity.

Earth's Most Productive Habitats

Coral reefs are one of the few places where animals are more abundant than plants. Despite the lack of conspicuous plants on reefs, these habitats are the most productive places on earth. Their productivity is about twice that of agricultural fields. Single-cell algae called zooxanthellae that live within the tissues of coral animals are responsible for the coral's dark coloration and much of the photosynthesis that contributes to high reef productivity. The relationship between corals and zooxanthellae is mutualistic, meaning both organisms benefit. The zooxanthellae protect corals from ultraviolet (UV) radiation and provide them with energy-rich sugars, while the corals offer zooxanthellae a place to grow and fertilizer from the corals' metabolic wastes.

"Bleaching" is a process by which corals expel their zooxanthellae and take on a white coloration. Coral bleaching has been happening with greater frequency in recent years and may be a signal that environmental changes are harming the health of corals.

Small, inconspicuous seaweeds are responsible for the remainder of reef productivity. The basal portions of these seaweeds are protected from herbivores by being located in the abundant crevices of reefs. As soon as the top portions grow beyond this protective shelter, however, they are rapidly consumed by fishes, sea urchins, or other herbivores. Scientists have counted as many as 150,000 bites per square meter per day being taken out of some areas of reefs. This intense herbivory benefits the reef by maintaining seaweeds in a fast growing state with little seaweed biomass. When herbivores are removed from reefs, the seaweeds grow to large size and smoother the corals within weeks. This problem has been observed on reefs where too many fish have been harvested.

Types of Coral Reefs

Coral reefs are generally grouped into one of four catagories: fringing, barrier, atoll, and patch reef. When coral animals colonize a new volcanic island and form a coral reef along the fringes of that island, it is called a fringing reef. The island slowly sinks into the ocean as it cools, while the reef continues to grow upward through the building activity of the coral animals. After millions of years, the original fringing reef is then located away from the shores of the existing island because the island has, in essence, shrunk. At this point, the reef is considered a barrier reef because there is now a lagoon between the reef and the island. Barrier reefs get their name because they serve as a barrier to waves heading toward land. Eventually, the island will sink completely below the water's surface and the atoll is the ring of reef left behind. This is how atolls form, and because the Indo-Pacific is the oldest ocean, atolls are most common here. If the oceanic island sinks at a rate greater than the reef can grow, then the reef will eventu-

Coral reefs offer habitats for many varieties of marine life. (Digital Stock)

ally die as the water becomes too deep and dark to support the hermatypic corals. Scientists have found deep undersea reefs on top of seamounts that apparently grew too slowly to remain near the water surface. Finally, the fourth category of reefs is patch reefs, which are small patches of corals that dot lagoons.

Not all barrier reefs are found off islands. Barrier reefs that grow along continents are the largest reefs, with the Great Barrier Reef along eastern Australia being over two thousand kilometers long and up to seventy kilometers wide. The second largest reef is a barrier reef located off the coast of Belize. These reefs can be used to describe different zones of a reef that one encounters as they move from the sea toward land. Approaching the reef from the sandy ocean floor, the reef slope is the first zone encountered. The reef slope is the steepest part of the reef and can be fifty meters deep. It extends from the base of the reef to the reef crest, located at about five meters depth.

Most waves crash near the reef crest, protecting shoreward zones from severe wave energy. The slope of the reef changes drastically at the reef crest, resulting in a gradually sloping zone, a reef flat that extends from the reef crest to the lagoon. Shoreward from the reef flat is a lagoon of shallow, calm water. The bottom of the lagoon is typically covered with fine, white calcium carbonate sand, with an occasional patch reef providing a hard structure for organisms. The lagoon extends to the shore where white beaches usually occur. The calm, crystal clear blue water and white beaches that make these locations popular destinations for tourists are the result of coral reef ecosystems.

—Greg Cronin

See also: Chaparral; Ecosystems; Food chains and food webs; Forests, coniferous; Forests, deciduous; Grasslands and prairies; Habitats and biomes; Lakes and rivers; Marine biology; Mountains; Rain forests; Savannas; Tidepools and beaches; Tundra.

Bibliography

Birkeland Charles, ed. *Life and Death of Coral Reefs*. New York: Chapman and Hall, 1997. A recent, comprehensive volume written and edited by top coral reef scientists. It covers the accretion and erosion of corals, reef ecology, the impacts of environmental stresses on reefs, and ideas for managing these diverse ecosystems.

Cousteau, Jacques-Yves, and Philippe Diolé. *Life and Death in a Coral Sea*. Translated by J. F. Bernard. Garden City, N.Y.: Doubleday, 1971. A colorful travel log of the *Calypso* as the world-famous explorer studies the reefs of the Indian Ocean and Red Sea. Full of interesting observations and photographs.

Greenberg, Jerry, and Idaz Greenberg. *The Living Reef*. Miami: Seahawk Press, 1972. A colorful, photographic guide to many of the organisms encountered on Caribbean reefs.

Keating, Barbara H., et al., eds. *Seamounts, Islands, and Atolls*. Washington, D.C.: American Geophysical Union, 1987. A collection of papers on the abundance, distribution, and geophysical processes of deep oceanic volcanoes, seamounts, islands, and atolls.

Köhler, Annemarie, and Danja Köhler. *The Underwater Explorer: Secrets of a Blue Universe*. New York: Lyons Press, 1997. A colorful divers' guide to the behavior, biology, ecology, and taxonomy of reef organisms.

National Geographic Society. *Jewels of the Caribbean Sea: A National Geographic Special*. Washington, D.C.: Author, 1994. A videotape with extraordinary footage of organisms living on and around coral reefs, from the smallest plankton to giant sperm whales.

REFLEXES

Type of animal science: Ethology
Fields of study: Biophysics, neurobiology

Reflexes are simple, unlearned, yet specific behavioral responses to particular stimuli. Virtually all animals, from protozoa to primates, exhibit reflexes.

Principal Terms

AXON: that part of a nerve cell through which impulses travel away from the cell body

EFFECTOR: that part of a nerve which transmits an impulse to an organ of response

FIXED ACTION PATTERN (FAP): a complex motor act involving a specific, temporal sequence of component acts

INTERNEURON: central neurons often interposed between the sensory and motor neurons

MOTOR NEURON: a nerve cell body and its processes which carry impulses from the central nervous system to a muscle, producing motion

NEURONS: the structural and functional units of the nervous system, each consisting of the nerve cell body and all its processes, as the dendrites and axon

RECEPTOR: a nerve ending specialized for the reception of stimuli

REFLEX ARC: the entire nerve path involved in a reflex action

SPINDLE: in a muscle, the bundle of nuclear fibers formed during one stage of mitosis

SYNAPSE: the point of contact between adjacent neurons, where nerve impulses are transmitted from one to the other

That there are immediate motor responses to sensory stimulation is a fact that has been apparent in the thinking and writing of many scholars for centuries. Yet the term "reflex" actually did not appear until the eighteenth century. Georg Prochaska of Vienna, one of the first to use the term (1784), stated that "the reflexion of sensorial into motor impressions . . . takes place in the *sensorium commune* (common sensory center). . . . This reflexion may take place either with consciousness or without."

The word "reflex," then, originates from the idea that nerve impulses are "reflected" in the central nervous system. Current use of the term dates from the earliest experimental investigations of the role of the spinal cord in mediating muscle responses to sensory stimuli. These studies, conducted in the 1820's by Charles Bell in England and François Magendie in France, first established that sensory fibers are contained in the dorsal (upper) roots and motor fibers in the ventral (underside) roots of the spinal cord.

Response to Stimuli

Reflexes are types of action consisting of relatively simple segments of behavior usually occurring as direct and immediate responses to particular stimuli uniquely correlated with them. In addition to other, more complex stimulus-bound responses such as fixed action patterns, reflexes account for many of the behavioral patterns of invertebrates. In higher animals, such as primates, learned behavior dominates; however, reflexes remain a significant component of total behavior.

Reflexes are genetically determined. No effort is needed to acquire reflexes; they simply occur automatically, usually taking place beneath the conscious level. Although heart rate and ventilation rate, for example, are constantly regulated by reflexes, most people are unaware of these modifications. In like manner, pupillary diameter and blood pressure are regulated reflexively, without conscious knowledge. Other reflex responses, however, such as perspiring, shivering, blinking, and maintaining an upright position, are more apparent, although they also occur without conscious intervention.

Only one neuron (or even an absence of neurons) is required for the simplest known reflexes. For example, ciliated protozoa, single-celled animals having no neurons, exhibit what appear to be reflex actions. When a paramecium collides with an object, it reverses the stroke of the cilia (hairlike outgrowths), backs away a short distance, turns, and again moves forward. When touched caudally (at or near the tail), the animal moves forward. In this instance, the animal's cell membrane itself serves as the receptor of the stimulus, and the cilia act as effectors for directed movement. Very simple reflexes also occur in higher animals. For example, when human skin is injured sufficiently to stimulate a single pain neuron, an unknown substance is released, causing small local blood vessels to dilate. In contrast to these simple responses, however, most reflexes require a vast sequence of neurons.

In most reflexes, the neurons involved are connected by specific synapses to form functional units in the nervous system. Such a sequence begins with sensory neurons and ends with effector cells such as skeletal muscles, smooth muscles, and glands, all controlled by motor neurons. Interneurons, or central neurons, are often interposed between the sensory and motor neurons. This sequence of neurons is called a reflex arc. The sensory aspect of the reflex arc conveys specificity regarding the particular reflex to be activated. That is, the sensory cells themselves determine which environmental change is sensed, either inside or outside the body. The remainder of the reflex response is regulated by the specific synaptic connections that lead to the effector neurons.

Types of Reflexes

One visible and familiar reflex is the knee-jerk or stretch reflex. This reflex involves the patellar (kneecap) tendon and a group of upper leg muscles. While other muscle groups exhibit similar reflexes, this one is described here because of its widespread clinical use. This reflex involves a relatively simple reflex arc in which the terminals of the sensory cells synapse directly on the effector neurons. It is thus an example of a monosynaptic reflex. When the patellar tendon is tapped, a brief stretch is imposed on the attached muscles (quadriceps femoris), including the muscle spindles that are embedded within the quadriceps. This stretch causes the sensory neurons of the spindles to fire impulses, which return to the spinal cord. Within the spinal cord, excitatory synapses are then activated on the specific motor neurons supplying the same muscles that were stretched (the quadriceps). A series of impulses occurs in the motor neurons, which activate a brief contraction of the quadriceps, which in turn extends the leg. Another component of this reflex causes the central inhibition of the flexor motor neurons. A possible function of stretch reflexes is to oppose sudden changes in the length of skeletal muscle.

The stretch reflex, as well as other more complex reflexes, involves the inhibition of antagonistic muscle groups. This activity requires the activation of inhibitory interneurons in the spinal cord, which synapse upon the motor neurons of antagonistic muscles. In other complex reflexes, there is a sweep of activity to higher and lower levels of the spinal cord and to the opposite side of the body. Examples of reflexes with these characteristics include the flexor reflex and accompanying crossed extensor reflex. These reflexes are commonly initiated by a painful stimulus. The returning sensory signals excite the motor neurons of the leg flexor muscles, which inhibit those of the leg extensor muscles on the same side as the stimulus. This behavior assures that the foot is quickly removed from the harmful stimulus. Balance, however, must also be maintained, which is accomplished by exciting the leg extensors of the opposite limb—thus the term "crossed extensor reflex." The flexor reflex is a prime example of a protective response, and it takes place so rapidly that the pain is felt only after the withdrawal response is complete. Speed of response is particularly important when the severity of the injury is time-dependent, such as in response to a chemical exposure or burn.

Many thousands of neurons are usually involved in flexor and extensor responses. From only a few to many sensory neurons may activate several hundred to several thousand motor neurons, depending upon the reflex involved. Each motor neuron, in turn, branches to synapse on as many as a thousand muscle cells. In addition, in reflexes with bilateral effects, another thousand or more interneurons may become involved.

The Functions of Reflexes

The functions of reflexes are numerous and varied. As previously mentioned, reflexes adjust many important biological functions rapidly and efficiently without conscious effort, while other reflexes are largely protective. For example, the eye and the ear, the most delicate and sensitive sensory systems, can be damaged or destroyed by overstimulation or by accident. The amount of light admitted to the retina is controlled by the pupillary reflex, which in humans can effect a change in pupillary diameter from approximately eight millimeters in darkness to two millimeters in bright light. A sudden flash of intense light, for example, can evoke the reflex closing of the eyelids, further protecting the delicate retina. The eyeball itself is protected from drying by the blink reflex and from mechanical injury by the eyelid closure reflex. The latter reflex is triggered when an object approaches the eye or when the lashes or cornea are touched. The ear is also protected from potentially damaging sounds through the reflex contraction of middle-ear muscles in response to loud noise. This reflex functions to lower the efficiency with which sound vibrations are transmitted through the bones of the middle ear and thus reduces the possibility of damage to the delicate hair cell receptor of the inner ear.

Another category of protective reflexes exhibited by many animals is stereotyped escape responses. For example, a startled squid takes evasive action by contracting its mantle muscles (the membranous flap or folds of the body wall), forcing a jet of water through its siphon. Fish respond to vibrations carried through water by contracting the muscles on one side of their bodies; the reflex contraction occurs on the side opposite the source of vibration, and the result is a sudden move away from potential danger.

Equally essential for survival are the numerous feeding reflexes exhibited by animals. For example, flies as well as many other insects possess chemoreceptors located on their feet, mouth parts, and antennae. Thus, when a hungry fly walks on a surface moistened with nutrients, a set of reflexes is triggered. A reflex extension of the proboscis occurs. If the proboscis receptors are favorably stimulated, then the animal begins to drink. Drinking continues until the crop is sufficiently distended to stimulate its stretch receptors. Finally, this stimulus initiates the reflex termination of feeding.

Instinctive or innate behaviors, such as courtship rituals, nest-building, aggression, and territorial behaviors, demonstrate many similarities to reflexes. Although generally more complex, they are, like reflexes, unlearned, species-specific, genetically determined, and stereotypic in nature. Importantly, fixed action patterns such as these are similar to reflexes in that they are initiated by a specific stimulus, called a sign stimulus or a releaser. Generally, both forms of behavior are also comparable in that they are thought to be controlled by specific sets of neurons that underlie each behavior. Like the more complex learned behaviors, however, the neural basis of instinctive behaviors remains largely unknown.

Modification of Response

Normally, the degree of reflexive response depends upon the intensity of the stimulus; many reflexes tend to weaken gradually if the stimulus is applied repeatedly. This phenomenon, termed habituation, allows an animal to ignore a familiar or repeated stimulus. If a strong, unexpected stimulus is introduced along with a reflex-evoking stimulus, however, the strength of the reflex is often enhanced. This process, called sensitization, causes the animal to detect a potentially threatening situation by responding forcefully to its environmental cues.

Similarly, reflexes may be modified by classic (Pavlovian) conditioning, a form of associative learning. For example, in dogs, the sight and smell of food causes the reflexive secretion of saliva. A light or the sound of a bell does not. Ivan Pavlov (1849-1936) demonstrated that linking the sight and smell of food (unconditioned stimuli) with a light or bell caused the animal to associate the light or bell with food. Thus, after several training sessions, introduction of the light or bell alone produced salivation. Whether classical conditioning contributes to a significant portion of learned behavior is still a controversial issue.

Finally, reflexes are often modified during voluntary movement or locomotion. Motion itself stimulates many sensory receptors and often elicits reflexes that oppose the intended movement. Such reflexes are thought to be overridden or suppressed by commands from the brain in order for the desired movements to be executed.

Early Reflex Studies

Seventeenth century French philosopher, mathematician, and scientist René Descartes (1596-1650) combined the physiological discoveries of Galen (a Greek physician of the second century C.E.) with a conception of the body as a machine to provide the first notion of reflex action. Descartes thought that within a nerve there were thin threads attached at one end to the sense organs. External stimuli pulled on the threads to open small gates to the ventricles, thereby allowing *pneuma* (the breath of life; soul or spirit) to flow back out of the ventricles (reflected) through the same hollow nerves. This activity then caused movement by the pineal gland (a small, cone-shaped gland in the brain of all vertebrates having a cranium), extending from the midline into the ventricles. He observed that in animals this process was strictly mechanical. In humans, who un-

like animals have souls, however, the soul was thought to interact with the body at the pineal gland and thus could influence the flow of the pneuma to the muscles.

Thomas Willis (1621-1675), anatomist, physician, and Oxford professor, advanced Galen's idea one step further and related it to actual brain structures. His *Cerebri Anatome*, illustrated by Sir Christopher Wren, was the most complete description of the brain to date. Sense impressions, Willis speculated, were carried by *pneuma* within the nerves to the *sensus communis* in the corpus striatum (two striated ganglia in front of the thalamus in each half of the brain) and then on to the corpus callosum (a mass of white, transverse fibers connecting the cerebral hemispheres in higher mammals) and the cerebral cortex (a layer of gray matter over most of the brain), where they were perceived and remembered. Some were "reflected" back to the muscles via the cerebellum (the section of the brain behind and below the cerebrum, regarded as the coordinating center for muscular movement). Voluntary movement was thus controlled by the cerebrum (upper, main part of the brain of vertebrates, consisting of two equal hemispheres—the largest part of the human brain and believed to control both conscious and voluntary processes) and involuntary, or "reflex" movement, by the cerebellum.

The first actual experiments on neural mechanisms of reflexes were conducted by Robert Whytt (1714-1766) of Edinburgh. Using frogs, he demonstrated that only a segment of the spinal cord was necessary for reflex responses to skin stimulation. He also showed that the pupillary reflex was contingent upon the midbrain. More than Descartes, he stressed the protective function of reflexes. While movement was strictly mechanical for Descartes, Whytt believed that it was dependent on the "sentient principle," even when involuntary or reflex, an idea that persisted into the nineteenth century.

Modern Reflex Studies

The modern concept of a reflex largely began with the nineteenth century English physiologist and physician, Marshall Hall (1790-1857). Hall used Charles Bell and François Magendie's distinction of sensory and motor roots to develop the notion of the reflex arc. Reflexes were, by then, by definition dependent on the spinal cord, independent of the brain, and strictly unconscious and involuntary. Hall also described the excitation or inhibition of reflex movements by various drugs. Finally, he was the first to use reflexes in medical diagnosis and treatment.

The next significant development was advanced by Charles Scott Sherrington in approximately 1890. Sherrington's work on reflexes (which led to the concept of the synapse) was built upon two foundations: He conducted a meticulous anatomical analysis of the nerves to various muscles and then used this knowledge to analyze quantitatively the reflex properties of specific nerves and muscle groups. This painstaking work provided the means for obtaining the first clear conception of the reflex as a combined structural and functional unit, and it established the reflex arc as a subject for further anatomical and physiological analysis by many twentieth century scientists. In addition, Sherrington emphasized the importance of the reflex as an elementary unit of behavior and thus laid one of the cornerstones for the modern studies of animal behavior.

Sherrington's first studies involved a correlation of anatomical tracing of sensory and muscle nerves with careful observations of various reflex behaviors. He introduced methods for cutting across the brain stem of a cat at the level of the midbrain, which produced a great enhancement of tone in the extensor muscles of the limbs (those muscles responsible for keeping an animal in the standing position). In order to study the reflex basis of this activity, Sherrington and his collaborators, in 1924, began to analyze the responses to passive stretch of an extensor muscle. The results indicated that stretch of the muscle by only a few millimeters produces a large increase in tension, as measured by a strain gauge. If the muscle nerve is cut, the tension developed is low because it results only from the passive elastic properties inherent in the muscle and its tendon. Thus, the

greater tension depends on a reflex pathway that passes through the spinal cord. The reflex activity produces contractions of the stretched muscle. Since the reflex feeds back specifically to the stretched muscle, it is called a "myotatic reflex" or "stretch reflex." This is the familiar knee-jerk reflex evoked by a tap on the tendon of the knee. While most muscles, invertebrate and vertebrate, demonstrate this type of reflex, working extensor muscles best demonstrate it. Although the feedback to the muscle stretched is excitatory, there is, in addition, an inhibitory effect on muscles with antagonistic actions at a joint; thus, when a knee flexor is stretched, some of the tension in the knee extensor dissipates. This action illustrates the principle of reciprocal innervation of the muscles to a joint.

Next, it was necessary to analyze the nervous pathways involved in these and other types of reflex activity. David Lloyd at Rockefeller University began these studies in about 1940. The experiments required laborious dissection of individual peripheral nerves as well as removal of the laminae (thin layer of tissue) of the vertebral bones to expose the spinal cord so that electrodes could be placed on the dorsal and ventral roots. A single burst could then be set up in a peripheral nerve, and the response of motor neurons could be recorded in terms of the compound action potential of their axons in the ventral root. He found that stimulation of a muscle nerve produces a short-latency, brief volley in the ventral root. The input from muscles is thus carried over large, rapidly conducting axons, and there is only one, or at most two or three, synaptic relays in the spinal cord. In contrast, the ventral root response to a volley in a skin nerve has a long latency and duration. This response suggests the involvement, in skin reflexes, of shower-conducting fibers, polysynaptic pathways, and prolonged activity in the neurons in these pathways.

Reflexes and Neurology

Since reflex responses are essentially invariant in healthy individuals, their examination can often provide valuable information about neurologic disorders. Relatively simple tests, such as pupillary responses or auditory perceptions, can reveal information about complex brain functions. Scratching the skin of the foot, an activity normally leading to flexion of the toes, can demonstrate the condition of the spinal circuits that control them. For example, if the toes extend and spread in response to scratching the sole of the foot (called the Babinski response), a spinal lesion in the pyramidal tracts descending from the motor centers of the cerebral cortex is indicated. When observations of these and many other reflexes are combined with data on paralyses, spasms, or losses of sensation, it is often possible to localize the site of the lesion.

In addition to the diagnostic functions of reflexes, the reflex concept has exerted a great influence on psychological thinking and initially led to premature attempts to develop a psychology based on reflexes. Pavlov's innovative work led to extensive research in the early twentieth century on the physiology of behavior; for some time, the conditioned reflex provided the best technique for enabling at least a part of the learning process to be investigated quantitatively and to be subjected to an exact analysis. The principles proposed by such behaviorists as Edwin Ray Guthrie, Clark Leonard Hull, and B. F. Skinner to explain psychological actions as conditioned or learned responses to external and internal stimuli were based in part on earlier reflex notions and upon the fundamental model of the conditioned reflex as demonstrated by Pavlov. It is now generally recognized, however, that the reflex relationship between stimulus and response is not nearly as simple as was previously thought. The use of the conditioned reflex as a model for learning in classical-conditioning experiments artificially isolates, to an extreme degree, part of the total learning process in higher animals and is by itself inadequate in attempting to analyze the complex physiological and mental interactions that ultimately determine the behavior of humans and other mammals.

—*Genevieve Slomski*

See also: Ethology; Habituation and sensitization; Nervous systems of vertebrates.

Bibliography

Asratian, Ezras A. *Compensatory Adaptations, Reflex Activity, and the Brain*. Translated by Samuel A. Corson. Elmsford, N.Y.: Pergamon Press, 1965. The studies documented in this book, complete with charts, graphs, and photographs, are based on the work of Ivan Pavlov. The author summarizes the data compiled by himself and his collaborators on the neural regulation of the organism and on the conditioned reflex.

Creed, R. S., Derek Denny-Brown, J. C. Eccles, E. G. T. Liddell, and Charles Scott Sherrington. *Reflex Activity of the Spinal Cord*. Oxford, England: Oxford University Press, 1932. Although dated, this brief but classic work by the foremost researchers of the time is directed at the medical student or anyone interested in the physiology of the nervous system. It is a concise account of the basic features of the reflex mechanism, as illustrated particularly by the mammalian spinal cord. The work contains a bibliography as well as charts, graphs, and illustrations.

Denny-Brown, Derek. *The Cerebral Control of Movement*. Springfield, Ill.: Charles C Thomas, 1966. This often-cited classic, a series of lectures given to commemorate Charles Sherrington, a former pupil, attempts to bridge the gap in relating the large corpus of knowledge of detailed interaction of nerve cells to the neuronal effects that underlie motor behavior and its disorders.

Fulton, J. F. *Muscular Contraction and the Reflex Control of Movement*. Baltimore: William & Wilkins, 1926. This dated but well-written and lengthy work is primarily concerned with the activity of skeletal muscle: its mode of contraction and the nature of its reflex control. The author studies muscle as a means of motion rather than as a machine for transforming energy. Aside from charts and graphs, the book contains a very extensive, although dated, bibliography.

Kandel, Eric R., and James H. Schwartz. *Principles of Neural Science*. 4th ed. New York: McGraw-Hill, 2000. A textbook of neurobiology aimed at students of behavior, biology, and medicine. Contains chapters on spinal reflexes and on the brain stem, reflexive behavior, and the cranial nerves.

Pavlov, Ivan Petrovich. *Lectures on Conditioned Reflexes: Twenty-five Years of Objective Study of the Higher Nervous Activity (Behavior) of Animals*. New York: Classics of Medicine Library, 1995. A reprint of Pavlov's classic studies on conditioned reflexes, or so-called Pavlovian responses.

Sechenov, Ivan M. *Reflexes of the Brain*. Translated by S. Belsky. Cambridge, Mass.: MIT Press, 1965. In this brief but classic work, the author analyzes and defines the ways in which muscular movements originate in the brain. Chapter 1 is devoted to involuntary movements; chapter 2 discusses voluntary movements. The work also contains a postscript, notes, and a biographical sketch of the author.

Shepherd, Gordon M. *Neurobiology*. 3d ed. New York: Oxford University Press, 1994. This excellent college textbook treats molecular neurobiology as the foundation for the study of the nervous system, and provides an orientation to these methods and a summary of their application to a model synapse. An extensive bibliography as well as numerous illustrations, charts, and graphs are included in the text.

REGENERATION

Type of animal science: Development
Fields of study: Cell biology, embryology, immunology

Regeneration is a phenomenon by which the cells surviving from damaged tissue grow to replace missing tissue. The process is common in tissues of lower animals (such as starfish and insects), but it is rare in mammals. Regeneration is important for replacing damaged organs and for understanding organismal development.

Principal Terms

CHEMOTAXIS: a process by which cells are attracted to a chemical, moving from low to high concentrations of the chemical, until the cells cluster

DETERMINATION: an event in organismal development during which a particular cell becomes committed to a specific developmental pathway

DIFFERENTIATION: the process by which a determined cell specializes or assumes a specific function

FATE MAP: a map of determined, but undifferentiated, tissue by which specific cell regions can be identified as giving rise to specific adult structures

IMAGINAL DISK: a determined, undifferentiated tissue in fruit fly larvae that gives rise to a specific adult structure

POSITIONAL INFORMATION: a concept by which differentiating cells organize themselves to produce a particular tissue type based upon cell-to-cell interactions

STEM CELL: a determined, undifferentiated cell that is hormonally activated and changes into a specific cell type

TRANSDETERMINATION: an event by which a determined, undifferentiated cell changes its determination, thereby giving rise to a different tissue type

Regeneration is a process by which some organisms replace damaged or missing tissue using living cells adjacent to the affected area. The phenomenon is not well understood, but several animal systems have enabled developmental geneticists to develop strong models describing the process.

The replacement of tissue is a common occurrence in fungi and plants. Regeneration can also occur in animals, although the capacity for regeneration progressively declines with increasing complexity in the animal species. Among primitive invertebrates (animal species lacking an internal skeleton), regeneration frequently occurs. A planarium can be split symmetrically into right and left halves. Each half will regenerate its missing mirror-image, resulting in two planaria, each a clone (exact genetic copy) of the other.

In higher invertebrates, regeneration occurs in echinoderms (such as starfish) and arthropods (such as insects and crustaceans). A radially symmetrical starfish can regenerate one or several of its five arms. Regeneration of appendages (limbs, wings, and antennae) occurs in insects such as cockroaches, fruit flies, and locusts. Similar processes operate in crustaceans such as lobsters, crabs, and crayfish. Limb regeneration extends even to lower vertebrate species (species having an internal skeleton) such as amphibians and reptiles, although on a very limited basis.

In amphibians, the newt can replace a lost leg. In reptiles, some lizards can lose their tails when captured by a predator, thus assisting their escape. If the lizard is young, the tail can regenerate.

The tail breaks because of a breakage plane in the tail which severs upon hormonal activation. The glass-snake lizard is such a species.

Principles of Regeneration

Regeneration in these organisms is based upon two principles: the symmetrical organization of cells in the organism, and the reversal of determination and differentiation in the surviving cells, termed blastema, adjacent to the missing tissue. These two factors are fundamental to the development of the organism.

In animal systems, two major body symmetries emerge—radial symmetry and bilateral symmetry. In radially symmetric organisms, including plants and primitive invertebrates such as hydra, jellyfish, and starfish, body tissues are arranged in a circular orientation about a central axis. Appendages may also be present that likewise orient in a circular pattern of cells. In bilaterally symmetrical organisms, including animal species such as planaria, arthropods, fish, amphibians, reptiles, birds, and mammals, the body is oriented into mirror-image halves about a central body plane, resulting in its having right and left equivalent structures along each half.

Body symmetry is critical for tissue regeneration because of positional information. The cells of an individual organize into a specific pattern during development. Cell-to-cell interactions and chemical messengers between cells provide the cells of a given tissue with information signals directing the cells to grow in a particular direction or pattern. The loss of a tissue portion may stimulate the remaining cells to carry out a programmed growth, that is, to complete a specified pattern.

Determination and differentiation both play an important role in the genetic basis for development. All cells of an organism contain the same genetic material; that is, they all contain the same deoxyribonucleic acid (DNA), the same genes. In a specialized organism (one having different tissue types—eyes, ears, skin, and nerves), these cells must behave differently, even though they contain the same genetic information. The process by which identical cells with the same genetic infor-

mation give rise to different tissues is termed differentiation.

What causes similar cells to differentiate to form different tissues is a process called determination. Prior to differentiation, cells become determined, meaning that some genes in these cells are turned on, making certain proteins, while other genes are turned off, not making other proteins. All cells of a specific tissue have the same genes that are turned on or off and therefore make the same proteins (for example, all red blood cells manufacture hemoglobin). Cells of other tissue types have different genes that are turned on or off and make other proteins (for example, epidermis cells manufacture keratin, not hemoglobin). How cells determine and differentiate depends upon chemical signals (hormones). Hormones signal different cells that receive chemically coded information based upon their location in the organism, which is based upon the organism's symmetry.

Cockroaches, Newts, and Fruit Flies

Three principal animal regenerative systems have been studied: cockroach limb regeneration, newt limb regeneration, and fruit fly imaginal disk regeneration. In all animal systems, regeneration occurs primarily in younger individuals undergoing metamorphosis, which is a change in development involving considerable alterations in body size and physical appearance. Adult regeneration is incomplete or does not occur.

Severing a cockroach limb results in distal regeneration; the remaining leg part regenerates the lost portion. If the middle portion of a leg is removed and the remaining leg parts are grafted together, complete regeneration of the missing middle portion occurs precisely between the grafted parts, but grafting requires correct orientation of the body parts. If the leg parts are grafted backward, regeneration will be distorted, resulting in a malformed limb, sometimes with multiple leg structures sprouting from one limb. Virtually identical results have been obtained for newt limb regeneration. Furthermore, limb regeneration in the presence of certain chemicals such as retinol palmitate causes complete limb regeneration, including undamaged regions. The net result is a severely deformed limb.

For the fruit fly *Drosophila melanogaster*, the period from egg to adult is approximately ten days at 25 degrees Celsius. The period includes roughly seven days during which the organism proceeds through three larval (maggot) stages, followed by a three-day immobile pupal stage, during which metamorphosis occurs. Metamorphosis involves the replacement and modification of larval body structures with adult body structures (eyes, legs, and wings). The cells that are to become the adult structures are present, but dormant, in the larval stages. These special cells, called imaginal disks, are determined to become specific adult structures, but remain undifferentiated until activated by the hormone ecdysone during metamorphosis.

There is one imaginal disk for each future body structure (two eye imaginal disks, six leg imaginal disks, for example). Gerold Schubiger and other geneticists have determined "fate maps" for each imaginal disk. They have determined what each cell group on each disk will become in the adult. For example, the male genital disk has been mapped so that specific cell groups are associated with the formation of the specific adult structures of, for example, the heart, penis, and testes.

Experiments by Peter J. Bryant and Schubiger focused upon removing parts of leg and wing imaginal disks. The remaining surviving cells either regenerated the missing tissue or duplicated themselves. Whether regeneration or duplication occurred depended upon the amount of tissue lost and the position of the surviving imaginal disk tissue. If a small portion of a disk or a particular region of the disk was lost, then the remaining disk cells regenerated the missing part, thus producing a complete disk and ultimately a normal adult structure. If a large section of a disk or a sensitive region of it was lost, then the remaining cells du-

Starfish are capable of regenerating broken-off arms. (PhotoDisc)

plicated a mirror-image of themselves, giving rise to a useless adult structure.

Models of Regeneration

These collective studies, especially those involving the *Drosophila* imaginal disks, have produced two comprehensive regeneration models: the gradient regeneration model and the polar coordinate model. The gradient regeneration model, proposed by Victor French, explains the regenerative capacity of a given tissue by arranging the cells of the tissue along a gradient of regenerative capacity. Cells are arranged in order of high regenerative capacity to low regenerative capacity. This high-to-low regenerative gradient is directly correlated to the positional information of each cell. Cells located proximal (near) to the main body axis have high regenerative capacity. Cells located distally (far) from the main body axis have progressively lower regenerative capacities. Removal of distal cells results in their replacement by regeneration of the proximal highly regenerative cells. The removed distal cells, which lack regenerative information because they are at the low end of the gradient, cannot regenerate the proximal cells.

The gradient regeneration model can best be visualized as a right triangle with its hypotenuse (longest side) being a downward slope. Highly regenerative cells (proximal to the main body axis) are at the top of the slope. They contain positional information for themselves plus information for all cells below them on the slope. Higher cells can replace lost lower cells (located distal to the main body axis). Distal cells low on the slope have considerably less positional information and therefore can only duplicate themselves.

The polar coordinate model, proposed by French with Peter J. Bryant and Susan V. Bryant, is a more elaborate version of the gradient model that explains not only the *Drosophila* imaginal-disk experiments but also the cockroach and newt regeneration experiments. This model is a three-dimensional gradient that covers regeneration not only in a proximate to distal direction but also from the exterior to interior. The polar coordinate model can best be visualized as a cone. The circular end of the cone represents proximal tissue, whereas the tapered tip represents distal tissue, thus simulating the proximal-to-distal regeneration gradient.

On the circular (proximal) base of the cone, imagine a bull's-eye. The outermost circle represents exterior tissue, whereas the circle center (bull's-eye) represents the most interior tissue. There is thus a three-dimensional regeneration gradient—proximal-to-distal and exterior-to-interior. The circle is furthermore subdivided clockwise into twelve regions, completing the polar coordinate model of tissue regeneration capacity.

The tissue pattern of regeneration will again favor those cells located at high gradient positions, namely proximal (cone base) and exterior (outside circle). These cells possess positional information for regeneration of lower gradient tissue. Lower gradient cells, located distally (cone tip) and interiorly (circle center), will have limited positional information and will be capable only of duplicating themselves.

For clockwise regeneration, the polar coordinate model operates by the shortest intercalation route; that is, if a small tissue section is lost, the remaining large section will regenerate the lost piece based upon positional information. If a large tissue section is lost, however, the remaining small section will lack sufficient regenerative information and will be capable only of duplicating a mirror image of itself.

The French, Bryant, and Bryant polar coordinate model is a three-dimensional regenerative capacity gradient intended to model a tissue based on cell position. The model really boils down to one principle: a large, proximal (or exterior) group of cells can regenerate missing tissue; a small, distal (or interior) group of cells cannot.

Studying Regeneration

Developmental geneticists have studied regeneration in a variety of ways. Among the principal experimental techniques have been fate map determination of imaginal disks and limb regeneration, already discussed above, transdetermination of imaginal disks, and studies of simple organismal development.

Geneticists have found that under special circumstances, an imaginal disk or a portion of an imaginal disk can change its pattern of determination, that is, it transdetermines. A wing occasionally grows from an eye, for example, or a leg from a wing. Cells that are programmed to follow one developmental route follow another route instead. The cause of transdetermination is unknown, but the process does follow specific patterns. For example, a genital imaginal disk can transdetermine to form an antenna or leg, but not vice versa. An antenna disk can transdetermine to produce an eye, wing, or leg, but the wing and leg disks cannot transdetermine to an antenna.

Further regenerative studies involve the model developmental systems, including the cellular slime mold *Dictyostelium discoideum*. In the presence of adequate food, this exists as single, amoeba-like cells. If the cells are starved, they release a chemical attractant (chemotaxic) substance, cyclic AMP, that attracts the cells to one another. The resulting cellular mass moves as a single unit until the organism finds a suitable food source, upon which the cells differentiate and specialize to produce and release spores, each of which subsequently gives rise to a new amoeba-like stage. Such studies are necessary because regeneration ultimately involves changes in the determination and differentiation of cells.

Future Prospects

Regeneration research presents two opportunities for further development: an understanding of higher cell differentiation and growth, and prospects for replacing lost or damaged human tissues and organs. The polar coordinate model for tissue regeneration indicates that tissue replacement depends upon blastemas at the damaged area replacing tissue using positional information. Future research includes genetic and molecular studies to identify intercellular chemotaxic molecules and other information molecules that mediate cell-to-cell communication and thereby control how cells develop and grow in specified patterns. Such research will unravel important clues to cellular development and regeneration.

Scientists' understanding of cellular differentiation and growth is currently limited to more primitive species (such as *Dictyostelium discoideum* and *Drosophila melanogaster*). The most effective breakthroughs have been with the *Drosophila* imaginal disk studies and cockroach and newt limb regeneration experiments. Much more work remains to be done, particularly with respect to genetic and molecular analysis. The action of specific steroid and protein hormones on cellular growth and differentiation is a further avenue of research.

While regeneration research has been pursued for many decades, it is still in its infancy. Further research will allow the understanding of organismal development.

—David Wason Hollar, Jr.

See also: Aging; Amphibians; Cleavage, gastrulation, and neurulation; Determination and differentiation; Development: Evolutionary perspective; Fertilization; Growth; Insects; Sexual development.

Bibliography

Alberts, Bruce, Dennis Bray, Julian Lewis, Martin Raff, Keith Roberts, and James D. Watson. *Molecular Biology of the Cell*. 3d ed. New York: Garland, 1994. As an intermediate-level undergraduate textbook, this outstanding work is a unique approach to biology in that it interweaves genetics and cell biology. Chapters on cellular mechanisms of development and differentiated cells and the maintenance of tissues are among the best discussions of developmental biology found in an undergraduate biology text. The book provides excellent photographs and diagrams.

Goodenough, Ursula. *Genetics*. 3d ed. New York: Holt, Rinehart and Winston, 1984. As a beginning undergraduate genetics text, this book provides excellent coverage of genetics, including molecular genetics, that maintains the importance of classical genetics experiments. The text cites extensive experimental evidence with numerous references. The chapter on control of gene expression in eukaryotes is a very clear presentation of gene control in cellular determination and differentiation.

Klug, William S., and Michael R. Cummings. *Concepts of Genetics*. 6th ed. Columbus, Ohio: Charles E. Merrill, 2000. This is a strong undergraduate genetics textbook that is clearly written and provides hundreds of illustrations and examples. The chapter on the role of genes in development includes a very good discussion of fate maps for *Drosophila* imaginal disks.

Lewin, Benjamin. *Genes VII*. New York: Oxford University Press, 1999. This advanced textbook is a comprehensive, yet very readable, summary of current advances in molecular biology. Lewin has complete command of all important breakthroughs in genetics and cell biology. The text includes numerous chapters on gene regulation in development, including a discussion of mutations which alter *Drosophila* determination and development.

Sang, James H. *Genetics and Development*. London: Longman, 1984. This is suitable as both a graduate text and a reference source for developmental geneticists. It clearly and precisely presents current theories of development, determination, differentiation, and regeneration. Numerous animal and plant systems are presented, with excellent diagrams and photographs. Several hundred references are cited for further research.

Starr, Cecie, and Ralph Taggart. *Biology: The Unity and Diversity of Life*. 9th ed. Pacific Grove, Calif.: Brooks/Cole, 2001. This is a popular beginning biology textbook for undergraduate science and nonscience majors. It is clearly written and includes excellent illustrations and photographs. The chapter on principles of reproduction and development discusses development and differentiation in animal systems.

Wallace, Robert A., Jack L. King, and Gerald P. Sanders. *Biosphere: The Realm of Life*. 2d ed. Glenview, Ill.: Scott, Foresman, 1988. This is an excellent introductory biology textbook for both science and nonscience majors. It is clearly written and comprehensive in its coverage, although it is not too detailed for the layperson. The chapter on animal origins and the lower invertebrates provides a discussion of radial and bilateral symmetry in primitive animals. Also includes a good description of organismal development.

REINDEER

Types of animal science: Anatomy, classification, ecology, reproduction
Fields of study: Anatomy, zoology

Reindeer are deer native to subarctic and arctic regions. Males and females have antlers. Reindeer migrate in herds of up to 200,000. Imported into Alaska and Canada, they are valuable resources.

Principal Terms

DOMESTICATED: trained and raised by humans for specific uses.
GESTATION: term of pregnancy
HERBIVORE: an animal that eats only plants
RUMINANT: a herbivore that chews and swallows plants, which enter its stomach for partial digestion, are regurgitated, chewed again, and reenter the stomach for more digestion

---

Reindeer are large deer, native to subarctic and arctic regions of northern Europe and Asia. They are related to North American caribou, as both are variants of the species *Rangifer tarandus*. Reindeer can be domesticated and have long been valuable possessions of humans in those regions of the world. They yield meat, cheese, butter, clothes, and draft animals able to carry heavy burdens.

Many Eurasian reindeer still run wild and are trapped for domestication. Whether wild or domesticated, reindeer are herbivores, eating only plants. Their diet is grass, moss, leaves, twigs, and lichens. They often obtain food by scraping snow cover with their antlers and hooves. Reindeer are diurnal, meaning that they are active only during the day. They spend most of their time seeking food. Their preferred habitats are barren, open plains (tundra), forests, grasslands, and mountains.

Physical Characteristics of Reindeer

Reindeer differ from most deer in having large, deeply cleft hooves, hairy muzzles to help to keep them warm, and antlers on both males and females. Reindeer have long bodies and legs. Their hooves are broad, to provide footing on snow and ice. Male reindeer are four feet tall at shoulder height and weigh up to six hundred pounds. Females are shorter but reach similar maximum weights. Both genders grow up to seven feet long. Their thick, waterproof fur is brown in summer and gray-brown in winter. White fur covers their rumps, tails, and the lower portions of their legs. Males have white neck manes during mating season. Reindeer do not see well, but they have an excellent sense of smell.

Reindeer antlers have pointed branches (points). In females, they grow to two-foot lengths, while males' antlers reach five-foot lengths. Very large male antlers have forty points. Those of females only have a few points. As in other deer, reindeer antlers are shed and regrown each year. Males lose their antlers in winter and females lose their antlers in late spring. The antlers that grow back are larger than those replaced. Antlers are important during mating season, when males fight for mates. Fights can damage antlers, so if they were not shed and regrown each year, many males would be unable to fight well, lose fights, and be unable to mate.

Reindeer are also ruminants, animals that chew and swallow their food more than once. After a little while, food that was swallowed reenters the ruminant mouth from the stomach. Reindeer and other ruminants chew the food, swallow it again, and the food enters a different stomach for additional digestion. The process, also called cud chewing, helps reindeer to get maximum

In the late nineteenth century, the U.S. Office of Education imported reindeer into Alaska and instituted a successful program to teach the Inuit how to raise them, as the Lapps of Finland had long done. (AP/Wide World Photos)

amounts of nutrients and vitamins from their difficult-to-digest food.

Reindeer are social animals. They live in groups of about 20 most of the year. The groups consist of a male, his mates, and their young. Reindeer migrate great distances each fall and spring to feeding and mating grounds, travelling in herds of up to 100,000 and migrating about twenty miles per day.

Reindeer mate mostly in October. Gestation is about eight months long. The female leaves the herd to give birth to one calf, in May or June. The calf weighs up to twenty pounds. Mother and calf then rejoin the herd and the calf nurses for six months. A calf can mate when three years old. The life span of reindeer is up to fifteen years.

North American Reindeer Imports

Reindeer are excellent sources of food, clothes, and draft animals, as the Laplanders of Finland

---

Reindeer Facts

Classification:
Kingdom: Animalia
Subkingdom: Bilateria
Phylum: Chordata
Subphylum: Vertebrata
Class: Mammalia
Order: Artiodactyla
Family: Cervidae (deer)
Genus and species: Rangifer tarandus (reindeer and caribou)
Geographical location: Northern Europe, Asia, Canada, and Alaska
Habitat: Open plains (tundra), forests, grasslands, and mountainous areas
Gestational period: Eight months
Life span: Twelve to fifteen years
Special anatomy: Antlers, ruminant stomach

have long known. In order to provide a reliable food source for the Inuit of Alaska, who live in a comparable environment with a similar social structure, the U.S. Office of Education imported thirteen hundred reindeer from Siberia near the end of the nineteenth century. Several million reindeer are now found throughout Alaska. In 1935, the Canadian government set up a herd of reindeer in the Yukon Territory to benefit Native Americans and Inuit. This herd also flourished and Native Americans and Inuit now own all reindeer herds in North America. The deer satisfy many of their basic needs, becoming a valuable North American resource.

—*Sanford S. Singer*

See also: Cattle, buffalo, and bison; Deer; Domestication; Elk; Herbivores; Herds; Horns and antlers; Moose; Ruminants.

Bibliography

Gerlach, Duane, Sally Atwater, and Judith Schnell, eds. *Deer*. Mechanicsburg, Pa.: Stackpole, 1994. This book discusses deer habits, habitats, species, and much more.

Lepthein, Emilie U. *Reindeer*. Chicago: Children's Press, 1994. This brief book, aimed at juveniles, covers basic aspects of the natural history of reindeer and caribou.

Russell, H. John. *The World of the Caribou*. San Francisco: Sierra Club Books, 1998. This book discusses caribou and reindeer habits, habitats, similarities, and differences.

Syroechkovskii, E. E. *Wild Reindeer*. Translated and edited by David R. Klein. Washington, D.C.: Smithsonian Institution Press, 1995. This in-depth book discusses many aspects of wild reindeer natural history, including food, reproduction, geographic distribution, hunting, and ecology.

REPRODUCTION

Type of animal science: Reproduction
Fields of study: Ethology, genetics, reproduction science

Reproduction is a prerequisite for life. Sexual organisms reproduce when the gametes of two adults unite. Similarities and differences in how these adults maximize their own lifetime offspring production influence the diversity of mating behaviors observed in nature.

Principal Terms

GAMETE: a sexual reproductive cell that must fuse with another cell to produce an offspring: a sperm or egg

MATE CHOICE: the tendency of members of one sex to mate with particular members of the other sex

MATE COMPETITION: competition among members of one sex for mating opportunities with members of the opposite sex

NATURAL SELECTION: the process that occurs when inherited physical or behavioral differences among individuals cause some individuals to leave more offspring than others

REPRODUCTIVE SUCCESS: the number of offspring produced by one individual relative to other individuals in the same population

SEXUAL DIMORPHISM: an observable difference between males and females in morphology, physiology, and behavior

SEXUAL SELECTION: the process that occurs when inherited physical or behavioral differences among individuals cause some individuals to obtain more matings than others

The ability to reproduce is central to the existence of any organism. Simple one-celled organisms reproduce asexually by duplicating their genetic material and dividing in half. For reproduction in sexual species the participation of two individuals is essential. Mating partners have a common interest: production of successful offspring. They also have conflicting interests. Appreciating the diversity of reproductive behaviors that occur in different organisms requires understanding of the reproductive conflicts that exist between mates even more than their mutual reproductive interests.

Early sexual species most likely consisted of individuals that produced gametes of similar size. Except that one set of individuals produced smaller gametes that would be called sperm and another set produced larger gametes called eggs, males and females did not exist. Yet, reproduction was sexual. Through time it is thought that distinct sexes evolved because some individuals obtained reproductive advantages by producing smaller than average gametes whose greater motility increased their likelihood of fertilizing other gametes, while other individuals obtained reproductive advantages by producing larger than average gametes whose greater stores of nutrients increased the survival prospects of their young. Such evolution of gamete dimorphism would then lead to the many specializations in appearance and behavior of present-day males and females.

Reproductive Success

For species whose parents do not provide care for their young, the maximum number of offspring that an individual female can produce is determined by the number of eggs that she can manufacture. For species in which parents do provide care, the number of young a female can produce might be limited more by the number of young she can raise than by the number of eggs she can make. For both these types of species, females that copulate with more than one male produce roughly the same number of young each season as females that mate with only one male. Thus, the quality of an individual mate, or the resources of paternal care that he can provide, should affect the reproductive success of females more than the number of mates they can obtain. As a result, females of most species are expected to be selective about which male fertilizes their eggs.

In contrast, the reproductive success of males of most species is not limited by the number of sperm they can produce. In species lacking parental care, males that mate with the most females usually leave the most offspring. In parental species, the reproductive consequences of mating with more than one female are slightly more complex. If one parent can provide sufficient care to raise young to independence, males that mate with multiple females usually leave more offspring than males that mate with only one female. If both parents are necessary to raise young, males that mate with only one female should leave more offspring than males with multiple mates. Given the reproductive advantages of mating with multiple females and a tendency for breeding males to outnumber breeding females, competition for mating opportunities is more pronounced in males than in females for many species. Furthermore, males should usually be less selective in mate choice than females.

Mate Competition and Mate Choice
Charles Darwin realized that the struggle to obtain mates could be an important process affecting the evolution of organisms. He deemed this process "sexual selection" to distinguish it from the struggle for existence, or "natural selection," and suggested two components of sexual selection: mate competition and mate choice. The importance of mate competition is rarely questioned by biologists because it often involves frequent and conspicuous aggressive interactions and distinctive weapons (for example, antlers of various ungulate species, horns of bovids, and enlarged canine teeth of primates). Mate choice is another matter. This exceedingly subtle behavior is difficult to document in any species. Although females of most species are demonstrably more reluctant to mate than males and are remarkably good at rejecting the mating attempts of males of other species as well as those of closely related males of their own species, few studies can convincingly show that females actively choose particular types of males as mates. The lack of direct evidence for the pervasiveness of mate choice in nature has caused many biologists to doubt its significance ever since Darwin first suggested it. Mate choice is also often misunderstood. The decision process it embodies is not thought to result from conscious deliberation; rather, individuals might simply react more favorably to some members of the opposite sex than to others.

Fighting for Mates
Similarities and differences in the details of mate competition and mate choice of different species can be exemplified by comparing three North American vertebrates: bullfrogs, sage grouse, and elephant seals. Male bullfrogs fight each other for small areas in ponds that females use as egg deposition sites. The wrestling matches that occur between males do not involve weapons as such, but being larger than an opponent almost always confers success. During nighttime choruses, females move among the territorial males, apparently assessing features of the male and his territory. Pairing begins when a female approaches a particular calling male and touches him. The male clasps the female, and within an hour, the female releases up to twenty thousand eggs, which the male fertilizes externally. Neither the male nor the female provides parental care for their young. Each year, roughly half of the males in a population obtain mates, and the most successful male may mate with six or seven different females.

Male elephant seals can weigh as much as three thousand kilograms and are highly aggressive.

Rather than fight for territories, they fight directly for groups of females, called harems, that haul out on land to give birth. Males that monopolize large groups of females might mate with ninety or more females each year. One study revealed that less than 10 percent of the males sire all the pups in a breeding colony. Success in male competition not only involves being large but also involves having formidable canine teeth to use as weapons. During a fight, males rear up to half their length, slam themselves against the opponent, and bite him on the neck. Skin on the chests of males is highly cornified; these "shields" provide some protection against such onslaughts, but injury is still common. Mate choice by females is limited to vocalizing before and during copulation. If the male attempting to mate with her is a subordinate individual, the dominant male quickly responds to the call and attacks the copulating male.

Male sage grouse often congregate, or "lek," in traditional areas where they display and fight to control small territories. The territories function as courtship sites and places to copulate. Males provide neither resources nor parental care for their young; yet, females initiate mating and appear to be highly selective in mate choice. Near unanimity in preferred mates by the females in the population results in only a few males obtaining all the matings.

In all three of these species, many males compete for mates by fighting for territories; however, some males employ different tactics to obtain mates. Small, young bullfrog males remain silent near large, calling males and attempt to intercept any female attracted by the calling male. Small, young elephant seal males lurk about on the surf and grab females as they leave land to feed in the ocean. These males force females to copulate. Some sage grouse males attempt copulation with females away from the display arena. In another lekking bird species, the ruff, two types of males exist: territorial males and satellite males. The latter males are nonaggressive and appear to capitalize on the ability of territorial males to attract females. Unlike the case of bullfrog males, however, genetic differences of male ruffs produce the strik-

ing differences in plumage and behavior, rather than their ages.

Mate choice by females is well developed in bullfrogs and sage grouse. The benefit that female bullfrogs obtain by choosing particular males is relatively straightforward: Chosen males tend to control superior egg deposition sites that increase offspring survival. Benefits that female sage grouse obtain from mate choice are unknown. Mate choice in such lekking species continues to pose a significant question for biologists.

Competition and Reproductive Success

Differential success of males in mate competition and mate attraction may translate to large differences in reproductive success. In contrast, variation in reproductive success among females is usually low because most females mate and produce at least some offspring. The relative amount of variation in reproductive success within each sex can influence the evolution of sexual traits. When only a few adults produce most of the offspring in a population, genes affecting the traits that underscore their success will be passed on to their offspring and quickly become the predominant characteristics of future generations. In contrast, if the most successful individual produces only slightly more offspring than other individuals, genes from all these parents will be present in roughly similar numbers in subsequent generations.

A consequence of greater variation in reproductive success of males relative to females is the evolution of elaborate sexual characteristics that are expressed only in males. These traits can be morphological, physiological, or behavioral. The extent of sexual dimorphism is predicted to be related to the relative variation in reproductive success in the sexes. Thus, species in which one or a few males sire most of the offspring produced in the population would tend to be species with considerable phenotypic differences between the sexes.

Field Research and Laboratory Studies

Studies of the reproductive behavior of organisms usually involve observation and experimentation of male and female interactions in nature or the laboratory. Early studies were mostly observational and cataloged the most typical behavior patterns observed in each sex. These studies ignored differences in behaviors among individuals. Because such differences can have significant consequences in terms of reproductive success, more recent studies usually involve marking males and females for individual recognition and recording various features of their morphology, behavior, and reproductive success.

Quantifying reproductive success in nature is a difficult task, and various methods have been used for different organisms. For all studies, the identity of each individual must be known. For some species, researchers can assess only the number of copulations that individuals obtain; for other species, they can count the number of young born; in yet others, they can determine the number of young that survive to independence or even sexual maturation.

Laboratory and field experimentation has been used to study a plethora of questions concerning the acquisition, function, and evolutionary significance of a variety of reproductive behaviors. Early studies of bird song investigated not only the role of song in attracting mates but also how individuals acquired their species-typical song. Choice experiments on female frogs using either naturally calling males or playbacks of recorded male calls revealed the call characteristics that females use in species and mate recognition. Crossing different species that varied in reproductive behaviors and noting the characteristics of their

Many female fish deposit their eggs on the rocky floor in shallow water, where they are externally fertilized by the male. (Joerg Boetel/Photo Agora)

hybrid offspring provided some insights into the genetic basis of various behaviors. Staging aggressive interactions between males that differ physically in some regard demonstrated the significance of various male characteristics.

Researchers investigate general trends in reproductive behaviors by using the comparative method. This method usually works best when fairly closely related taxa (for example, species within the same genera or family) are considered. Using the comparative method, researchers can look for the relationships between the degree of sexual dimorphism in some characteristic and sex-specific differences in reproductive success variation, or the ecological and social conditions that affect male behaviors such as territoriality or female behaviors such as the amount of maternal care provided to young.

A thorough understanding of reproductive behaviors requires the creative use of all three methods of investigation: observations of individuals in nature to document normal behavioral patterns, precise experimentation on behaviors under controlled conditions to understand mechanisms of behavior and the stimuli that produce them, and comparison of trends in closely related species to gain insights into the evolutionary history of behavioral traits.

Reproduction is one of the defining attributes of life itself, and for most organisms, reproduction is sexual. Yet, despite the universality of sexual reproduction, the behavior patterns associated with it in different organisms are exceedingly diverse. For solitary organisms, sexual reproduction may be the only form of social behavior. For highly social species, individual interactions can be much more complex but still usually influenced by sex in some manner. Biologists seek to find some order to the variety of reproductive behaviors observed in different organisms. A unifying theme for this diversity is an evolutionary one: How do the sexes differ in maximizing the number of offspring they can produce? More specifically: How do males or females maximize the number of offspring they produce given the behavior patterns of the other sex and the various ecological factors that affect them? Thus, current research on reproductive behaviors has gone well beyond the point of merely describing what animals do to reproduce to determining why they do what they do.

—*Richard D. Howard*

See also: Adaptations and their mechanisms; Asexual reproduction; Birth; Copulation; Courtship; Embryology; Ethology; Evolution: Historical perspective; Fertilization; Hormones and behavior; Infanticide; Lactation; Marsupials; Monotremes; Parthenogenesis; Reproductive strategies; Reproductive systems of female mammals; Reproductive systems of male mammals; Sex differences: Evolutionary origin; Sex differences: Evolutionary origin; Sexual development.

Bibliography

Alcock, John. *Animal Behavior: An Evolutionary Approach.* 7th ed. Sunderland, Mass.: Sinauer Associates, 2001. An extremely up-to-date textbook, utilizing case studies to illustrate topics. Covers mating and mating behavior in the context of natural selection.

Clutton-Brock, Tim H., F. E. Guinness, and S. D. Albon. *Red Deer: Behavior and Ecology of Two Sexes.* Chicago: University of Chicago Press, 1982. Provides an in-depth analysis of a fifteen-year study of the reproductive behavior of male and female red deer. This study has produced some of the very best information on the reproductive biology of a long-lived vertebrate.

Ferraris, Joan D., and Stephen R. Palumbi, eds. *Molecular Zoology: Advances, Strategies, and Protocols.* New York: Wiley-Liss, 1996. Discusses the uses and methods of molecular biology as related to a wide variety of zoological topics, including reproduction.

Gould, James L., and Carol Grant Gould. *Sexual Selection*. 2d ed. New York: Scientific American Library, 1997. Summarizes some of the best research to date on the reproductive behaviors of a variety of animals. It is well written and well illustrated. Provides a small but select bibliography for more detailed reading.

Halliday, Tim. *Sexual Strategy*. Chicago: University of Chicago Press, 1980. Covers a range of topics related to sexual behavior in a clear, informative manner.

Krebs, John R., and Nicholas B. Davis. *An Introduction to Behavioural Ecology*. 4th ed. Sunderland, Mass.: Sinauer Associates, 1997. Covers various aspects of reproductive behavior and develops some of the theoretical approaches used by biologists.

Short, R. V., and E. Balaban, eds. *The Differences Between the Sexes*. New York: Cambridge University Press, 1994. A collection of symposium papers covering a wide variety of topics related to sex differences in animals.

Wiley, R. Haven. "Lek Mating System of the Sage Grouse." *Scientific American* 238 (May, 1978): 114-125. Gives additional information on the reproductive behavior of this fascinating species and on the general phenomenon of lekking behavior.

REPRODUCTIVE STRATEGIES

Types of animal science: Ecology, reproduction
Fields of study: Ethology, genetics, reproduction science

Reproductive strategies are a set of attributes involved in an organism's maximizing its reproductive success. Theoretical and experimental studies of reproductive strategies reveal why various reproductive patterns have evolved.

Principal Terms

BET-HEDGING: a reproductive strategy in which an organism reproduces on several occasions rather than focusing efforts on a single or few reproductive events

K STRATEGY: a reproductive strategy typified by low reproductive output; common in species living in areas having limited critical resources

LITTER SIZE: the number of offspring produced per birth; also referred to as "clutch size"

POPULATION DENSITY: number of individuals per unit of area, especially when it pertains to the group's reproductive potential

R STRATEGY: a reproductive strategy involving high reproductive output; found often in unstable or previously unoccupied areas

REPRODUCTIVE STRATEGY: a set of traits that characterizes the successful reproductive habits of a group of organisms

---

The term "reproductive strategies" is probably something of a misnomer. A strategy implies that an organism has had conscious forethought in determining how to proceed with its reproductive events, that some planning has occurred. With the exception of humans, who can plan aspects of their parenthood, this is virtually impossible. Having a reproductive strategy implies only that an organism has evolved a pattern that maximizes its success in the production of offspring.

The concept of reproductive strategies is closely related to that of natural selection. Natural selection results in the more fit individuals within a population, under a given set of environmental circumstances, being more likely to pass on their genes to future generations. By this process, the gene pool (genetic makeup) of the population is altered over time. An organism's fitness can be assessed by evaluating two key characteristics—survival and reproductive success. The organism's reproductive strategy, then, is that blend of traits enabling it to have the highest overall reproductive success. Application of the term "reproductive strategy" has also been extended to describe patterns beyond individual organisms: the population, species—even entire groups of similar species, such as carnivorous mammals.

Examination of reproductive strategies is part of the larger study of life-history evolution, which attempts to understand why a given set of basic traits has evolved. These traits include not only those pertaining to reproduction but also those such as body size and longevity. To consider a reproductive strategy appropriately, one must view it within the context of the organism's overall life history, precisely because these traits (particularly body size) often affect reproductive traits. One should also evaluate the role that the organism's ancestry plays in these processes. A species' evolutionary history can have a profound effect on its current attributes.

Traits and Behaviors

A reproductive strategy consists of a collection of basic reproductive traits, including litter, or "clutch," size (the number of offspring produced per birth), the number of litters per year, the number of litters in a lifetime, and the time between litters, gestation, or pregnancy length. The age of the mother's first pregnancy is also a consideration. Another trait is the degree of development of the young at birth. In different species, mothers put varying levels of time and energy into the production of either relatively immature, or altricial, offspring or offspring that are well developed, or precocial.

Reproductive strategies also consist of behavioral elements, such as the mating system and the amount of parental care. Mating systems include monogamy (in which one male is mated to one female) and polygamy (in which an individual of one sex is mated to more than one from the other). The type of polygamy when one male mates with several females is called polygyny; the reverse is known as polyandry.

Finally, physiological events such as those involved in ovulation (what happens when the egg or eggs are shed from the ovary) may also be used to characterize a reproductive strategy. Some mammals are spontaneous ovulators. Females shed their eggs during the reproductive cycle without any physical stimulation. Other mammalian species are induced ovulators—a female ovulates only after being physically stimulated by a male during copulation. These patterns, induced and spontaneous ovulation, may be regarded as alternate reproductive strategies, each enabling a type of species to reproduce successfully under certain conditions.

The overall effectiveness of a reproductive strategy is important to consider with respect to the relative success of the offspring (even those in future generations) in leaving their own descendants. A sound reproductive strategy results in increased fitness. An organism's fitness as it affects the population's gene pool may not be adequately assessed until several generations have passed.

The r and K Selection Model

The model of r and K selection is the most widely cited description of how certain reproductive traits are most effective under certain environmental conditions. To appreciate this model, an understanding of elementary population dynamics is needed. At the early stages of a population's growth, the rate of addition of new individuals (designated r) tends to be slow. After a sufficient number of individuals is reached, the growth rate can increase sharply, resulting in a boom phase. In most environments, however, unrestrained growth cannot continue indefinitely. Critical resources—food, water, and protective cover—become more scarce as the environment's carrying capacity (K) is approached. Carrying capacity is the maximal population size an area can support. When the population approaches this level, growth rate slows, as individuals now have fewer resources to convert into the production of new offspring.

This pattern is defined as density-dependent population growth—the density or number of individuals per area that influences its growth. This description of population dynamics is also referred to as logistic growth and was conceived by the Belgian mathematician Pierre-François Verhulst in the early nineteenth century. It has successfully described population growth in many species.

The r and K selection model was presented by Robert H. MacArthur and Edward O. Wilson in their influential book, *The Theory of Island Biogeography* (1967). They argued that in the early phase of a population's growth, individuals should evolve traits associated with high reproductive output. This enables them to take advantage of the relatively plentiful supply of food. The evolution of such traits is called r selection, after the high population growth rates occurring during this phase. They also suggested that, as the carrying capacity was approached, individuals would be selected that could adjust their lives to the now reduced circumstances. This process is called K selection. Such individuals should be more efficient in the conversion of food into offspring, producing fewer young than those living in the population's early phase. In a sense, a shift from productive to efficient individuals occurs as the population grows.

Other biologists, most notably Eric Pianka, have extended this concept of r and K selection to entire species rather than only to individuals at different stages of a population's growth. Highly variable or unpredictable climates commonly create situations in which population size is first diminished but then grows rapidly. Species commonly occurring in such environments are referred to as r strategists. Those living in more constant, relatively predictable climates are less likely to go through such an explosive growth phase. These species are considered to be K strategists. According to this scheme, an r strategist is characterized by small body size, rapid development, high rate of population increase, early age of first reproduction, a single or few reproductive events, and many small offspring. The K strategist has the

opposite qualities—large size, slower development, delayed age of first reproduction, repeated reproduction, and fewer, larger offspring.

Various combinations of r and K traits may occur in a species, and few are entirely r- or K-selected. Populations of the same species commonly occupy different habitats during their lives or across their geographic ranges. An organism might thus shift strategies in response to environmental changes—it may, however, be constrained by its phylogeny or ancestry in the degree to which its strategies are flexible.

Criticism of the Model

Because the r and K model of reproductive strategies seems to explain patterns observed in nature, it has become widely accepted. It has also met with considerable criticism. Charges against it include arguments that the logistic population-growth model (on which the r and K strategies model is based) is too simplistic. Another is that

Seals congregate in huge numbers for mating. (PhotoDisc)

cases of r and K selection have not been adequately tested. Mark Boyce, an ecologist, has persuasively argued that for the r and K model to be most useful it must be viewed as a model of how population density affects life-history traits. Within this framework, also called density-dependent natural selection, the concept of r and K selection remains true to the one that MacArthur and Wilson originally proposed. Boyce suggests that the ability of r and K selection to explain reproductive strategies will have the best chance of being realized when approached in this fashion.

In addition to the r and K model, there are many other ways of describing reproductive strategies. For example, some species, such as the Chinook salmon, are semelparous: They reproduce only once before dying. The alternate is to be iteroparous—having two or more reproductive events over the organism's life. If juvenile death rates are high, an individual might be better off reproducing on several occasions rather than only once. (This reproductive strategy is referred to as "bet-hedging.") Finally, it has also been useful to evaluate reproductive strategies based on the proportion of energy that goes into reproduction relative to that devoted to all other body functions. This mode of analysis addresses such considerations as reproductive effort and resource allocation.

Studying Reproductive Strategies

Initially, one who studies the reproductive strategy of an organism should attempt to characterize its reproduction fully. The sample examined must be representative of the population under consideration—it should account for the variability of the traits being measured. Studies can involve any of several approaches. Short-term laboratory studies can uncover some hard-to-observe features, but there is no substitute for long-term field research. By studying an organism's reproduction in nature, a biologist has the best chance of determining how its reproduction is shaped by an environment. If the research is performed over several seasons or years, patterns of variability can be

better understood. This is important in determining how the physical environment influences reproductive traits.

After data have been systematically collected, it might then be possible to characterize a reproductive strategy. Imagine that a mouse population becomes established in a previously uninhabited area and that the population has a high reproductive rate (it produces large litters). The young develop quickly and produce many young themselves. Because of this combination of circumstances, one might consider the reproductive strategy to be r-selected since the population has a high reproductive output in an unexploited area. Though the concept of r and K strategies is problematic, it still is common to typify a strategy as r- or K-selected based upon this approach.

Because a reproductive strategy needs to be seen as part of an organism's overall life history, however, other things should be measured to understand it fully. These may include the life span and population attributes such as survival patterns. Values should be taken for different age groups to characterize the population's strategy. Correlational analysis is a statistical procedure that is used to evaluate reproductive strategies. Through such a methodology, one assesses the degree of association between two variables or factors. This may involve relationships between two reproductive variables or between a reproductive and an environmental variable—for example, to determine whether there is a significant correlation between litter size and decreasing body size in mammals. If one were found to occur, the conclusion that smaller species typically have larger litters might be drawn, which is, in fact, true. Such an analysis enables the characterization of a change in reproductive strategy based on body size. Simply establishing a correlation does not prove that causation has occurred—it does not automatically mean that one factor is responsible for the expression of the other.

Multivariate statistical procedures are also used to analyze reproductive strategies. These allow the determination of how groups of reproductive traits are associated and of how they can be explained by several factors. One might determine that a certain bird species produces its greatest number of young, and that the young grow most rapidly, at northern locations having high snow levels. Such an approach is often needed in dealing with reproductive strategies—a combination of traits typically requires explanation.

Reproduction and Survivial

The characterization of an organism's reproductive strategy involves more than an understanding of reproductive traits. There is a successful process by which offspring are produced, and reproductive success is one of the two principal measures of fitness—the other is survival. Because a successful reproductive strategy ultimately results in high fitness, any discussion of these strategies bears directly on issues of natural selection and evolution.

An organism's reproductive strategy represents perhaps the most significant way in which the organism adapts to its environment. A successful reproductive strategy represents a successful mode of passing genes on to the next generation, so traits associated with a reproductive strategy are under intense natural selection pressure. If environmental conditions change, the original strategy may no longer be as successful. To the extent that an organism can shift its reproductive strategy as circumstances change, its genes will persist.

The study of reproductive strategies has helped scientists understand why certain modes of reproduction occur, based upon observations of a species itself and of its environment. An understanding of reproductive strategies may also be of some practical use. An organism's reproduction directly influences its population dynamics.

If an animal has small litters and is at an early age at first reproduction, its population should grow at a concomitantly high rate. These and other components of reproduction may strongly affect a species' population growth. A knowledge of how reproduction influences population dynamics can be important in wildlife management activities, which can range from strict preservation efforts to overseeing trophy hunting.

—*Samuel I. Zeveloff*

See also: Adaptations and their mechanisms; Courtship; Displays; Evolution: Historical perspective; Mating; Population Growth; Reproduction; Reproductive systems of female mammals; Reproductive systems of male mammals; Sex differences: Evolutionary origin; Tool use.

Bibliography

Austin, C. R., and R. V. Short, eds. *The Evolution of Reproduction.* New York: Cambridge University Press, 1976. This book is part of an excellent series on mammal reproduction designed for undergraduate biology students. It is a multiauthored volume addressing how evolution has shaped various aspects of mammal reproduction. The chapter "Selection for Reproductive Success" is important background reading on reproductive strategies. Bibliography, index, illustrations.

Boyce, Mark S. "Restitution of r- and K-Selection as a Model of Density-Dependent Natural Selection." *Annual Review of Ecology and Systematics* 15 (1984): 427-447. Boyce shows why r and K selection is most appropriate as a model of density-dependent natural selection. Some reading is challenging, but there are useful sections for beginners on the history and meaning of r and K selection. Bibliography.

Clutton-Brock, T. H., F. E. Guiness, and S. D. Albon. *Red Deer: Behavior and Ecology of Two Sexes.* Chicago: University of Chicago Press, 1982. This book details perhaps the best-known and most widely respected case study of a species' reproductive strategy. Of particular note is its discussion of how reproductive success in stags (males) is much more variable than that of the hinds (females). Contains a bibliography, an index, and charts.

Ferraris, Joan D., and Stephen R. Palumbi, eds. *Molecular Zoology: Advances, Strategies, and Protocols*. New York: Wiley-Liss, 1996. Discusses the uses and methods of molecular biology as related to a wide variety of zoological topics, including reproduction.

MacArthur, Robert H., and Edward O. Wilson. *The Theory of Island Biogeography*. Princeton, N.J.: Princeton University Press, 1967. Though rather advanced, this book is noteworthy as the foundation for the theory of r and K selection. All references to this theory should include this classic text. It also introduces another important topic of modern ecology, island biogeography, which establishes the principles of speciation and extinction on islands. Bibliography, index, and illustrations.

Pianka, Eric R. *Evolutionary Ecology*. 6th ed. New York: Harper & Row, 2000. This text provides an excellent overview of the ecological bases for evolutionary events. Its population ecology section covers various aspects of reproductive strategies including reproductive effort, r and K selection, and bird clutch-size patterns. Though a college-level book, much of it is accessible to those who have taken high school biology. Contains a bibliography, an index, and illustrations.

Wrangham, Richard W., W. C. McGrew, Frans B. M. De Waal, and Paul G. Heltne, eds. *Chimpanzee Cultures*. Cambridge, Mass.: Harvard University Press, 1996. A collection of essays on chimpanzee behavior, including reproductive strategies.

REPRODUCTIVE SYSTEM OF FEMALE MAMMALS

Type of animal science: Reproduction
Fields of study: Developmental biology, embryology, physiology

The mammalian female reproductive system is a complex system of organs and hormones that functions to produce offspring. The female also provides a protective environment in which the offspring develop until birth.

Principal Terms

ANTERIOR PITUITARY GLAND: the front portion of the pituitary gland, which is attached to the base of the brain; the source of luteinizing hormone (LH) and follicle-stimulating hormone (FSH)

ESTRUS CYCLE: hormonally controlled changes that make up the female reproductive cycle in most mammals; ovulation occurs during the estrus (heat) period

EXTERNAL GENITALS: the external reproductive parts of the female

GONAD: the primary reproductive organ (the ovary in females and the testes in males), which produces sex cells (gametes) and sex hormones

MENSTRUAL CYCLE: a series of regularly occurring changes in the uterine lining of a nonpregnant primate female that prepares the lining for pregnancy

OVARY: the female gonad, which produces ova and the hormones estrogen and progesterone

OVUM (pl. OVA): the female reproductive cell (gamete); a mature egg cell

UTERUS: the hollow, thick-walled organ in the pelvic region of females that is the site of menstruation, implantation, development of the fetus, and labor

---

The function of the mammalian female reproductive system, in cooperation with the male reproductive system, is to produce offspring. The role of the female is very complex: She must produce gametes (sex cells) called ova (singular, ovum) or eggs, provide the site for the combination of ova with sperm from the male (fertilization), and nourish and protect a developing fetus during pregnancy. She also must provide for the delivery of offspring from her body to the outside. These functions are carried out by a group of organs or structures and a number of chemicals called hormones. The major organs of the system include ovaries, which produce ova and hormones; uterine (Fallopian) tubes that transport the ovum and provide the site of fertilization; the uterus, which houses the developing offspring; the vagina, which receives the male penis and sperm during sexual intercourse and also functions as a birth canal; and the external genitals. Hormones important to the function of the female reproductive system include estrogen and progesterone from the ovaries, and follicle-stimulating hormone (FSH) and luteinizing hormone (LH) from the anterior pituitary gland. Release of FSH and LH is under control of chemicals called releasing factors, which are produced by a small region of the brain called the hypothalamus.

The Ovaries

The paired ovaries are the primary sex organ, or gonad, of the female. They are analogous to the testes in the male reproductive system and actually develop from the same tissue. The size and shape of the ovary depend on the age and size of the female and whether the female usually has a single offspring or several at one time. Before birth, small groups of cells called follicles are formed in each ovary. In the center of each follicle is a single large cell called an oocyte, which is able to mature into an ovum. In other words, all the oocytes a female will ever produce were already in place before she was born. The ova are special cells; they are formed by a process (meiosis) that results in a cell with only half of the chromosomes found in other body cells. The only other cells that divide by this process are those that form sperm in males. When an ovum and sperm unite, then, each cell contributes half the necessary chromosomes to make a new complete cell. This new cell, the first cell of an offspring, will have characteristics of each parent. The follicles develop in response to the hormone FSH from the anterior pituitary gland. A mature follicle releases its mature ovum through the wall of the ovary into the pelvic cavity. This process is called ovulation and is controlled by LH and FSH.

The ovary also produces the female sex hormones—estrogen and progesterone. Estrogen is produced by the maturing follicle cells. In addition to causing growth of the sex organs at puberty and stimulating growth of the uterine lining each month, estrogen is responsible for the appearance of female secondary sex characteristics. The follicle cells that are left behind following ovulation form a structure called the corpus luteum, which produces both estrogen and progesterone. The most important function of progesterone is to stimulate the lining of the uterus to complete its preparation for pregnancy.

The Accessory Organs
The rest of the internal structures of the reproductive system are called accessory organs. The first of these is a pair of uterine (Fallopian) tubes, or oviducts, which extend from each ovary into the uterus. They are frequently shaped like funnels, with fingerlike ends, called fimbria, that partially surround each ovary. Movements of the fimbria sweep the ovum and some attached cells into the uterine tube following ovulation. If fertilization is to take place, it will be in the uterine tube.

The uterus in most mammals consists of two horns and a body, although much variation occurs. Marsupials, mammals that have pouches, such as the opossum, have two completely separate uteri, each opening to the outside through a separate vagina. Rats, mice, and rabbits have uteri with two horns. Primates have simple uteri with no horns. The uterus has an amazing ability to expand during pregnancy. In all cases, the wall of the uterus is thick and muscular. This muscle layer, the myometrium, is able to contract rhythmically and powerfully to move the young down the birth canal and out of the mother's body during the birth process. The lining of the uterus is called the endometrium.

The uterus narrows down into a muscular, necklike region called the cervix. This structure acts like a valve to keep the opening into the uterus closed most of the time. This prevents bacteria and other harmful objects from entering. The final internal accessory structure is the vagina, a thin-walled muscular tube. The vagina surrounds the cervix of the uterus at its anterior end and extends to its opening to the outside of the body. It allows for childbirth, sexual intercourse, and, in primates, menstrual flow. The walls of the vagina normally touch one another and have deep folds that allow for stretching without damage. A thin fold of tissue called the hymen partially covers the external opening of the vagina. This structure has no function and varies considerably in different mammals.

The external structures of the female reproductive system are called the external genitals. These include the labia majora, labia minora, and clitoris. Two thick, hair-covered folds of skin, the labia majora, protect and enclose other structures. In some mammals, two smaller hair-free folds of skin are located within the labia majora. These folds, the labia minora, are very prominent in primates but small in most other mammals and completely lacking in some. They enclose a region called the vestibule. Within the vestibule are located the clitoris, the external opening from the urinary system (the urethra), and the external opening of the vagina. The clitoris is a small structure almost covered by the anterior ends of the labia minora. It is very sensitive, being richly supplied with nerve endings and blood vessels.

Sexual Maturity

Reproduction can occur only after females reach sexual maturity. In mammals, this requires the full development of the reproductive structures. The point at which maturity is attained is ultimately under the control of the hypothalamus, as it controls the release of FSH and LH. Many factors, such as attainment of a particular body weight, temperature, day length, and climate may influence the release of hormones.

After the female reaches maturity, reproductive activities are cyclic. In mammals, there are two different kinds of reproductive cycles. Most mammals have an estrus cycle in which females will mate with a male only if they are "in heat," which happens at certain restricted times. An estrus cycle is divided into stages: an inactive phase, called anestrus, which last for days, weeks, months, or years; proestrus, during which the follicles are developing; estrus, when ovulation occurs; and metestrus, when the ova are moving into the oviduct. Females mate, and may become very aggressive about finding a mate, during estrus only. Usually ovulation is triggered by LH from the pituitary gland. In some mammals, including cats and rabbits, ovulation does not occur until the animal mates. Many females signal that they are in estrus. The signals may be chemical—a special scent which carries for a long distance, for example—or visual. Chimpanzees, for example, develop pink swollen skin on the external genitals during estrus.

The Menstrual Cycle

Primates have a menstrual cycle instead of an estrus cycle. The menstrual cycle is coordinated by estrogen and progesterone from the ovary. These hormones, in turn, are controlled by FSH and LH from the anterior pituitary gland, so all the functions of the reproductive structure are coordinated and synchronized. The three stages of the menstrual cycle are the menses, proliferation, and secretion stages.

In menses, the thick endometrium is sloughed off and flows out of the uterus and out of the body through the vagina. This is also called the menstrual flow or menstrual period. The menstrual fluid consists of roughly equal parts of blood and other accumulated bodily fluids. In the proliferation phase, the endometrium again grows thick. Ovulation occurs in the ovary at the end of this stage, following a sudden increase in the release of LH from the anterior pituitary gland. In the secretion stage, the endometrium becomes very thick and cushiony and prepared to nourish a developing embryo if fertilization has occurred. If fertilization did not occur, the endometrial cells die and the cycle begins again. These stages are controlled by estrogen and progesterone from the ovary. Female primates will mate throughout the entire menstrual cycle.

Studying the Female Reproductive System

Detailed examination of individual reproductive tissues is performed using a variety of very thin tissue slices, various dyes and stains, and microscopes. Frequently, preserved tissue is used. Electron microscopes have made it possible to magnify single cells, or parts of cells, several thousand times to observe minute details of structure. Fresh tissue is also examined. It is possible to freeze a small tissue sample quickly, slice it very thin, and then expose the tissue to chemicals, which can add to researchers' understanding of the function of particular cells.

The study of reproductive hormones and the understanding of their function demand the use of many different methods. Again, much information comes from nonhuman studies. The procedures vary widely but a typical laboratory experiment may involve removing the ovaries from a female rat and then injecting small amounts of estrogen or progesterone to observe the response of the endometrium. It is also possible to use chemicals that block, or inhibit, one or more specific hormones. By creating an abnormal, controlled situation and observing the results, an understanding about the role of individual hormones within a complex interrelated system can be obtained.

It is also frequently necessary to measure how much hormone is present in some bodily fluid, either for research to gain understanding of normal function or for medical diagnosis. This is very difficult, as most hormones occur in very minute concentrations. Procedures called radioimmunoassay (RIA) techniques, introduced during the late 1950's and early 1960's, represent a very important advance in the study of hormone concentrations. These procedures, which use special recognition molecules for each hormone, plus certain hormones that have been purified and made radioactive, make it possible to measure levels of hormones as low as one trillionth of a gram (a picogram).

In a system as complex in its function as the female reproductive system, it is not surprising that information is obtained from a variety of sources. Each technique has made a contribution to an understanding of the whole system.

Understanding the Female Reproductive System

The reproductive system is unique among all body systems. It is the only system not called upon to function continuously for the well-being of the individual. It is nonfunctional during the early part of the female's life, then is activated by chemical messages from the anterior pituitary gland. Its primary function is not, after all, the well-being of one individual but rather the continued existence of the species. It is also unique in that it must interact with another individual, a male, in order to fulfill this function. Throughout the reproductive years, all the functions of the female reproductive systems are directed toward pregnancy.

—*Frances C. Garb*

See also: Breeding programs; Cloning of extinct or endangered species; Copulation; Endocrine systems of vertebrates; Fertilization; Gametogenesis; Hormones in mammals; Lactation; Pregnancy and prenatal development; Reproductive systems of male mammals; Sex differences: Evolutionary origins; Sexual development; Wildlife management.

Bibliography

Banks, William J. *Applied Veterinary Histology*. 3d ed. St. Louis: Mosby-Year Book, 1993. A good general reference on animal physiology and organ systems.

Berne, Robert M., and Matthew N. Levy. *Principles of Physiology*. 3d ed. St. Louis: C. V. Mosby, 2000. A clearly written textbook at the college level. Emphasis is on cellular mechanisms and regulation. The overview of reproductive function includes an excellent description of interactions between brain, pituitary gland, and gonads. The female system is covered separately. Excellent summary charts and figures. References provided for each chapter. Index.

Hayssen, Virginia, Ari van Tienhoven, and Ans van Tienhoven, eds. *Asdell's Patterns of Mammalian Reproduction: A Compendium of Species-Specific Data*. Ithaca, N.Y.: Comstock, 1993. An encyclopedic reference work listing information on the reproductive characteristics of mammalian species, including size and mass of newborns, litter size, age at sexual maturity, length of estrus, gestation, and lactation, and details of frequency and seasons of reproduction.

Hickman, Cleveland P., Jr., Larry Roberts, and Frances Hickman. *Integrated Principles of Zoology*. 11th ed. St. Louis: Mosby College Publishing, 2001. A comprehensive textbook, richly illustrated. Chapters on mammals, the reproductive process, and principles of development survey reproductive strategies. A very readable work. A historical perspective and references are included with each chapter. The book has an index, a glossary, and appendices.

Rijnberk, A., and H. W. De Vries. *Medical History and Physical Examination in Companion Animals*. Translated by B. E. Belshaw. Boston: Kluwer, 1995. A handbook for veterinarians, includes coverage of the reproductive systems of animals that are commonly found as pets.

REPRODUCTIVE SYSTEM OF MALE MAMMALS

Type of animal science: Reproduction
Field of study: Physiology

The male reproductive system is a group of organs that function together to produce sperm and carry them to the outside of the male body. Sperm carry the male's genetic information to his offspring. Continued survival of the species is assured by the proper functioning of the male and female reproductive systems.

Principal Terms

CHROMOSOME: a molecule of deoxyribonucleic acid (DNA) that contains a string of genes, which consist of coded information essential for all cell functions, including the creation of new life

EJACULATION: the process of expelling semen from the male body

ENDOCRINE GLANDS: glands that produce hormones and secrete them into the blood

ERECTION: the process of enlargement and stiffening of the penis because of increased blood volume within it

FERTILIZATION: the union of a sperm with an ovum; fertilization is the first step in the creation of a new individual

GAMETE: a reproductive cell—sperm in the male, ovum in the female; produced in the gonads, gametes contain a set of chromosomes from the adult male or female

GONAD: the organ responsible for production of gametes—the testis in the male, the ovary in the female

GONADOTROPIN: a hormone that stimulates the gonads to produce gametes and to secrete other hormones

SEMEN: the sperm-containing liquid that is expelled from the male body

The reproductive systems of all male mammals have the same basic design. The reproductive organs produce sperm and deliver it to the outside of the body. The sperm can be regarded as packages of chromosomes that the animal passes on to his offspring. Hormones and nerves control and coordinate the functions of the reproductive organs.

The Brain and Reproduction

Although the brain is not usually considered to be a component of the reproductive system, part of the brain is, in fact, essential to the function of the reproductive organs because of the hormones produced there. This part of the brain is the hypothalamus, a relatively small area that acts without conscious control. The hypothalamus is located in the lower middle of the brain; it contains centers that control eating, drinking, body temperature, and other essential functions.

Hypothalamic control over reproduction in the male is primarily by way of the hormone called gonadotropin-releasing hormone (GnRH). GnRH is released from the hypothalamus to enter blood vessels that carry it to the pituitary gland, a small gland suspended just below the hypothalamus. When GnRH arrives at the pituitary, it stimulates the pituitary to produce and release two more hormones, follicle-stimulating hormone (FSH) and luteinizing hormone (LH). FSH and LH in the male are identical to hormones of the same names in the female. The names of these hormones describe their functions in the female. Like other hormones, FSH and LH are released into the blood and circulate throughout the body. FSH and LH are called gonadotropin hormones: gonadotropin means "gonad stimulating." These are the hormones that stimulate the gonads (testes in the male, ovaries in the female) to produce sperm or eggs and to secrete gonadal hormones. In the male, the gonadal hormones are primarily testosterone and related hormones. There is a chain of hormonal commands, with GnRH from the hypothalamus at the top of the chain. GnRH stimulates the pituitary to secrete FSH and LH, which in turn stimulate the testes to produce sperm and testosterone.

In addition to the chain leading from the brain to the pituitary to the testes, information is sent back to the brain from the testes, a checks-and-balances system using principles of negative feedback to ensure that the hormones are produced in the appropriate quantities. If, for example, the hormone system gets slightly out of balance, leading to too much testosterone being produced, this excess of testosterone will be sensed by the hypothalamus. It will cause a temporary shutdown of GnRH production, leading to the system's correcting itself, because then a little less testosterone will be produced. If testosterone levels fall too low, the opposite will happen: GnRH, and then FSH and LH, and then finally testosterone, will all increase, again resulting in a correction of the original aberration. The hormonal system is a delicately balanced network that ensures the proper functioning of the testes.

The Testes

The testes are the sites of sperm production. Within the testes are hundreds of tiny tubes, the seminiferous tubules, that are responsible for sperm production. The sperm develop gradually from round cells called spermatogonia, which are located in the walls of the seminiferous tubules. As a sperm matures, it develops a long, whiplike tail attached to an oval head. The head of the sperm contains chromosomes, the genetic information of the male that will be passed on to his offspring. The sperm of some mammals can be distinguished under the microscope by characteristic differences in their appearance.

Between the seminiferous tubules are clusters of hormone-producing cells, the interstitial or Leydig cells. The Leydig cells produce testosterone and related hormones. Testosterone is essential for proper sperm development. In addition, testosterone is responsible for the development of male body features, including, in most species, a large muscle mass, and for the growth of the reproductive organs during puberty. In some animals, testosterone is also linked to aggressive and reproductive behaviors.

The testes of most mammals are located in the scrotum, a pouch of skin and muscle that is suspended outside the abdomen. In some animals, the testes may be withdrawn into the abdomen when the animal is startled or when it is not in the breeding condition.

The function of the scrotum is to maintain the temperature of the testes at a few degrees lower than average body temperature. The capability to maintain this temperature of the scrotum is rooted in the fact that the muscles within the scrotum are responsive to temperature. Under warm conditions, the scrotum relaxes, allowing the testes to move away from the body and lose heat. In cool temperatures, the opposite occurs: The scrotum wrinkles, pulling the testes closer to the body and allowing them to stay warmer. The reduced temperature maintained by the scrotum is mandatory for the production of normal fertile sperm. Fever or other situations that raise the temperature of the scrotum can interfere with sperm production, even resulting in temporary infertility. In a few large mammals (such as elephants, whales, and dolphins), the testes are not located within a scrotum, but instead occupy a position in the abdomen. It is not known why these species apparently do not require a temperature lower than that of the body for sperm production.

The Epididymus and Vas Deferens

Sperm are removed from the testes by a system of tubes that lead out of the body. Located next to each testis within the scrotum is the epididymis, a highly coiled tube that is directly connected to the seminiferous tubules of the testes. The epididymis serves two functions: sperm maturation and sperm storage. The epididymis is drained by a long, thin tube called the vas deferens, which carries sperm out of the scrotum through the inguinal canal into the abdomen. The inner end of the vas deferens is a widened area that may serve as a site of storage for mature sperm.

The vas deferens passes in a loop next to and under the bladder, the sac that stores urine until it can be removed from the body. Immediately beneath the bladder, the vas deferens is connected by a short tube, the ejaculatory duct, to the urethra. The urethra is the long, fairly straight tube that carries either urine from the bladder or sperm from the reproductive system. A valve located in the urethra below the bladder opens and closes to prevent sperm and urine from mixing, so that only one type of fluid is in the urethra at a time.

From their site of production in the testes, sperm pass through the epididymis, the vas deferens, the ejaculatory duct, then finally the urethra to the outside of the body. As sperm are expelled from the body along this route, they are mixed with seminal fluid to produce semen. Seminal fluid is secreted into the tubes by three sets of glands: the seminal vesicles, the prostate, and the bulbourethral (Cowper's) glands. The sperm never enter these glands; fluid is squeezed out of them into the tubes where the sperm are located.

The Penis

The penis is designed to deliver sperm to the female system. The penis consists of a long shaft with an enlarged head, the glans. The skin of the penis, especially the glans, is extremely sensitive to touch. In some species, the penis is withdrawn into a sheath of skin except during sexual arousal.

Internally, the penis contains the outer segment of the urethra, as well as erectile tissue. This erectile tissue is designed like a sponge. The many blood vessels in the erectile tissue are capable of greatly expanding and increasing the quantity of blood that they contain. When this happens, the erectile tissues swell, and the entire penis increases in length and width and becomes stiff. This process, called erection, is an involuntary reflex: It cannot be consciously prevented or caused. Erection can result from direct stimulation of the penis, as during sexual contact, or from erotic sights or sounds. In some animals, a bone within the penis, the baculum, assists in maintenance of the erection.

Continued sexual stimulation will eventually result in an ejaculation, with semen being forced out of the body by contractions of muscles in the fluid-producing glands and along the tube system. Ejaculation is coordinated by nerves that arise in the spinal cord. The normal volume of fluid ejaculated varies from species to species. In man, it is usually two to six milliliters; it may be up to one hundred milliliters in pigs. The ejaculate of most animals contains many millions of sperm per milliliter of fluid.

Studying Male Reproduction

The hormonal system that controls the male reproductive system is the subject of much research. The most straightforward type of hormonal research is simply descriptive: The scientist seeks to describe the levels of the reproductive hormones when the animal of interest is in different physiological states. The hormones can be measured in blood samples taken from the animals. Obtaining a blood sample from an experimental animal may pose difficulties: Some large animals may be difficult to restrain, and some small animals may not have veins large enough for an easy puncture. Another consideration is how often blood samples should be taken. Endocrinologists have become increasingly aware of the importance of the pattern of hormone release over time. In particular, it now appears that fluctuations in hormone levels within a time frame of minutes or hours may be critical in regulating the responses of hormone target sites. To obtain blood samples with such a high frequency, researchers usually implant a cannula into a vein of the animal; the cannula can be left in place for repeated blood sampling with very little stress to the animal.

Scientists interested in hormonal feedback may examine the roles of specific hormones by removing one of the endocrine glands from the system, and then examining the effects on the remaining hormones. For example, the testes (as the site of testosterone production) can be removed from an experimental animal. Blood samples after the surgery can then be assayed to determine the circulating levels of LH, FSH, and GnRH. The endocrine glands may be left in place, but the researcher may administer hormones either by injection or by implanting timed-release capsules containing the hormone under the skin. Then, blood samples taken from the animal will reveal how levels of hormones produced by the animal's own endocrine glands have changed as a result of the exposure to the added hormone.

A technique that is widely used to study males of seasonally breeding species is to subject the animals to carefully controlled environmental conditions. Length of exposure to light, temperature, rainfall, nutrients in the diet, and other factors can be controlled in the laboratory to determine which acts as the cue for seasonal reproduction. The status of the reproductive system can be determined by various methods. The testes can be measured: Inactive testes are usually smaller and lighter in weight. Hormone levels in the blood can be measured: Testosterone and other hormones may decrease when the animal is reproductively quiescent, or the male can be exposed to a female to determine whether he will show mating behavior.

For some types of research, the most revealing experiments may not use the entire animal (referred to as in vivo research), but will instead focus on specific organs. Living samples of organs can be maintained in the laboratory for such in vitro experimentation. For the in vitro approach, a small piece of living tissue can be removed from an animal and the cells suspended in a liquid that contains the nutrients necessary for their life. Under these isolated conditions, the scientists can investigate a number of areas such as which hormones tissues produce and the hormones that make the tissue itself respond. Organs respond optimally to a particular pattern of hormonal stimulation, and this is another important area of research. By combining the results of in vivo and in vitro experiments, scientists are able to piece together a complete picture of how the reproductive system functions.

Controlling Reproduction

Knowledge of how the male reproductive system functions has allowed scientists to develop technologies for controlling reproduction to enhance or curtail fertility in domestic animals. Knowledge of male reproductive physiology has been applied to the management of domestic breeding populations. Hormone measurements and sperm counts can be used to determine the optimum age at which to begin breeding young stock. Techniques for collecting and storing semen can be combined with artificial insemination of females to increase the number of offspring produced by valuable males, thus resulting in improvement of the population. These methods are particularly valuable to breeders of large animals because maintaining large numbers of males of these species (such as stallions and bulls) can be costly and difficult because of the aggressive behaviors that these males may exhibit.

The study of seasonal breeding has also been of value in agriculture. Scientists now know much about the environmental conditions that are responsible for promoting reproductive activity in many domestic species. Farmers can apply this knowledge to their breeding stock to increase production throughout the year. Another area in which reproductive studies are of vital importance is the enhancement of the breeding of captive animals that are endangered in the wild. Zoos, once considered merely spectacles for entertainment, are now seen by many as the last hope of saving many species on the verge of extinction. Knowledge of the conditions necessary for successful breeding of exotic animals will help to increase their numbers and, perhaps, to return them to the wild.

—*Marcia Watson-Whitmyre*

See also: Breeding programs; Cloning of extinct or endangered species; Copulation; Endocrine systems of vertebrates; Fertilization; Gametogenesis; Hormones in mammals; Hydrostatic skeletons; Lactation; Pregnancy and prenatal development; Reproductive systems of female mammals; Sex differences: Evolutionary origins; Sexual development; Wildlife management.

Bibliography

Carter, Carol Sue, I. Izja Lederhendler, and Brian Kirkpatrick, eds. *The Integrative Neurobiology of Affiliation.* Cambridge, Mass.: MIT Press, 1999. A collection of papers from a symposium, with coverage of the neurobiology of affiliation associated with reproductive behaviors.

Knobil, Ernst, and Jimmy D. Neill, eds. *The Physiology of Reproduction.* 2 vols. 2d ed. New York: Raven Press, 1994. Covers the entire scope of mammalian reproduction, looking at both human and nonhuman species.

Marshall Graves, J. A., R. M. Hope, and D. W. Cooper, eds. *Mammals From Pouches and Eggs: Genetics, Breeding, and Evolution of Marsupials and Monotremes.* Canberra, Australia: CSIRO, 1990. A collection of papers on the often-overlooked topic of marsupial and monotreme reproduction.

Nalbandov, A. V. *Reproductive Physiology of Mammals and Birds: The Comparative Physiology of Domestic and Laboratory Animals and Man.* San Francisco: W. H. Freeman, 1976. This book is written for beginning students of reproductive physiology, and it focuses on the essentials. As the title implies, the author takes a comparative perspective, surveying reproduction in different species.

Setchell, B. P. *The Mammalian Testis.* Ithaca, N.Y.: Cornell University Press, 1978. Although written for a scientific audience, the work is noteworthy because of its completeness. Contains detailed information about all aspects of the testis, scrotum, spermatogenesis, and hormonal control, including historical perspectives of the research in male reproduction. The book is also useful because it covers all mammals, not only humans, so the reader is able to appreciate similarities and differences among the mammals.

Van Tienhoven, Ari. *Reproductive Physiology of Vertebrates.* 2d ed. Ithaca, N.Y.: Cornell University Press, 1983. This is a good reference for anyone interested in reproduction among nonmammalian vertebrates. The text covers many subjects not often found in one volume, including effects of the environment, immunological aspects of reproduction, puberty, and animals with intersex characteristics.

REPTILES

Type of animal science: Classification
Fields of study: Herpetology, systematics (taxonomy), zoology

Living reptiles are vertebrates that lay eggs or bear live young. Included in this class are turtles, the tuatara, lizards and snakes, and crocodilians. The study of reptiles provides insight into a group of often very successful animals and into the characteristics of their ancestors, including groups that gave rise to birds and mammals.

Principal Terms

ANAPSIDA: a group of reptiles in which the temporal region of the skull lacks openings

CHELONIA (TESTUDINES): a living order of reptiles composed of turtles and tortoises

CLEIDOIC EGG: a shelled egg equipped with internal membranes that make terrestrial reproduction possible

CROCODYLIA: a living order of reptiles that includes crocodiles and alligators

DIAPSIDA: a group of reptiles in which the temporal region of the skull is characterized by two openings

EURYAPSIDA: an extinct group of reptiles in which the temporal region of the skull is characterized by a single opening situated high on the side of the skull

RHYNCHOCEPHALIA: a living order of reptiles represented by a single species, the tuatara

SQUAMATA: a living order of reptiles composed of lizards and snakes

SYNAPSIDA: an extinct group of reptiles in which the temporal region of the skull is characterized by a single opening; this group gave rise to mammals

VENOM: a toxic substance that must be injected in order to elicit damaging effects

---

Reptiles are a class of vertebrates characterized by their ability to produce cleidoic eggs (which are similar to bird eggs and were their evolutionary precursors). The development of this egg, protected by an impervious shell, was a historic step, as it allowed animals to exploit terrestrial habitats. The egg and the reptiles' dry, horny scales differentiate all living reptiles from amphibians. Skeletal features (single bones for sound conduction in the middle ear and jaws composed of several bones) and the lack of feathers and hair differentiate reptiles from birds and mammals, respectively. One must realize, however, that it is the combination of features that characterizes reptiles and that, since soft tissues such as reproductive tracts and skin do not fossilize well, distinguishing extinct forms from other closely related vertebrate groups is often dependent on a single characteristic and may not be precise. For example, *Archaeopteryx*, a primitive bird, would have been classified as a reptile had not feathers been adventitiously preserved.

Early Reptiles, Turtles, and Crocodiles

Reptiles arose from amphibians roughly 315 million years ago, during the Carboniferous period. These "stem-reptiles" gave rise to all other groups. The "cheek," or temporal region, is very important in reptilian classification. In early forms, the region was solid (lacked openings). These forms are placed in the subclass Anapsida (without recesses or openings). Along with some of these earliest reptilian fossils are some from a distinctly different group, with a single temporal opening. These animals are placed in the subclass Synapsida. Referred to as "mammal-like reptiles," they gave rise to mammals prior to their extinction. During the Permian period (280-215 million years ago), another reptilian group appeared. This group, characterized by two temporal recesses, is placed in the subclass Diapsida (two openings), to which the majority of reptiles, living and extinct, belong. Another subclass, the Euryapsida, with a single opening high on the side of the skull, became extinct near the end of the Mesozoic era (approximately 175 million years ago). It included large marine (sea-dwelling) forms, such as fishlike ichthyosaurs and flat-bodied, long-necked plesiosaurs.

Anapsids include not only the stem-reptiles, but also a surviving order, the Chelonia or Testudines, composed of turtles and tortoises. Primitive turtles appear almost fully formed in the fossil record, and their relationship to the first reptiles is uncertain. Their principal feature is a shell, composed of flattened ribs fused to layers of bony tissue, usually covered by large, flat scales called scutes. The upper portion, to which vertebrae are fused, is the carapace; the underside is the plastron. They are connected by bridges. Shoulders and hips have been modified and are located within the rib cage, a unique arrangement. All lay eggs on land. The order contains almost 250 species in 75 genera and 13 families. Although it is easy to fall into the trap of thinking that "if it has a shell, it is a turtle" and "a turtle is a turtle is a turtle," the group is quite diverse. The most common (and probably primitive) body plan is associated with marsh dwellers. Characterized by somewhat flattened shells (streamlined for locomotion through water) and webbed feet for propulsion, these surprisingly agile swimmers include the familiar sliders that drop into the water from logs or the bank when approached too closely. Though adept in water, they are no match for sea turtles; these animals have reduced, flattened shells to minimize resistance and limbs modified into paddles with which they fly through the water using movements almost identical to those of birds. Though they lay their eggs on land, they are prac-

tically helpless there. Graceful in the water, they often swim considerable distances, guided by a very effective navigational sense. Sea turtles include the largest living reptiles, the leatherbacks, which may exceed a ton in weight. Other turtles are bottom dwellers; ambush predators or scavengers, they lie in wait or slowly crawl along bottoms of ponds and streams. They possess webbed feet and flat, often rough shells. Algae growing on their shells serves as camouflage, hiding them from prey and predators. A final body plan characterizes land turtles, which include tortoises. A firm, generally high-domed shell minimizes surface area through which water might be lost (critical in terrestrial animals) and also resists attacks by predators to which a land dweller is exposed. Some very large forms are quite long-lived; one documented record exceeds 150 years.

The term "dinosaur" is often used to describe any large, extinct reptile, including mammal-like synapsids, marine euryapsids, and diapsids such as flying reptiles and some large lizards. Used properly, however, it refers to only one group of diapsids, the Archosauria, or "ruling reptiles," which gave rise to birds. Other close relatives are in the living order Crocodylia, which now contains only twenty-two species in eight genera and three families. These animals share some very advanced features that cause some authorities to place them in a distinct class, the Dinosauria. All have fully partitioned hearts, allowing separate circuits for oxygenated blood to be carried to the body and deoxygenated blood to the lungs. Recent evidence indicates that many dinosaurs may have possessed birdlike capabilities for temperature regulation, allowing levels of activity beyond that of other reptiles. Modern crocodilians, some exceeding 7.5 meters in length, are quite aquatic and feed principally on fish or animals ambushed as they drink. They are restricted to tropical and subtropical zones. All are egg-layers.

Lepidosauria and Squamata

All other living reptiles are in the diapsid group Lepidosauria (scaly reptiles). The tuatara, a lizardlike reptile up to sixty centimeters long, is the only surviving member of the order Rhynchocephalia, a diverse assembly that coexisted with dinosaurs. Restricted to roughly thirty small islands off the coast of New Zealand and well adapted to a cool climate, it demonstrates considerable longevity (approximately 120 years) but also has a low reproductive rate. It feeds primarily on insects and eggs and the young of sea birds or other tuataras, with which it shares burrows.

Arguably the most successful reptilian group, extinct or living, is the order Squamata. All are equipped with efficient vomeronasal organs with which they "smell" by sampling air or substrate with their tongues. They may lay eggs or give live birth. There are approximately 3,750 species of lizards, suborder Sauria or Lacertilia, in almost 400 genera and some 16 families. They are found from north of the Arctic Circle to the southern tip of South America; such range is perhaps attributable to their exceedingly efficient capacity for thermoregulation. Some lizards at below-freezing temperatures may maintain body temperatures near 20 degrees Celsius. It is difficult to characterize lizards because of their tremendous diversity. Some are legless—an adaptation for burrowing or living in dense grass. Others have the capacity for gliding. Feet may be modified for running (many run on their hind legs, in one instance so rapidly that the lizard can run on water for considerable distances) or climbing, with digits equipped with claws and/or adhesive pads. Teeth may be used for grasping, cutting, or crushing food. Tails may be prehensile (capable of grasping), may be used for balancing while climbing or running, may come equipped with spines or knobs for defense, may be capable of fat storage, and may even break off if grasped by a predator (often to regenerate rapidly). Some lizards are excellent swimmers; the marine iguana of the Galápagos Islands feeds primarily on seaweed. Two species are venomous. The smallest lizards are only a few centimeters long; the largest may exceed three meters.

Snakes

Snakes, suborder Serpentes or Ophidia, are distinct from lizards in that they lack external ears, eyelids, and limbs (at least one of which most lizards have—worm lizards lack these features and have been treated as snakes or even placed into a separate suborder). Despite these constraints, snakes are quite diverse: There are almost 2,400 species in more than 400 genera and 11 families, ranging from the Arctic Circle to the southern tip of South America. Leglessness was a primitive adaptation for burrowing, but modern snakes also swim, crawl, climb, and, in one case, even glide adeptly. The elongated body form that accompanies limblessness requires that paired internal organs be arranged longitudinally, with one often degenerating. Digestive tracts are short and straight, resulting in all snakes being carnivorous, as meat is more easily digested by snakes than plant material. Locomotion is surprisingly varied. The familiar serpentine movement works well either on land or in water, but heavy-bodied snakes often use rectilinear locomotion, pushing their bodies in straight lines by alternately raising and retracting their large belly scales. In tight quarters, snakes anchor their necks and pull their bodies forward or, alternately, push off using anchored tails (many are equipped with spines for this purpose). A few snakes, especially on loose substrates such as sand, sidewind, pushing down to prevent sliding while lifting loops of their bodies laterally.

Snakes swallow their prey whole. Jaws, which are loosely attached to the skull and to each other, alternately slide forward and pull back on food with recurved teeth. Prey may be swallowed alive, killed by constriction, or killed with venom. Venom injection may accompany a bite or may be facilitated by special fangs in the rear or front of each jaw. In vipers, the bones to which fangs are attached rotate, so that very long fangs can be folded back when not in use. Burrowing snakes tend to be slender and small, with smooth scales and rigid heads. Aquatic snakes are usually stout, with rough scales to prevent slipping through water. Arboreal snakes (climbers) are often extremely slender. Active hunters are usually more slender than ambush predators, which eat more rarely but can consume much larger items. Some snakes have temperature-sensitive pits with which they find prey in the dark. Sizes range from a few centimeters to almost ten meters.

Studying Reptiles

Methods used to study reptiles are determined by the nature of the particular investigation. Historical studies rely on paleontological methods. The discovery of fossils, followed by recovery, preservation, reconstruction, and analysis, leads to an understanding of the structure and function of prehistoric animals and provides information about both how they lived and conditions in which they existed. Comparisons, especially of structures, with other fossils and with animals alive today constitute much of the field of comparative anatomy and lend insights into relationships between various living and extinct forms. These studies, in turn, lead into the discipline of systematics, which attempts to reconstruct relationships and build classification schemes accordingly. The actual naming of various groups is called taxonomy. Since fossil records are typically incomplete, however, other methods must be used to establish fully the nature of relationships.

Similarities and differences between living forms may be established on the basis of detailed anatomical studies or various biochemical techniques. In the latter case, the analysis of the molecular structure of the deoxyribonucleic acid (DNA) and proteins produced by different species (or even by different populations of the same species) allows determinations of how closely related certain forms may be. These studies also have considerable evolutionary implications, providing insights not only into methods that might have

The Komodo dragon is the largest living lizard species, with strong jaws that can sever the muscles of most vertebrates. (Corbis)

resulted in evolutionary changes among reptiles but also into the processes that were responsible for the origins of birds and mammals. One phenomenon that was first discovered in reptiles is parthenogenesis, the development of an individual from an unfertilized egg, a process that leads to all-female populations. In lizards these often result from hybridization between two species, the offspring of which are distinctive. Analysis of mitochondrial DNA, which is passed to descendants through the egg (never the sperm), allowed determination of which hybridizing parental species was maternal.

Another field of study that uses biochemical techniques and that focuses on reptiles concerns the venoms produced by some snakes and two lizard species. Knowing the composition of venoms is important in determining their effectiveness as devices to kill prey, but it is also significant in that some substances in these venoms have been shown to possess functional traits that have important medical implications.

Reptiles have also been used widely in ecological and behavioral studies. Methods include both field and laboratory techniques. Studies in natural situations often involve extensive observations and, as such, require organisms that are easily observed. Laboratory studies, in which environmental factors are often simulated and then modified, demand subjects that are small and readily maintained in captivity. Many of these studies also have physiological implications. Studies monitoring such factors as body temperatures, food intake, and foraging strategies often rely on reptiles, especially lizards, as they are ideally suited to these observational and experimental investigations. Considerable work in reproductive physiology also relies on reptiles; their eggs are accessible, and they demonstrate developmental patterns similar to those in birds and mammals. This aptitude as a subject for studies has also resulted in reptiles becoming the focus of many biogeographical studies, especially on islands, where (as in deserts) they often dominate the fauna. For many of the same reasons, environmental studies of endangered species or altered habitats frequently use reptiles as models.

Reptiles and Other Animals

The study of reptiles not only increases scientists' knowledge of this fascinating group of animals, but also has many other applications. Historically, reptiles were the first group of fully terrestrial vertebrates; they dominated the earth for many millions of years and gave rise to both birds and mammals. Thus, studies of fossil forms, with additional insights from investigations of living species, provide insights into the conditions that prevailed on the earth during prehistoric times. They also lead to theories regarding relationships between animal groups and lend understanding to the origins and nature of early birds and mammals. Studies of this type need not be restricted to arcane facts relevant only to times long past; they are also significant in understanding long-term biological processes, such as those that reflect climatic cycles and periods of mass extinction (both problems of considerable interest today). Also, since many

reptilian groups are sufficiently old to predate the breakup of Pangaea (the single landmass that existed historically and which has since broken up into the continental areas of today), or at least its subsequent parts, applications can be made to the areas of biogeography and even geology.

Physiological investigations of reptiles have been invaluable in developmental and reproductive studies and in increasing knowledge of how animals interact with their environments. Lizards, especially, have been widely studied in regard to their ability to thermoregulate effectively using environmental sources of heat. These studies have numerous applications to broader investigations of homeostasis and adaptations to cold and hot environments by many animals (including man). Lizards also have been widely used as models in behavioral and ecological studies. Especially in the tropics, they are abundant, diverse, easily observed, and often remarkably well-adapted to their environment. Like birds, much of their behavior is quite stereotypical, that is, innate and consistent. As a result, the recognition of patterns is much easier than it is in secretive mammals, for example, where the problems are further magnified by frequent modifications of instinctive mechanisms by learned behaviors. This same visibility and ease of observation lends itself to ecological investigations. Lizards have been more widely utilized in niche partitioning studies than has any other animal. Niche partitioning studies seek to investigate how limited resources are used by animal communities. They often center on the hypothesis that food habits are critical, but microhabitat preferences, activity cycles, and other aspects of environmental impact are also involved.

—*Robert Powell*

See also: *Allosaurus; Apatosaurus; Archaeopteryx; Brachiosaurus;* Chameleons; Circulatory systems of vertebrates; Cold-blooded animals; Crocodiles; Dinosaurs; Evolution: Historical perspective; Extinction; Fish; Hadrosaurs; Ichthyosaurs; Lizards; Pterosaurs; Sauropods; Scales; Snakes; Systematics; Stegosaurs; *Triceratops;* Turtles and tortoises; *Tyrannosaurus;* Thermoregulation; Velociraptors.

Bibliography

Bakker, Robert T. *The Dinosaur Heresies: New Theories Unlocking the Mystery of the Dinosaurs and Their Extinction*. New York: Kensington, 1986. A marvelously entertaining book that treats dinosaurs as homeotherms. Well-written text and good line drawings shed a whole new light on dinosaurs as active, dynamic animals. Provides more information than most other dinosaur books that glut the market. Bibliography. Written for a general audience.

Conant, Roger. *A Field Guide to Reptiles and Amphibians of Eastern and Central North America*. 3d ed. Boston: Houghton Mifflin, 1998. Along with Stebbins's guide to western forms, the best of many identification manuals to North American amphibians and reptiles. Numerous illustrations indicate diagnostic characteristics. Some discussion of biological aspects of different groups. Glossary and guide to additional references. Written for the general reader.

Halliday, Tim R., and Kraig Adler. *The Encyclopedia of Reptiles and Amphibians*. New York: Facts on File, 1998. An excellent combination of technical competence and a writing level suitable for the interested layperson. Well-illustrated, the diagrams and photographs alone justify the cost of the book. Coverage includes general aspects of biology, ancestry, relationships, and taxonomy of each major group. Probably the best single reference for an amateur herpetologist. Glossary and bibliography.

Hickman, Cleveland P., Larry S. Roberts, and Frances M. Hickman. *Integrated Principles of Zoology*. 10th ed. St. Louis: Times Mirror/Mosby, 2001. A textbook written with exceptional clarity, one of the best in a field of many good general zoology books. General coverage of all aspects of reptilian biology. Numerous photographs and clear diagrams illustrate representative types and examples of pertinent concepts. Selected references are given at the end of the chapter. Glossary. College level, but suitable for advanced high school students.

Kirshner, David S. *Reptiles and Amphibians*. Washington, D.C.: National Geographic Society, 1996. A comprehensive and detailed survey of reptiles and amphibians, with chapters all written by experts. Illustrated with high-quality photographs and paintings.

Spellerberg, Ian F. *Mysteries and Marvels of the Reptile World*. New York: Scholastic, 1995. Written for children aged nine to twelve. Focuses on odd facts and the more unusual aspects of reptile life.

Stebbins, Robert C. *A Field Guide to Western Reptiles and Amphibians*. 2d ed. Boston: Houghton Mifflin, 1985. Along with Roger Conant's guide to eastern forms (above), the best of many identification manuals to North American amphibians and reptiles. Numerous illustrations indicate diagnostic characteristics. Some discussion of biological aspects of different groups. Glossary and guide to additional references. Written for the general public.

Vitt, Laurie J., Janalee P. Caldwell, and George R. Zug. *Herpetology: An Introductory Biology of Reptiles and Amphibians*. 2d ed. San Diego, Calif.: Academic Press, 2001. An outstanding introductory college-level textbook. Covers evolution, classification, development, reproduction, population, and environmental issues, and introduces the history of the field.

RESPIRATION AND LOW OXYGEN

Type of animal science: Physiology
Fields of study: Anatomy, biochemistry, physiology

Nearly all animals require free oxygen in order to convert food to living tissue and energy, yet many animals face periods when oxygen supply is diminished. Adaptation to low oxygen involves developing mechanisms to compensate for the reduced oxygen supply.

Principal Terms

HYPERVENTILATION: an increase in the flow of air or water past the site of gas exchange (lung, gill, or skin)

HYPOXIA: from two Latin words, *hypo* and *oxia*, meaning "low oxygen"

METABOLISM: the sum of all of the reactions that take place in an animal allowing it to move, grow, and carry out body functions

RESPIRATORY PIGMENT: a protein that "supercharges" the body fluid (blood) with oxygen; the oxygen can bind to the pigment and then be released

RESPIRATORY SURFACE: the gill, lung, or skin site at which oxygen is taken up from the air or water into the animal, with the release of carbon dioxide at the same time and site

SYSTEMIC: referring to a group of organs that function in a coordinated and controlled manner to accomplish some end, such as respiration

VENTILATION: the movement, often by pumping, of air or water to the site of gas exchange; commonly thought of as breathing

---

Adaptation to low oxygen refers to a number of different changes in metabolism or body function, or both, that animals use to survive low-oxygen conditions. Low-oxygen conditions mean a reduction in the amount of oxygen available in relation to the need or demand for oxygen by the cells, or tissues. Low oxygen, or hypoxia, therefore, can result from either a decrease in the supply at constant demand, or an increase in demand at a constant supply. The former, a reduction in oxygen supply, is the focus of the present discussion. Low oxygen resulting from increased oxygen demand usually is referred to as tissue hypoxia and is discussed only briefly here.

Oxygen is required by animal cells in order to produce energy used for growing, moving around, or simply maintaining normal body functions. At times when less oxygen is available, animals must either move to some place where there is sufficient oxygen, or change some internal function or process. A change in internal function or process is an adaptation that allows the animal to live with less oxygen or that will be a means of keeping the supply of oxygen to the tissues great enough to meet the needs of the cell.

External or environmental hypoxia results from one of two conditions: a greater utilization of oxygen by plants and animals than can be renewed by natural processes, or a lower density of air (at high elevations). It is necessary that animals cope with a decrease in oxygen in some way, in part because of the consequences of a decrease in internal oxygen supply. If oxygen supply to the tissues and cells falls, then the functions that require oxygen will fail or at least be reduced. The functions that fail include the "maintenance functions" of a cell, apart from growing or producing specialized chemicals. The types of low-oxygen conditions that have received the most attention from researchers are high altitude, diving by air breathers, and oxygen depletion in water. Some of these are temporary; others, such as high altitude, can last for a lifetime for animals that do not migrate.

Increase in Ventilation

One common adaptation to hypoxia is an increase in ventilation—the amount of air or water that the animal breathes. This increase is referred to as hyperventilation, and it makes up for the reduced amount of oxygen in the air or water by breathing a greater volume. This response is especially common in mammals that move to high altitude, fish and crabs, and some other water-breathing animals. A simple mathematical example will show how the response is effective. If an animal normally breathes one liter of air per minute and removes half the oxygen (air contains 209 milliliters of oxygen per liter), then it is taking in 104.5 milliliters of oxygen per minute. If the amount of oxygen in the air falls to only 104.5 milliliters per liter, and the animal still uses half of that (52.25 milliliters), then it must breathe two liters of air per minute to keep taking up 104.5 milliliters of oxygen per minute. Such an increase in ventilation can be accomplished by an increase in the frequency of the respiratory pump, or by increasing the volume of water or air that is moved with each "breath." For this response to occur, the nervous system must sense the reduction in oxygen and provide a nerve impulse to the brain, which then stimulates the ventilatory pump(s) to increase activity.

It may seem that this adaptation is all that would be needed for animals to survive hypoxia, but there are some limitations to this adaptation. First, hyperventilation causes increased muscular activity and an increase in oxygen used to move the respiratory muscles. The greater ventilation volume is a benefit, but the cost is a greater demand for oxygen. For animals that breathe air, the increase is rather small, but for animals that breathe water, the increase in the muscular activity causes a substantial increase in the oxygen used to pump the water, so that the "cost" may be greater than the "benefit" when the oxygen falls to

low levels. Another problem exists for air breathers. Air breathers are in danger of losing water in the air that is exhaled (desert animals, such as camels, have elaborate mechanisms to conserve this respiratory water loss). Hyperventilation increases the water loss and requires the animal to drink more water. A final problem for both air and water breathers is that carbon dioxide is lost from the same respiratory surface where oxygen is taken up. Hyperventilation thus increases the loss of carbon dioxide, changing the chemical balance of the body as a whole.

Blood Flow and Oxygen Delivery

In response to an internal hypoxia, many animals also exhibit an increase in the flow of blood to the tissues. This response is similar to that described for the ventilation system. An increased rate of flow compensates for a smaller amount of oxygen delivered for a given volume of blood (or respiratory medium in the case of ventilation). As with ventilation, blood flow can be elevated by increasing heart rate or by increasing the volume of blood pumped with each beat. There are numerous limitations to the effectiveness of this response, and it is only short term. The limitations center on the critical role of blood flow and blood pressure in the function of other systemic body functions. An excellent example is how kidney filtration rate increases with blood pressure.

Long-term adaptations to low oxygen often increase the ability of systemic respiratory functions to maintain oxygen delivery to the tissues. In the case of internal oxygen transport, this can be accomplished by increasing the amount of oxygen carried by the blood. A higher concentration of respiratory pigment accomplishes this, increasing either the number of red blood cells or the concentration of respiratory pigment in the blood. This adaptation requires the synthesis of new proteins and possibly new cells. Not surprisingly, many days or even weeks may be needed to increase respiratory pigment levels. Another way in which oxygen transport by the respiratory pigment may be improved is by increasing the concentration of a chemical that affects oxygen binding. This adaptation requires a change in metabolism and is discussed below.

Metabolic Alteration

One final type of adaptation to low oxygen is an alteration in the basic metabolism of the animal. Metabolic changes can take one of several forms. First, a simple reduction in metabolism will lower the need and demand for oxygen by the cells. To be effective, this must occur before the oxygen has been exhausted, so as not to impair normal functions. A few animals show this type of adaptation, which is thought to result from the metabolic reactions being limited by the availability of oxygen. Second, the chemical reactions involved in metabolic pathways (a series of chemical reactions) may be altered in low-oxygen conditions so that different reactions take place to maintain energy production. The nature of these adaptations is that an alternative metabolic pathway requires different enzymes and perhaps different chemicals in the reactions. Last, a metabolic adaptation may yield a product that has an enhancing effect on oxygen transport. An enhancement of oxygen transport occurs when certain chemicals increase the ability of the respiratory pigment to bind oxygen or cause the respiratory pigment to bind oxygen at lower oxygen levels; this is called an increase in oxygen affinity. The change in metabolism at low oxygen thus results in an improvement in the supply of oxygen to the tissues. This response is seen in both vertebrates and invertebrates.

An excellent example of an animal that shows nearly all of the adaptations to low oxygen is the blue crab, the common commercial crab found throughout the Gulf Coast of North America and on the East Coast from Florida to New York. To compensate for low oxygen, the blue crab increases the flow of water over the gills, thereby keeping the amount of oxygen that actually passes over the gills nearly constant. Blue crabs also increase the heart rate, thereby increasing the rate of blood flow in the gills and to the muscles and organs of the body. This increase helps maintain the oxygen supply. If the period of hypoxia is brief, only a few hours, then these reactions may

be all that is required for the animal to survive. If, however, the hypoxia continues for days or even weeks, then other responses come into play. There are changes in metabolism and in the way in which oxygen is transported in the body. Metabolism actually decreases so that less oxygen is needed by the animal. When that happens, then there must be some activity, such as swimming, that the animal gives up for lack of energy. The other change is an improvement in the way oxygen is transported to the tissues by the respiratory pigment, hemocyanin—a certain kind of protein, dissolved in the blood, that binds oxygen at the gills and can release the oxygen at the tissues where it is used by the cells. This improvement takes the form of increasing the level of a chemical in the blood that changes the binding of oxygen to hemocyanin. The hemocyanin then works as well when the animal is in hypoxic water. In addition, the crab can, and does, make the hemocyanin in a new form, so that it works better in the hypoxic conditions.

Understanding the Respiratory and Circulatory Systems

Adaptation to low oxygen (either high demand or reduced supply) has been studied with the idea of understanding the functional capabilities of the respiratory and circulatory systems that supply oxygen to the tissues. Many different experimental protocols and procedures are used to assess the balance between oxygen uptake and oxygen demand when the external supply of oxygen is limited and demand remains constant. One approach to the study of low-oxygen conditions has been to compare animals that live at sea level in high-oxygen habitats with those living in habitats in which oxygen levels are low. A comparison of water breathers and air breathers is, strictly speaking, within the realm of consideration. Water holds much less oxygen than air, and is therefore a low-oxygen condition. Freshwater at room temperature contains about 0.8 milliliter of oxygen per 100 milliliters of water; there are 20.9 milliliters of oxygen in the same volume of air. To obtain the same amount of oxygen, an animal must thus take all the oxygen from either 2,600 milliliters of water or 100 milliliters of air. Consequently, air breathers have much lower ventilation rates, at the same temperature, than do water breathers of the same size. The lower ventilation rate of air breathers is considered functional by reducing the loss of water from the respiratory surface.

Such adaptation is principally evolutionary and involves the transition from water breathing to air breathing in the evolutionary transition to land. There are a great many morphological as well as physiological consequences of this transition.

Adaptation to short-term hypoxia has been studied under controlled conditions in the laboratory in a variety of animals. Short-term conditions may mean anything from a few hours to weeks or even months. The length of the low-oxygen exposure generally depends on the animal used, its ability to withstand low oxygen, and the nature of the inquiry into the responses. Some clams, for example, are able to live in the absence of oxygen for several weeks. These experiments require careful monitoring of the animal and the conditions to ensure that oxygen neither rises nor falls too low and that the animals will survive.

A method that has been used to study adaptation to low oxygen in mammals is to conduct field studies in which the subjects are temporarily moved to high elevations. Mountain-climbing expeditions have been involved in some of these experiments in areas throughout the world. Additionally, experimental stations have been established at certain locations for the purpose of conducting these research projects. In this way, medical researchers are able to bring in appropriate equipment and supplies necessary to make complex and precise measurements.

All approaches to the study of low-oxygen adaptation require measurements of respiratory function or metabolic processes, or both. These measurements assess the uptake and transport of oxygen and the transport and excretion of carbon dioxide. The specific measurements are of the rate of oxygen uptake, ventilation volume and rate, blood flow, heart rate, oxygen transport proper-

ties of the blood, and oxygen uptake. In long-term monitoring studies of free-ranging animals, the animals are frequently fitted with implanted electrodes and blood sampling tubes. In this way, measurements can be made routinely over long periods without disturbing the animals.

A common measure of respiratory function is the total amount of oxygen used by an animal in a given period of time. The rate of excretion of carbon dioxide is another measure of overall function. The ratio of carbon dioxide loss to oxygen uptake is used to determine the nature of the metabolic pathways at a given time. Different metabolic pathways have characteristic oxygen uptake and carbon dioxide excretion ratios, and these are used in a predictive or diagnostic fashion. Some of the methods do not impair normal activity and can be used in low-oxygen experiments. The technique of placing small animals in respiratory chambers is used in these types of experimentation. Measurement of single organ function is used more often with larger animals, such as humans.

Evolution, Metabolism, and Ecology

Adaptation to low oxygen has been studied to understand three concepts better: evolutionary changes associated with oxygen availability, cell metabolism, and ecology of hypoxic habitats. Literally every aspect of oxygen uptake, transport, and utilization has received some attention.

Some of the evolutionary changes during the transition from water to land and from low to high altitudes have been studied as problems related to low oxygen. Results indicate that air breathers have lower ventilation rates than do water breathers of the same size. The lower ventilation rates are possible because of the higher oxygen levels, but they result in higher internal carbon dioxide levels. Animals such as insects, reptiles, and mammals have respiratory structures that are internalized and are inpocketings of the body wall. This arrangement aids in water conservation and helps keep the respiratory surfaces moist.

Just as important as research on the transition to land has been the information gained about the evolution of life in high-oxygen environments as compared to the low-oxygen conditions that are believed to have occurred in the ancient oceans. From this research, it is clear that the major advances in respiratory systems are present in invertebrates and probably evolved quite early in the history of life on earth. Marine worms possess closed circulatory systems, respiratory pigments, red blood cells, special gas exchange structures, gills, and alternate metabolic pathways.

Biologists interested in metabolism and the factors that cause metabolic rate to change have examined the relationship between metabolic rate and other physiological functions. Specifically, oxygen supply, carbon dioxide removal, and glucose supply have been examined because all three are directly involved in aerobic metabolism. Imposing a limitation on external oxygen supply has therefore been used as an experimental tool to probe the limits and capabilities of cellular metabolism.

One of the observations that biologists have made over the years is that animals tend to find a way to live in places that are in any way habitable, and they tend to adapt to occupy new habitats. Understanding the physiological mechanisms required or used in adaptations to low-oxygen habitats, such as stagnant pools of water, has allowed explanation of some evolutionary changes.

There are several habitat types that undergo hypoxia routinely, and the utilization of the natural resources of those habitats, as well as the effective preservation of the habitats, generally dictates that attention be paid to the effects on the animals. One of the bodies of water that undergoes low-oxygen conditions is the Chesapeake Bay, and the effects on the animals there have been studied.

—*Peter L. deFur*

See also: Circulatory systems of invertebrates; Circulatory systems of vertebrates; Gas exchange; Heart; Lungs, gills, and tracheas; Metabolic Rates; Mountains; Respiration; Respiration in birds.

Bibliography

Bicudo, J. Eduardo P. W., ed. *The Vertebrate Gas Transport Cascade: Adaptations to Environment and Mode of Life.* Boca Raton, Fla.: CRC Press, 1993. A collection of symposium papers related to gas transport and respiratory mechanisms.

Bryant, Christopher, ed. *Metazoan Life Without Oxygen.* New York: Chapman and Hall, 1991. A collection of essays addressing the question of how ancient animals dealt with low oxygen or interruptions in oxygen supply to vital organs. Traces the evolution of animal species to adapt to the accumulation of oxygen in earth's atmosphere.

Dejours, Pierre. "Mount Everest and Beyond: Breathing Air." In *A Companion to Animal Physiology*, edited by C. Richard Taylor, Kjell Johansen, and Liana Bolis. New York: Cambridge University Press, 1982. This chapter is a scholarly paper that Dejours delivered at a symposium. It is clearly written and has illustrations. Basic principles of respiration are explained; topics such as how animals deal with conditions of low oxygen are discussed. The chapter does not require an extensive science background.

Graham, Jeffrey B. *Air-Breathing Fishes: Evolution, Diversity, and Adaptation.* San Diego, Calif.: Academic Press, 1997. Comprehensive coverage of the puzzling existence of air-breathing fishes. Easy to read for nonspecialists with a background in biology. Numerous tables and illustrations.

Hill, R. W., and G. A. Wyse. *Animal Physiology.* New York: Harper & Row, 1989. The chapter on the physiology of diving in birds and mammals gives an excellent explanation of the problems of limited oxygen supply and metabolic function under such conditions. The chapter discusses responses of two groups of animals that have been well studied. This text also has background material in related topics. The text is understandable, the information is current, and the figures are illustrative.

Hochachka, Peter, and George Somero. *Strategies of Biochemical Adaptation*. Philadelphia: W. B. Saunders, 1973. This book is a monograph that treats the general concepts of biochemical adaptations and gives the logic behind the authors' present explanations. Although the book is more than twenty-five years old, the explanations are first rate; it is clearly written, with sound reasoning. It is not filled with masses of equations, complex graphs, and numerous references, but concentrates on the concepts. The first two chapters are directly related to this topic, and there are other references to low oxygen in the book.

Paganelli, Charles V., and Leon E. Farhi, eds. *Physiological Function in Special Environments*. New York: Springer-Verlag, 1988. This book presents the papers from a symposium on environmental physiology held in Buffalo, New York. The chapters are written by experts in the field for presentation to scientists. The papers are well written and informative, and are suited to the person who is able to master the other entries here and still wishes to pursue the topic. Chapters 1 to 4 on adaptation to altitude, and chapters 12 and 13, on comparative physiology, deal with physiological adaptations related to low-oxygen conditions.

RESPIRATION IN BIRDS

Type of animal science: Physiology
Fields of study: Anatomy, biochemistry, evolutionary science, ornithology, physiology, zoology

Respiration in birds presents a physiological challenge that has been solved by the evolution of a unique and very efficient respiratory system.

Principal Terms

ADENOSINE TRIPHOSPHATE (ATP): the energy currency of cell metabolism in all organisms

AEROBIC: requiring free oxygen; any biological process that can occur in the presence of free oxygen

DIFFUSION: the net movement of molecules from an area of high concentration to one of lower concentration as a result of random molecular movements

HEMOGLOBIN: a protein in vertebrate red blood cells that carries oxygen and carbon dioxide

METABOLISM: the sum of all chemical processes occurring within a cell or living organism

RESPIRATION: the utilization of oxygen; in air-breathing vertebrates, the inhalation of oxygen and the exhalation of carbon dioxide

Flying is an activity that demands an extraordinary amount of metabolic energy. During flight, the active muscles of birds require and produce very large amounts of the cellular "energy currency" known as adenosine triphosphate (ATP) to power their contractions. To accomplish this, the flight muscles must be well supplied with oxygen extracted from the atmosphere. The flight muscles furthermore generate large amounts of carbon dioxide as a waste product of their metabolism, and this must be eliminated from the blood.

These facts make it tempting to propose that birds need a specially designed respiratory system because their need for oxygen must be greater than that of comparably sized mammals. Many textbooks say exactly this. Studies have shown that resting birds and comparably sized resting mammals both consume nearly identical amounts of oxygen. The argument says that flying birds must consume more oxygen than mammals exercising at high levels of performance. Further studies have shown that flying birds consume ten to fifteen times more oxygen than when at rest. This is no greater than the increase in oxygen consumption that any well-conditioned human athlete can attain.

The primary reasons that birds require a special respiratory system become clear when the structural limitations of size and weight imposed by flight and the extreme environments encountered during high-altitude flight (such as low oxygen availability) are taken into consideration. The respiratory system of a bird needs to be relatively light, compact, and efficient in order to maximize the animal's chances of survival. The avian respiratory system has evolved successfully, which is evident if one observes birds in steady flight at altitudes so high that the scarcity of oxygen would cause most mammals to become comatose even when totally inactive.

The Avian Respiratory System

This sort of respiratory performance has been achieved, within the imposed size and weight restrictions, through the evolution of the avian respiratory system into a unique and very efficient structure. Typical mammalian lungs expand and collapse inside the body with each breath (the lungs have a variable volume), and there is a bidirectional flow of gas into and then back out of them. In contrast, the bird's two lungs have a nearly constant volume through which most of the inhaled gas flows in one direction. This is possible because of the presence of several large, thin-walled air sacs that join the lungs and assist in their ventilation, much in the manner of a system of bellows to push air through the lungs.

The inhaled air initially follows an anatomic pathway in birds that resembles that of mammals—through a trachea (windpipe) that then divides into two tubes called the primary bronchi, each of which enters a lung. At this point, the anatomy of the bird becomes different as the primary bronchi actually pass completely through the avian lung to terminate in paired abdominal air sacs. These air sacs also have connections to several hollow bones in the bird's legs. Near the posterior end of the primary bronchi, furthermore, there is a pair of posterior thoracic air sacs that are connected to the primary bronchi via laterobronchi. Joined to the primary bronchi via one of several secondary tubes called the ventrobronchi are paired anterior thoracic air sacs and, depending upon the bird species, usually several other anteriorly located air sacs.

Very soon after entering the lung, four ventrobronchi branch from a primary bronchus. Farther along the primary bronchus, there are as many as ten additional secondary bronchi, called the dorsobronchi, which also leave the primary bronchus. The dorsobronchi and ventrobronchi both divide many times as they penetrate the lung tissue. Eventually the dorsobronchi become connected with the ventrobronchi by thousands of tiny parabronchi, each of which is only about one millimeter in diameter.

The parabronchial walls are like a lattice, perforated extensively with pockets called atria, which themselves have indentations known as infundibula. Interconnecting the infundibula are numerous tiny air capillaries approximately five micrometers in diameter. The lattice itself is made up of

the capillary blood vessels into which the oxygen and out of which the carbon dioxide diffuse during the exchange of these gases between the air and the blood. This region of parabronchi and the many tiny air passages extending from them, all of which are enmeshed by the very dense network of blood capillaries, make up the gas exchange area of the avian lung.

Inhalation and Exhalation
The bronchi, primary and secondary, and the air sacs do exchange gas with the blood during avian respiration. Their only function is to help move the air through the actual gas exchange organs, the lungs. To achieve the unidirectional flow of air through the lungs, the inhaled, oxygen-rich air passes through the primary bronchi directly into the posterior air sacs, which are being expanded as a result of the movements of inhalation. While this is happening, the air that is already in the gas exchange areas of the lungs is being pulled out of them and into the anterior air sacs, which are also experiencing expansion during inhalation. Replacing this air in the lungs is a portion of the oxygen-rich air from the primary bronchi that does not enter the posterior air sacs, but flows into the posterior lung regions and is directed forward to the anterior areas.

During exhalation, both the posterior and anterior air sacs are compressed and their contents emptied. The air from the posterior air sacs enters the gas exchange areas of the lungs, not yet having lost any of its oxygen, and flows in a posterior to anterior direction. The air from the anterior air sacs, loaded with the carbon dioxide which had diffused out of the blood capillaries, enters the primary bronchi and is exhaled by the animal. The cycle then repeats itself with the next inhalation.

During the cycles of inhalation and exhalation, the air is always passing through the gas exchange areas of the avian lungs in a single posterior to anterior direction. The air passing through the gas exchange areas of the lung also always contains a relatively constant and high percentage of oxygen that can diffuse into the blood inside the lung's blood capillaries. The blood entering these capil-

The structural limitations of size and weight as well as the low oxygen conditions encountered at high altitudes have caused birds to evolve exceptionally efficient respiratory systems. (Corbis)

laries is rich in carbon dioxide, which diffuses out of the blood and into the air as it passes through the lung on its way to the anterior air sacs, and eventually out of the animal.

Blood and Oxygen

It is necessary here to address the possibility that avian blood might have some special properties that allow it to extract and transport oxygen much better than mammalian blood. The blood of birds, however, is actually not significantly better in this respect. The primary oxygen transporter is the avian hemoglobin molecule, and it does not perform significantly better than the hemoglobin found in mammals, nor is there any real difference in the amount of hemoglobin contained in avian blood. Avian blood has no special properties that would contribute to the remarkable performance of the avian respiratory system.

The flow of the blood with respect to the gas flow through the gas exchange regions is a crucial factor. The flow of blood through the pulmonary (lung) blood capillaries is in a direction that permits the efficient extraction of the air's oxygen. While it was previously believed that there was a countercurrent flow of air and blood, with the air moving through the lung in a posterior to anterior direction and the blood flowing through the lung in an anterior to posterior direction, it is now known to be a crosscurrent system. This crosscurrent system has the air still passing in a posterior to anterior direction, but the blood flows in a combination of the opposite direction and at a right angle to the air flow. This critical design feature, combined with those mentioned in the next paragraph, permits the avian lung to be the most efficient of any known gas exchange organ for an air breathing animal.

To comprehend the compact and efficient design of the avian lung, it is necessary only to measure the ratio of the surface area it uses for gas exchange to the lung volume. This value is approximately ten times greater than the ratio found in mammals, allowing more blood per unit of volume of lung to be present at any time. This is really of importance only when it is noted that avian lungs have more blood in their capillaries per unit of volume of lung than do mammalian lungs. Since only the blood in the capillaries can pick up oxygen and get rid of carbon dioxide, the value of such features is obvious: small and very efficient lungs.

Avian Anatomy and Physiology

The primary techniques utilized in the study of the avian respiratory system come from the fields of anatomy and physiology. Modern work involves the use of electron microscopes, miniature gas flow meters, and even tiny radio transmitters that can send signals from sensors implanted in a freely moving animal.

The connection of the airways with the large air sacs was first reported in the seventeenth century. In the eighteenth century, pioneering studies were performed which showed the interesting fact that some hollow bones of the bird were connected to the respiratory system. It is, in fact, possible for a bird to breathe through one of these bones if its normal airway is blocked and the hollow center of the bone is opened to the atmosphere.

The rate at which birds consume oxygen can be measured in much the same way as it is in other animals. This can involve the use of a respirometer, a device to record the volumes of air inhaled and exhaled during breathing. It can also indicate how much oxygen is retained by the animal for its survival. Other methods are also used if the experimental demands preclude the use of such a device.

Scientists have even sampled the air at different places in the respiratory system of birds. To do this, it is necessary to insert small tubes into the regions of interest (for example, into the different air sacs) and to withdraw small samples of the gas found within these structures at different times during the breathing cycle. The samples are then tested for their oxygen and carbon dioxide content. The results can reveal whether the gas has already experienced exchanges of oxygen and carbon dioxide with the blood, and whether the gas is being replaced at regular intervals or is only being stored there. The latter question involves using a marker gas mixture that the animal first inhales for one or more breaths and that is then replaced by normal air for inhalation. The rate of disappearance of the marker gas indicates how quickly the gas content of the region being studied is replaced.

Similarly, miniature gas flow meters have been developed for investigating the direction of the gas flow within the various regions of the respiratory system. Inserting these into the airways of larger birds shows clearly that gas flows through the bird lung in a unidirectional way. The results of such investigations have allowed scientists to discard incorrect hypotheses about the manner in which the inhaled air is distributed throughout the bird's respiratory system and about the functions of various respiratory structures.

Biochemical studies of avian blood measure the same factors as are measured in other animal blood tests. The quantities of interest are the hemoglobin content of the blood (important because it carries essentially all the oxygen transported by the blood), the ability of avian hemoglobin to carry oxygen, the effects of various factors on the hemoglobin-oxygen interactions (such as the normally occurring acids of the blood, carbon dioxide, and temperature), and the ability of the hemoglobin to transport carbon dioxide. It is important to learn about these factors because of the intimate relationships between the respiratory organs and the blood (as transporter of the oxygen and carbon dioxide between the lungs and the other tissues of the body). These studies have revealed no dramatic differences between avian and mammalian blood with respect to the blood's respiratory functions.

These and many other methods are used to investigate avian respiratory systems. As a result of these studies, scientists know that a bird's ability to maintain sustained flight at high altitudes results not from any special properties of its blood, but is based on the very special design of its respiratory organs.

Altitude and Oxygen Demand

Regardless of the animal, there is one common need for survival—oxygen. Oxygen must be delivered at a rate high enough to permit the organism's cellular functions to take place, most of which require a continuous and large supply of adenosine triphosphate (ATP). The energy-rich ATP molecules are made in large quantity by the processes of aerobic cellular respiration (oxygen-dependent cellular respiration). Without sufficient oxygen, rates of aerobic cellular respiration sufficient for maintaining life are impossible. Correctly interpreting the value of the avian respiratory system requires an understanding of the characteristics and demands of avian life. Because of the high activity levels of flying birds, but especially because of the elevations at which this activity occurs and the consequent reduction of oxygen availability, birds require very efficient respiratory systems. To provide them with the required oxygen within the context and demands of their lifestyle, they have evolved a distinctive respiratory system. The anatomical, physiological, behavioral, and chemical components of the bird are all exquisitely adapted to these demands, not as independent entities but as an integrated system.

The unique structure of the avian respiratory system endows it with the distinction of being the most efficient air breathing organ among the vertebrates. It is able to extract oxygen from the air so efficiently that prolonged flight is possible at altitudes where the low availability of oxygen is disabling to most vertebrates even at rest.

Knowledge of how this is accomplished may eventually prove useful to engineers in the design and development of high-efficiency oxygen extraction devices. Large-scale applications of this sort of device could serve to enrich the atmosphere with oxygen for the survival of animals in hostile environments. Regardless of the possible practical applications of the knowledge of the structure and function of avian respiratory systems, they have provided a remarkable example of a creative solution to environmental demands which resulted from the process of natural selection.

—John V. Urbas

See also: Adaptations and their mechanisms; Birds; Circulatory systems of vertebrates; Flight; Gas exchange; Lungs, gills, and tracheas; Respiration; Respiratory system; Wings.

Bibliography

Baumel, Julian J., ed. *Handbook of Avian Anatomy: Nomina Anatomica Avium.* 2d ed. Cambridge, Mass.: Nuttall Ornithological Club, 1993. Covers all aspects of avian anatomy, including the respiratory system.

Eckert, Roger, David Randall, and George Augustine. *Animal Physiology.* 4th ed. New York: W. H. Freeman, 1997. This college-level text has a detailed analysis of the physiology of respiration, including the avian respiratory system. There is a good list of suggested readings and a list of technical references cited within the text. A glossary and detailed index are also included.

Farner, Donald S., James R. King, and Kenneth C. Parkes, eds. *Avian Biology.* London: Academic, 1997. A general text on avian biology, including respiration.

Hildebrand, Milton. *Analysis of Vertebrate Structure.* 5th ed. New York: John Wiley & Sons, 2001. This classic textbook is for college-level readers. Includes material relevant to the avian respiratory system. The index is very complete, while the glossary is only adequate. The list of references for each chapter is good, but some are of a quite advanced nature.

Proctor, Noble S. *Manual of Ornithology: Avian Structure and Function.* New Haven, Conn.: Yale University Press, 1993. Focuses exclusively on structure, with terms clearly identified and defined. Good index.

Schmidt-Nielsen, Knut. *How Animals Work*. New York: Cambridge University Press, 1972. This is an excellent short text designed for the general reader. Much of the text deals with various aspects of avian respiration. The description of many pivotal experimental results allows the reader to follow the deductive processes involved in the solution of many scientific problems. Most of the references are quite technical. The index is adequate.

_____. "How Birds Breathe." *Scientific American* 225 (December, 1971): 72-79. An informative article presenting some of the history of the study of avian respiratory systems. The very clear diagrams and explanations of techniques used by the author's research group make this appropriate for the high school reader.

Welty, Joel Carl. *The Life of Birds*. 4th ed. Philadelphia: W. B. Saunders, 1988. A general college-level textbook that describes the anatomy and physiology of the elements of the avian respiratory system in a clear manner. There are several good suggestions for further reading, and there is a very extensive list of references, most of which are quite technical. The index is excellent.

Woakes, A. J., M. K. Grieshaber, and C. R. Bridges, eds. *Physiological Strategies for Gas Exchange and Metabolism*. New York: Cambridge University Press, 1991. A general book on gas exchange and metabolism, with specific chapters covering the gas transport systems of fish, crustaceans, amphibians and reptiles, and birds.

MAGILL'S ENCYCLOPEDIA OF SCIENCE

ANIMAL LIFE

ALPHABETICAL LIST

Volume 1

Volume 2

Volume 3

Volume 4

CATEGORY LIST

Amphibians
Amphibians
Cold-blooded animals
Frogs and toads
Metamorphosis
Salamanders and newts
Vertebrates

Anatomy
Anatomy
Antennae
Beaks and bills
Bone and cartilage
Brain
Cell types
Claws, nails, and hooves
Digestive tract
Ears
Endoskeletons
Exoskeletons
Eyes
Feathers
Fins and flippers
Fur and hair

Heart
Horns and antlers
Hydrostatic skeletons
Kidneys and other excretory
    structures
Lungs, gills, and tracheas
Muscles in invertebrates
Muscles in vertebrates
Noses
Scales
Sense organs
Shells
Skin
Tails
Teeth, fangs, and tusks
Tentacles
Wings

Arthropods
Arachnids
Arthropods
Centipedes and millipedes
Cold-blooded animals
Crabs and lobsters

Crustaceans
Exoskeletons
Horseshoe crabs
Invertebrates
Scorpions
Spiders
Vertebrates

Behavior
Adaptations and their
    mechanisms
Altruism
Camouflage
Cannibalism
Carnivores
Communication
Communities
Competition
Copulation
Courtship
Death and dying
Defense mechanisms
Displays
Domestication

Neutral mutations and
  evolutionary clocks
Nonrandom mating, genetic
  drift, and mutation
Nutrient requirements
Osmoregulation
pH maintenance
Regeneration
Reproduction
Sense organs
Water balance in vertebrates

Classification
Aardvarks
Allosaurus
American pronghorns
Amphibians
Animal kingdom
Antelope
Ants
Apatosaurus
Arachnids
Archaeopteryx
Armadillos, anteaters, and
  sloths
Arthropods
Baboons
Bats
Bears
Beavers
Bees
Beetles
Birds
Brachiosaurus
Butterflies and moths
Camels
Cats
Cattle, buffalo, and bison
Centipedes and millipedes
Chameleons
Cheetahs
Chickens, turkeys, pheasant,
  and quail
Chimpanzees
Chordates, lower
Clams and oysters

Cockroaches
Coral
Crabs and lobsters
Cranes
Crocodiles
Crustaceans
Deer
Dinosaurs
Dogs, wolves, and coyotes
Dolphins, porpoises, and
  toothed whales
Donkeys and mules
Ducks
Eagles
Echinoderms
Eels
Elephant seals
Elephants
Elk
Fish
Flamingos
Flatworms
Flies
Foxes
Frogs and toads
Geese
Giraffes
Goats
Gophers
Gorillas
Grasshoppers
Grizzly bears
Hadrosaurs
Hawks
Hippopotamuses
Hominids
Horses and zebras
Horseshoe crabs
Hummingbirds
Hyenas
Hyraxes
Ichthyosaurs
Insects
Invertebrates
Jaguars
Jellyfish

Kangaroos
Koalas
Lampreys and hagfish
Lemurs
Leopards
Lions
Lizards
Lungfish
Mammals
Mammoths
Manatees
Meerkats
Mice and rats
Moles
Mollusks
Monkeys
Moose
Mosquitoes
Mountain lions
Neanderthals
Octopuses and squid
Opossums
Orangutans
Ostriches and related
  birds
Otters
Owls
Pandas
Parrots
Pelicans
Penguins
Pigs and hogs
Platypuses
Polar bears
Porcupines
Praying mantis
Primates
Protozoa
Pterosaurs
Rabbits, hares, and pikas
Raccoons and related
  mammals
Reindeer
Reptiles
Rhinoceroses
Rodents

Roundworms
Ruminants
Salamanders and newts
Salmon and trout
Sauropods
Scorpions
Seahorses
Seals and walruses
Sharks and rays
Sheep
Shrews
Skunks
Snails
Snakes
Sparrows and finches
Spiders
Sponges
Squirrels
Starfish
Stegosaurs
Storks
Swans
Systematics
Tasmanian devils
Termites
Tigers
Triceratops
Turtles and tortoises
Tyrannosaurus
Ungulates
Velociraptors
Vertebrates
Vultures
Wasps and hornets
Weasels and related mammals
Whale sharks
Whales, baleen
White sharks
Woodpeckers
Worms, segmented
Zooplankton

Ecology
Adaptations and their
    mechanisms
Biodiversity

Bioluminescence
Breeding programs
Camouflage
Carnivores
Chaparral
Cloning of extinct or
    endangered species
Communities
Competition
Deep-sea animals
Demographics
Deserts
Ecological niches
Ecology
Ecosystems
Endangered species
Food chains and food webs
Forests, coniferous
Forests, deciduous
Grasslands and prairies
Habitats and biomes
Herbivores
Lakes and rivers
Marine animals
Marine biology
Mimicry
Mountains
Omnivores
Paleoecology
Plant and animal interactions
Poisonous animals
Pollution effects
Population analysis
Population fluctuations
Population genetics
Population growth
Rain forests
Reefs
Ruminants
Savannas
Scavengers
Swamps and marshes
Symbiosis
Tidepools and beaches
Tundra
Urban and suburban wildlife

Wildlife management
Zoos

Evolution
Adaptive radiation
Apes to hominids
Clines, hybrid zones, and
    introgression
Coevolution
Convergent and divergent
    evolution
Evolution: Historical
    perspective
Extinction
Extinctions and evolutionary
    explosions
Fossils
Gene flow
Genetics
Hardy-Weinberg law of
    genetic equilibrium
Heterochrony
Hominids
Homo sapiens and human
    diversification
Human evolution analysis
Isolating mechanisms in
    evolution
Multicellularity
Mutations
Natural selection
Neutral mutations and
    evolutionary clocks
Nonrandom mating, genetic
    drift, and mutation
Phylogeny
Prehistoric animals
Punctuated equilibrium and
    continuous evolution

Fields of Study
Anatomy
Biogeography
Biology
Demographics
Ecology

Category List

Embryology
Ethology
Human evolution analysis
Marine biology
Paleoecology
Paleontology
Phylogeny
Physiology
Population analysis
Population genetics
Systematics
Veterinary medicine
Zoology

Fish
Cold-blooded animals
Deep-sea animals
Eels
Fins and flippers
Fish
Lakes and rivers
Lampreys and hagfish
Lungfish
Marine animals
Marine biology
Reefs
Salmon and trout
Scales
Seahorses
Sharks and rays
Tidepools and beaches
Vertebrates
Whale sharks
White sharks

Genetics
Gene flow
Genetics
Hardy-Weinberg law of
   genetic equilibrium
Homo sapiens and human
   diversification
Human evolution analysis
Morphogenesis
Multicellularity
Mutations

Natural selection
Neutral mutations and
   evolutionary clocks
Nonrandom mating, genetic
   drift, and mutation

Geography
Biogeography
Chaparral
Deep-sea animals
Deserts
Fauna: Africa
Fauna: Antarctica
Fauna: Arctic
Fauna: Asia
Fauna: Australia
Fauna: Caribbean
Fauna: Central America
Fauna: Europe
Fauna: Galápagos Islands
Fauna: Madagascar
Fauna: North America
Fauna: Pacific Islands
Fauna: South America
Forests, coniferous
Forests, deciduous
Grasslands and prairies
Habitats and biomes
Lakes and rivers
Marine animals
Mountains
Rain forests
Reefs
Savannas
Swamps and marshes
Tidepools and beaches
Tundra
Urban and suburban wildlife

Habitats
Chaparral
Deep-sea animals
Deserts
Forests, coniferous
Forests, deciduous
Grasslands and prairies

Habitats and biomes
Lakes and rivers
Marine animals
Mountains
Rain forests
Reefs
Savannas
Swamps and marshes
Tidepools and beaches
Tundra
Urban and suburban wildlife

Herbivores
American pronghorns
Antelope
Apatosaurus
Beavers
Brachiosaurus
Camels
Cattle, buffalo, and bison
Deer
Donkeys and mules
Elephants
Elk
Food chains and food webs
Giraffes
Goats
Gophers
Hadrosaurs
Herbivores
Herds
Hippopotamuses
Horns and antlers
Horses and zebras
Hyraxes
Kangaroos
Koalas
Mammoths
Manatees
Mice and rats
Moles
Moose
Pandas
Pigs and hogs
Platypuses
Porcupines

Category List

Monkeys
Monotremes
Moose
Mountain lions
Neanderthals
Orangutans
Otters
Pandas
Pigs and hogs
Placental mammals
Platypuses
Polar bears
Porcupines
Primates
Rabbits, hares, and pikas
Raccoons and related
    mammals
Reindeer
Reproductive system of
    female mammals
Reproductive system of male
    mammals
Rhinoceroses
Rodents
Ruminants
Seals and walruses
Sheep
Shrews
Skunks
Squirrels
Tasmanian devils
Tigers
Ungulates
Vertebrates
Warm-blooded animals
Weasels and related mammals
Whales, baleen

Marine Biology
Clams and oysters
Coral
Crabs and lobsters
Crustaceans
Deep-sea animals
Dolphins, porpoises, and
    toothed whales

Echinoderms
Eels
Elephant seals
Fins and flippers
Fish
Horseshoe crabs
Ichthyosaurs
Jellyfish
Lakes and rivers
Lampreys and hagfish
Lungfish
Manatees
Marine animals
Marine biology
Mollusks
Octopuses and squid
Otters
Penguins
Reefs
Salmon and trout
Scales
Seahorses
Seals and walruses
Sharks and rays
Sponges
Starfish
Tentacles
Tidepools and beaches
Turtles and tortoises
Whale sharks
Whales, baleen
White sharks
Zooplankton

Marsupials
Fauna: Australia
Kangaroos
Koalas
Marsupials
Opossums
Reproduction
Reproductive strategies
Reproductive system of
    female mammals
Tasmanian devils

Omnivores
Baboons
Bears
Chimpanzees
Food chains and food webs
Gorillas
Hominids
Lemurs
Monkeys
Omnivores
Orangutans
Raccoons and related mammals
Shrews
Skunks
Weasels and related mammals

Physiology
Adaptations and their
    mechanisms
Aging
Asexual reproduction
Bioluminescence
Birth
Camouflage
Circulatory systems of
    invertebrates
Circulatory systems of
    vertebrates
Cleavage, gastrulation, and
    neurulation
Cold-blooded animals
Determination and
    differentiation
Digestion
Digestive tract
Diseases
Endocrine systems of
    invertebrates
Endocrine systems of
    vertebrates
Endoskeletons
Estrus
Exoskeletons
Fertilization
Gametogenesis
Gas exchange

Hormones and behavior
Hormones in mammals
Imprinting
Infanticide
Lactation
Life spans
Marsupials
Mating
Metamorphosis
Molting and shedding
Monotremes
Morphogenesis
Nesting
Offspring care
Pair-bonding
Parthenogenesis
Pheromones
Placental mammals
Pregnancy and prenatal
    development
Regeneration
Reproduction
Reproductive strategies
Reproductive system of
    female mammals

Reproductive system of male
    mammals
Sex differences: Evolutionary
    origins
Sexual development
Vocalizations

Reptiles
Allosaurus
Apatosaurus
Archaeopteryx
Brachiosaurus
Chameleons
Cold-blooded animals
Crocodiles
Dinosaurs
Hadrosaurs
Ichthyosaurs
Lizards
Pterosaurs
Reptiles
Sauropods
Scales
Snakes
Stegosaurs

Triceratops
Turtles and tortoises
Tyrannosaurus
Velociraptors
Vertebrates

Scientific Methods
Breeding programs
Cloning of extinct or
    endangered species
Demographics
Fossils
Hardy-Weinberg law of
    genetic equilibrium
Human evolution analysis
Mark, release, and recapture
    methods
Paleoecology
Paleontology
Population analysis
Population genetics
Systematics
Veterinary medicine